T0190079

 Birkhäuser

Contemporary Mathematicians

Gian-Carlo Rota
Joseph P.S. Kung

Editors

For further volumes:
http://www.springer.com/series/4817

Peter Duren   Lawrence Zalcman
Editors

Menahem Max Schiffer: Selected Papers Volume 1

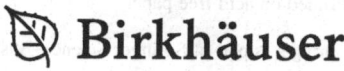 Birkhäuser

Editors
Peter Duren
Department of Mathematics
University of Michigan
Ann Arbor, Michigan, USA

Lawrence Zalcman
Department of Mathematics
Bar-Ilan University
Ramat-Gan, Israel

ISBN 978-1-4939-3698-4   ISBN 978-0-8176-8085-5 (eBook)
DOI 10.1007/978-0-8176-8085-5
Springer New York Heidelberg Dordrecht London

Mathematics Subject Classification (2010): 30-XX, 31-XX, 35-XX, 49-XX, 76-XX, 20-XX, 01-XX

Printed on acid-free paper

Springer is part of Springer Science+Business Media (www.birkhauser-science.com)

Max Schiffer in his Stanford office, ca.1976

Preface

These two volumes contain some four dozen articles selected from over 130 papers published by Menahem Max Schiffer, the dominant figure in geometric function theory during the second half of the twentieth century. The papers are accompanied by commentaries which aim variously to place them in their historical context, elaborate on the circumstances of their composition, and provide indications of subsequent developments. Also included are a complete bibliography of Schiffer's publications, a list of his doctoral students with the dates and titles of their dissertations, a biographical chronology, and a number of personal reminiscences of Schiffer, the man and the mathematician, by collaborators and admirers.

The idea of publishing a selection of Schiffer's papers was proposed by Gian-Carlo Rota, editor of the *Contemporary Mathematicians* series, around 1990. Schiffer himself endorsed the project, providing advice and encouragement. Paul Garabedian and Dennis Hejhal were also involved in the early planning and participated in the difficult process of selecting papers to be reprinted. However, over the next 20 years, progress was sporadic at best; and it was only at the beginning of 2012 that work on the current volumes began in earnest.

Limitations of space precluded the reprinting of several highly important but very long papers, including [27] (with Gabor Szego) and [53] (with George Pólya). Fortunately, both of these joint papers have been reprinted in the collected works of their respective coauthors and are also readily available online.

Special thanks are due to the following individuals for their valuable assistance:

Dinah Singer (née Schiffer), for providing biographical materials relating to her father and for her good-natured patience in the face of long delays;

Edwin F. Beschler, who accompanied this project from its very inception and, on his own initiative, secured the necessary permissions from the copyright holders to reprint the selected articles;

Brad Osgood of Stanford University, who worked diligently to extract relevant information from the Stanford archives;

Ofer Tzemach, Director of the Central Archive of the Hebrew University, and his colleagues, for making available Schiffer's personal file as a member of the academic staff of that institution; and, finally,

Miriam Beller, who oversaw the technical aspects of producing the commentaries and related material.

Our hope is that these volumes will enable a new generation of analysts to become acquainted with and appreciate the enduring achievements of one of the truly great modern masters.

Peter Duren   *Lawrence Zalcman*
Ann Arbor, Michigan, USA   Jerusalem, Israel

Contents

Publications of M. M. Schiffer

1. Ein neuer Beweis des Endlichkeitssatzes für Orthogonalinvarianten. *Math. Z.* 38 (1934), 315 322.
2. Sur un principe nouveau pour l'évaluation des fonctions holomorphes. *Bull. Soc. Math. France* 64 (1936), 231 240.
3. Un calcul de variation pour une famille de fonctions univalentes. *C. R. Acad. Sci. Paris* 205 (1937), 709 711.
4. Sur un problème d'extrémum de la représentation conforme. *Bull. Soc. Math. France* 66 (1938), 48 55.
5. A method of variation within the family of simple functions. *Proc. London Math. Soc.* (2) 44 (1938), 432 449.
6. On the coefficients of simple functions. *Proc. London Math. Soc.* (2) 44 (1938), 450 452.
7. Sur les domaines minima dans la théorie des transformations pseudoconformes. *C. R. Acad. Sci. Paris* 207 (1938), 112 115.
8. Sur un théorème de la représentation conforme. *C. R. Acad. Sci. Paris* 207 (1938), 520 522.
9. (with S. Bergmann) Familles bornées de fonctions de deux variables complexes dans des domaines avec une surface remarquable. *C. R. Acad. Sci. Paris* 207 (1938), 711 713.
10. Sur la variation de la fonction de Green de domaines plans quelconques. *C. R. Acad. Sci. Paris* 209 (1939), 980 982.
11. Sur la variation du diamètre transfini. *Bull. Soc. Math. France* 68 (1940), 158 176.
12. On the subadditivity of the transfinite diameter. *Proc. Cambridge Philos. Soc.* 37 (1941), 373 383.
13. Variation of the Green function and theory of the p-valued functions. *Amer. J. Math.* 65 (1943), 341 360.
14. The span of multiply connected domains. *Duke Math. J.* 10 (1943), 209 216.
15. (with S. Bergman) Bounded functions of two complex variables. *Amer. J. Math.* 66 (1944), 161 169.
16. Sur l'équation différentielle de M. Löwner. *C. R. Acad. Sci. Paris* 221 (1945), 369 371.
17. Hadamard's formula and variation of domain-functions. *Amer. J. Math.* 68 (1946), 417 448.
18. On the modulus of doubly-connected domains. *Quart. J. Math. Oxford Ser.* 17 (1946), 197 213.
19. The kernel function of an orthonormal system. *Duke Math. J.* 13 (1946), 529 540.
20. (with S. Bergman) A representation of Green's and Neumann's functions in the theory of partial differential equations of second order. *Duke Math. J.* 14 (1947), 609 638.
21. (with S. Bergman) On Green's and Neumann's functions in the theory of partial differential equations. *Bull. Amer. Math. Soc.* 53 (1947), 1141 1151.
22. An application of orthonormal functions in the theory of conformal mapping. *Amer. J. Math.* 70 (1948), 147 156.
23. (with S. Bergman) Kernel functions in the theory of partial differential equations of elliptic type. *Duke Math. J.* 15 (1948), 535 566.

24. (with R. von Mises) On Bergman's integration method in two-dimensional compressible fluid flow, in *Advances in Applied Mechanics* (R. von Mises and T. von Kármán, editors), Academic Press, New York, 1948, pp. 249 285.

25. Faber polynomials in the theory of univalent functions. *Bull. Amer. Math. Soc.* 54 (1948), 503 517.

26. (with P. R. Garabedian) Identities in the theory of conformal mapping. *Trans. Amer. Math. Soc.* 65 (1949), 187 238.

27. (with G. Szego) Virtual mass and polarization. *Trans. Amer. Math. Soc.* 67 (1949), 130 205.

28. (with A. C. Schaeffer and D. C. Spencer) The coefficient regions of schlicht functions. *Duke Math. J.* 16 (1949), 493 527.

29. (with P. R. Garabedian) On existence theorems of potential theory and conformal mapping. *Ann. of Math.* (2) 52 (1950), 164 187.

30. (with D. C. Spencer) The coefficient problem for multiply-connected domains. *Ann. of Math.* (2) 52 (1950), 362 402.

31. Various types of orthogonalization. *Duke Math. J.* 17 (1950), 329 366.

32. (with S. Bergman) Various kernels in the theory of partial differential equations. *Proc. Nat. Acad. Sci. U.S.A.* 36 (1950), 559 563.

33. (with S. Bergman) Some linear operators in the theory of partial differential equations. *Proc. Nat. Acad. Sci. U.S.A.* 36 (1950), 742 746.

34. Some recent developments in the theory of conformal mapping, in R. Courant, *Dirichlet s Principle, Conformal Mapping, and Minimal Surfaces*, Interscience, New York, 1950, Appendix, pp. 249 323.

35. (with S. Bergman) Kernel functions and conformal mapping. *Compositio Math.* 8 (1951), 205 249.

36. (with S. Bergman) Kernel functions and partial differential equations. I. Boundary value problems in the theory of non-linear partial differential equations of elliptic type. *J. Analyse Math.* 1 (1951), 375 386.

37. (with S. Bergman) A majorant method for non-linear partial differential equations. *Proc. Nat. Acad. Sci. U.S.A.* 37 (1951), 744 749.

38. (with D. C. Spencer) A variational calculus for Riemann surfaces. *Ann. Acad. Sci. Fenn. Ser. A I* 93 (1951), 9 pp.

39. (with D. C. Spencer) On the conformal mapping of one Riemann surface into another. *Ann. Acad. Sci. Fenn. Ser. A I* 94 (1951), 10 pp.

40. Variational methods in the theory of conformal mapping, in *Proceedings of the International Congress of Mathematicians, Cambridge, Mass., 1950*, American Mathematical Society, Providence, R.I., 1952, Vol. 2, pp. 233 240.

41. (with K. S. Miller) On the Green's function of ordinary differential systems. *Proc. Amer. Math. Soc.* 3 (1952), 433 441.

42. (with K. S. Miller) Monotonic properties of the Green's function. *Proc. Amer. Math. Soc.* 3 (1952), 948 956.

43. (with F. S. Bodenheimer) Mathematical studies in animal populations. I. A mathematical study of insect parasitism. *Acta Biotheoretica Ser. A* 10 (1952), 23 56.

44. (with P. R. Garabedian and H. Lewy) Axially symmetric cavitational flow. *Ann. of Math.* (2) 56 (1952), 560 602.

45. (with S. Bergman) Potential-theoretic methods in the theory of functions of two complex variables. *Compositio Math.* 10 (1952), 213 240.

46. (with D. C. Spencer) Some remarks on variational methods applicable to multiply connected domains, in *Construction and Applications of Conformal Maps: Proceedings of a Symposium* (E. F. Beckenbach, editor), National Bureau of Standards, Appl. Math. Series 18, U.S. Government Printing Office, Washington, D.C., 1952, pp. 193 198.

47. (with S. Bergman) Theory of kernel functions in conformal mapping, in *Construction and Applications of Conformal Maps: Proceedings of a Symposium* (E. F. Beckenbach, editor), National Bureau of Standards, Appl. Math. Series 18, U.S. Government Printing Office, Washington, D.C., 1952, pp. 199 206.

48. (with P. R. Garabedian) Variational problems in the theory of elliptic partial differential equations. *J. Rational Mech. Anal.* 2 (1953), 137 171.

49. (with P. R. Garabedian) Convexity of domain functionals. *J. Analyse Math.* 2 (1953), 281 368.

50. Variational methods in the theory of Riemann surfaces, in *Contributions to the Theory of Riemann Surfaces* (L. Ahlfors *et al.*, editors), Annals of Math. Studies 30, Princeton University Press, Princeton, N.J., 1953, pp. 15 30.

51. (with S. Bergman) *Kernel Functions and Elliptic Differential Equations in Mathematical Physics*, Academic Press, New York, 1953.

52. The Applied Mathematics Laboratory at Stanford University, in *Proceedings of a Conference on Training in Applied Mathematics* (1953), pp. 37 40.

53. (with G. Pólya; appendix by H. Helfenstein) Convexity of functionals by transplantation. *J. Analyse Math.* 3 (1954), 245 346.

54. Variation of domain functionals. *Bull. Amer. Math. Soc.* 60 (1954), 303 328.

55. (with D. C. Spencer) *Functionals of Finite Riemann Surfaces*, Princeton University Press, Princeton, N.J., 1954.

56. (with P. R. Garabedian) On estimation of electrostatic capacity. *Proc. Amer. Math. Soc.* 5 (1954), 206 211.

57. (with S. Bergman) Properties of solutions of a system of partial differential equations, in *Studies in Mathematics and Mechanics Presented to Richard von Mises*, Academic Press, New York, 1954, pp. 79 87.

58. (with P. R. Garabedian) On a double integral variational problem. *Canad. J. Math.* 6 (1954), 441 446.

59. (with P. R. Garabedian) A coefficient inequality for schlicht functions. *Ann. of Math.* (2) 61 (1955), 116 136.

60. (with P. R. Garabedian) A proof of the Bieberbach conjecture for the fourth coefficient. *J. Rational Mech. Anal.* 4 (1955), 427 465.

61. Partial differential equations of elliptic type, in *Modern Mathematics for the Engineer* (E. F. Beckenbach, editor), McGraw-Hill, New York, 1956, pp. 110 144.

62. The Fredholm eigen values of plane domains. *Paci"c J. Math.* 7 (1957), 1187 1225.

63. Sur les rapports entre les solutions des problèmes intérieurs et celles des problèmes extérieurs. *C. R. Acad. Sci. Paris* 244 (1957), 2680 2683.

64. Sur la polarisation et la masse virtuelle. *C. R. Acad. Sci. Paris* 244 (1957), 3118 3121.

65. Problèmes aux limites et fonctions propres de l'équation intégrale de Poincaré et de Fredholm. *C. R. Acad. Sci. Paris* 245 (1957), 18 21.

66. Partial differential equations of the elliptic type, in *Lecture Series of the Symposium on Partial Differential Equations, University of California, Berkeley, 1955* (N. Aronszajn and C. B. Morrey, Jr., editors), University of Kansas Press, Lawrence, Kansas, 1957, pp. 97 149.

67. Applications of variational methods in the theory of conformal mapping, in *Calculus of Variations and its Applications*, Proc. Symposia Appl. Math. 8 (L. M. Graves, editor), McGraw-Hill, New York, 1958, pp. 93 113.

68. Fredholm eigen values of multiply-connected domains. *Paci"c J. Math.* 9 (1959), 211 269.

69. (with G. Pólya) Sur la représentation conforme de l'extérieur d'une courbe fermée convexe. *C. R. Acad. Sci. Paris* 248 (1959), 2837 2839.

70. Extremum problems and variational methods in conformal mapping, in *Proceedings of the International Congress of Mathematicians, Edinburgh, 1958*, Cambridge University Press, New York, 1960, pp. 211 231.

71. Analytical theory of subsonic and supersonic flows, in *Handbuch der Physik*, Springer-Verlag, Berlin, 1960, Vol. 9, Part 3, pp. 1 161.

72. (with Z. Charzyński) A new proof of the Bieberbach conjecture for the fourth coefficient. *Arch. Rational Mech. Anal.* 5 (1960), 187 193.

73. (with Z. Charzyński) A geometric proof of the Bieberbach conjecture for the fourth coefficient. *Scripta Math.* 25 (1960), 173 181.

74. (with N. S. Hawley) Connections and conformal mapping. *Acta Math.* 107 (1962), 175 274.

75. (with P. L. Duren) A variational method for functions schlicht in an annulus. *Arch. Rational Mech. Anal.* 9 (1962), 260 272.

76. (with J. Siciak) Transfinite diameter and analytic continuation of functions of two complex variables, in *Studies in Mathematical Analysis and Related Topics; Essays in Honor of George Pólya* (G. Szego, editor), Stanford University Press, Stanford, California, 1962, pp. 341 358.

77. (with P. L. Duren) The theory of the second variation in extremum problems for univalent functions. *J. Analyse Math.* 10 (1962/63), 193 252.

78. Fredholm eigenvalues and conformal mapping. *Rend. Mat.* (5) 22 (1963), 447 468.

79. (with E. Reich) Estimates for the transfinite diameter of a continuum. *Math. Z.* 85 (1964), 91 106.

80. (with R. Adler and M. Bazin) *Introduction to General Relativity*, McGraw-Hill, New York, 1965.

81. (with B. Epstein) On the mean-value property of harmonic functions. *J. Analyse Math.* 14 (1965), 109 111.

82. (with G. Springer) Fredholm eigenvalues and conformal mapping of multiply connected domains. *J. Analyse Math.* 14 (1965), 337 378.

83. (with O. Tammi) On the fourth coefficient of bounded univalent functions. *Trans. Amer. Math. Soc.* 119 (1965), 67 78.

84. (with O. Tammi) The fourth coefficient of a bounded real univalent function. *Ann. Acad. Sci. Fenn. Ser. A I* 354 (1965), 32 pp.

85. (with L. Sario and M. Glasner) The span and principal functions in Riemannian spaces. *J. Analyse Math.* 15 (1965), 115 134.

86. (with P. R. Garabedian and G. G. Ross) On the Bieberbach conjecture for even *n*. *J. Math. Mech.* 14 (1965), 975 989.

87. (with N. S. Hawley) Half-order differentials on Riemann surfaces. *Acta Math.* 115 (1966), 199 236.

88. Half-order differentials on Riemann surfaces. *SIAM J. Appl. Math.* 14 (1966), 922 934.

89. (with O. Tammi) A method of variations for functions with bounded boundary rotation. *J. Analyse Math.* 17 (1966), 109 144.

90. A variational method for univalent quasiconformal mappings. *Duke Math. J.* 33 (1966), 395 411.

91. (with P. R. Garabedian) The local maximum theorem for the coefficients of univalent functions. *Arch. Rational Mech. Anal.* 26 (1967), 1 32.

92. (with N. S. Hawley) Riemann surfaces which are doubles of plane domains. *Paci"c J. Math.* 20 (1967), 217 222.

93. (with O. Tammi) On the fourth coefficient of univalent functions with bounded boundary rotation. *Ann. Acad. Sci. Fenn. Ser. A I* 396 (1967), 26 pp.

94. Univalent functions whose *n* first coefficients are real. *J. Analyse Math.* 18 (1967), 329 349.

95. On the coefficient problem for univalent functions. *Trans. Amer. Math. Soc.* 134 (1968), 95 101.

96. (with O. Tammi) On bounded univalent functions which are close to identity. *Ann. Acad. Sci. Fenn. Ser. A I* 435 (1968), 26 pp.

97. (with O. Tammi) On the coefficient problem for bounded univalent functions. *Trans. Amer. Math. Soc.* 140 (1969), 461 474.

98. (with J. A. Hummel) Coefficient inequalities for Bieberbach-Eilenberg functions. *Arch. Rational Mech. Anal.* 32 (1969), 87 99.

99. Some distortion theorems in the theory of conformal mapping. *Atti Accad. Naz. Lincei Mem. Cl. Sci. Fis. Mat. Natur. Sez. Ia* (8) 10 (1970), 1 19.

100. (with R. N. Pederson) Further generalizations of the Grunsky inequalities. *J. Analyse Math.* 23 (1970), 353 380.

101. (with H. G. Schmidt) A new set of coefficient inequalities for univalent functions. *Arch. Rational Mech. Anal.* 42 (1971), 346 368.

102. (with O. Tammi) A Green's inequality for the power matrix. *Ann. Acad. Sci. Fenn. Ser. A I* 501 (1971), 15 pp.

103. (with G. Schober) An extremal problem for the Fredholm eigenvalues. *Arch. Rational Mech. Anal.* 44 (1971/72), 83 92.

104. (with R. N. Pederson) A proof of the Bieberbach conjecture for the fifth coefficient. *Arch. Rational Mech. Anal.* 45 (1972), 161 193.

105. Inequalities in the theory of univalent functions, in *Inequalities III* (O. Shisha, editor), Academic Press, New York, 1972, pp. 311 319.

106. (with G. Schober) A remark on the paper "An extremal problem for the Fredholm eigenvalues." *Arch. Rational Mech. Anal.* 46 (1972), 394.

107. (with R. J. Adler, J. Mark, and C. Sheffield) Kerr geometry as complexified Schwarzschild geometry. *J. Mathematical Phys.* 14 (1973), 52 56.

108. Article in *How to Write Mathematics*, American Mathematical Society, Providence, R.I., 1973, pp. 49 61.

109. (with J. Hersch and L. E. Payne) Some inequalities for Stekloff eigenvalues. *Arch. Rational Mech. Anal.* 57 (1975), 99 114.

110. (with R. Osserman) Doubly-connected minimal surfaces. *Arch. Rational Mech. Anal.* 58 (1975), 285 307.

111. (with G. Schober) Coefficient problems and generalized Grunsky inequalities for schlicht functions with quasiconformal extensions. *Arch. Rational Mech. Anal.* 60 (1975/76), 205 228.

112. (with G. Schober) A distortion theorem for quasiconformal mappings, in *Advances in Complex Function Theory, Maryland 1973/74* (W. E. Kirwan and L. Zalcman, editors), Lecture Notes in Math. 505, Springer-Verlag, Berlin, 1976, pp. 138 147.

113. (with S. Friedland) Global results in control theory with applications to univalent functions. *Bull. Amer. Math. Soc.* 82 (1976), 913 915.

114. (with G. Schober) Representation of fundamental solutions for generalized Cauchy-Riemann equations by quasiconformal mappings. *Ann. Acad. Sci. Fenn. Ser. A I Math.* 2 (1976), 501 531.

115. (with S. Friedland) On coefficient regions of univalent functions. *J. Analyse Math.* 31 (1977), 125 168.

116. (with J. A. Hummel) Variational methods for Bieberbach-Eilenberg functions and for pairs. *Ann. Acad. Sci. Fenn. Ser. A I Math.* 3 (1977), 3 42.

117. (with G. Schober) A variational method for general families of quasiconformal mappings. *J. Analyse Math.* 34 (1978), 240 264.

118. (with H. Samelson) Dedicated to the memory of Stefan Bergman. *Applicable Anal.* 8 (1978/79), 195 199 (1 plate).

119. (with E. Netanyahu) On the monotonicity of some functionals in the family of univalent functions. *Israel J. Math.* 32 (1979), 14 26.

120. (with G. Schober) An application of the calculus of variations for general families of quasiconformal mappings, in *Complex Analysis, Joensuu 1978* (I. Laine, O. Lehto, and T. Sorvali, editors), Lecture Notes in Math. 747, Springer-Verlag, Berlin, 1979, pp. 349 357.

121. (with J. A. Hummel and B. Pinchuk) Bounded univalent functions which cover a fixed disc. *J. Analyse Math.* 36 (1979), 118 138.

122. (with G. Schober) The dielectric Green's function and quasiconformal mapping. *J. Analyse Math.* 36 (1979), 233 243.

123. (with A. Chang and G. Schober) On the second variation for univalent functions. *J. Analyse Math.* 40 (1981), 203 238.

124. Stefan Bergman (1895 1977): in memoriam. *Ann. Polon. Math.* 39 (1981), 5 9 (1 plate).

125. Fredholm eigenvalues and Grunsky matrices. *Ann. Polon. Math.* 39 (1981), 149 164.

126. (with D. Aharonov and L. Zalcman) Potato kugel. *Israel J. Math.* 40 (1981), 331 339.

127. (with P. L. Duren and Y. J. Leung) Support points with maximum radial angle. *Complex Variables Theory Appl.* 1 (1982/83), 263 277.

128. (with L. Bowden) *The Role of Mathematics in Science*, Mathematical Association of America (New Mathematical Library, Vol. 30), Washington, D. C., 1984.

129. George Pólya (1887 1985). *Math. Mag.* 60 (1987), 268 270.

130. Pólya's contributions in mathematical physics, *Bull. London Math. Soc.* 19 (1987), 591 594. [Part of obituary of George Pólya, *ibid.*, pp. 559 608.]

131. (with P. L. Duren) Grunsky inequalities for univalent functions with prescribed Hayman index. *Paci"c J. Math.* 131 (1988), 105 117.

132. (with P. L. Duren) Conformal mappings onto nonoverlapping regions, in *Complex Analysis* (J. Hersch and A. Huber, editors), Birkhäuser Verlag, Basel, 1988, pp. 27 39.

133. (with P. L. Duren) Sharpened forms of the Grunsky inequalities. *J. Analyse Math.* 55 (1990), 96 116.

134. (with P. L. Duren) Goluzin inequalities and minimum energy for mappings onto nonoverlapping regions. *Ann. Acad. Sci. Fenn. Ser. A I Math.* 15 (1990), 133 150.

135. (with P. L. Duren) Univalent functions which map onto regions of given transfinite diameter. *Trans. Amer. Math. Soc.* 323 (1991), 413 428.

136. (with P. L. Duren) Robin functions and energy functionals of multiply connected domains. *Paci"c J. Math.* 148 (1991), 251 273.

137. (with P. L. Duren) Robin functions and distortion of capacity under conformal mapping. *Complex Variables Theory Appl.* 21 (1993), 189 196.

138. Issai Schur: Some personal reminiscences, in *Mathematik in Berlin*, Vol. II, (H. Begehr, editor), Shaker Verlag, Aachen, 1998, pp. 177 181.

Editors Note. Two items, which appear in lists of Schiffer's publications prepared in the 1940s but are absent from later lists, have been omitted:

Lösung der Aufgabe 128, *Jber. Deutsch. Math.-Verein.* **42** (1933), Abt. 2, 118 119.

(with S. Sambursky) Static universe and nebular red shift. II, *Phys. Rev.* (2) **53** (1938), 256 263.

Doctoral Students of M. M. Schiffer

The following list gives the names of the 20 students at Stanford for whom Schiffer served as doctoral advisor and the titles of their dissertations.

Year	Name	Title of Dissertation
1954	Ralph H. Pennington	Surface instabilities on pulsating gas bubbles
1963	Hubert Halkin	On the necessary condition for optimal control of nonlinear systems
1963	Frank B. Thiess	An analytic representation of the invariant distributions of quantum field theory
1965	John L. Troutman	Eigenvalues and eigenfunctions of a class of potential operators in the plane
1965	James W. Daniel	The conjugate gradient method for nonlinear operator equations
1965	Donald R. Snow	Reachable regions and optimal controls
1967	James W. Burrows	Estimates for Fredholm eigenvalues of plane domains
1967	Gerald S. Goodman[1]	Univalent functions and optimal control
1969	Richard G. Kellner	Some eigenvalue problems connected with potential theory

---

[1] Dissertation written under Charles Loewner; Schiffer became official advisor on Loewner's death.

Year	Name	Title of Dissertation
1970	Duane W. DeTemple	Generalizations of the Grunsky-Nehari inequalities
1970	Barnabas B. Hughes	The De Numeris Datis of Jordanus De Nemore: A critical edition, analysis, evaluation and translation
1970	Keith A. Rose	On a class of extremal elliptic operators
1972	Dennis A. Hejhal	Theta-functions, kernel functions and Abelian integrals
1973	Douglas J. Nelson	Extremal problems in the class of Bieberbach-Eilenberg functions
1975	William H. Barker	Plane domains with hyperelliptic double
1976	Andrew N. Harrington	Some extremal problems in conformal and quasiconformal mapping
1977	Yusuf Avc	Quadrature identities and the Schwarz function
1977	Donald K. Krecker	Extremal problems in quasiconformal mappings with side conditions
1979	Daniel B. Pouquet-Barthez	Extremal problems in the class of bounded univalent functions with side conditions
1985	John A. Velling[2]	Spherical geometry and the Schwarzian differential equation

---

[2]N.S. Hawley was principal advisor; Schiffer is listed as second advisor.

Chronology of Menahem Max Schiffer

1911	Born on September 24 in Berlin, the eldest of three children of Chaim and Miriam (Alpern) Schiffer
1917 1930	Primary and secondary education at Hindenburg-Oberrealschule zu Berlin-Wilmersdorf
1930 1933	Studies at Friedrich-Wilhelms-Universität in physics under Erwin Schrödinger, Walther Nernst, and Max von Laue, and in mathematics under Issai Schur, Erhard Schmidt, Ludwig Bieberbach, and Richard von Mises; first contacts with Stefan Bergman
1933	Leaves Germany for Mandatory Palestine in response to rise of Nazis
1933 1934	Studies at the Hebrew University of Jerusalem
1934	First paper [1] published in *Mathematische Zeitschrift* Awarded M.A. in mathematics by Hebrew University for thesis on invariant theory
1934 1939	Research student in mathematics at Hebrew University under direction of Michael Fekete; contacts with Otto Toeplitz
1937	Marries Fanya Rabinovich
1938	Research stay in Paris; further contacts with Bergman Method of boundary variation [5] published in *Proceedings of the London Mathematical Society*
1939	Receives first Ph.D. in mathematics awarded by Hebrew University, with thesis *Conformal representation and univalent functions*
1939 1943	Junior assistant in Department of Theoretical Physics, Hebrew University
1943	Method of interior variation [13] published in *American Journal of Mathematics*

1943 1946	Instructor in Department of Theoretical Physics, Hebrew University
1946 1949	Research Lecturer at Harvard University; work with Bergman, von Mises, and Garabedian
1947	Promotion (in absentia) to Lecturer in Theoretical Physics, Hebrew University
1949 1950	Visiting Professor of Mathematics, Princeton University; work with D.C. Spencer
1950	Invited lecture "Variational methods in the theory of conformal mapping" [40] at International Congress of Mathematicians, Cambridge, Massachusetts
1950 1951	Professor of Applied Mathematics, Hebrew University
1951 1952	Visiting positions at Harvard and Stanford; membership in Stanford's Applied Mathematics and Statistics Lab
1952	Appointed Professor of Mathematics, Stanford University
1953	*Kernel Functions and Elliptic Differential Equations in Mathematical Physics* [51], with Stefan Bergman, Academic Press, published
1954	*Functionals of Finite Riemann Surfaces* [55], with D.C. Spencer, Princeton University Press, published
1954 1959	Executive Head of Stanford Mathematics Department
1955	Proof of Bieberbach Conjecture for $n = 4$ [60], with P.R. Garabedian, published in *Journal of Rational Mechanics and Analysis*
1958	Plenary lecture "Extremum problems and variational methods in conformal mapping" [70] at International Congress of Mathematicians, Edinburgh
1965	*Introduction to General Relativity* [80], with Ronald Adler and Maurice Bazin, McGraw-Hill, published
1967	Appointed Robert Grimmett Professor of Mathematics (first endowed chair in mathematics at Stanford)
1968	Elected to American Academy of Arts and Sciences
1970	Elected to U.S. National Academy of Sciences
1973	Awarded honorary doctorate by Technion Israel Institute of Technology Elected foreign member of Finnish Academy of Sciences

1976	Recipient of Dean's Award for Teaching in the Stanford School of Humanities and Sciences
1977	Retires from Stanford as Professor Emeritus
1978 1983	Appointment as Visiting Professor in the Faculty of Mathematics of the Technion; home base remains at Stanford
1984	*The Role of Mathematics in Science* [128], with Leon Bowden, Mathematical Association of America, published
1993	Last research paper [137] (with Peter Duren) published in *Complex Variables: Theory and Application*
1997	Dies on November 8 in Palo Alto, California

Personal Reminiscences

Recollections of Menahem Max Schiffer

Paul R. Garabedian

Max Schiffer was a truly admirable mathematician. Already as an undergraduate at Brown, I read some of his earliest papers about variational methods in conformal mapping. Right after George Springer and I arrived at Harvard to become graduate students there, George discovered Schiffer over in Stefan Bergman's office with Richard von Mises, and he urged me to go introduce myself. I quickly made friends with Max and we started a fruitful collaboration that lasted for twenty years. We soon wrote a joint paper that preceded my Ph.D. thesis, which was stimulated by the activities in Bergman's office. I learned how to write mathematical papers from Schiffer and he taught me his generous outlook on joint work. His advice has guided me through all the remaining phases of my career.

Schiffer's variational method for conformal mapping is one of the most original and effective tools of modern analysis and applied mathematics. The generalizations to fluid dynamics and plasma physics have had a significant effect on computer codes that have turned out to be very successful in those fields. The theory is a delightful combination of complex analysis and partial differential equations. In his work Schiffer made remarkable contributions to both pure and applied mathematics. His good taste in elegant proofs and his solid grasp of mathematical physics were a source of inspiration for those with the good fortune to become his collaborators. His optimism and imagination made it a joy to go explain to him a theorem whenever one had the good luck to find something new.

P. Duren and L. Zalcman (eds.), *Menahem Max Schiffer: Selected Papers Volume 1,*
Contemporary Mathematicians, DOI 10.1007/978-0-8176-8085-5_1,
© Springer Science+Business Media New York 2013

Memories of Menahem Schiffer

Robert Finn

Although Menahem (Max) Schiffer was the person who formally appointed me at Stanford and who was "Executive Head" of the Mathematics Department when I arrived there in 1959, I had little direct professional contact with him during his lifetime. But his mathematical stature, inner dignity, and personal nobility were unmistakable, and I profited in profound ways from them during my tenure at Stanford. The occasional general lectures he gave were major events not to be missed. He saw each topic in active context with the whole of science and had an amazing talent for developing the material in all its ramifications, displaying insights and interconnections with a clarity that left me with a sense of awe and revelation. He did this invariably without use of lecture notes; it simply flowed from within him, providing the listener a clear awareness of the majestic achievements that arise in great science.

At Stanford as elsewhere, there were differences of view on practical decisions, and Max's perspectives did not always prevail. He felt very strongly that all colleagues should be treated equally and that they should be allowed full independence in deciding their focus of activity, their frequency of publication, and their relations with students. In the developing world of salary competition for "recognized" investigators and lecturers, that was not to happen.

Max was also intent on providing support for individual activity in every possible way. I recall a particular instance during my initial year at Stanford, when a coauthor came to visit me while we prepared a paper. Max observed our collaboration and offered him a position in the department. That put me into a difficult situation, as this person was trained as an engineer and knew little formal mathematics. As it turned out, the matter could be happily resolved by letting the "Aero and Astro" Department know of his potential availability; he became a valuable and productive member of that department until his death a few years ago.

In another instance, I recall Max pleading at a department meeting (with tears in his eyes) that we retain a nontenured colleague, who was a fruitful and original thinker but had difficult relations with some department members and also with administrators. I had personally gained scientific insight from my contacts with that colleague, and I felt a deep relief and gratitude that the decision came out favorably.

It was shortly after my arrival at Stanford that Max sought an appointment for Felix Browder, with a view to filling a gap in functional analysis in our course offerings. The department was united on the choice; but that was during the McCarthy era, and Felix's father was then head of the U.S. Communist Party. Felix volunteered that he was not politically active beyond being a registered Democrat, but the Stanford administration nevertheless refused to accept him. Max suffered the decision with dignity and grace, and Felix went on to a distinguished career at other universities, receiving among many honors the National Medal of Science.

P. Duren and L. Zalcman (eds.), *Menahem Max Schiffer: Selected Papers Volume 1*,
Contemporary Mathematicians, DOI 10.1007/978-0-8176-8085-5_2,
© Springer Science+Business Media New York 2013

Max's tenure at Stanford was marked by a pronounced expansion in the composition of the Mathematics Department. He had himself been appointed by Gabor Szego, who also brought us Stefan Bergman and Charles Loewner among others. Despite being rebuffed on the Browder appointment, Max continued the trend and was instrumental in a succession of appointments bringing strong support to those already present and adding still further stature. Stanford developed into one of the world's few major centers for classical analysis. And during his tenure as department head, Max also found time to complete his proof with Garabedian of the basic result $|a_4| \leq 4$ for the celebrated "coefficient problem."

Max had many informal personal contacts with people closely connected to developing political events in his time. He was able to observe the unfolding events, with a cognizance and perspective available to few others. I thought often and also suggested to him that he write his memoirs, which could have provided a unique insight into the political and social changes of the period, and the motivations of those responsible for major decisions. I continue to consider it a loss for reason in our universe that he did not do so.

Menahem Schiffer embodied for me the highest personal qualities, of profound scientific attainment joined with human dignity and with consideration for the needs and respect for the achievements of others. He was generous in every sense and incapable of personal dispute or of vindictive motivation. And he was also the model scientist of my idealized youthful dreams, who was animated by a compelling need to comprehend and describe the elusive physical laws that govern our universe. He remains for me a central and convincing example that the naive dreams of one's youth can in fact come to fruition on this generally more "realistic" and "practical" earth.

Working with Max Schiffer

Peter Duren

The first few years after graduate school can be a fragile time in a mathematician's career. Looking back, I see how fortunate I was to have had Max Schiffer as a postdoctoral mentor and early collaborator. In order to describe the circumstances that brought us together, I have to begin with a few words of personal history.

As a graduate student at MIT in the late 1950s, I specialized in functional analysis and was especially attracted by its interplay with problems of classical analysis. Among my teachers in those days were Norman Levinson, Alberto Calderón, Joseph Walsh (at Harvard), Klaus Roth (visiting from London), Elias Stein, Jürgen Moser, and my doctoral adviser Gian-Carlo Rota. Through their inspirational teaching and my own reading (for instance of Zeev Nehari's book *Conformal Mapping*), I developed a passion for analysis. I hoped for a postdoctoral position at a place strong in analysis and was delighted when Stanford offered me a two-year instructorship starting in the fall of 1960.

Stanford in 1960 was a major center of classical analysis, with a faculty that included George Pólya, Gabor Szego, Charles Loewner, Stefan Bergman, Max Schiffer, Bob Finn, David Gilbarg, Gordon Latta, and Sam Karlin. Younger members were Paul Garabedian (just leaving for NYU when I arrived), Halsey Royden, Bob Osserman, and Karel de Leeuw. Ralph Phillips and Paul Cohen came in 1961. That year I shared an office with Jaap Korevaar, a visitor who gave some fascinating talks on Tauberian theorems in

Szego's seminar. In 1960 1961, I audited Szego's course on orthogonal polynomials and Schiffer's course on extremal problems for univalent functions. Both were tremendously inspiring and opened new worlds of mathematics to me. I was much impressed by Schiffer's polished style of lecturing. Before a lively audience of 20 or 30 students and faculty in the Colloquium room, he developed and applied the elegant variational methods that he had pioneered. His presentation was formal but always clear and enthusiastic. He lectured without a scrap of notes, undeterred by long complicated formulas which he wrote rapidly and accurately on the blackboard. Two of the students produced lecture notes and distributed them to the class. Schiffer hoped those notes could be developed into a book, and for years afterward he worked sporadically with one of the note-takers toward that end, but ultimately the project was abandoned.

At some point in Schiffer's course, he used a variational method to derive the elegant inequality $|b_2| \leq \frac{2}{3}$. I was intrigued by the result and managed to derive it in a more elementary way from the Grunsky inequalities. I mentioned this to Schiffer after one of his lectures, and as I recall he asked me to come to his office and show him the details. This I believe was how we first became acquainted. He encouraged me to generalize and publish my little observation and proposed that we work together. As it turned out, this was the beginning of a collaboration that would continue for thirty years, resulting in ten

P. Duren and L. Zalcman (eds.), *Menahem Max Schiffer: Selected Papers Volume 1*,
Contemporary Mathematicians, DOI 10.1007/978-0-8176-8085-5_3,
© Springer Science+Business Media New York 2013

joint papers. It was also the beginning of what eventually became a close friendship.

We began on a project that Schiffer proposed, to work out the second variation and apply it to the Bieberbach conjecture. It was a wonderful opportunity for me to learn variational techniques from the master, and our eventual publication [77] was an important asset in the early stages of a professional career. We worked together in Schiffer's office in the old Sequoia Hall, two specified days per week from 10:00 a.m. to noon, as I recall. My new mentor, I soon learned, was well organized and accustomed to working on a regular schedule. But when Dieter Gaier came through to give a Colloquium talk, he made an interesting conjecture about conformal mappings of an annulus. It was an extremal problem which Schiffer believed a variational method could handle, and he proposed that we interrupt our work on the second variation to pursue that idea. We then devised a suitable variational technique which solved Gaier's problem and had other applications. Our work appeared in the paper [75].

In 1962 I moved to Michigan and pursued other mathematical interests, especially problems related to H^p spaces, but Max and I met frequently at conferences on complex analysis. In the late 1960s he invited me back to Stanford to work on a new idea he had conceived, but that project proved intractable and eventually we had to admit defeat. During the 1970s I began working again in geometric function theory, even writing a book on univalent functions, but Max and I were not collaborating then. His main collaborator during that period was Glenn Schober, with whom I had also begun to work.

Then came the Schiffers' visit to Michigan. Max had retired from Stanford in 1977. We invited him to spend the fall semester of 1980 in Ann Arbor, teaching a graduate course on topics of his choice. His wife Fanya came willingly but without enthusiasm, apparently expecting Ann Arbor to be something of a cultural desert. The Schiffers were housed in an apartment near campus, a short walk from the Mathematics Department in Angell Hall, where Max lectured on applications of variational methods to a variety of extremal problems in mathematics and physics. Our mathematical collaboration got off to a good start when we discussed the problem of maximum radial angle of the arc omitted by a support point in the standard class S of univalent functions. The upper bound of $\pi/4$ had been known for several years, and the sharpness of that bound had just been shown by a simple example with radial angle $\pi/4$ at the tip of its omitted arc, actually a half line in that case. The problem was then to describe all such extremal functions. Max quickly realized that a support point with maximum radial angle would solve two independent extremal problems, so that its omitted arc would be a trajectory of two quadratic differentials whose quotient would yield a purely algebraic relation. Following this idea, we found the unexpected result that the known example was essentially unique, leading to the paper [127]. Y. J. Leung was added as a coauthor when he came to town after Max had left and we managed to uncover additional information.

Meanwhile, my wife Gay and I were getting to know Max and Fanya better than ever before. Since the Schiffers were in Ann Arbor without a car, Gay was taking Fanya along on trips to the supermarket, where Fanya would complain that the fruit was cheaper in California, and better. But on the brighter side, Fanya was finding the public library a good source of reading material, and she discovered to her delight that world-class musicians were performing in Hill Auditorium, a large concert hall very close to their apartment. Together with Max and Fanya that fall, we attended concerts by Rostropovich, Vladimir Horowitz, and Rudolph Serkin, among others. Fanya was thrilled, and she had to revise any notion that the town was a cultural desert. Max seemed to enjoy the outings, but we got the impression that concerts were more Fanya's passion than his. Otherwise, we spent many an evening together, often in the company of other local mathematicians, conversing on topics of common interest and learning more about each others' personal lives. For instance, we learned of Max's passion for collecting early editions of books important in the history of science. Later I was privileged to be shown some of that collection at his home in Palo Alto.

Following his visit to Ann Arbor, Max and I entered a period of intense collaboration. Max was working steadily, searching for new applications of his variational methods. When he hit upon an especially promising idea, he would telephone me and describe it in broad terms, and when I expressed interest he suggested that I visit him at Stanford to work on it. Then he would write out a more detailed preliminary account, in a familiar scrawl that I had learned to decipher, and send it by mail prior to my visit. Upon my arrival in California, we spent a social evening together, then Max and I would agree to meet in his office at 10:00 the next morning. Our working sessions typically involved Max doing calculations at the blackboard while I raised questions, made suggestions, and took careful notes. After a series of straightforward calculations, Max would pause to gather his thoughts, then proclaim in his thick German accent, "Now comes zee trick." And indeed the "trick" was always an ingenious argument.

Upon return to Michigan, my job was to work through the notes, supply missing details, and produce a first draft of our paper. I soon became aware of a simple recipe for the success that Max enjoyed with variational methods. Generally speaking, a variational method will provide information that an extremal function maps onto the complement of a network of analytic arcs which are trajectories of a quadratic differential whose form depends on the functional to be maximized. That information can be effectively parlayed to quantitative information when the quadratic differential involves a perfect square. In that case, at least in principle, the square root can be integrated to produce a specific sharp inequality. It became clear to me that Max was a wizard at devising interesting functionals whose variational calculus would generate quadratic differentials with perfect squares. Examples can be found, for instance, in our papers [131, 132, 133, 134]. The paper [135] is an exception, but the functional considered there leads to standard elliptic integrals.

I cannot recall my time at Stanford without thinking of Charlotte Austin, later Charlotte Crabtree. Charlotte was already installed as Schiffer's secretary when I arrived in 1960, and she typed our two early papers. When I first began working with Max, she welcomed me warmly and took me under her wing like a mother hen. She was completely devoted to Max and his work, and she did him great service until her retirement some time in the 1980s. In return, Max and Fanya treated her like a member of their family. Soon after retirement, Charlotte suffered a severe stroke which left her partly paralyzed and unable to speak. Full of compassion, the Schiffers drove to her home every Sunday for a long visit.

During the 1970s and 1980s a series of very stimulating conferences on complex analysis were held in the research institute at Oberwolfach, in the rural Black Forest area of Germany. I had the good fortune to attend some of those conferences, but Max was never there. Finally the young organizers, knowing of our connection, asked me to intervene. "We keep inviting Schiffer to come," they said, "but he always says he has a conflict. We know the real reason he never comes. Will you please tell him that none of us was born when all of that happened?" The next time I saw Max, I relayed their message. He made clear that his refusal to attend was not a protest, but his memories of Germany were just so painful that he couldn't bear to return.

At that point I understood why he almost never used the German language. Only his first paper was written in German. I had noticed that at international conferences he rarely conversed in German, even when talking to native Germans. However, when he lectured in English, little German words such as "aber" would often creep into his speech.

Our last collaboration came in 1989, when I spent the winter and spring quarters at Stanford. During that visit we completed the papers [133, 134, 136], and we laid the groundwork for the paper [137] on Robin capacity, as we called it, the last and arguably the best of our ten joint papers. I wish Max had lived to see the subsequent developments. Following publication of our paper, the concept of Robin capacity has been embraced and studied by a number of mathematicians.

In September of 1991 I made a short visit to Stanford to participate in the Mathematics Department's celebration of Max's 80th birthday. I greeted Max and asked how he had been. His reply was a total surprise. "I'm fine," he said. "Fanya and I went to Berlin." When I expressed astonishment, he explained, "Fanya had a frequent flyer ticket, and she looked to see what was the farthest place she could go with it. That turned out to be Berlin, so we went." He then told me that on arrival in Berlin they had found a large convention in progress and no hotel rooms were available. A local agent asked if they would mind staying in a private home. They agreed and were assigned by chance to the home of an old woman who lived in the neighborhood where Max had grown up. She knew some of the same people Max had known before the war and could tell him what had happened to them. For Max this visit was a wonderful healing experience. Had it occurred earlier, I believe he would have returned to Germany, perhaps for conferences at Oberwolfach. But Max was finally beginning to show his age. Painfully, he told me he had decided to stop doing mathematical research. "Let it be," he said. "Let it be."

Memories of Max Schiffer

Lawrence Zalcman

I first met Max Schiffer in the summer of 1968, when I arrived at Stanford. I was a fresh Ph.D. from MIT, trained mainly in functional analysis, but with a burgeoning interest in complex function theory. After seven years in New England, California had beckoned like the Garden of Eden; but the decision to go to Stanford was based largely on its exceptional strength in complex analysis. It was immediately apparent that even in so distinguished a department as Stanford's, Schiffer stood out as someone special. (Today, we would say he was one of the department's two "superstars"; the other, of course, was Paul Cohen.) In addition to Max's exceptional intellectual power, there was a great breadth and depth of learning to back it up. His very extensive knowledge of classical mathematics (including algebra, which he had studied under Schur) was complemented by a professional competence in physics; and he was extremely widely read in the history and philosophy of science, as well. In this connection, I recall the following incident, which is representative. I had been reading René Thom's *Structural Stability and Morphogenesis*, a heady blend of mathematics, biology, and pre-Socratic philosophy; the experience left me in a state of excited confusion. I turned to Max for guidance, lending him the book for the weekend. By the following Monday, he had read it, assimilated it, and was able to give me some excellent advice on how to distinguish the wheat from the chaff.

During our years at Stanford, Max and Fanya showed us many kindnesses, opening their home to us on more than one occasion to meet distinguished visiting mathematicians. Max also played an important role in smoothing my entry into the community of complex analysts. I had seen an announcement for a symposium on complex analysis to be held during the summer of 1973 at the University of Kent in Canterbury. This promised to be the event of the year in function theory; virtually everyone who was anyone in complex variables was scheduled to attend. I had some results I was eager to present, but there was a problem: I had no invitation. When I turned to Max for advice, he suggested that I contact the organizers and mention that I was writing at his suggestion. It worked like a charm! In the event, the symposium exceeded all expectations, with a number of talks presenting results that are now considered classical. Nevanlinna pronounced it the best conference *he* had ever attended. My lecture, which to this day I view as my real debut as a complex analyst, turned out to be even more successful than I dared hope; and the broad smile and enthusiastic reception with which Max greeted me afterwards remain cherished memories.

At Stanford, Max and I ran a weekly seminar, at which graduate students and faculty presented lectures based on the current literature in complex variables. We also discussed mathematics on a regular basis and even worked on some problems together, though we did not publish any joint papers at that time. (Ultimately, this was to be remedied; more on that later.) Eventually, I left

P. Duren and L. Zalcman (eds.), *Menahem Max Schiffer: Selected Papers Volume 1*, Contemporary Mathematicians, DOI 10.1007/978-0-8176-8085-5_4, © Springer Science+Business Media New York 2013

Stanford for the University of Maryland, but that did not cut off our contact with the Schiffers. In fact, at Maryland, we finally had the opportunity to reciprocate some of their hospitality. We also ran into them regularly during (their and our) frequent visits to Israel. Once we moved to Israel for good in 1985, our personal contacts became more limited, though we did see them a couple of times on their visits there. An additional professional connection developed in 1987, when I became editor of *Journal d Analyse Mathématique*. Max was then the senior member of the editorial board, having been associated with *Journal d Analyse* since its founding in 1951.

It was during one of our earlier stays in Israel, in 1980, that Max and I finally managed to write a joint paper. Our coauthor was Dov Aharonov of the Technion. Let me describe the main result very briefly. According to a famous result of Isaac Newton, a solid homogeneous ball exerts on other bodies a gravitational attraction identical to that exerted by a point mass (of total mass equal to that of the ball) situated at the center of the ball. Now suppose that one has a solid homogeneous body P whose gravitational attraction on other bodies is identical to that of some point mass. Must P be a ball? This problem was posed by Lee Rubel, who called P a "potato." Max, Dov, and I proved that the answer is "yes," thus establishing the converse to a 300 year old theorem of Newton. This result extends (in various ways)

to collections ("sacks") of disjoint potatoes, to not necessarily homogeneous (though appropriately regular) potatoes, to unbounded potatoes, and even to rotten potatoes. We called the paper "Potato Kugel" since our initial result asserts that the potato must be a Kugel (German for "ball"). Of course, the title is a bit of a joke, as potato kugel is also the name of a traditional eastern European Jewish delicacy, a kind of potato pudding. Further sidelights on this paper appear in the commentary on [126].

Finally, something should be said about Max's optimism. It was real, and it was contagious; and it definitely had a beneficial effect on my approach to mathematics. I'm sure that many others can say the same.

Max Schiffer died on November 8, 1997. His wife Fanya and daughter Dr. Dinah S. Singer (Chief of the Molecular Regulation Section of the Experimental Immunology Branch of the Center for Cancer Research and Director of the Division of Cancer Biology at the National Cancer Institute) generously donated his large personal library to Bar-Ilan University. Ensconced there in a separate room of the Mathematics Library, the Schiffer Collection, with its extensive holdings in mathematics, physics, philosophy, and history, attracts scholars and researchers from within the university community and beyond. It is a fitting memorial to one of the true giants of twentieth-century complex analysis.

Some Reminiscences of My Thesis Advisor, Max Schiffer

Dennis Hejhal

Analytic functions were my first real passion in mathematics. As an undergraduate at the University of Chicago, I think I first encountered the name "M. M. Schiffer" early in 1968 in connection with a research project that I was pursuing aimed at trying to explicate a somewhat nonstandard type of kernel function associated with multiply connected plane domains and circular slit mappings.

Two years later, while attending my first meeting in complex analysis (at Michigan State), I had the very good fortune of having Larry Zalcman approach me to strike up a conversation about the type of work I liked to do. This led to a seminar invitation from Stanford, where, in May 1970, I was thrilled to meet not only Max Schiffer but also Stefan Bergman and Halsey Royden. The warmth and excitement that I felt during that visit prompted me to rethink my earlier plans and decide to attend Stanford (instead of Chicago) for my graduate work, starting that September.

I spent two years at Stanford. Building on the work I had already done at Chicago, I was able to begin doing research at Stanford virtually from day one. At the same time, I was encouraged to continue the reading project of some years' standing that I had undertaken on the advice of Lars Ahlfors so as to deepen my knowledge of complex function theory. Though for the most part I liked to work independently, I have fond memories of the many meetings I had with both Professors Schiffer and Royden (as I addressed them then) wherein I'd first give a progress report on my latest work, following which we'd chat about a variety of new problems, angles, or ideas.

Max had more of a formal, "old-world" style than did Royden. As such, my meetings with him would generally take place in his office at a preset time every 2 weeks or so. Max's broad interests and extensive knowledge ensured that there was never any shortage of things to talk about even when progress had been a bit slow. On a personal level, Max was always cordial and invariably upbeat.

During my time at Stanford, I obtained results in perhaps a half-dozen different directions. The result of most interest to Max grew serendipitously out of some reading in several complex variables that I had been doing in early 1971. In working through the second volume of Osgood's *Lehrbuch der Funktionentheorie*, I was struck by the development of the Klein prime form $\Omega(t, \tau)$ in the setting of Fuchsian uniformization. Its transformation law resembled that of a mixture of a $-\frac{1}{2}$ order differential and a Riemann theta function $\theta[u_\alpha(t) - u_\alpha(\tau) - g_\alpha]$ (an object discussed by Osgood a few chapters hence). Several months earlier, Max had given me a reprint of his 1966 *Acta* paper with Newton Hawley on half-order differentials; upon reviewing that, I immediately began to suspect that their Szego kernel function $\Lambda(p, q)$ admitted a simple representation in terms of θ and $\Omega(t, \tau)$.

Not being overly knowledgeable about theta functions at that point, I naturally turned to the library. There, after looking first at Fricke and

P. Duren and L. Zalcman (eds.), *Menahem Max Schiffer: Selected Papers Volume 1*, Contemporary Mathematicians, DOI 10.1007/978-0-8176-8085-5_5, © Springer Science+Business Media New York 2013

Klein's two-volume classic on automorphic forms, I soon encountered a somewhat dilapidated, dust-covered copy of H. F. Baker's 1897 treatise, *Abel s Theorem and the Allied Theory of Theta Functions*. Within that, I was surprised to find the explicit development of a variant of both $\Omega(t, \tau)$ and $\Lambda(p, q)$ for a broad class of Schottky groups, together with a link to theta functions. By adapting Baker's discussion and reworking part of Osgood's, I was readily able to verify the validity of the formula that I had begun to suspect must hold for $\Lambda(p, q)$. Max became excited when I showed him my result. He immediately suggested trying to use the formula to determine the constants a_{jk} in the classical kernel function identities

$$4\pi \widehat{K}(z; \bar{\xi})^2 = K(z; \bar{\xi}) + \sum_{j,k} a_{jk} w'_j(z) \overline{w'_k(\xi)};$$

$$4\pi \widehat{L}(z; \xi)^2 = L(z; \xi) + \sum_{j,k} a_{jk} w'_j(z) w'_k(\xi).$$

In this, however, the road remained blocked until, after several weeks' fumbling, it finally hit me (in June 1971) that one needed to "back up" and view the foregoing relations as a limiting case of a more basic identity involving theta functions and a third variable η. (Max's comment to me that Eq. (III 26) in the *Acta* paper "had to be significant" was, in effect, the "push" that I needed.)

Max was very pleased with the final result, namely, a formula for a_{jk} as a quotient of theta constants. He suggested that preparing a comprehensive write-up of the topic would form an excellent thesis. I concurred; and, in this way, Max became my Ph.D. advisor.

My thesis defense took place early in the summer of 1972. As occasionally happens at such, the event turned out to be something of a "nervous affair" (hardly my finest hour). I do still smile, though, at one of the "curves" tossed my way during the question segment. Out of the blue, Max decided to ask me about a technique that I had used some years earlier in a work aimed at proving the real-analytic dependence of certain domain functionals. Somehow I drew a blank on what he was seeking and found myself standing there in embarrassed silence for 25 30 seconds.

George Pólya was sitting next to Max. (I lived in a small apartment attached to Pólya's house during my Stanford years; he and I often talked.) Pólya was 84 at the time and becoming hard of hearing. Wondering why I had stopped, he leaned over to Max and whispered, "Max, Max, what was the question?" In a louder whisper, readily audible to me, Max replied: "I am *asking* him for the Schwarz alternating principle." I, of course, instantly continued, "Ah, yes, the Schwarz . . . ."

Max Schiffer was indeed a very kind man.[1]

After leaving Stanford, I tended to see Max chiefly during the summer months on my occasional return visits to the Bay Area. In addition to any mathematical discussions with him, these visits would sometimes include the always-welcome treat of enjoying Fanya and Max's warm hospitality over a dinner in their home near campus. (Max's friendliness quickly made it impossible for me to continue my student habit of addressing him as "Professor.") I especially remember a dinner, a couple years after Stefan Bergman died, at which Max shared with us a number of extremely funny stories about Bergman's Harvard and early Stanford days. Max had a role in several of these; he was clearly very fond of Bergman.

Though, in time, our areas of primary mathematical interest began to diverge, Max always took an interest in whatever I was doing. For my part, as I look back, I am struck by the number of papers of mine whose origin can be traced back to ideas (sometimes embryonic) that are already present in the Schiffer Hawley *Acta* paper from 1966: things involving monodromy groups, accessory parameters, Poincaré series, moduli, etc. Their paper also played a big role

---

[1] I would be remiss if I failed to mention here a second curve one that arrived about a year later. Following a lecture in Maryland, I was surprised to learn that, at approximately the same time as I, John Fay had independently found a formula for a_{jk} equivalent to mine. Fay's work utilized a deeper analysis of theta functions, which enabled him to get hold of the a_{jk} in the setting of a broader, much more versatile identity than the (straightforward, 3-variable) one that I had employed for this purpose. See Fay's 1973 Springer Lecture Notes and also Mumford's *Tata Lectures on Theta II*, p. 224.

in kindling my nascent interest in automorphic forms, including on the historical side of things (e.g., by inspiring me to take a serious look at the second volume of Poincaré's *Oeuvres*).

While, on one level, Max represented for me "the classical, old-style, European professor," Max was also a man with a ready sense of both humor and optimism. I really liked that about him. I still smile, for instance, at his habit of closing his New Year's greeting cards to me with the words "Good Luck and Good Theorems!" I also remember the time (it must have been around 1979) that Max and I found ourselves booked on the same flight from San Francisco to Ann Arbor, where there was a meeting of some sort. We agreed to meet near the airline counter so we could arrange to try to sit together. When Max arrived, I noticed that he was not attired in a coat and tie (as he normally was). I casually remarked, "Max, don't you usually wear a tie? You're a professor." He looked at me and, without skipping a beat,

pronounced with an air of authority, "When you travel, it's OK to go in *civilian* clothing."

In 1990, I visited the Palo Alto area for about a month; Max let me use his office in the department. One day, when he and I were both there, we got to talking about "life's ups and downs" and I asked him if he had some secret to maintaining his zest for research work, indeed in multiple areas, over all these years. (He was then 78.) I was genuinely struck by Max's response. He said, "You know, in everything I've ever done, I've always viewed myself as an amateur. This has enabled me to continually try new things and never worry about what I come up with." Max gave me something to seriously think about that afternoon.

Though formally I was "with" Max for only two years, the extent of his influence on me extends far beyond that. I will always be grateful to him for the example he set and for what he taught me about being a mathematician.

Max Schiffer and the Technion

Dov Aharonov

Max Schiffer's contributions to Israeli mathematics in general, and to the Technion and its Faculty of Mathematics in particular, deserve special mention. Schiffer visited the Technion frequently, from the 1950s into the 1980s. In 1973, he was awarded an honorary doctorate by the Technion; and in 1978, after his retirement from Stanford, he received a long-term appointment in the Technion's Department of Mathematics, which enabled him to visit the department whenever he pleased. (Paul Erdos also held such a position, and Haim Brezis currently enjoys such an appointment.)

During his long-term visits, Schiffer delivered lecture series on various subjects. The ones I remember best dealt with differential equations, integral equations, calculus of variations, fluid dynamics, and general relativity. The lectures were, quite simply, wonderful. Max's rare combination of detailed knowledge on an unusually wide front, together with his deep understanding of both mathematics and physics, made him a great teacher. He was also a fantastic lecturer. Students and professors from all the different faculties came to hear him (and many, dazzled by his exceptional erudition, stayed on to ask him questions concerning their own problems in mathematics, physics, and even engineering). I still recall Harold Shapiro's reaction when he learned about the great impact of Schiffer's lectures at the Technion: "I think you should translate the lecture notes into English and publish them properly." As Harold put it, "Schiffer belongs to the world." (It can still be done, I believe, and would be well worth the effort.)

In 1976, a special volume of *Journal d Analyse Mathématique* was published on the occasion of Max Schiffer's 65th birthday. It is, I think, a remarkable volume, with contributions from Agmon, Ahlfors, Garabedian, Gehring, Hayman, Hörmander, Lehto, Nehari, Pfluger, and Pólya, among many others. I am particularly proud that my first paper with Harold Shapiro on quadrature domains appears among the papers of that volume.

P. Duren and L. Zalcman (eds.), *Menahem Max Schiffer: Selected Papers Volume 1*,
Contemporary Mathematicians, DOI 10.1007/978-0-8176-8085-5_6,
© Springer Science+Business Media New York 2013

M. M. Schiffer, Explorer

Steven R. Bell

My favorite objects of complex analysis and potential theory for a finitely connected region in the plane are the Bergman kernel and projection, the Szego kernel and projection, the Garabedian kernel, the Poisson kernel, the Ahlfors map, the zeroes of the Szego kernel, solution operators to the inhomogeneous Cauchy Riemann equations, the harmonic measure functions and their derivatives, Green's function and operator for the Laplacian, and the Neumann kernel and operator for the Laplacian.

I have often felt envious of the pleasure that Max Schiffer must have taken in uncovering relationships between these objects. He and his collaborators were the first mathematicians to set foot on this continent of geometric analysis, and they very quickly charted out the main attractions. Name any two of the objects above, and you will find that Schiffer had a hand in finding a simple and beautiful relationship between them and then had the further pleasure of extending it all to finite Riemann surfaces.

After spending years looking for hidden valleys that Schiffer and his co-explorers might have missed, I think I know what drove Schiffer to come up with so many of his gems of analysis. The tantalizing relationships between these objects have often led me to feel just inches away from getting a stranglehold on one of them, that there is a formula for the object almost as simple and concrete as on the unit disc; but, alas, terms involving at least one abstractly defined object appear in the formula, so there is more to do. The origins of these results always trace back to papers of Schiffer and clearly follow in his tradition. In fact, I am now fairly certain that Schiffer was on the same quest that I have followed so fruitlessly these past years, namely, to lay hands concretely on Green's function of a multiply connected domain in the plane.

I never met Max Schiffer, but I have read so many of his papers and books that I feel as if I know the man. I wish he were here today. I'd like to be able to tell him about exciting developments in the theory of quadrature domains and double quadrature domains which lead me to believe that success in our long-term quest for a formula which expresses Green's function in finite terms may, finally, lie just around the corner.

P. Duren and L. Zalcman (eds.), *Menahem Max Schiffer: Selected Papers Volume 1*,
Contemporary Mathematicians, DOI 10.1007/978-0-8176-8085-5_7,
© Springer Science+Business Media New York 2013

Selected Papers

[1] Ein neuer Beweis des Endlichkeitssatzes für Orthogonalinvarianten

[1] Ein neuer Beweis des Endlichkeitssatzes für Orthogonalinvarianten. *Math. Z.* **38** (1934), 315–322.

Ein neuer Beweis des Endlichkeitssatzes für Orthogonalinvarianten.

Von

Max Schiffer in Berlin.

---

§ 1.

Hat man eine Form k-ten Grades in n Variablen $x_1, x_2, \ldots, x_n$, und unterwirft man diese einer linearen Substitution, die in bekannter Weise als $x = s(x')$ bezeichnet wird, aus einer Gruppe $\mathfrak{G}$, so ergibt sich das Formenproblem der Invariantentheorie folgendermaßen. Ist die Form als $f(a, x)$ gegeben, wobei a die Gesamtheit der Koeffizienten der Form repräsentiert, so ordne man die Form nach den x_r' um, so daß man $f(a, x) = f(a^s, x')$ erhält. Hierbei stellen die a^s eine lineare Substitution der a dar, die der Substitution s eindeutig zugeordnet ist. Man suche nun alle homogenen rationalen Funktionen $J(a)$ auf, für die gilt

$$J(a^s) = \gamma_s J(a),$$

wobei γ_s nur von s abhängt. Dies sind die Invarianten der Form in bezug auf die Gruppe $\mathfrak{G}$, auf deren Bestimmung es ankommt.

Wählt man als Substitution $x = s'(x')$, wobei s' die zu s transponierte Matrix ist, so unterliegen die a einer Gruppe $\mathfrak{Q}$ von linearen Substitutionen $Q(s)$, welche die Gruppe $\mathfrak{G}$ darstellen. Zu jeder Substitution s aus $\mathfrak{G}$ gibt es also genau eine zugeordnete Substitution $Q(s)$ aus $\mathfrak{Q}$, so daß, wenn $Q(s)$ und $Q(t)$ den Matrizen s und t aus $\mathfrak{G}$ entsprechen, ihr Produkt $Q(s) \cdot Q(t)$ der Matrix $s \cdot t$ aus $\mathfrak{G}$ zugeordnet ist. Eine solche Darstellung werde im folgenden als Homomorphismus zu $\mathfrak{G}$ bezeichnet. Die Gruppe $\mathfrak{Q}$ ist also der Gruppe $\mathfrak{G}$ homomorph.

Man kann dann von vornherein folgende Frage aufwerfen. Man suche alle homogenen rationalen Funktionen irgendeiner Variablenreihe a, die invariant bleiben, wenn man die a einer Substitution $L(s)$ aus einer zu $\mathfrak{G}$ homomorphen Gruppe $\mathfrak{L}$ unterwirft. Dies Problem enthält dann als Spezialfall das Formenproblem. Im folgenden soll nur noch von den zu einem solchen Homomorphismus gehörenden Invarianten gesprochen werden.

Im Vordergrunde bei allen diesen Betrachtungen steht nun die Frage nach einer endlichen Basis. Unter dieser ist in der Invariantentheorie eine Gesamtheit von endlich vielen Invarianten zu verstehen, aus denen sich alle anderen ganz und rational mit konstanten Koeffizienten aufbauen lassen. Aus dem Hilbertschen Formensatz ist die Existenz einer endlichen Basis sofort herzuleiten, wenn man einen Operator O mit folgenden Eigenschaften herleiten kann:

1. $O[f(a)] = F(a)$ ist entweder Null oder eine Invariante vom gleichen Grade wie $f(a)$.

2. $O[f(a) + g(a)] = O[f(a)] + O[g(a)]$.

3. Ist $J(a)$ eine Invariante, so ist $O[J(a) \cdot g(a)] = J(a) \cdot O[g(a)]$.

4. $O[J(a)] = c J(a)$, wobei c eine von Null verschiedene Konstante ist.

Einen solchen Operator gewann z. B. Hilbert für die projektive Gruppe mit Hilfe des Ω-Prozesses.

§ 2.

Nun hat Hurwitz[1]) einen Integrationsprozeß entwickelt, der bei jeder abgeschlossenen kontinuierlichen Gruppe $\mathfrak{G}$ in bezug auf den Homomorphismus aller $Q(s)$ anwendbar ist und der einen Operator der eben beschriebenen Art liefert. Diese Methode läßt sich auch ohne weiteres auf alle stetigen Homomorphismen dieser Gruppen anwenden. Der wichtigste Fall, an dem auch Hurwitz seine Methode entwickelt, ist der Fall der reellen Drehungsgruppe. Deren stetige Homomorphismen sind übrigens alle rational, wie Herr Schur[2]) gezeigt hat.

Bei der reellen Drehungsgruppe geht Hurwitz folgendermaßen vor. Er beschränkt sich nur auf absolute Invarianten, da eine Erweiterung auf beliebige Invarianten dann unschwer erfolgen kann. Er betrachtet den n^2-dimensionalen Raum aller reellen Matrizen mit n^2 Elementen und in ihm den Unterraum $\mathfrak{R}$ aller orthogonalen Matrizen. Geht nun a^s durch Anwendung irgendeines Homomorphismus $L(s)$ der Drehungsgruppe aus a hervor, ist ferner $f(a)$ eine beliebige Funktion der Variablen $a_1, \ldots, a_m$, so wende man folgenden Operator H an:

$$H[f(a)] = \int_{\mathfrak{R}} f(a^s) \, d s = F(a).$$

---

[1]) Über die Erzeugung der Invarianten durch Integration. Göttinger Nachrichten 1897, S. 71.

[2]) Neue Anwendung der Integralrechnung auf Probleme der Invariantentheorie. Sitzungsber. d. Berl. Akad. 1924, S. 297.

Hierbei ist ds das Volumenelement in $\Re$, dessen Bestimmung das Kernstück der Untersuchung von Hurwitz ist.

Man erkennt, daß der Operator H den Bedingungen 2, 3, 4 ohne weiteres genügt, wenn unter $J(a)$ absolute Invarianten gegenüber dem betreffenden Homomorphismus $L(s)$ verstanden werden. Aber auch 1 ist erfüllt. Denn es ist

$$F(a^t) = \int_{\Re} f(a^{s\,t})\, ds,$$

wenn auch t eine Drehung ist. Nun ist im n^2-dimensionalen Matrizenraum die Multiplikation aller Matrizen mit einer festen orthogonalen Matrix nur eine Drehung des Raumes, bei der überdies der Unterraum $\Re$ in sich übergeht. Dann ist bekanntlich nach den Regeln der Integralrechnung $ds = d(s \cdot t)$, und somit ist

$$F(a^t) = \int_{\Re} f(a^{s\,t})\, d(s \cdot t) = \int_{\Re} f(a^s)\, ds = F(a).$$

Damit ist die Invarianz von $F(a)$ nachgewiesen.

Ebenso wie hier

$$\int_{\Re} f(a^{s\,t})\, ds = \int_{\Re} f(a^s)\, ds$$

bewiesen wurde, kann man auch ohne Mühe folgern:

$$\int_{\Re} f(a^{t\,s})\, ds = \int_{\Re} f(a^s)\, ds.$$

Diese Tatsache wird später eine wesentliche Rolle spielen.

§ 3.

Bei der rechnerischen Durchführung des Hurwitzschen Prozesses kommt es vor allem darauf an, eine Parameterdarstellung der Drehungsgruppe zu finden. Dazu bedarf es einer näheren Untersuchung dieser Gruppe. Es ist daher von Interesse, daß ein anderer Integrationsprozeß existiert, der mit dem Hurwitzschen Prozeß im wesentlichen identisch ist und der rein formal zu entwickeln ist, ohne nähere Eigenschaften der orthogonalen Gruppe zu gebrauchen. Besonders bemerkenswert ist es, daß die Abgeschlossenheit dieser Gruppe nicht benutzt wird. Einen ähnlichen Prozeß benutzt schon Herr Schur[3]) im projektiven Fall.

---

[3]) Neue Anwendungen der Integralrechnung auf Probleme der Invariantentheorie I. Sitzungsber. d. Berl. Akad. 1924, S. 189.

Zur Bildung des neuen Prozesses braucht man einen Hilfssatz, der sich schon bei Kelland und Tait[4]) und allgemeiner bei Frobenius[5]) findet, und auf den mich mein Freund, Herr W. Ledermann, aufmerksam machte. Ist nämlich τ eine beliebige Matrix von nicht verschwindender Determinante, so ist

$$\sigma(\tau) = (\tau\,\tau')^{-1/2}\,\tau$$

eine orthogonale Matrix.

Man sieht sofort, daß rein formal wirklich $\sigma(\tau)\,(\sigma(\tau))' = E$ ist, wenn E die Einheitsmatrix bedeutet. Denn es ist

$$\sigma(\tau)\cdot(\sigma(\tau))' = (\tau\,\tau')^{-1/2}\,\tau\,\tau'\,(\tau\,\tau')^{-1/2} = E.$$

Hierbei ist unter $(\tau\,\tau')^{-1/2}$ jene Wurzelbildung verstanden, die schon Sylvester und Frobenius benutzten und die folgendermaßen definiert ist:

Es sei P eine symmetrische Matrix mit positiven Eigenwerten (was ja für $(\tau\,\tau')$ erfüllt ist); man kann dann eine reelle orthogonale Matrix Q so angeben, daß $Q\cdot P\,Q^{-1}$ in der Diagonalform

$$Q\,P\,Q^{-1} = \begin{pmatrix} \omega_1 & 0 & 0 & \dots & 0 \\ 0 & \omega_2 & 0 & \dots & 0 \\ \vdots & & \ddots & & \vdots \\ 0 & & & \ddots & 0 \\ 0 & 0 & 0 & \dots & \omega_n \end{pmatrix}$$

erscheint. Man nehme nun an, daß kein $\omega_\nu = 0$ wird und daß alle Wurzeln verschieden sind. Dann ist die Interpolationsaufgabe

$$a_0 + a_1\omega_i + \dots + a_{n-1}\omega_i^{\,n-1} = \omega_i^{-1/2} \quad \text{für } i = 1, 2, \dots, n$$

lösbar, und zwar erscheinen dann die a_ν in folgender Gestalt:

$$a_\nu = \frac{G_\nu(\sqrt{\omega_1}, \dots, \sqrt{\omega_n})}{\sqrt{\omega_1} \cdots \sqrt{\omega_n}\, V(\omega_1, \dots, \omega_n)}$$

Hierbei ist G_ν eine ganze rationale Funktion von $\sqrt{\omega_1}, \dots, \sqrt{\omega_n}$, und V ist die Vandermondesche Determinante $\displaystyle\prod_{\alpha < \beta}(\omega_\alpha - \omega_\beta)$. Bildet man nun

$$f(P) = a_0 + a_1 P + \dots + a_{n-1} P^{n-1} = Q^{-1} \begin{pmatrix} \omega_1^{-1/2} & 0 & & \dots & 0 \\ 0 & \omega_2^{-1/2} & & & \vdots \\ \vdots & & \ddots & & 0 \\ 0 & \dots & 0 & & \omega_n^{-1/2} \end{pmatrix} Q,$$

so ist das jedenfalls eine Funktion, die der Gleichung

$$f(P) \cdot f(P) = P^{-1}$$

---

[4]) Quaternions, 1873, Chapter 10.
[5]) Über die cogredienten Transformationen der bilinearen Formen. Sitzungsber. d. Berl. Akad. 1896, S. 7.

27

genügt. Man kann also $f(P) = P^{-1/2}$ setzen, und es ist bekannt, daß hierbei $f(P)$ nicht von der Wahl der transformierenden Matrix Q abhängt. Man hat damit also $P^{-1/2}$ eindeutig definiert, falls nur alle Werte ω_ν untereinander und von Null verschieden sind. Man kann nun aber (nach einer mündlichen Bemerkung von Herrn Schur) die a_ν in eine Gestalt bringen, an der man das Verhalten dieser Funktion studieren kann, wenn mehrere Wurzeln einander gleich werden oder wenn eine Wurzel verschwindet. Für $\sqrt{\omega_i} = \sqrt{\omega_k}$ verschwindet sowohl G_ν als auch V in der obigen Darstellung für a_ν; daher sind V und G_ν durch die Vandermondesche Determinante $\prod\limits_{\alpha < \beta} (\sqrt{\omega_\alpha} - \sqrt{\omega_\beta})$ teilbar. Man kürze in den a_ν aus, und man erhält im Nenner von a_ν abgesehen von dem Wurzelprodukt $\prod\limits_\alpha \sqrt{\omega_\alpha}$ nur die „Geminante" $\prod\limits_{\alpha < \beta} (\sqrt{\omega_\alpha} + \sqrt{\omega_\beta})$. Man ersieht daraus erstens, daß die oben definierte Bildung $f(P)$ auch noch Sinn hat, wenn zwei Wurzeln gleich werden und daß auch dann $f(P) \cdot f(P) = P^{-1}$ wird. Zweitens bemerkt man aber auch, daß $|P|^l f(P)$ für einen hinreichend großen Exponenten l eine stetige Funktion der Matrix wird, die für $P| \to 0$ gegen den Wert Null konvergiert. Denn in $|P|^{l-1} a_\nu$ treten ja nur Faktoren der Form $\dfrac{\sqrt{\omega_\alpha} \, \sqrt{\omega_\beta}}{\sqrt{\omega_\alpha} + \sqrt{\omega_\beta}}$ auf, die wegen der Ungleichung

$$\frac{ab}{a+b} < a, \quad \frac{ab}{a+b} < b \qquad \text{für } a > 0,\ b > 0$$

beschränkt sind, so daß $|P|^l a_\nu$ bereits gegen Null geht. $\big($Man sieht leicht ein, daß es genügt $l \geqq \left[\dfrac{n}{2}\right] + 2$ zu wählen, wenn $\left[\dfrac{n}{2}\right]$ die größte ganze Zahl unterhalb von $\dfrac{n}{2}$ ist.$\big)$ Dieses Verhalten von $|P|^l f(P)$ wird im folgenden eine wesentliche Rolle spielen. Übrigens hat Frobenius das Verhalten von $f(P)$ für den Fall studiert, daß $|P| \to 0$ geht. Er wies nach, daß $f(P)$ konvergiert und somit im Limes eine orthogonale Matrix darstellt.

§ 4.

Mit den eben abgeleiteten Hilfsmitteln kann man nun einen Integrationsprozeß definieren, der die in § 1 verlangten Eigenschaften hat. Ist wieder $f(a)$ eine beliebige Form, so wende man folgenden Operator O_l an:

$$O_l[f(a)] = \int\limits_{-\infty}^{+\infty} e^{-\sum\limits_{i,k}^{1 \ldots n} t_{ik}^2} |\tau|^{2l} f(a^{\sigma(\tau)})\, d\varrho = G_l(a).$$

Hierbei bedeutet $d\varrho$ das Volumenelement $d\xi_{11}\ldots d\xi_{nn}$. Es ist dabei über alle n^2 Elemente der Matrix $\tau = (\xi_{ik})$ von $-\infty$ bis $+\infty$ zu integrieren. $\sigma(\tau)$ ist dabei die in § 3 definierte Matrix, die zu τ eindeutig zugeordnet ist, und l ist ein Exponent, der ausreicht, um $|\tau|^{2l} f(a^{\sigma(\tau)})$ zu einer beschränkten Funktion der ξ_{ik} zu machen, was ja nach § 3 stets möglich ist. Wie in § 2 geht $a^{\sigma(\tau)}$ aus a hervor, indem man auf a die $\sigma(\tau)$ entsprechende Substitution aus irgendeinem stetigen Homomorphismus der Drehungsgruppe anwendet.

Auch bei diesem Operator sind die Eigenschaften 2., 3., 4. augenfällig. Denn $O_l(f)$ ist linear und homogen, und ist $J(a)$ eine Invariante gegen den betreffenden Homomorphismus, so ist

$$O_l[J(a) U(a)] = J(a) \cdot O_l[U(a)],$$

da $J(a^{\sigma(\tau)}) = J(a)$ ist. (Es werden hier wie bei Hurwitz nur absolute Invarianten untersucht.) Speziell ist

$$O_l[J(a)] = J(a) \int\limits_{-\infty}^{+\infty} e^{-\Sigma \xi_{ik}^2} |\tau|^{2l}\, d\varrho = J(a) \cdot c_l.$$

Hierbei ist der Faktor $c_l > 0$. Nachzuweisen bleibt also nur, daß $G_l(a)$ eine Invariante ist. Dazu bilde man

$$G_l(a^t) = \int\limits_{-\infty}^{+\infty} e^{-\Sigma \xi_{ik}^2} |\tau|^{2l} f(a^{\sigma(\tau)t})\, d\varrho.$$

Hierbei stellt t eine orthogonale Matrix dar. In diesem Fall ist aber

$$\sigma(\tau) \cdot t = (\tau\tau')^{-1/2}\tau t = (\tau t \cdot t'\tau')^{-1/2}\,\tau t = \sigma(\tau \cdot t).$$

Führt man ferner die Komponenten η_{ik} der Matrix $\lambda = \tau t$ ein, so gilt

$$\sum_{i,k}^{1\ldots n} \xi_{ik}^2 = \sum_{i,k}^{1\ldots n} \eta_{ik}^2;\quad |\tau| = |\lambda|;\quad d\varrho = d\eta_{11}\ldots d\eta_{nn} = d\mu.$$

Somit folgt

$$G_l(a^t) = \int\limits_{-\infty}^{+\infty} e^{-\Sigma \eta_{ik}^2} |\lambda|^{2l} f(a^{\sigma(\lambda)})\, d\mu = G_l(a).$$

Damit ist gezeigt, daß auch die Bedingung 1 erfüllt ist.

§ 5.

Nun ist es noch von Interesse, den Nachweis zu führen, daß der Hurwitzsche Prozeß mit dem Prozeß $O_l[f(a)]$ bis auf einen konstanten Faktor übereinstimmt. Der Beweisgedanke ist der gleiche wie der in dem analogen Fall der projektiven Gruppe von Herrn Schur[6]) entwickelte. Dort wurde nachgewiesen, daß jeder invariantenerzeugende Prozeß, der mit dem Hilbertschen Operator vertauschbar ist, mit diesem bis auf einen konstanten Faktor übereinstimmt.

Man bilde zum Beweis

$$H\left[G_l(a)\right] = \int\limits_{\Re}\left[\int\limits_{-\infty}^{+\infty} e^{-\Sigma \xi_{ik}^2} \dots d\varrho\right] ds,$$

wobei wieder $H\left[f(a)\right]$ den Hurwitzschen Operator bedeuten soll. Da der Integrand stetig ist und das uneigentliche Integral absolut und gleichmäßig konvergiert (jetzt kann ja die Abgeschlossenheit der Drehungsgruppe benutzt werden), sind die Integrationen vertauschbar. Also ist

$$H\left[G_l(a)\right] = \int\limits_{-\infty}^{+\infty} e^{-\Sigma \xi_{ik}^2} |\tau|^{2l}\left\{\int\limits_{\Re} f(a^{u(\tau)\cdot s})\, ds\right\} d\varrho.$$

Nun wurde schon in § 2 hervorgehoben, daß auch

$$\int\limits_{\Re} f(a^{ts})\, ds = \int\limits_{\Re} f(a^s)\, ds$$

ist, wofern t eine orthogonale Matrix ist. Also ist

$$H\left[G_l(a)\right] = \int\limits_{-\infty}^{+\infty} e^{-\Sigma \xi_{ik}^2} |\tau|^{2l}\left\{\int\limits_{\Re} f(a^s)\, ds\right\} d\varrho$$

und folglich

$$H\left[G_l(a)\right] = c_l\, H\left[f(a)\right].$$

Andererseits ist $G_l(a)$ eine Invariante und $H\left[f(a)\right]$ erfüllt die Bedingung 4. Also ist

$$H\left[G_l(a)\right] = h\, G_l(a),$$

wobei

$$h = \int\limits_{\Re} ds$$

ist. Also folgt

$$O_l\left[f(a)\right] = \frac{c_l}{h}\, H\left[f(a)\right].$$

Damit ist die Identität der beiden Operationen nachgewiesen, was zu einer bemerkenswerten Integralidentität Anlaß gibt. Dieselben Überlegungen zeigen auch, daß Prozesse der Art

$$\int\limits_{-\infty}^{+\infty} e^{-\left(\Sigma \xi_{ik}^2\right)^\nu} |\tau|^{2l}\, \varphi(|\tau|)\, f(a^{\sigma(\tau)})\, d\varrho$$

---

[6]) Neue Anwendungen der Integralrechnung usw. I, S. 194.

mit $\nu = 1, 2, \ldots$, und beliebigen Funktionen $\varphi(|\tau|)$, für die das Integral noch konvergiert, ebenfalls bis auf einen konstanten Faktor mit dem Hurwitzschen Prozeß identisch sind.

Rechnet man im binären Fall den Prozeß $O_t[f(a)]$ wirklich durch, so kann man die Integralidentität direkt rechnerisch bestätigen; man findet, daß das Integral von $O_t[f(a)]$ sich durch geeignete Transformationen in ein Produkt von drei Integralen zerlegen läßt. Davon führen zwei Integrale auf Konstanten, die nur von l abhängen. Das dritte Integral ist gerade $H[f(a)]$.

Berlin-Siemensstadt, den 11. März 1933.

(Eingegangen am 13. März 1933.)

Commentary on

[1] *Ein neuer Beweis des Endlichkeitssatzes für Orthogonalinvarianten*, Math. Z. **38** (1934), 315 322.

This is Schiffer's very first paper (and the only one he published under the name "Max Schiffer"), written while he was a student at Friedrich-Wilhelms-Universität (now Humboldt-Universität zu Berlin) and submitted shortly before he left Germany for Palestine. The story of how it came to be written is related in [138].

The subject matter of [1] belongs to algebra, more specifically, to classical invariant theory, for which see Weyl's magisterial exposition [W]. Hilbert had shown that for certain classical groups G, the ring of invariants of G is finitely generated. His proof was based on the existence of an appropriate projection operator from functions to invariants, which he constructed using a method known as Cayley's Ω-process. Subsequently, Hurwitz [H] showed that integration over a compact Lie group also yields such an operator. In [1], Schiffer treats the important case of the special orthogonal group $SO(n)$ and develops a different approach, also involving integration, which avoids some of the technical difficulties in [H], while generating essentially the same operator.

Perhaps surprisingly, Schiffer never published his most enduring achievement in invariant theory; that fell to no less a worthy than Hermann Weyl! Shortly before departing for Palestine in 1933, Schiffer deposited with his teacher Issai Schur a manuscript detailing his latest research results. A Hebrew version of this paper was accepted as Schiffer's master's thesis in mathematics by the Hebrew University of Jerusalem in 1934. Meanwhile, Schur had submitted the manuscript on Schiffer's behalf to *Compositio Mathematica*; however, it was rejected as being too abstract. Eventually, it came to the attention of Weyl, who, describing its contents as "a particularly straightforward and powerful process for the generation of invariants" [W, ix], published (with Schiffer's permission) an account of it under the title "A modified proof of the main theorem on invariants" as Supplement C in the second edition of *The Classical Groups* [W, pp. 300 303].

References

[H] A. Hurwitz, *Über die Erzeugung der Invarianten durch Integration*, Nachrichten von der Gesellschaft der Wissenschaften zu Göttingen, Mathematisch-Physikalische Klasse, 1897, 71 90.

[W] Hermann Weyl, *The Classical Groups: Their Invariants and Representations*, second edition, Princeton University Press, 1946.

LAWRENCE ZALCMAN

[2] Sur un principe nouveau pour l'évaluation des fonctions holomorphes

[2] Sur un principe nouveau pour l'évaluation des fonctions holomorphes. *Bull. Soc. Math. France* **64** (1936), 231–240.

SUR UN PRINCIPE NOUVEAU
POUR L'ÉVALUATION DES FONCTIONS HOLOMORPHES ;

Par Menahem Schiffer
à Jérusalem.

Dans la présente Note je veux montrer l'importance d'une nouvelle conception qui permet d'évaluer beaucoup de propriétés des fonctions holomorphes. J'obtiens d'ailleurs quelques théorèmes concernant les domaines, obtenus par la représentation du cercle-unité à l'aide des fonctions analytiques.

Nous nous occuperons de la famille des fonctions

$$(1) \qquad w = f(z) = z + a_2 z^2 + a_3 z^3 + \ldots$$

holomorphes dans le cercle-unité, et nous étudierons les ensembles-images qu'on obtient de ce cercle à l'aide de ces fonctions. Ce sont des domaines, éventuellement couverts plusieurs fois et de connection multiple, mais jamais ne contenant l'infini dans leur intérieur. Ainsi nous obtenons pour chaque fonction $f(z)$ un domaine D dans le plan w qui possède une frontière, éventuellement composée de plusieurs continus. Parmi eux il y en a un, que nous appellerons la « frontière extérieure » dont on peut relier es points avec l'infini sans passer par des points de D. C'est la conception de la frontière extérieure $\mathcal{R}$ qui joue un rôle important dans nos recherches.

1. L'ensemble continu $\mathcal{R}$ définit un domaine $\hat{D}$ de connection simple (contenant l'origine) dont il est la frontière, et ce domaine peut être représenté simplement sur le cercle-unité à l'aide d'une fonction univalente $\varphi(z)$, normée de manière que son inverse est de la forme

$$(2) \qquad F(z) = \gamma(z + \alpha_2 z^2 + \ldots) \qquad \text{avec} \qquad \gamma > 0.$$

La fonction (2) représente le cercle-unité simplement sur le domaine $\dot{D}$. Une méthode de conclusion intéressante dont M. Rogosinski s'est spécialement servi avec succès ([1]), fournit

$$(3) \qquad \gamma \geqq 1$$

parce que

$$(4) \qquad F^{-1}\{f(z)\} = \frac{1}{\gamma}z + \left(\frac{a_2}{\gamma} - \frac{\alpha_2}{\gamma^2}\right)z^2 + \ldots$$

représente le cercle-unité sur lui-même et que l'inégalité de Cauchy exige

$$\frac{1}{\gamma} \leqq 1.$$

Nous relions ce fait au théorème de Koebe-Faber sur les fonctions univalentes dans le cercle-unité. Ce théorème dit ([2]) que la frontière du domaine fourni par une fonction univalente dans le cercle-unité

$$(1') \qquad f(z) = \gamma z + \beta z^2 + \ldots$$

ne s'approche jamais de l'origine jusqu'à une distance moindre de $\frac{\gamma}{4}$. Nous obtenons ainsi le théorème I :

Théorème I. — *La frontière extérieure d'un domaine dans lequel une fonction (1) transforme le cercle-unité reste toujours à une distance plus grande que $\frac{1}{4}$ de l'origine* ([3]).

Nous nous servirons aussi de l'énoncé plus précis

$$(5) \qquad |\alpha| \geqq \frac{\gamma}{4}$$

([1]) Rogosinski, *Über den Wertevorrat einer analyt. Funktion. Schriften der Königsberger gelehrten Gesellschaft, Naturwissensch. Klasse* (8, Jahr, *Heft*, 1, 1931, p. 2, Satz A.).

([2]) P. Koebe, *Über die Uniformisierung der algebraischen Kurven II, Math. Annalen*, 69, 1910. S. 46. — G. Faber, *Neuer Beweis eines Koebe-Bieberbachschen Satzes...* (*Sitz. Per. d. math. phys. Kl. d. Kgl. Bayr-Akad. d. Wiss.*, 1916, S. 39-42).

([3]) M. Rogosinski a eu l'obligeance de m'informer que ce résultat a été déjà trouvé par lui-même et par M. Dieudonné.

si α est un point quelconque de la frontière extérieure de D. Nous verrons que la valeur de γ est importante pour beaucoup de propriétés du domaine D et ainsi indirectement aussi de la fonction $f(z)$. La frontière extérieure à son tour donne une évaluation par excès pour ce coefficient et de là l'importance de cette dernière.

2. Prenons par exemple la fonction

$$(4) \qquad F^{-1}\{f(z)\} = \frac{1}{\gamma}z + \left(\frac{a_2}{\gamma} - \frac{\alpha_2}{\gamma^2}\right)z^2 + \cdots$$

qui satisfait pour $|z| \leqq 1$ à l'inégalité

$$(6) \qquad |F^{-1}\{f(z)\}| \leqq 1.$$

La fonction

$$h(z) = \frac{\dfrac{F^{-1}\{f(z)\}}{z} - \dfrac{1}{\gamma}}{1 - \dfrac{F^{-1}\{f(z)\}}{z} \cdot \dfrac{1}{\gamma}}$$

$$= \frac{\left(\dfrac{a_2}{\gamma} - \dfrac{\alpha_2}{\gamma^2}\right)z + \cdots}{1 - \dfrac{1}{\gamma^2} - \left(\dfrac{a_2}{\gamma^2} - \dfrac{\alpha_2}{\gamma^3}\right)}z - \cdots = \frac{\left(\dfrac{a_2}{\gamma} - \dfrac{\alpha_2}{\gamma^2}\right)}{1 - \dfrac{1}{\gamma^2}}z + \cdots$$

satisfait de manière analogue pour $|z| \leqq 1$ à l'inégalité

$$|h(z)| \leqq 1.$$

L'inégalité de Cauchy exige donc

$$\left|\frac{a_2}{\gamma} - \frac{\alpha_2}{\gamma^2}\right| \leqq 1 - \frac{1}{\gamma^2}$$

ou

$$(7) \qquad |a_2| \leqq \gamma - \frac{1}{\gamma} + \frac{|\alpha_2|}{\gamma}.$$

D'après un théorème de M. Bieberbach sur les coefficients de fonctions univalentes ([1]) on a l'inégalité

$$(8) \qquad |\alpha_2| \leqq 2.$$

([1]) L. Bieberbach, *Über die Koeffizienten derj. Potenzreihen, welche eine schlichte Abbildung des Einheitskreises vermitteln* (*Berl. Ber.*, 940-955, 1916, p. 946).

Ainsi nous obtenons à l'aide de (5) et (7)

$$| a_2 | \leqq 4 | \alpha | + \frac{| \alpha_2 | - 1}{\gamma}.$$

Pour se servir de (3) il faut distinguer deux cas.

a. $| \alpha_2 | \geqq 1$. Alors (3) et (8) fournissent :

(9) $$| a_2 | \leqq 4 | \alpha | + 1.$$

b. $| \alpha_2 | \leqq 1$. Dans ce cas (9) vaut également.

Donc nous avons prouvé le théorème suivant :

THÉORÈME II. — *Soit* α *un point de la frontière extérieure d'un domaine donné par une fonction* (1)*; alors on a l'inégalité*

$$| a_2 | \leqq 4 | \alpha | + 1.$$

Ce théorème est le plus exact qu'on peut trouver, parce que pour

$$f(z) = \frac{z}{(1-z)^2} = z + 2 z^2 + 3 z^3 + \ldots$$

on a, en effet,

$$\alpha = - \frac{1}{4} \quad \text{et} \quad 2 = 4 . \left| - \frac{1}{4} \right| + 1.$$

On peut évaluer de la même manière les autres coefficients à l'aide de la frontière extérieure. A cette fin on se servira de l'algorithme de M. Schur pour les coefficients de fonctions bornées [1] dont nous avons fait ici le premier pas. Ces évaluations sont naturellement d'intérêt spécial, si la fonction $f(z)$ n'est pas bornée, mais possède une frontière extérieure. Au lieu du point le plus éloigné de l'origine du domaine, c'est-à-dire du module-maximum de la fonction dont se sert l'inégalité de Cauchy, nous nous servons dans l'évaluation d'un point choisi à volonté sur la frontière extérieure, même du plus rapproché.

D'un autre côté l'inégalité (9) fournit une évaluation intéressante pour α à l'aide de a_2. Nous constatons à l'aide de (9) qu'une valeur β, que la fonction (1) ne prend pas, doit être plus grande

[1] I. SCHUR, *Über Potenzreihen, die im Innern des Einheitskreises beschränkt sind* (*Journal für die reine und angew. Math.*, 147, 1916, p. 206-232).

que $\dfrac{|a_2|-1}{4}$ ou sera point d'un trou dans la représentation du cercle-unité, fourni par $f(z)$.

3. Nous nous occuperons de la même manière de l'inégalité de Jensen. Supposons que 0, β_1, $\ldots$, β_n soient des zéros de $f(z)$ se trouvant dans le cercle-unité. Dans ce cas ils seront aussi zéros de la fonction (4). Appliquons à cette fonction dont le module pour $|z| < 1$ est toujours < 1 l'inégalité de Jensen

$$(10) \qquad \left| \frac{1}{\beta_1 \beta_2 \ldots \beta_n} \right| \leqq \gamma \leqq 4\,|\,\alpha\,|.$$

Donc nous avons :

THÉORÈME III. — *Soient* 0, β_1, $\ldots$, β_n *des zéros d'une fonction* (1) *et* α *un point de la frontière extérieure correspondante; alors on a*

$$\left| \frac{1}{\beta_1 \beta_2 \ldots \beta_n} \right| \leqq 4\,|\,\alpha\,|.$$

4. Au lieu de l'inégalité de Schwarz qui se sert du maximum, nous trouverons une inégalité analogue également à l'aide de la frontière extérieure. Appliquons l'inégalité de Schwarz à la fonction (1) donnant pour $|z| < 1$,

$$(11) \qquad |\,F^{-1}\{f(z)\}\,| \leqq |\,z\,|.$$

Nous en tirons pour

$$f(z) = F\,[F^{-1}\{f(z)\}]$$

à l'aide de l'inégalité de M. Bieberbach

$$|\,F(\eta)\,| \leqq \frac{\gamma\,|\,\eta\,|}{(1-|\,\eta\,|)^2}$$

pour $|\,\eta\,| < 1$ et pour des fonctions univalentes (1′)

$$|\,f(z)\,| \leqq \frac{\gamma\,|\,z\,|}{(1-|\,z\,|)^2}$$

pourvu que $|\,z\,| \leqq 1$. En vertu de (5) nous avons donc :

THÉORÈME IV. — *Soit* $f(z)$ *une fonction de la famille* (1) *et α un point de la frontière extérieure correspondante; alors on a*

(12) $$|f'(z)| \leqq \frac{4|\alpha||z|}{(1-|z|)^2}$$

5. Transformons enfin en vue de la conception de la frontière extérieure l'inégalité

$$|g'(z)| \leqq |\frac{1-|g(z)|^2}{1-|z|^2},$$

valable dans le cercle-unité pour la dérivée d'une fonction (1) satisfaisant à $|g(z)| \leqq 1$. Appliquons cette inégalité à la fonction (4) ayant la dérivée

$$[F^{-1}\{f(z)\}]' = \frac{f'(z)}{F'\{F^{-1}[f(z)]\}}.$$

Nous obtenons

(13) $$|f'(z)| \leqq \frac{1}{1-|z|^2} \cdot |F'\{F^{-1}[f(z)]\}|.$$

D'après un théorème de MM. Bieberbach-Pick sur les fonctions univalentes ([1]), on a pour $|\eta| < 1$

$$|F'(\eta)| \leqq \frac{\gamma(1+|\eta|)}{(1-|\eta|)^3}.$$

Maintenant d'après l'inégalité de Schwarz nous avons

$$|F^{-1}[f(z)]| \leqq |z|.$$

Ainsi on a

$$|F'\{F^{-1}[f(z)]\}| \leqq \frac{\gamma(1+|z|)}{(1-|z|)^3},$$

et d'après (13) et (5)

(14) $$|f'(z)| \leqq \frac{4|\alpha|}{(1-|z|)^4}.$$

Donc :

THÉORÈME V. — *Soit $f(z)$ une fonction de la famille* (1) *et α un point de la frontière extérieure, alors on a*

$$|f'(z)| \leqq \frac{4|\alpha|}{(1-|z|)^4}.$$

([1]) PICK, *Über den Koebeschen Verzerrungssatz, Leipz. Ber.*, 68, 1916, p. 58-64.
BIEBERBACH, 1. c., p. 946.

On voit déjà quel est le principe commun à toutes ces évaluations et aussi qu'on peut trouver ainsi beaucoup d'autres inégalités. Mais nous quittons ces considérations pour nous occuper maintenant de propriétés géométriques de la frontière extérieure et de sa relation avec les autres continus de la frontière.

6. Soient α et β deux points de la frontière de D qu'on peut relier sans passer par des points de D. Nous construisons la fonction

$$(15 \qquad f_1(z) = \frac{f(z)}{1 - \frac{f(z)}{\alpha}} = z + b_2 z^2 + \ldots$$

appartenant aussi à la famille (1). Pour elle

$$\beta' = \frac{\beta}{1 - \frac{\beta}{\alpha}}$$

est un point de la frontière extérieure, parce que la transformation

$$(15') \qquad \omega_1 = \frac{\omega}{1 - \frac{\omega}{\alpha}}$$

fait correspondre β' à β, et ∞ à α, et à la courbe continue entre les deux points donnés une courbe semblable reliant les deux points nouveaux. Du théorème I nous déduisons

$$|\beta'| \geqq \frac{1}{4}.$$

Donc nous en tirons

THÉORÈME VI. — *Si α et β sont deux points d'un même trait continu de la frontière de D on a*

$$(16) \qquad \left| \frac{1}{\alpha} - \frac{1}{\beta} \right| \leqq 4.$$

Ce théorème est aussi valable pour toute paire de points de la frontière d'un trou de D et dans ce cas il présente un intérêt spécial. Interprétons-le géométriquement; supposons que β soit le point du trou le plus rapproché de l'origine. On a

$$(17) \qquad |\alpha - \beta| \leqq 4 |\alpha| \cdot |\beta|$$

pour tous les points α du trou. Cela nous donne une borne supérieure pour $|\alpha|$

$$(18) \qquad |\alpha| \leqq \frac{|\beta|}{1 - 4|\beta|}.$$

Donc

$$(19) \qquad |\alpha - \beta| \leqq \frac{4|\beta|}{1 - 4|\beta|}|\beta|.$$

Définissons l'angle ε par l'égalité

$$\sin\frac{\varepsilon}{2} = \frac{4|\beta|}{1 - 4|\beta|},$$

alors nous tirons de l'inégalité (17) que tous les points α se trouvent dans un cercle autour de β qu'on voit de l'origine sous un angle moindre de ε. Nous pouvons donc associer à tout nombre ε un rayon ρ de manière que nous voyons le trou de l'origine sous un angle moindre de ε, si un seul point du trou se trouve plus prochain de l'origine que ρ. M. Valiron a étudié plus profondément tous ces problèmes concernant les valeurs exceptionnelles ([1]).

7. Nous établirons enfin une relation intéressante entre les points des frontières différentes de D. Nous construisons la fonction automorphe

$$(20) \qquad g(z) = \gamma(z + a_2 z_2 + \ldots), \qquad \gamma > 0,$$

définie pour $|z| < 1$, qui représente son domaine fondamental contenant l'origine simplement sur le domaine $\hat{D}_1$ qu'on déduit de $\hat{D}$ en y supprimant le domaine T qui était un trou de D. L'inégalité de Schwarz fournit

$$(3') \qquad \gamma \geqq 1,$$

parce que

$$g^{-1}\{f(z)\} = \frac{1}{\gamma}z + \ldots$$

est holomorphe dans le cercle-unité et le représente sur lui-même. Le groupe de la fonction $g(z)$ est composé de translations non euclidiennes et possède un seul paramètre.

([1]) G. VALIRON, *Sur un théorème* de MM. KOEBE et LANDAU [*Bull. sc. Math.* (2), 51, p. 34-42].

La fonction

$$(21) \qquad G(z) = -\rho \log \left\{ 1 - \frac{g(z)}{\rho} \right\}, \qquad G(0) = 0,$$

est holomorphe dans le cercle-unité, si ρ est un point du trou. En des points différents homologues pour la fonction $g(z)$, $G(z)$ prend des valeurs qui se distinguent par des multiples de $2\pi i\rho$ différents de zéro. Ainsi nous voyons que la fonction

$$(22) \qquad G(z) = \gamma z + \left(\frac{\gamma^2}{2\rho} + \gamma a_2 \right) z^2 + \dots$$

est univalente. D'après l'inégalité (8) de Bieberbach nous trouvons

$$\left| \frac{\gamma}{2\rho} + a_2 \right| \leqq 2,$$

$$(23) \qquad \left| \frac{\gamma}{2\rho} \right| \leqq 2 + |a_2|.$$

De (3) et (9) suit

$$\left| \frac{1}{2\rho} \right| \leqq 4 \left(|\alpha| + \frac{3}{4} \right),$$

si α est un point de la frontière extérieure. Si nous appelons tout trait continu de la frontière qui n'est pas la frontière extérieure, une frontière intérieure, nous pouvons énoncer le théorème suivant :

THÉORÈME VII. — *Soit α un point de la frontière extérieure de g et ρ un point d'une frontière intérieure; alors ils satisfont à l'inégalité*

$$(24) \qquad \frac{1}{8} \leqq |\rho| \left(|\alpha| + \frac{3}{4} \right).$$

En combinant cette inégalité avec (5) nous obtenons le théorème suivant :

THÉORÈME VIII. — *Soit α un point de la frontière extérieure et ρ un point d'une frontière intérieure de D; alors ils satisfont à l'inégalité*

$$(25) \qquad |\alpha| \cdot |\rho| \geqq \frac{1}{32}.$$

En effet (25) suit immédiatement de (5), si $|\rho| \geqq \frac{1}{8}$, tandis que autrement (24) fournit

$$|\rho\alpha| \geqq \frac{1}{8} - \frac{3}{4} |\rho| \geqq \frac{1}{32}.$$

Les théorèmes VII et VIII montrent qu'un point très rapproché de l'origine d'une frontière intérieure force toute la frontière extérieure de rester à grande distance de l'origine. D'après VIII α et ρ se comportent réciproquement. Maintenant nous verrons que $|\alpha|$ se comporte asymptotiquement comme $|\rho| e^{\frac{\text{const.}}{|\rho|}}$.

8. Pour voir cela prenons la fonction (21). Si α est un point de la frontière extérieure de D, alors

$$-\rho \log\left\{ 1 - \frac{\alpha}{\rho} \right\}, \qquad \left| \mathcal{J} \log\left\{ 1 - \frac{\alpha}{\rho} \right\} \right| \leq \pi$$

sera un point de la frontière extérieure du domaine, obtenu du cercle-unité à l'aide de la fonction (21). Ainsi nous obtenons du théorème I :

THÉORÈME IX. — *Soit α un point de la frontière extérieure de D, ρ un point d'une frontière intérieure : alors ils satisfont à l'inégalité*

$$(26) \qquad \left| \rho \log\left\{ 1 - \frac{\alpha}{\rho} \right\} \right| \geq \frac{1}{4}$$

Pour $|\rho| \leq \frac{1}{8}$ nous sommes certains que

$$\left| 1 - \frac{\alpha}{\rho} \right| \geq \left| \frac{\alpha}{\rho} \right| - 1 \geq 1$$

parce que $|\alpha| \geq \frac{1}{4}$. Donc, pour des $|\rho|$ très petits,

$$\log\left| \left\{ 1 - \frac{\alpha}{\rho} \right\} \right|$$

n'est pas très grand que pour des $|\alpha|$ très grands. Pour l'ordre de grandeur de $|\alpha|$ nous tirons de (26) l'information suivante :

$$|\rho| \log |\alpha| \geq \frac{1}{4} + O\left(|\rho| \log \frac{1}{|\rho|} \right).$$

Toutes ces évaluations faites ici peuvent être améliorées ou extendues. Mais dans cette Note nous avons voulu seulement montrer les relations générales et attirer l'attention sur l'importance de la frontière extérieure.

Commentary on

[2] *Sur un principe nouveau pour l évaluation des fonctions holomorphes*, Bull. Soc. Math. France **64** (1936), 231 240.

In this early paper, Schiffer applies the Schwarz lemma and the classical growth and distortion theorems of geometric function theory to deduce similar inequalities for a general analytic function $f(z) = z + a_2 z^2 + \cdots$ in the unit disk $\mathbb{D}$. Let Ω denote the set of values assumed by f, and let $\widehat{\Omega}$ be the simply connected domain bounded by the outer boundary C of Ω. (Loosely speaking, $\widehat{\Omega}$ is obtained from Ω by filling in the holes.) Then by the Riemann mapping theorem there is a uniquely determined function $F(z) = \gamma z + A_2 z^2 + \cdots$, where $\gamma > 0$, which maps $\mathbb{D}$ conformally onto $\widehat{\Omega}$. Schiffer shows, for instance, that

$$|\alpha| \geq \frac{\gamma}{4} \geq \frac{1}{4}, \qquad |a_2| \leq 4|\alpha| + 1,$$

and $|f(z)| \leq \dfrac{4|\alpha||z|}{(1 - |z|)^2}$

for each point $\alpha \in C$ and all $z \in \mathbb{D}$.

Schiffer cites a 1931 paper of Rogosinski for what is now known as the method of subordination, but curiously fails to mention Rogosinski's major paper [R] of 1933 on the same topic. In later years, the technique of subordination was finely honed at the hands of Rogosinski, Goluzin, and others to produce a large body of interesting results. Details and further references can be found in the books [G, D].

References

[D] Peter L. Duren, *Univalent Functions*, Springer-Verlag, 1983.

[G] G. M. Goluzin, *Geometric Theory of Functions of a Complex Variable*, second edition, Izdat. "Nauka", Moscow, 1966; English transl., American Mathematical Society, 1969.

[R] Werner Rogosinski, *Zum Majorantenprinzip der Funktionentheorie*, Math. Z. **37** (1933), 210 236.

PETER DUREN

[4] Sur un problème d'extrémum de la représentation conforme

[4] Sur un problème d'extrémum de la représentation conforme. *Bull. Soc. Math. France* **66** (1938), 48–55.

SUR UN PROBLÈME D'EXTRÉMUM
DE LA REPRÉSENTATION CONFORME

PAR M. MENAHEM SCHIFFER

(Jérusalem).

1. Soit

$$(1) \qquad z = f(\zeta) = \zeta + \frac{\alpha_1}{\zeta} + \frac{\alpha_2}{\zeta^2} + \cdots$$

holomorphe et univalente dans le domaine $|\zeta| > 1$; alors on a pour le domaine D représenté par $f(\zeta)$ le théorème classique :

Si a et b sont deux points quelconques de la frontière C de D, on a toujours $|a - b| \leqq 4$.

Pour la fonction

$$(2) \qquad p(\zeta) = \zeta + \frac{1}{\zeta},$$

cette borne est atteinte en effet aux points $a = 2$, $b = -2$.

2. On est amené facilement à la généralisation suivante de cette question. On considère sur la frontière C n points $a_1, a_2, \ldots, a_n$ et l'on forme le produit des différences

$$(3) \qquad \mathcal{P}_n(C; a_1, a_2, \ldots, a_n) = \prod_{\nu < \mu} |a_\nu - a_\mu|.$$

Le maximum de cette fonction des a_i pour une frontière donnée C

$$(3') \qquad \mathcal{P}_n(C) = \operatorname*{Max}_{a_i \subset C} \left\{ \prod_{\nu < \mu} |a_\nu - a_\mu| \right\}$$

dépend de C seulement et par là de $f(\zeta)$. $\mathcal{P}_n(C)$ est borné pour n fixe, et, puisque la famille (1) est normale et fermée, il y a pour chaque n une fonction $f_n(\zeta)$ de la famille (1), pour laquelle $\mathcal{P}_n(C)$ atteint en effet sa borne supérieure. Nous appelons la fonction $f_n(\zeta)$ une fonction extrémale du problème du maximum de $\mathfrak{T}_n(C)$.

Dans la présente Note nous montrerons quelques propriétés géométriques de la frontière C_n, appartenant à une fonction extrémale $f_n(\zeta)$. Nous démontrerons que C_n est une courbe de Jordan, composée d'arcs analytiques, qui satisfont tous à un certain type d'équations différentielles. Nous ferons aussi une première application au problème des coefficients de la famille (1); en outre nous croyons aussi la méthode employée de quelque intérêt en elle-même.

3. Nous choisissons une frontière fixe C_n et désignons sur elle un point O, qui sera l'origine du plan des z. Nous coupons sur C_n, autour de O, une section $\mathfrak{S}$, de diamètre transfini ρ. C'est-à-dire qu'il existe une fonction

$$(4) \qquad \zeta = \varphi(z) = \rho\, \psi\left(\frac{z}{\rho}\right) = z + k\rho + \frac{\mathcal{B}_1 \rho^2}{z} + \frac{\mathcal{B}_2 \rho^3}{z^2} + \ldots,$$

qui représente l'extérieur de $\mathfrak{S}$ simplement sur le domaine $|\zeta| > \rho$. On voit aisément que la fonction $\psi(t)$ représente l'extérieur d'un continu de diamètre transfini 1 sur l'extérieur du cercle-unité. $\psi(t)$ possède donc pour fonction inverse une fonction de la famille (1). Nous remarquons que $\psi(t)$ et par là aussi K, $\mathcal{B}_1$; $\mathcal{B}_2, \ldots$ dépendent du choix de $\mathfrak{S}$. Si l'on choisit une suite de section $\mathfrak{S}(\rho)$ de manière que $\mathfrak{S}(\rho) \subset \mathfrak{S}(\rho')$ pour $\rho < \rho'$, ρ étant le diamètre transfini de $\mathfrak{S}(\rho)$, on peut supposer que $\psi(t)$ dépend d'un paramètre ρ et dans ce sens nous écrirons plus exactement $\psi_\rho(t)$, $K(\rho)$, $\mathcal{B}_1(\rho)$, $\mathcal{B}_2(\rho), \ldots$ s'il est nécessaire.

La fonction inverse de (4)

$$(5) \qquad z = \varphi^{-1}(\zeta) = \rho\, \psi^{-1}\left(\frac{\zeta}{\rho}\right) = \zeta - k\cdot\rho - \frac{\mathcal{B}_1 \rho^2}{\zeta} - \frac{(\mathcal{B}_2 + k\mathcal{B}_1)\rho^3}{\zeta^2} - \ldots$$

représente le domaine $|\zeta| > \rho$ simplement sur l'extérieur de $\mathfrak{S}$. Puisque $\psi^{-1}(\tau)$ appartient à la famille (1), on a par le théorème des aires

$$(6) \qquad 1 \geqq |\mathcal{B}_1|^2 + 2\,|\mathcal{B}_2 + k\mathcal{B}_1|^2 + \ldots.$$

On a en particulier $|\mathcal{B}_1| \leqq 1$.

D'un autre côté nous trouverons pour chaque α donné ($|\alpha| \leqq 1$) toujours des fonctions univalentes dans le domaine $|\zeta| > \rho$ et possédant un développement

$$(7) \qquad z = \zeta + \frac{\alpha \rho^2}{\zeta} + \ldots$$

Prenons par exemple la fonction

$$(7') \qquad z = q_\alpha(\zeta) = \rho \sqrt{\alpha}\, p\left(\frac{\zeta}{\rho \sqrt{\alpha}}\right) = \zeta + \frac{\alpha \rho^2}{\zeta}$$

qui est holomorphe et univalente pour $|\zeta| > \rho$ à cause de $|\alpha| \leqq 1$.
La fonction

$$(8) \qquad q_\alpha[\varphi(z)] = z + k\rho + \frac{(\alpha + \mathcal{B}_1)\rho^2}{z} + \ldots$$

est, pour tout $|\alpha| \leqq 1$, holomorphe et univalente dans l'extérieur de $\mathcal{S}$ et représente donc tout le domaine D_n (de frontière C_n) sur un domaine D'_n (de frontière C'_n), qu'on obtient aussi de $|\zeta| > 1$ à l'aide d'une fonction (1). Le domaine D_n appartenait à une fonction extrémale; nous trouvons donc

$$(9) \qquad \mathcal{P}_n(C'_n) \leqq \mathcal{P}_n(C_n)$$

ou

$$(9') \qquad \mathcal{P}_n(C'_n; a'_1, \ldots, a'_n) \leqq \mathcal{P}_n(C_n; a_1, \ldots, a_n)$$

si C'_n, a'_i résulte de C_n, a_i par la représentation (8) et si les a_i étaient l'ensemble de points, qui donnait à $\mathcal{P}_n(C_n; a_1, \ldots, a_n)$ son maximum sur C_n. Ainsi nous obtenons

$$(10) \qquad \prod_{\nu < \mu} \left| a_\nu - a_\mu + (\alpha + \mathcal{B}_1)\rho^2\left(\frac{1}{a_\nu} - \frac{1}{a_\mu}\right) + \rho^3(\ldots) \right| \leqq \prod_{\nu < \mu} |a_\nu - a_\mu|$$

ou

$$(10') \qquad \prod_{\nu < \mu} \left| 1 - (\alpha + \mathcal{B}_1)\rho^2 \frac{1}{a_\nu a_\mu} + \rho^3(\ldots) \right| \leqq 1,$$

$$(10'') \qquad \left| 1 - (\alpha + \mathcal{B}_1)\rho^2 \sum_{\nu < \mu} \frac{1}{a_\nu a_\mu} + \rho^3(\ldots) \right| \leqq 1.$$

L'inégalité $(10'')$ est valable pour tout ρ. Faisons converger ρ vers zéro; selon (6), $\mathcal{B}_1(\rho)$ est borné et il existe donc au moins une valeur limite $\mathcal{B}_0$ des $\mathcal{B}_1(\rho)$, et pour elle nous tirons de $(10'')$

$$(11) \qquad \mathcal{R}\left\{(\alpha + \mathcal{B}_0) \sum_{\nu < \mu} \frac{1}{a_\nu a_\mu}\right\} \geq 0.$$

$\mathcal{R}(x)$ désignant la composante réelle de x. Le coefficient α est encore quelconque soumis à la condition $|\alpha| \leq 1$. Puisqu'on a aussi $|\mathcal{B}_0| \leq 1$, (11) peut être satisfait seulement pour $|\mathcal{B}_0| = 1$. Soit donc $\mathcal{B}_0 = e^{i\theta}$. Un choix correspondant des α donne à la somme $\alpha + \mathcal{B}_0$ tous les arguments de l'intervalle ouvert

$$\left(-\frac{\pi}{2} + \theta, \frac{\pi}{2} + \theta\right).$$

Grâce à (11) nous avons

$$(12) \qquad \arg \sum_{\nu < \mu} \frac{1}{a_\nu a_\mu} = -\theta$$

ou

$$(12') \qquad \mathcal{B}_0 \sum_{\nu < \mu} \frac{1}{a_\nu a_\mu} \geq 0.$$

Il suit de (12), que l'argument θ de la valeur limite $\mathcal{B}_0$ est déterminé sans ambiguïté $\left(\text{excepté le cas } \sum_{\nu < \mu} \frac{1}{a_\nu a_\mu} = 0\right)$. Donc

$$\lim_{\rho \to 0} \mathcal{B}_1(\rho) = e^{i\theta}$$

existe.

4. De $\lim \mathcal{B}_1(\rho) = e^{i\theta}$ et (6) on tire

$$(13) \qquad \lim_{\rho \to 0} \left\{ \psi_\rho^{-1}(\tau) + k(\rho) \right\} = \tau - \frac{e^{i\theta}}{\tau} = p_\theta(\tau).$$

Cette fonction limite représente $|\tau| > 1$ sur l'extérieur d'un segment de droite, qui a la direction $\sigma = \pm e^{i\frac{\theta + \pi}{2}}$. Pour tout ε et des ρ suffisamment petits les points de frontière du domaine, représenté à l'aide de $\psi_\rho^{-1}(\tau)$ de $|\tau| > 1$, se trouvent donc tous à une distance moindre de ε d'une droite de direction σ, et puisque cette frontière contient toujours l'origine, ils se trouvent aussi à une distance moindre de 2ε de la droite d par l'origine avec la direction σ. Pour $\rho \to 0$ nous avons $\varepsilon(\rho) \to 0$. La fonction (5) représente donc pour des ρ suffisamment petits le domaine $|\zeta| > \rho$ sur un domaine dont les points de frontière se trouvent à une distance $\leq 2\rho\varepsilon(\rho)$ de la droite d. Construisons maintenant autour de O un cercle de rayon ρ: il contient une section de C_n dont le diamètre transfini est au plus ρ. Toute intersection de cette section avec le cercle $z = \rho$ apparaît de O sous un angle ω par rapport à d, qui est borné par l'inégalité

(14) $$\sin \omega \leq 2\,\varepsilon(\rho).$$

On voit donc que d est la tangente à C_n en O (¹).

5. Puisque O était un point quelconque sur C_n, nous avons le théorème suivant :

La frontière C_n obtenue à l'aide d'une fonction extrémale $f_n(\zeta)$ est une courbe continue $z = z(s)$ (s un paramètre réel), qui possède toujours une tangente, excepté en $2(n-1)$ points, et qui satisfait à l'équation différentielle

(15) $$\left(\frac{dz}{ds}\right)^2 \sum_{\nu < \mu}^{1\ldots n} \frac{1}{(a_\nu - z)(a_\mu - z)} = 0.$$

L'équation (15) vient de (12') en posant $\mathcal{B}_0 = e^{i\theta}$ et $\sigma^2 = -e^{i\theta}$. Elle est en effet une équation différentielle pour la courbe

$$z(s) = x(s) + i\,\gamma(s).$$

Car, en écrivant l'équation de cette courbe sous la forme $\gamma = \gamma(x)$, nous trouvons

(15') $$\frac{d\gamma}{dx} = \frac{\mathfrak{I}\,\{\,z'(s)\,\}}{\mathcal{R}\,\{\,z'(s)\,\}};$$

on obtient le second membre de (15') aisément de (15). Les $2(n-1)$ points critiques de l'équation différentielle sont les $(n-2)$ zéros et les n pôles de la somme du premier membre de (15). Si l'on connaissait les points $a_1, a_2, \ldots, a_n$ on aurait déterminé complè-

---

(¹) Pour une formulation générale du principe appliqué ici et une recherche plus profonde à l'aide de la théorie des ensembles, voir une Note *A method of variation within the family of simple functions* qui va paraître dans les *Proceedings of the London Math. Society*.

tement une fonction extrémale $f_n(\zeta)$. En tout cas les a_i sont soumis à la restriction que si l'on résout pour eux l'équation (15), il faut qu'une courbe existe, qui satisfait à (15) et qui passe par tous ces points a_i. Ceci est en général seulement possible pour certains n-tuples de a_i.

6. Maintenant nous nous servirons du résultat précédent pour évaluer le coefficient $\mathcal{O}_2$ dans (1). Dans le cas $n = 3$ nous obtenons pour C_3 l'équation différentielle

$$(16) \qquad \left(\frac{dz}{ds}\right)^2 \frac{z - \dfrac{a_1 + a_2 + a_3}{3}}{(z - a_1)(z - a_2)(z - a_3)} \leqq 0.$$

On voit aisément que, dans le cas d'extrémum, les points a_i ne se trouvent pas sur une droite et que dans ce cas la courbe résolvante de (16) est composée de trois arcs analytiques, qui se rencontrent au centre de gravité $g = \dfrac{a_1 + a_2 + a_3}{3}$ sous des angles de 120°. Faisons g origine du plan des z et construisons autour de g un cercle de rayon r. Dans celui-ci se trouve, pour r suffisamment petit, une section étoilée E de C_n; si son diamètre transfini est ρ, alors on a $\dfrac{r}{4} \leqq \rho \leqq r$, tel que ρ et r convergent en même temps vers zéro. Prenons la suite des sections E(ρ), qu'on obtient en faisant converger $r \to 0$. C_n possède en g trois tangentes, qui forment l'une avec l'autre des angles de 120°; donc E(ρ) tend (avec une approximation d'ordre supérieur en ρ) vers une étoile de trois segments rectilignes égaux, qui forment des angles de 120°.

Nous représentons E(ρ) sur un cercle $|\zeta| > \rho$ à l'aide de

$$(4') \qquad \zeta = \varphi(z) = \rho\, \psi\left(\frac{z}{\rho}\right) = z + k \cdot \rho + \frac{\mathcal{B}_1 \rho^2}{z} + \frac{\mathcal{B}_2 \rho^3}{z^2} + \dots$$

La fonction inverse $\psi_\rho^{-1}(\tau)$ de $\psi_\rho(t)$ représente l'extérieur du cercle-unité sur l'extérieur d'une étoile E$'(\rho)$, qui s'obtient de E(ρ) par une dilatation de facteur $\dfrac{1}{\rho}$. E$'(\rho)$ converge comme E(ρ) vers une étoile avec des segments égaux et des angles égaux; appelons-le E$'_0$. Plus exactement : E$'(\rho)$ converge vers E$'_0$ dans ce sens, qu'un domaine contenant E$_0$ dans son extérieur, contient aussi tous les $E'(\rho)$ dans l'extérieur pour ρ suffisamment petit. L'extérieur de E'_0 est le noyau de tous les extérieurs des $E'(\rho)$; car on voit sans difficulté que E'_0 est le plus grand domaine avec la propriété énoncée. D'après un théorème classique de la théorie de la représentation conforme la suite $\varphi_\rho(t)$ converge, dans tout domaine contenant E'_0 dans l'extérieur, uniformément vers la fonction univalente $\psi_0(t)$, qui représente l'extérieur de E'_0 sur l'extérieur du cercle-unité. Puisque tous les $\psi_\rho(t)$ étaient normalisés au développement $t + K + \frac{\mathcal{B}_1}{\cdot t} + \frac{\mathcal{B}_2}{t^2} + \ldots$, ceci est valable également pour $\psi_0(t)$. Ce fait et la propriété de représentation déterminent $\psi_0(t)$ essentiellement. Toutes les fonctions normalisées, représentant le cercle-unité sur une étoile E'_0, ont la forme

$$(17) \qquad g_1(\tau) = \tau \sqrt[3]{1 + \frac{2\lambda^3}{\tau^3} + \frac{\lambda^6}{\tau^6}} \qquad |\lambda| = 1.$$

Cette fonction représente le domaine $|\tau| > 1$ sur l'extérieur de l'étoile rectiligne $\left(0, \frac{1}{\lambda}\sqrt[3]{4}, \frac{1}{\lambda}e^{\frac{2\pi i}{3}}\sqrt[3]{4}, \frac{1}{\lambda}e^{\frac{4\pi i}{3}}\sqrt[3]{4}\right)$. Les $\psi_\rho^{-1}(\tau)$ convergent vers un certain $g_1(\tau)$, λ dépendant de l'orientation de E'_0. Nous trouvons donc pour $(4')$

$$(18) \qquad \lim_{\rho \to 0} k(\rho) = \lim_{\rho \to 0} \mathcal{B}_1(\rho) = 0, \qquad \lim_{\rho \to 0} \mathcal{B}_2(\rho) = -\frac{2}{3}\lambda^3.$$

Si

$$(19) \qquad r(\zeta) = \zeta + \frac{\mathcal{A}_1 \rho^2}{\zeta} + \frac{\mathcal{A}_2 \rho^3}{\zeta^2} + \ldots$$

est une fonction holomorphe et univalente dans $|\zeta| > \rho$, alors

$$(20) \quad r[\varphi(z)] = z + k \cdot \rho + \frac{(\mathcal{B}_1 + \mathcal{A}_1)\rho^2}{z} + \frac{(\mathcal{B}_2 - k\mathcal{A}_1 + \mathcal{A}_2)\rho^3}{z^2} + \ldots$$

est holomorphe et univalente à l'extérieur de E. Concluons comme en (3) et considérons $a_1 + a_2 + a_3 = 0$; alors nous obtenons

$$(21) \qquad \left|1 + 3\rho^3(\mathcal{B}_2 - k\mathcal{A}_1 + \mathcal{A}_2)\frac{1}{a_1 a_2 a_3} + \rho^4(\ldots)\right| \leqq 1.$$

Si $\rho \to 0$, il suit de (18) et (21)

$$(22) \qquad \mathcal{R}\left\{\left(-\frac{2}{3}\lambda^3 + \mathcal{A}_2\right)\frac{1}{a_1 a_2 a_3}\right\} \leqq 0.$$

Cela n'est possible que si l'on a toujours $|\mathfrak{A}_2| \leqq \frac{2}{3}$. Ainsi nous trouvons :

Soit $f(\zeta) = \zeta + \dfrac{\mathfrak{A}_1}{\zeta} + \dfrac{\mathfrak{A}_2}{\zeta^2} + \dots$ holomorphe et univalente pour $|\zeta| > 1$, alors on a $|\mathfrak{A}_2| \leqq \frac{2}{3}$ et cette limite est atteinte pour

$$(17') \qquad f(\zeta) = \zeta \sqrt[3]{1 + \frac{2}{\zeta^3} + \frac{1}{\zeta^6}}.$$

[5] A method of variation within the family of simple functions

[5] A method of variation within the family of simple functions. *Proc. London Math. Soc.* **44** (1938), 432–449.

A METHOD OF VARIATION WITHIN THE FAMILY OF SIMPLE FUNCTIONS

By MENAHEM SCHIFFER.

[Received 12 October, 1937.—Read 18 November, 1937.]

1. *Introduction.* In the theory of conformal representation various problems of extremum have been stated and solved. There are two possible ways of investigating such problems. In the first method, all conformal representations of the type considered are discussed, and inequalities for them are established. Then it is shown that, for a certain representation, the bound of the inequality is actually attained, and this representation solves the corresponding extremum problem. An example of this method is the proof of the "Verzerrungssatz" of Koebe. In the second method, on the other hand, the extremal representation is discussed with particular reference to the extremal character of its adjoint function. This method is similar to that used in the classical calculus of variations. It has been used by Grötzsch†, Marty‡, and others.

In the present paper a new and general method of the second type is described. It is a kind of calculus of variation applied to the family of all normalized functions which are simple (univalent) in a given domain. This method proves to be of great applicability to a large number of problems involving simple functions. With its aid, differential equations can be found, either for the functions of the family with certain extremal properties, or, at least, for the boundary curves of the domain conformally mapped with their aid from the given domain. In this way numerous problems in simple functions can be solved; in this paper, the method will be applied to the problem of the coefficients of simple functions, and will be used to develop a generalization for multiply connected domains of some theorems of growth of simple functions.

---

† H. Grötzsch, "Über ein Variationsproblem der konformen Abbildung", *Leipziger Berichte*, 82 (1930), 251–263.

‡ F. Marty, "Sur le module des coefficients de MacLaurin d'une fonction univalente", *Comptes rendus*, 198 (1934), 1569–1571.

2. *The fundamental lemma.*   Our method consists of a comparison of the extremal function with infinitesimally neighbouring functions of the same family.   For this purpose the following lemma is very useful:

LEMMA.   *There exists a set of positive constants*

$$K^*, \ A^* \leqslant 4^2, \ B^* \leqslant 4^3, \ \dots,$$

such that the following statements hold:

(a) *Let C be a bounded continuum in the z-plane, containing more than one point and possessing as complement the domain D. Then, for each $z_0 \subset C$ and for each value of the parameter $\rho > 0$, there exist functions*

$$(1) \qquad F_\rho(z, z_0) = z - z_0 + K\rho + \frac{A\rho^2}{z - z_0} + \frac{B\rho^3}{(z - z_0)^2} + \cdots,$$

which are simple and regular in D, the pole at infinity excepted, and for which the moduli of the coefficients $K, A, B, \dots$ are bounded by $K^*, A^*, B^*, \dots$.

(b) *Let the function $s(z)$ be analytic and not identically vanishing on C. Suppose that, at each fixed point $z_0 \subset C$, for each sequence of functions $F_{\rho'}(z, z_0)$ of the family (1) for which $\lim_{\rho' \to 0} A = a$ exists and is different from 0, we have $R\{as(z_0)\} \geqslant 0$. Then C is an analytic curve $z(t)$, satisfying the differential equation*

$$\left(\frac{dz}{dt}\right)^2 s(z) + 1 = 0.$$

Proof. I.   We first prove (a).   Without loss of generality we can suppose that $z_0 = 0$.   (This will be assumed throughout I–IV.)   Let p and q be two different points of C.   Then

$$(2) \qquad F(z) = \frac{p - q}{\log \dfrac{1 - p/z}{1 - q/z}} = z - \frac{p + q}{2} - \frac{(p - q)^2}{12z} - \frac{\Pi_3(p, q)}{z^2} - \frac{\Pi_4(p, q)}{z^3} \cdots$$

is simple in D and regular, the pole at ∞ excepted. $\Pi_\nu(p, q)$ denotes a homogeneous polynomial of degree ν in p and q.   Putting $\rho = \max(|p|, |q|)$, we can then write (2) in the form

$$(2') \qquad F_\rho(z) = z - \frac{p + q}{2\rho}\rho - \left(\frac{p}{\rho} - \frac{q}{\rho}\right)^2 \frac{\rho^2}{12z} - \Pi_3\left(\frac{p}{\rho}, \ \frac{q}{\rho}\right)\frac{\rho^3}{z^2} - \cdots,$$

and thus show that (2) is a function of the type (1) for the case considered $z_0 = 0$.   Further, ρ can be chosen arbitrarily small, because in each neighbourhood of O distinct points p, q can be found. This proves (a), if K^*, A^*, B^*, ... are taken as the bounds for the coefficients

$$\Pi_\nu\left(\frac{p}{\rho}, \frac{q}{\rho}\right)$$

of $(2')$.

But to prove (b) we shall have to consider also another class of functions (1) which are obtained in the following way. We choose a sub-continuum I_ρ of C containing the point O, the complement of which is a domain D_ρ, with the mapping radius $\rho > 0$. We map D_ρ conformally on $|\eta| > \rho$ by means of the simple function

$$(3) \qquad \eta = \zeta(z) = z + K(\rho)\,\rho + \frac{A(\rho)\rho^2}{z} + \frac{B(\rho)\rho^3}{z^2} + \frac{C(\rho)\rho^4}{z^3} + \dots.$$

The inverse function of (3),

$$(4) \qquad z = \zeta^{-1}(\eta) = \eta - K(\rho)\,\rho - \frac{A(\rho)\rho^2}{\eta} - \frac{\hat{B}(\rho)\rho^3}{\eta^2} - \frac{\hat{C}(\rho)\rho^4}{\eta^3} - \dots,$$

is simple and regular in $|\eta| > \rho$, the pole at infinity excepted. By the principle of areas (Flächensatz) of Bieberbach-Faber[†], we have therefore

$$(5) \qquad |A(\rho)|^2 + 2\,|\hat{B}(\rho)|^2 + 3\,|\hat{C}(\rho)|^2 + \dots \leqslant 1.$$

Hence all the coefficients $A(\rho)$, $\hat{B}(\rho)$, $\hat{C}(\rho)$, ... of (4) are bounded, independently of ρ and the choice of I_ρ; applying the same principle to the function $z = \sqrt{\{\rho\zeta^{-1}(\eta^2/\rho)\}}$, which is also regular and simple in $|\eta| > \rho$, we prove also that $|K(\rho)| \leqslant 2$; from this, the boundedness of the coefficients $B(\rho)$, $C(\rho)$, ... of (3) can be proved. In particular we find that $|A(\rho)| \leqslant 1$. On the other hand, the functions

$$(6) \qquad h_\gamma(\eta) = \eta + \frac{\gamma\rho^2}{\eta}$$

are simple in $|\eta| > \rho$ and regular for each $|\gamma| \leqslant 1$, the pole at ∞ excepted. Therefore the function

$$(7) \qquad F_\rho(z) = h_\gamma\big(\zeta(z)\big) = z + K(\rho)\,\rho + \frac{\big(A(\rho)+\gamma\big)\rho^2}{z} + \frac{\tilde{B}(\rho, \gamma)\rho^3}{z^2} + \dots$$

[†] See, for example, Bieberbach, *Lehrbuch der Funktionentheorie*, 2 (2. Aufl., 1931), 72-74, or Titchmarsh, *The theory of functions*, (1932), 209.

is simple in D_ρ and regular for each $|\gamma| \leqslant 1$, the point ∞ excepted. Its coefficients $\tilde{B}(\rho, \gamma)$, $\tilde{C}(\rho, \gamma)$, ... are bounded for all $\rho < 0$, $|\gamma| \leqslant 1$, independently of I_ρ. The function (7) therefore also belongs to the class (1), the constants K^*, A^*, B^*, ... being suitably chosen.

We now fix K^*, A^*, B^*, ... so that they give simultaneously upper bounds for the moduli of the coefficients of (2') and (7). It is obvious that we may take $A^* = 4^2$, $B^* = 4^3$, ... For the functions (2') and (7) are simple outside of the circle $|z| \leqslant 4\rho$, and by the principle of areas (5) we get the above upper bounds for the coefficients. We assert that part (b) of the lemma holds also for these constants.

II. We suppose henceforth that $s(0) \neq 0$, and we put

$$\operatorname{sgn} s(0) = \frac{s(0)}{|s(0)|} = e^{i\tau}.$$

We consider the function (2') for which $p \to 0$ and $q = 0$. Then

$$\rho = \max(|p|, |q|) = |p|,$$

and (2') can be written in the form

$$(2'') \qquad F_\rho(z) = z - \tfrac{1}{2}(\operatorname{sgn} p)\rho - \frac{(\operatorname{sgn} p)^2 \rho^2}{12z} - \frac{\Pi_3(\operatorname{sgn} p, 0)\rho^3}{z^2} - \dots.$$

We call the direction of $e^{i\kappa}$ a limit direction at O with respect to points of C, if a sequence of points $p_\nu \subset C$ with $p_\nu \to 0$ can be chosen for which

$$\lim_{\nu \to \infty} \operatorname{sgn} p_\nu = e^{i\kappa}.$$

Let us consider the functions (2'') for which $p = p_\nu$, $\rho = \rho_\nu = |p_\nu|$, and $\lim_{\nu \to \infty} \operatorname{sgn} p_\nu = e^{i\kappa}$. In consequence of the hypothesis of (b) we must have $\tfrac{1}{2}\pi \leqslant 2\kappa + \tau \leqslant \tfrac{3}{2}\pi$; for

$$\lim_{\nu \to \infty} \left[-\frac{\operatorname{sgn} p^2}{12} \right] = \frac{e^{i(2\kappa + \pi)}}{12} = a$$

is not zero. Hence κ must lie either in the closed interval

$$[\tfrac{1}{4}\pi - \tfrac{1}{2}\tau, \ \tfrac{3}{4}\pi - \tfrac{1}{2}\tau], \ \text{or in} \ [\tfrac{5}{4}\pi - \tfrac{1}{2}\tau, \ \tfrac{7}{4}\pi - \tfrac{1}{2}\pi].$$

All limit directions in O lie therefore in the two vertically opposite right angles which are bisected by the straight line l through O with the direction $e^{\frac{1}{2}i(\pi - \tau)}$. Hence, if $|p|$ is sufficiently small, all points $p \subset C$

2 F 2

have their arguments in the interior of the intervals

$$[\tfrac{1}{4}\pi-\tfrac{1}{2}\tau-\epsilon,\ \tfrac{3}{4}\pi-\tfrac{1}{2}\tau+\epsilon],\quad [\tfrac{5}{4}\pi-\tfrac{1}{2}\tau-\epsilon,\ \tfrac{7}{4}\pi-\tfrac{1}{2}\tau+\epsilon],$$

for each given $\epsilon>0$.

We show, in addition, that 0 is a boundary point of C.

III. We consider the function (7) which is simple in the domain D_ρ. Since $|A(\rho)|\leqslant 1$ and since γ may take any value such that $|\gamma|\leqslant 1$, it follows from the hypothesis of (b) that $\lim_{\rho\to 0}|A(\rho)|=1$ for each sequence of continua I_ρ. For, in the contrary case, a sub-sequence I_{ρ_ν} could be chosen, satisfying $\lim_{\nu\to\infty}|A(\rho_\nu)|<1$; and, consequently, by a proper choice of γ_ν, the argument of $a=\lim_{\nu\to\infty}\{A(\rho_\nu)+\gamma_\nu\}$ could be fixed arbitrarily, contradicting the hypothesis of (b).

We now choose a sub-sequence I_{ρ_ν} yielding a convergent sub-sequence $A(\rho_\nu)$ with $\lim_{\nu\to\infty}A(\rho_\nu)=e^{i\vartheta}$. Also, by choosing γ_ν properly, we can make the argument of a take an arbitrarily given value in the open interval $(\vartheta-\tfrac{1}{2}\pi,\ \vartheta+\tfrac{1}{2}\pi)$. By the hypothesis of (b), all the values of

$$(\tau+\vartheta-\tfrac{1}{2}\pi,\ \tau+\vartheta+\tfrac{1}{2}\pi)$$

must lie in the interval $[-\tfrac{1}{2}\pi,\ +\tfrac{1}{2}\pi]$. This is only possible when these two intervals of length π coincide, i.e. $\tau+\vartheta=0$. This condition determines ϑ uniquely; and therefore we have

$$(8)\qquad\qquad\qquad \lim_{\rho\to 0}A(\rho)=e^{-i\tau}.$$

Thus, the functions (4) mapping $|\eta|>\rho$ conformally on D_ρ necessarily have coefficients $A(\rho)$ tending to $e^{-i\tau}$ as $\rho\to 0$. The functions

$$(4')\qquad\qquad z'=\eta'-K(\rho)-\frac{A(\rho)}{\eta'}-\frac{\hat{B}(\rho)}{\eta'^2}-\cdots$$

map the circle $|\eta'|>1$ on domains D_ρ' which is obtained from D_ρ by the magnification $z'=z/\rho$. In virtue of (8) and (5) $\hat{B}(\rho)$ and $\hat{C}(\rho)$, all tend to 0 as $\rho\to 0$; and so $(4')$ can be written in the form

$$(4'')\qquad\qquad z'=\eta'-K(\rho)-\frac{e^{-i\tau}}{\eta'}+\delta(\rho)H_\rho\left(\frac{1}{\eta'}\right).$$

Here $|H_\rho(1/\eta')|\leqslant M_\epsilon$ in each domain $|\eta'|\geqslant 1+\epsilon>1$, M_ϵ being independent of I_ρ, while $\delta(\rho)\to 0$ as $\rho\to 0$. Choosing ϵ sufficiently small and making a corresponding choice of ρ, we can make the image of the circle $|\eta'| = 1 + \epsilon$ lie in an arbitrary vicinity of a segment of a straight line

$$[-2e^{\frac{1}{2}i(\pi-\tau)} - K(\rho),\ +2e^{\frac{1}{2}i(\pi-\tau)} - K(\rho)],$$

which is obtained from $|\eta'| = 1$ by the conformal representation

$$z' = \eta' - K(\rho) - \frac{e^{-i\tau}}{\eta'}.$$

Thus, *a fortiori*, the boundary B_ρ' of D_ρ' also lies, in an arbitrary neighbourhood of this segment; for it lies in the interior of the image of $|\eta'| = 1 + \epsilon$, obtained by (4''). In other words, the boundary in question lies in an arbitrary neighbourhood of a straight line l_ρ with the direction $e^{\frac{1}{2}i(\pi-\tau)}$. On the other hand, B_ρ' contains the point O, this being a boundary point of C, and therefore also of D_ρ; hence, B_ρ' lies in an arbitrary neighbourhood of the straight line l with the direction $e^{\frac{1}{2}i(\pi-\tau)}$ *through the origin*. Therefore the distance of each point of B_ρ' from l is less than $d(\rho)$; and this converges with ρ to zero. The boundary points of D_ρ itself which also form the boundary of I_ρ have distances from l less than $\rho d(\rho)$, and the same holds for all interior points of I_ρ, if any such exist.

If we now describe round $z = 0$ a circle Γ_ρ with radius ρ, then, for ρ sufficiently small, at least one point $P(\rho)$ of C lies on Γ_ρ, and it can be connected with O by a continuum I_σ entirely interior to Γ_ρ, whose mapping radius is σ. Since the mapping radius is monotonic, we have $\sigma \leqslant \rho$. $P(\rho)$ has, therefore, as shown above, a distance less than $\sigma d(\sigma)$ from the straight line l; thus, the line $\overline{OP(\rho)}$ forms with l an angle $\omega(\rho)$ which satisfies the inequality $|\sin \omega(\rho)| < d(\sigma)$. Hence, $\lim_{\rho \to 0} \omega(\rho)$ is 0 or π, and the secants $\overline{OP(\rho)}$ converge to l as $\rho \to 0$. This shows that at least one of the directions $\pm e^{\frac{1}{2}i(\pi-\tau)}$ is a limit direction for C in $z = 0$.

IV. We now take the two vertically opposite angular domains containing all the points with arguments in the intervals $[\frac{1}{4}\pi - \frac{1}{2}\tau - \epsilon,\ \frac{3}{4}\pi - \frac{1}{2}\tau + \epsilon]$ and $[\frac{5}{4}\pi - \frac{1}{2}\tau - \epsilon,\ \frac{7}{4}\pi - \frac{1}{2}\tau + \epsilon]$ for a fixed value of $\epsilon > 0$, and cut two sectors out of these by means of a circle Γ_{ρ_0} with radius ρ_0 and centre 0. If ρ_0 is chosen sufficiently small, all points of C in the interior of Γ_{ρ_0} lie in these two sectors, by II. If, further, ρ_0 is chosen so small that not all points of C are inside Γ_{ρ_0}, there lies in at least one of the two sectors, say in S_1, and on Γ_{ρ_0} a point of C which can be connected with O by a continuum entirely in Γ_{ρ_0} and therefore also in S_1. On each circle Γ_ρ with $\rho < \rho_0$ there lies also at least one point $P(\rho)$ which can be connected with O by a continuum which lies in Γ_ρ and S_1. The mapping radius of the connecting continuum I_σ is $\sigma \leqslant \rho$. By the expansion in III the points $P(\rho)$ have distances less than $\sigma d(\sigma)$ from the straight line l; the secants $\overline{OP(\rho)}$ therefore form angles $\omega(\rho)$ with that ray l_1 of l which lies in S_1, $\omega(\rho)$ being such that $\lim_{\rho \to 0} \omega(\rho) = 0$. We choose for each ρ exactly one $P(\rho)$; and so we get a sequence of points which converge monotonically to O with ρ and for which the secants $\overline{OP(\rho)}$ converge to the ray l_1. Now it will be shown that each sequence of points Q_ν in the same sector S_1 which converge to O has the same property, viz. that the secants $\overline{OQ_\nu}$ converge to the ray l_1. It will thus be proved that the direction l_1 is the only limit direction in the sector S_1.

To prove this assertion, we consider the sequences of points

$$ Q_1, \quad Q_2, \quad \ldots, \quad Q_n, \quad \ldots \to 0 $$

and $P_1 = P(|Q_1|)$, $P_2 = P(|Q_2|)$, ..., $P_n = P(|Q_n|) \to 0$; we have $|P_\nu| = |Q_\nu|$ and $\lim_{\nu \to \infty} \overline{OP_\nu} = l_1$.

We construct the functions (2′) with $p = P_\nu$, $q = Q_\nu$, and $\rho = |P_\nu|$. Then

$$ (9) \quad F_\rho(z) = z - \frac{P_\nu + Q_\nu}{2\rho} \rho - \frac{(\operatorname{sgn} P_\nu - \operatorname{sgn} Q_\nu)^2 \rho^2}{12z} - \frac{\Pi_3(\operatorname{sgn} P_\nu, \operatorname{sgn} Q_\nu)\rho^3}{z^2} - \cdots $$

is a function of the form (1). For ν sufficiently large, the argument of P_ν is arbitrarily near to $\frac{1}{2}(\pi - \tau)$ (if S_1 lies in this direction from 0) and $\arg Q_\nu$ is in the interval $[\frac{1}{4}\pi - \frac{1}{2}\tau - \epsilon, \frac{3}{4}\pi - \frac{1}{2}\tau + \epsilon]$; the argument of $(\operatorname{sgn} P_\nu - \operatorname{sgn} Q_\nu)$ then lies, as is most easily seen from geometric considerations, in one of the intervals $[\pi - \frac{1}{2}\tau - \delta_\nu, \frac{9}{8}\pi - \frac{1}{2}\tau + \delta_\nu]$ or $[-\frac{1}{8}\pi - \frac{1}{2}\tau - \delta_\nu, -\frac{1}{2}\tau + \delta_\nu]$ with $\delta_\nu > 0$ and $\delta_\nu \to 0$. Therefore, the argument of $\{-(\operatorname{sgn} P_\nu - \operatorname{sgn} Q_\nu)^2 e^{i\tau}\}$ lies in the interval $[\frac{3}{4}\pi, \frac{5}{4}\pi]$. This and the hypothesis of (b) yield the result that

$$ (10) \qquad \lim_{\nu \to \infty} \{\operatorname{sgn} P_\nu - \operatorname{sgn} Q_\nu\} = 0; $$

for otherwise a sub-sequence Q_ν could be chosen for which

$$ \lim_{\nu \to \infty} \{-(\operatorname{sgn} P_\nu - \operatorname{sgn} Q_\nu)^2\} = a \neq 0 $$

without satisfying the condition $R\{as(z_0)\} \geqslant 0$. But (10) shows that the angle $P_\nu O Q_\nu$ converges to zero; therefore also $\lim_{\nu \to \infty} \overline{OQ_\nu} = l_1$, which was the assertion. An analogous property holds, if a sequence $P(\rho)$ in the sector S_2 is at our disposal.

V. Abandoning now the restriction $z_0 = 0$, we can summarize all the results proved hitherto in the following statement:

At each point z_0 with $s(z_0) \neq 0$, C possesses limit directions which can be classified into two groups. First a limit direction h which is one of the two directions $\pm e^{\frac{1}{2}i(\pi - \tau)} = \operatorname{sgn}\{\pm 1/\sqrt{-s(z_0)}\}$; second all the other directions, if any such exist. These lie in that right angle which is bisected by h produced backwards. The nearest limit direction different from h, therefore, forms with h an angle greater than or equal to $\frac{3}{4}\pi$. These geometric facts are invariant with regard to conformal representation; for such a representation transforms limit directions into limit directions and leaves the angles between different limit directions unchanged. Thus, at each point z_0, the system of limit directions is subjected only to a rigid rotation. We use this fact to obtain a simpler rule for the system of limit directions at different points of C.

Let $z_0 \subset C$, and let z_0 be not a zero of $s(z)$; then the function

$$(11) \qquad u = y(z) = \int_{z_0}^{z} \sqrt{-s(\zeta)}\, d\zeta$$

is regular and simple for $|z - z_0| \leqslant k$ and k sufficiently small. A subcontinuum T of C, which lies entirely in this circular domain, contains z_0 and does not reduce to this point, is mapped by (11) on a continuum T' in the u-plane which contains $u = 0$. At each point of T' the limit directions of the image points of C are separated into two groups. First into a limit direction parallel to the real axis, and secondly into limit directions which are separated from the first by an empty domain of aperture greater than or equal to $\frac{3}{4}\pi$.

For let z^* be an arbitrary point of T, and suppose that, for a sequence $z_\nu \to z^*$, $z_\nu \subset C$,

$$(12) \qquad \lim_{\nu \to \infty} \operatorname{sgn}(z_\nu - z^*) = e^{i\psi};$$

then, for $u_\nu = y(z_\nu)$ and $u^* = y(z^*)$, we have

$$(12') \qquad \lim_{\nu \to \infty} \operatorname{sgn}(u_\nu - u^*) = e^{i\psi}\operatorname{sgn}\sqrt{-s(z^*)}.$$

The limit direction $e^{i\psi} = \pm\operatorname{sgn}\{1/\sqrt{-s(z^*)}\}$ is therefore transformed into a real direction.

By a theorem of Haslam-Jones†, the projection of T' on the imaginary axis in the u-plane has measure zero. T' is a continuum and contains the

---

† Tangential properties of a plane set of points ", *Quart. J. of Math.*, 7 (1936), 116–123.

origin; therefore it lies entirely on the real axis. The corresponding part T of C in the z-plane therefore lies on the curve

$$\int_{z_0}^{z} \sqrt{-s(z)}\, dt = t,$$

with t a real number, which is got from the real u-axis by an analytical representation.

This holds for $|z - z_0| \leqslant k$; but this integral, along an arbitrary part of C which does not contain zeros of $s(z)$, can now be calculated by regarding this part as composed of sufficiently small pieces for which the above consideration holds. Finally, we can even integrate over the whole curve by regarding it as composed of parts which do not contain zeros of $s(z)$ in their interior. Along the whole curve C we therefore have

$$(13) \qquad\qquad \int_{z_0}^{z} \sqrt{-s(z)}\, dz = t,$$

a real number if $z_0 \subset C$. For an analytic branch of C, t can be chosen as the parameter of the curve. Differentiating (13), we get the differential equation

$$(14) \qquad\qquad \left(\frac{dz}{dt}\right)^2 s(z) + 1 = 0$$

for all points z which are not zeros of $s(z)$.

This is the assertion of the lemma.

From the proof it is obvious that C can have branch points or cusps only at the zeros of $s(z)$.

3. *The problem of the coefficients.* Let us now apply this lemma to the problem of the coefficients of a simple function. We imagine a given domain Δ of finite or infinite connectivity in the ζ-plane; it is supposed only that $\zeta = 0$ belongs to Δ. Now consider the family of all functions

$$(15) \qquad\qquad z = f(\zeta) = \zeta + a_2\zeta^2 + a_3\zeta^3 + \ldots + a_n\zeta^n + \ldots$$

which are regular and simple in Δ. This family is normal and therefore, for each n, there exists at least one function $f_n(\zeta)$ of the family for which $|a_n|$ attains its greatest value within the family. An important property of these functions may now be pointed out.

The function $f(\zeta)$ maps the domain Δ on a domain D in the z-plane the complements of which are certain continua R_ν and points P_ν. We choose a fixed continuum R among the R_ν (*i.e.* R contains more than one point) and select on it a point $z^* \neq \infty$. Round z^* we describe a circle Γ_r with radius $r < |z^*|$ and a concentric circle $\Gamma_{\frac{1}{3}r}$ with radius $\frac{1}{3}r$. Hence the origin lies outside Γ_r. Consider now the sub-continuum C of R which contains z^* and lies entirely inside $\Gamma_{\frac{1}{3}r}$. We apply the lemma to C: for each $\rho > 0$ and each $z_0 \subset C$ there exist functions (1) which are simple and regular in the whole z-plane, C and ∞ excepted. The expansion (1) converges for all points z outside Γ_r, ∞ excepted. Therefore the functions

$$(16) \quad F_\rho\{f(\zeta); z_0\} = f(\zeta) - z_0 + K\rho + \frac{A\rho^2}{f(\zeta) - z_0} + \frac{B\rho^3}{\{f(\zeta) - z_0\}^2} + \frac{C\rho^4}{\{f(\zeta) - z_0\}^3} + \cdots$$

are regular and simple in Δ and their expansion (16) converges uniformly in the vicinity of $\zeta = 0$. For, to $\zeta = 0$ there corresponds $z = 0$, and this point lies outside Γ_r. Further, the coefficients B, C, ... are bounded for all values of ρ in consequence of § 2, I; and so, in the neighbourhood of $\zeta = 0$, we may write

$$(16') \quad F_\rho\{f(\zeta); z_0\} = f(\zeta) - z_0 + K\rho + \frac{A\rho^2}{f(\zeta) - z_0} + O(\rho^3),$$

where $O(\rho^3)$ denotes terms of at least the third order in ρ. Normalizing this function according to (15), we get

$$(16'') \quad f^*(\zeta) = f(\zeta) - \frac{A\rho^2}{z_0} \left\{ \frac{f(\zeta)^2}{z_0^2} + \frac{f(\zeta)^3}{z_0^3} + \cdots \right\} + O(\rho^3).$$

This is a new normalized and simple function in Δ. We now put

$$(17) \quad [f(\zeta)]^r = \sum_{\nu=r}^{\infty} a_\nu^{(r)} \zeta^\nu = \zeta^r + r a_2 \zeta^{r+1} + \cdots.$$

Then we can write (16'') in the form

$$(18) \quad f^*(\zeta) = \sum_{n=1}^{\infty} \left\{ a_n - \frac{A\rho^2}{z_0} \sum_{k=2}^{\infty} \frac{a_n^{(k)}}{z_0^k} + O(\rho^3) \right\} \zeta^n.$$

This rearrangement of (13'') is legitimate in view of the uniform convergence of (16'') in the vicinity of the origin. (18) is a function of the family (15) which is regular and simple in Δ.

If now, in particular, $f(\zeta)$ is an extremal function $f_n(\zeta)$ with the coefficient a_n of the largest absolute value in the family, then for each z_0 we have the inequality

$$(19) \quad |a_n^*| = \left| a_n - \frac{A\rho^2}{z_0} \sum_{k=2}^{n} \frac{a_n^{(k)}}{z_0^k} + O(\rho^3) \right| \leqslant |a_n|;$$

hence, obviously, $a_n \neq 0$ and

(19′)
$$R\left\{A \frac{1}{a_n} \sum_{k=2}^{n} \frac{a_n^{(k)}}{z_0^{k+1}} + O(\rho)\right\} \geqslant 0.$$

Take any sequence $F_{\rho_\nu}(z; z_0)$ of functions (1), with $\rho_\nu \to 0$, for which

$$a = \lim_{\nu \to \infty} A_\nu$$

exists and is different from zero. For each A_ν the inequality (19′) holds, and so we get, in the limit when $\rho \to 0$,

(19″)
$$R\left\{a \frac{1}{a_n} \sum_{k=2}^{n} \frac{a_n^{(k)}}{z_0^{k+1}}\right\} \geqslant 0.$$

Thus, the supposition of part (b) of the lemma holds with

$$s(z) = \frac{1}{a_n} \sum_{k=2}^{n} \frac{a_n^{(k)}}{z_0^{k+1}}.$$

$s(z)$ does not vanish identically since $a_n^{(n)} = 1$. Therefore C is an analytic curve $z(t)$ satisfying the differential equation

(20)
$$\left(\frac{dz}{dt}\right)^2 \frac{1}{a_n} \sum_{k=2}^{n} \frac{a_n^{(k)}}{z(t)^{k+1}} + 1 = 0.$$

This holds for that part of R which lies inside $\Gamma_{\frac{1}{2}r}$. By regarding R as composed of parts for which (20) holds, we can show that R itself is also an analytic curve satisfying (20). The only critical points that the curve R can have are the zeros of $s(z)$ which lie on it. At them the curve can fork into different branches and it can possess cusps. We also see that D has no exterior points, *i.e.* it is a slit domain. Thus we have proved the following theorem:

THEOREM I. *Let Δ be a domain containing $\zeta = 0$. Among all functions (15) which are simple and regular in Δ, there exists at least one, $f_n(\zeta)$, with the n-th coefficient a_n largest in absolute value. This function maps Δ on a slit domain D, the boundary of which consists of curves satisfying the differential equation (20) and possibly single points.*

This theorem includes the theorem due to Marty[†], that each extremal function $f_n(\zeta)$ maps Δ on a slit domain; another special case of Theorem I was recently published by the writer[‡].

[†] F. Marty, *loc. cit.*

[‡] "Un calcul de variation dans une famille de fonctions univalentes", *Comptes rendus*, 205 (1937), 709–711.

4. *The case $n = 2$.* In the case $n = 2$ it follows from (20) that, for each boundary continuum of D which contains more than one point, the following differential equation holds:

$$(21) \qquad \left(\frac{dz}{dt}\right)^2 \frac{1}{a_2 z(t)^3} + 1 = 0.$$

Let z_1 be a fixed point in one of these continua; then, for every other point $z(t)$ of this same continuum, the integration of (21) yields

$$(22) \qquad z(t) = [z_1^{-\frac{1}{2}} \pm \tfrac{1}{2} it \sqrt{a_2}]^{-2}.$$

Thus, by means of a real parameter t which is restricted to a certain interval, we can represent each boundary continuum containing more than one point in the form

$$(22') \qquad z(t) = [z_\nu^{-\frac{1}{2}} + it \sqrt{a_2}]^{-2},$$

where z_ν is a point on the continuum considered. Now the extremal function $f_2(\zeta)$ maps Δ on a slit domain. On the other hand this function has no pole in Δ. Therefore, the point ∞ is always a boundary point and lies on one of the continua considered. This continuum is either a single point or a curve (22'). In the latter case this continuum must have the parametric representation $z = -(1/a_2)t$; for this is the only unbounded curve (22'). Thus we obtain the following

THEOREM II. *The extremal function whose coefficient a_2 has the largest absolute value in the family maps the domain Δ on a slit region which is bounded by curves (22'); if the point ∞ is not a single boundary point, it lies on the rectilinear slit $z = -(1/a_2)t$.*

It may be added that this result can also be derived from a theorem of de Possel[†] which we shall prove in §7. If the initial domain Δ is the interior of the unit circle, an obvious conclusion from (22') is the theorem of Bieberbach[‡]: $|a_2| \leqslant 2$.

5. *The case $n = 3$.* In this case, the differential equation

$$(23) \qquad \left(\frac{dz}{dt}\right)^2 \frac{1}{a_3} \left\{ \frac{1}{z(t)^4} + \frac{2a_2}{z(t)^3} \right\} + 1 = 0$$

† R. de Possel, " Zum Parallelschlitztheorem unendlich-vielfach zusammenhängender Gebiete ", *Göttinger Nach.* (1931), 199–202.

‡ L. Bieberbach, " Über die Koeffizienten derj. Potenzreihen, welche eine schlichte Abbildung des Einheitskreises vermitteln ", *Berliner Berichte* (1916), 940–955 (946).

must be considered.   It shows that if $a_2 \neq 0$ the continua of the boundary of D are analytic curves which can have singularities only at the points $z = -1/(2a_2)$ and $z = \infty$.   By the transformation $z = 1/u$ it is easily shown that, if ∞ is not a single point, the curve containing it converges toward it in the asymptotic direction $\mu = \arg(-a_2/a_3)$, it does not, however, pass beyond but retraces its track.   The point ∞ is, therefore, either a single point, or the endpoint of a boundary slit.   If, on the other hand, in the neighbourhood of $z_0 = -1/(2a_2)$ we put $u = z - z_0$, we have near $u = 0$ the differential equation

$$(23') \qquad \left(\frac{du}{dt}\right)^2 \frac{2a_2}{a_3}\frac{u}{(z_0+u)^4} + 1 = 0.$$

From this we infer that the curve under consideration can fork at the point z_0 into at most three analytic branches which meet at z_0 at angles of $2\pi/3$.   At all points except $-1/(2a_2)$ and ∞, the continua are regular analytic curves.   The parameters a_2, a_3 of the equation (23) are, of course, different for each given domain Δ.   We shall now continue the analysis of the special case in which the domain Δ is the interior of the unit circle. We shall show by this example how in many cases differential equations for the extremal function can be derived from the differential equations for the boundary.

Let $z = f_3(\zeta)$ be the extremal function considered.   From the above-mentioned properties of the boundary of D we infer that $f_3(\zeta)$ is regular everywhere on the periphery of the unit circle, except at the points which correspond to $z = \infty$ and to $z = -1/(2a_2)$.   To the point $z = \infty$ corresponds a double pole on the unit circle; to the point $z = -1/(2a_2)$ there correspond algebraic singularities of $f_3(\zeta)$, as is evident from the behaviour, in view of (23'), of the boundary curves in the neighbourhood of the critical point $-1/(2a_2)$.   It can be seen from (23') that the part of the boundary which lies in a sufficiently small neighbourhood of $-1/(2a_2)$ is obtained by a conformal representation from the real axis in the t-plane with the help of a mapping function of the local parameter $t^{\frac{2}{3}}$; thus the conformal representation of the periphery of the unit circle on this same piece of the boundary, which is furnished by $f_3(\zeta)$, must be expansible in the neighbourhood of that point ζ_0 which corresponds to $-1/(2a_2)$ in the form

$$(24) \qquad f_3(\zeta) = -\frac{1}{2a_2} + C_1(\zeta - \zeta_0)^{\frac{2}{3}} + \cdots.$$

Since $f_3(\zeta)$ is analytic on the unit circle, except at the points mentioned above, we can there write $\zeta = e^{is}$ and introduce this s as the real parameter of the boundary curve by means of the relation $z(s) = f(e^{is})$. s is obtained from the old parameter t of the curve by a real parameter transformation, and therefore, in place of (23), we can write

$$(25) \qquad g(\zeta) = \frac{1}{a_3} \frac{\zeta^2 f_3'(\zeta)^2}{f_3(\zeta)^4} \{2a_2 f_3(\zeta) + 1\} \geqslant 0 \qquad (\zeta = e^{i\vartheta}).$$

The function $g(\zeta)$ is regular in the interior of the unit circle except for a double pole at $\zeta = 0$. By (25) it can be continued without ambiguity over the whole plane by Schwarz's principle of reflection. Its only poles are $\zeta = 0$ and $\zeta = \infty$; for, in the unit circle, the simple function $f_3(\zeta)$ has no zeros and no poles except the above mentioned double pole. This does not, however, give rise to singularities of $g(\zeta)$. The branch points of $f_3(\zeta)$, also, do not give rise to singular points of $g(\zeta)$; for, on the Riemann surface of $f_3(\zeta)$, $g(\zeta)$ has poles only in the vicinity of the branch-points ζ_0; but from the expansion (24) results that $g(\zeta)$ is regular even at these points; finally, $g(\zeta)$ is uniform in the ζ-plane, so that ζ_0 is also not a branch-point for it; therefore $g(\zeta)$ is regular at ζ_0. Hence, $g(\zeta)$ is a rational function which has conjugate values at points symmetric with respect to the unit circle and possesses double poles at $\zeta = 0$ and $\zeta = \infty$. Therefore

$$(26) \qquad g(\zeta) = \frac{A}{\zeta^2} + \frac{B}{\zeta} + C + \bar{B}\zeta + \bar{A}\zeta^2,$$

with C real.

By comparing the coefficients, supposing that $a_3 > 0$, and using (15), we find that

$$(27) \qquad \frac{\zeta^2 f_3'(\zeta)^2}{f_3(\zeta)^4} \{2a_2 f_3(\zeta) + 1\} = \frac{1}{\zeta^2} + \frac{2a_2}{\zeta} + 2a_3 + 2\bar{a}_2\zeta + \zeta^2.$$

This is a differential equation for $f_3(\zeta)$ which still contains the two unknown parameters a_2 and a_3. All the other coefficients of $f_3(\zeta)$ are to be calculated successively from (27). But an algebraic relation connecting a_2 and a_3 is at once evident. To the finite endpoint of the boundary slit of D must correspond a zero of $f_3'(\zeta)$ on the unit circle. From (27) it follows that the polynomial $P(\zeta) = \zeta^4 + 2\bar{a}_2\zeta^3 + 2a_3\zeta^2 + 2a_2\zeta + 1$ possesses at least a double zero, and therefore its discriminant vanishes. We thus find that a_3 satisfies the equation

$$(28) \quad a_3^4 - R(a_2^2)a_3^3 - \left(5|a_2|^2 - \frac{|a_2|^4}{4} + 2\right)a_3^2$$

$$+ 9R(a_2^2)\left(1 + \frac{|a_2|^2}{2}\right)a_3 - (|a_2|^2 - 1)^3 - \frac{27}{4}R(a_2^2)^2 = 0.$$

The extremal function for the case $n = 3$ is already known; Löwner† has proved that the function

$$(29) \qquad p(\zeta) = \frac{\zeta}{(1-\zeta)^2} = \sum_{n=1}^{\infty} n\zeta^n$$

possesses the third coefficient largest in absolute value. In this paper, we have only shown that the function $f_3(\zeta)$ possesses some characteristic properties of (29). But on the other hand, our method can immediately be generalized from $n = 3$ to arbitrary n.

6. *On the conjecture* $|a_n| \leqslant n$: It has been conjectured that the function (29) is, for each n, the extremal function $f_n(\zeta)$ in the unit circle. We shall now prove that (29) satisfies the differential equation (20) for each n. Since the function (29) maps the periphery of the unit circle on the section of the negative axis from $-\frac{1}{4}$ to $-\infty$, it is sufficient to show that, for $z < -\frac{1}{4}$, we always have

$$(30) \qquad s_n(z) = \sum_{k=2}^{n} \frac{a_n^{(k)}}{z^{k+1}} \leqslant 0 \quad (n = 2, 3, \ldots),$$

the $a_n^{(k)}$ being the coefficients belonging to $p(\zeta)$.

By (16″), all $s_n(z)$ are obtained by means of the generating function

$$(31) \qquad \frac{p(\zeta)^2}{\{p(\zeta)-z\}z^2} = - \sum_{n=2}^{\infty} s_n(z)\,\zeta^n.$$

It must be shown therefore that the function (31) has positive coefficients in the expansion in powers of ζ for all $z < -\frac{1}{4}$. Disregarding a positive factor, the left-hand side of (31) is

$$(32) \qquad B(\zeta) = \frac{\zeta^2}{(1-\zeta)^2\{1-(2-t)\,\zeta+\zeta^2\}},$$

where $0 \leqslant t \leqslant 4$. Putting $2 - t = 2\cos\phi$, we find that

$$(33) \qquad B(\zeta) = \frac{\zeta^2}{(1-\zeta)^2} \sum_{\nu=0}^{\infty} \frac{\sin(\nu+1)\phi}{\sin\phi}\,\zeta^\nu = \sum_{n=2}^{\infty} t_n\zeta^n,$$

where

$$(34) \qquad t_n = \sum_{\nu=1}^{n} \frac{\sin\nu\phi}{\sin\phi}\,(n-\nu) = \frac{1}{4\sin^2\frac{1}{2}\phi}\left\{n - \frac{\sin n\phi}{\sin\phi}\right\}.$$

---

† K. Löwner, " Untersuchungen über schlichte konforme Abbildungen des Einheitskreises, I ", *Math. Annalen*, 89 (1923), 103–121 (121).

Now, disregarding a positive factor, t_n is equal to $-s_n(z)$. From (34), it follows that t_n is positive for all values of $z < -\frac{1}{4}$, and therefore $s_n(z)$ is negative. We have thus proved

THEOREM III. *The function* (29) *satisfies the differential equations* (20) *for each extremal function* $f_n(\zeta)$.

7. *Generalizations of the coefficients problem.* Hitherto we have used the lemma of § 2 to estimate the coefficients of simple functions with the normalization (15). It is evident that more difficult problems on extrema of simple functions can also be treated in the same way. For example: Let $P(a_2, a_3, ..., a_m)$ be a polynomial of arbitrary order in the variables $a_2, a_3, ..., a_m$. We seek that function $f(\zeta)$ which is simple in the domain Δ and for which $|P(a_2, a_3, ..., a_m)|$ becomes a maximum. By the above considerations, we obtain

THEOREM IV. *The function* (15) *which is regular and simple in* Δ *and for which* $|P(a_2, a_3, ..., a_m)|$ *has its maximum within the family, maps* Δ *on a domain D whose boundary continua satisfy a differential equation*

$$(35) \qquad \left(\frac{dz}{dt}\right)^2 \sum_{\nu=3}^{m+1} \frac{c_\nu}{\{z(t)\}^\nu} + 1 = 0,$$

where the c_ν *are certain constants depending on the domain* Δ *and on the polynomial P.*

The problem of the coefficients can also be set up for other normalizations than (15). For example : Let Δ be a domain in the ζ-plane containing the point at infinity, and consider the family of all functions which are simple in Δ and normalized in the form

$$(36) \qquad f(\zeta) = \zeta + \frac{a}{\zeta} + \frac{b}{\zeta^2} + \dots.$$

By the method of § 3, we construct from $f(\zeta)$ a new function

$$(37) \qquad F_\rho\{f(\zeta)\} = f(\zeta) + \frac{A\rho^2}{f(\zeta) - z_0} + O(\rho^3)$$

which is simple in Δ. This development is convergent in the vicinity of ∞, and therefore $F_\rho\{f(\zeta)\}$ admits the expansion

$$(38) \qquad F_\rho\{f(\zeta)\} = \zeta + \frac{a + A\rho^2 + O(\rho^3)}{\zeta} + \dots.$$

This is a correctly normalized function (36) and the consideration of §3 can be applied. For each extremal function a differential equation of the boundary of the mapped domain D can be found.

E.g., we seek that function (36) which is simple in Δ and for which $|a|$ assumes the largest possible value. For it the inequality

$$(39) \qquad R\left\{\frac{A}{-a}\right\} \geqslant 0$$

must hold at each point z_0 of the complement of D. By the lemma, D is therefore a slit domain whose slits satisfy the differential equation

$$(40) \qquad -\frac{1}{a}\left(\frac{dz}{dt}\right)^2 + 1 = 0.$$

We thus prove that the function (36), simple in Δ and with the coefficient a largest in absolute value among the family, maps Δ on a domain bounded by parallel slits with the common direction sgn $\sqrt{a}$.

This theorem is due to de Possel† who used it to establish the possibility of mapping an arbitrary domain on a domain bounded by parallel slits.

8. *On the growth of simple functions*: We content ourselves with the following remarks. We show that the lemma can also be used to prove certain theorems concerning the growth of simple functions. It is known that, among all functions (15) which are simple and regular in the interior of the unit circle, the function (29) has the quickest growth in the direction of the positive axis and assumes, at each point of it, values of larger modulus than those of every other function of the family. We investigate now the generalized problem which has the following form. Let Δ once more be an arbitrary domain in the ζ-plane not containing the point at infinity. Let us consider the family (15) of functions which are regular and simple in Δ. We fix a point ζ_0 in Δ, and seek that function of the family for which $|f(\zeta_0)|$ assumes the largest possible value within the family. Using the formulae (16′) and (16″) and concluding as in §3, we obtain

$$(41) \qquad \left(\frac{dz}{dt}\right)^2 \frac{f(\zeta_0)}{z^2\{z-f(\zeta_0)\}} + 1 = 0$$

as the differential equation for the extremal function, which must hold for each of the mapped boundary slits.

---

† R. de Possel, *loc. cit.*

If z_0 a fixed point on a given boundary slit in the image-plane, we get

$$(42) \qquad \int_{z_0}^{z} \frac{dx}{x \sqrt{(1-x/\gamma)}} = t,$$

a real number with $\gamma = f(\zeta_0)$. Carrying out the integration, we find that

$$(43) \qquad \log \left\{ \frac{1 - \sqrt{(1-z/\gamma)}}{1 + \sqrt{(1-z/\gamma)}} \right\} - \log \left\{ \frac{1 - \sqrt{(1-z_0/\gamma)}}{1 + \sqrt{(1-z_0/\gamma)}} \right\} = t,$$

or writing

$$C = \frac{1 + \sqrt{(1-z_0/\gamma)}}{1 - \sqrt{(1-z_0/\gamma)}},$$

and taking s as a real parameter instead of t

$$(44) \qquad z = f(\zeta_0) \frac{4sC}{(1+sC)^2}.$$

Here s varies in a certain real interval and (44) furnishes the corresponding points of the boundary continuum considered.

As in §§ 4 and 5, it is evident that one of the boundary continua must pass through ∞; but the only unbounded curve (44) is the straight line $z = f(\zeta_0) T$, T being another real parameter. Hence

THEOREM V. *If ζ_0 is a point in a given domain Δ, the function (15) which is regular and simple in Δ and assumes at ζ_0 the largest modulus within the family maps Δ on a domain bounded by curves with the parametric representation (44). In particular the straight line $z = f(\zeta_0) T$, with T varying in a certain real interval, is included among them.*

A completely analogous statement can be made concerning that function (15) which is simple in Δ and assumes in ζ_0 the least modulus within the family; the same statement is also valid for the simple function (15) which possesses the boundary point B whose distance from the origin is the least possible.

The Hebrew University,
    Jerusalem.

[6] On the coefficients of simple functions

[6] On the coefficients of simple functions. *Proc. London Math. Soc.* **44** (1938), 450–452.

ON THE COEFFICIENTS OF SIMPLE FUNCTIONS

By MENAHEM SCHIFFER.

[Received 24 February, 1938.—Read 17 March, 1938.]

1. A classical problem concerning simple functions may be formulated in the following way: Let Δ be a domain in the ζ-plane containing $\zeta = 0$, and let

(1) $$z = f(\zeta) = \zeta + a_2\zeta^2 + a_3\zeta^3 + \ldots a_n\zeta^n + \ldots$$

be regular and simple in Δ. The set of all normalized functions (1) forms a normal family, and there exist therefore functions $f_n(\zeta)$ of the type (1) for which $|a_n|$ possesses the largest value among the family. What is then this maximal value of $|a_n|$ and what can be stated of the extremal functions $f_n(\zeta)$ possessing the maximal $|a_n|$?

Each function (1) can be characterized by the domain D in the z-plane on which Δ is mapped with the aid of (1); and the problem is solved when the domains D_n belonging to the extremal functions $f_n(\zeta)$ are determined. By a method due to Marty* it is easily shown that the domains D_n are, for each n, slit domains. In a recent paper† it was shown that the boundary slits of D_n are analytic curves $z(t)$ satisfying the differential equation

(2) $$\left(\frac{1}{z}\frac{dz}{dt}\right)^2 S_n(z) + 1 = 0.$$

Here, $S_n(z)$ is a polynomial of the degree $n-1$ in $1/z$, the coefficients of which are formed in a certain manner from the coefficients of $f_n(\zeta)$. This

---

* F. Marty, "Sur le module des coefficients de MacLaurin d'une fonction univalente", *Comptes rendus*, 198 (1934), 1569.

† M. Schiffer, "A method of variation within the family of simple functions", *Proc. London Math. Soc.* (2), 44 (1938), 433–449.

polynomial may be obtained in the following way. We form the function

$$(3) \qquad F_z(\zeta) = \frac{f_n(\zeta)}{1 - f_n(\zeta)/z} = \sum_{\nu=1}^{\infty} a_\nu \{1 + S_\nu(z)\} \zeta^\nu \quad (a_1 = 1),$$

and in this development $S_n(z)$ is the polynomial* just mentioned.

The only possible critical points of the boundary curves of D_n are the zeros of $S_n(z)$ lying on these. So far as is known at present they may be branch points of these curves, and this paper is devoted to the investigation of these critical points.

2. If z is a boundary point of D_n the function (3) is simple and regular in Δ, and belongs therefore to the family (1). In view of the extremal property of $f_n(\zeta)$, we have

$$(4) \qquad |1 + S_n(z)| \leqslant 1.$$

If, at a boundary point $z_0 \neq \infty$, we have $S_n(z_0) = 0$, then equality holds in (4), and $F_{z_0}(\zeta)$ is also an extremal function. Therefore, all theorems on $f_n(\zeta)$ hold also for $F_{z_0}(\zeta)$ and in particular there exists an equation analogous to (2).

The image of Δ given by $F_{z_0}(\zeta)$ is obtained from D_n by the linear transformation

$$(5) \qquad y = \frac{z}{1 - z/z_0}.$$

Its boundary curves $y(\tau)$ satisfy the differential equation

$$(2') \qquad \left(\frac{1}{y}\frac{dy}{d\tau}\right)^2 T_n(y) + 1 = 0,$$

where $T_n(y)$ is obtained from the generating function

$$(3') \qquad G_y(\zeta) = \frac{F_{z_0}(\zeta)}{1 - F_{z_0}(\zeta)/y} = \sum_{\nu=1}^{\infty} a_\nu \left(1 + S_\nu(z_0)\right)\left(1 + T_\nu(y)\right) \zeta^\nu.$$

If z is connected with y by (5), we have also, by (3),

$$(3'') \qquad G_y(\zeta) = \frac{f_n(\zeta)}{1 - f_n(\zeta)/z} = \sum_{\nu=1}^{\infty} a_\nu \left(1 + S_\nu(z)\right) \zeta^\nu;$$

hence, comparing (3') with (3'') and assuming that $S_n(z_0) = 0$, we have

$$(6) \qquad T_n(y) = S_n(z).$$

---

* The author is much indebted to Miss M. L. Cartwright for calling his attention to this relation.

2 G 2

Substituting for ζ in terms of z in (2′) with the aid of (5) and (6), we get a new differential equation (involving a new parameter τ)

$$(7) \qquad \left(\frac{1}{z}\frac{dz}{d\tau}\right)^2 \frac{1}{(1-z/z_0)^2} S_n(z) + 1 = 0$$

for the boundary curves of D_n; and, since t and τ are both real, (2) and (7) together yield

$$(8) \qquad \left(1-\frac{z}{z_0}\right)^2 \geqslant 0$$

for all z on the boundary curves. This shows that

$$(9) \qquad z = Tz_0,$$

with real T. So we find that if $S_n(z_0) = 0$ holds for a finite boundary point z_0, all the boundary slits which contain more than one point lie on a ray through the origin.

Therefore the boundary curves of the domain D_n are analytic curves which have no finite critical points.

The Hebrew University,
     Jerusalem.

[8] Sur un théorème de la représentation conforme

[8] Sur un théorème de la représentation conforme. *C. R. Acad. Sci. Paris* **207** (1938), 520–522.

THÉORIE DES FONCTIONS. — *Sur un théorème de la représentation conforme.*
. Note de M. **Menahem Schiffer**, présentée par M. Paul Montel.

Soit B un domaine borné dans le plan z, contenant l'origine. Désignons
par $\mathcal{F}$ la famille des fonctions $f(z)$ holomorphes et univalentes dans B et
normées pour $z = o$: $f(o) = o$, $f'(o) = 1$. Cherchons une fonction de $\mathcal{F}$
possédant dans B le moindre module maximum $\mathfrak{M}$ de la famille. L'existence
d'une telle fonction extrémale $w = f(z)$ suit du fait que la famille $\mathcal{F}$ est
normale. $f(z)$ représente B sur un domaine D du plan w. Toute fonction $g(w)$ normée dans $w = o$, holomorphe et univalente dans D prend
dans $\overline{D}$ au moins une valeur $\geq \mathfrak{M}$. Soit maintenant w_0 un point extérieur
à D; la fonction

$$(1) \qquad g_\rho(w; w_0) = \left[w + \frac{e^{i\varphi}\rho^2}{w_0} + \frac{e^{i\varphi}\rho^2}{w - w_0} \right]\left[1 - \frac{e^{i\varphi}\rho^2}{w_0^2} \right]^{-1}$$

sera normée pour $w = o$, holomorphe et univalente dans D pour tout φ et
pour ρ suffisamment petit. Il y a donc dans $\overline{D}$ un w tel qu'on ait

$$(2) \qquad |g_\rho(w; w_0)| = |w|\left| 1 + \frac{e^{i\varphi}\rho^2 w}{(w - w_0)w_0^2} + (\rho^4) \right| \geq \mathfrak{M},$$

et de $|w| \leq \mathfrak{M}$ suit

$$(2') \qquad \mathcal{R}\left\{ \frac{e^{i\varphi}}{w_0^2} \frac{1}{1 - w_0/w} \right\} + (\rho^2) \geq \frac{1}{\rho^2}\left(\frac{\mathfrak{M}}{|w|} - 1 \right) \geq o,$$

w dépend encore de ρ, φ et w_0. Pour un w_0 fixe choisissons φ de manière
que $e^{i\varphi}w_0^{-2} < o$. Alors w ne dépend que de ρ. Considérons une suite $\rho_\nu \to o$.
    D'après $(2')$ on trouve $\lim |w(\rho_\nu)| = \mathfrak{M}$. De ces $w(\rho_\nu)$, extrayons
maintenant une suite partielle et convergente telle qu'on ait $\lim w(\rho_\nu) = w^*$,
$(|w^*| = \mathfrak{M})$. De $(2')$ suit, à la limite $\rho = o$,

$$(3) \qquad \mathcal{R}\{ -1 + w_0/w^* \} \geq o.$$

Cette inégalité n'est possible que si $|w_0| \geq |w^*| = \mathfrak{M}$. *Dans le cercle $|w| \leq \mathfrak{M}$
ne se trouvent donc que des points intérieurs ou frontières de* D.
    Soit w_1, $(|w_1| < \mathfrak{M})$, un point d'un continu frontière $\mathcal{C}$ de D. De $\mathcal{C}$
nous coupons un continu partiel $\mathcal{C}(\sigma)$ du diamètre transfini σ et contenant w_1. Considérons toutes les fonctions de la forme

$$(4) \qquad \chi(w) = w + \mathcal{A}\sigma^2(w - w_1)^{-1} + \mathcal{B}\sigma^3(w - w_1)^{-2} + \dots,$$

univalentes et méromorphes à l'extérieur de $\mathcal{C}(\sigma)$. La fonction

$$\psi(w) = \frac{[\chi(w) - \chi(0)]}{\chi'(0)}$$

a les mêmes propriétés et est normée pour $w = 0$. La fonction

$$h(w) = g_\rho\{\psi(w); w_0\},$$

enfin sera dans D holomorphe, univalente et normée pour ρ et σ suffi-samment petits et φ arbitraire, si $w_0 = \mathfrak{M}^2/\overline{w_1}$. Pour $\mathcal{A}$ et σ donnés, choisissons $\rho^2 = |\mathcal{A}| \mathfrak{M}^4 |w_1|^{-4} \sigma^2$ et $\varphi = \arg(-\mathcal{A})$. Alors on a le déve-loppement suivant, valable à l'extérieur de tout cercle contenant $\mathcal{C}(\sigma)$,

$$(5) \qquad h(w) = w + \frac{\mathcal{A}\sigma^2}{w_1^2}\left[\frac{w^2}{w - w_1} - \frac{w^2}{w - w_0}\right] + (\sigma^2).$$

De la propriété d'extrémum de D suit ici aussi l'existence d'un $w \subset \bar{\mathrm{D}}$ tel qu'on ait $|h(w)| \geqq \mathfrak{M}$. Pour une suite $\sigma \to 0$ choisissons une suite conver-gente $\mathcal{A}(\sigma) \to \alpha$ et $w(\sigma) \to w^*$. Encore une fois on trouve $|w^*| = \mathfrak{M}$. Parce que $w(w - w_1)^{-1} - w(w - \mathfrak{M}^2/\overline{w_1})^{-1} > 0$ pour tout $|w| = \mathfrak{M}$, on trouve de $|h(w)| \geqq \mathfrak{M}$ à la limite $\rho = 0$

$$(6) \qquad \mathcal{R}\{\alpha w_1^{-2}\} \geqq 0$$

pour toute suite convergente de coefficients $\mathcal{A}$ dans (4). Cela vaut pour tout point $w_1 \subset \mathcal{C}$. Comme j'ai démontré récemment [1], il suit de ce fait, que $\mathcal{C}$ est une courbe analytique $w(t)$, satisfaisant à l'équation diffé-rentielle

$$(7) \qquad \frac{1}{w^2}\left(\frac{dw}{dt}\right)^2 + 1 = 0.$$

Donc $w = ce^{it}$ est la seule représentation possible pour les fentes de D. D *est donc un cercle, coupé le long d'arcs de cercle concentriques.*

Il est évident qu'on obtient des résultats analogues en considérant, au lieu de $\mathscr{F}$, certaines familles de fonctions p-valentes, comme par exemple celles ayant le développement $f(z) = z^p + a_{p+1} z^{p+1} + \ldots$ autour de l'origine ou la famille $\mathscr{F}_p$ des $f(z) = z + a_2 z^2 + \ldots$ ne possédant d'autre zéro que l'origine. S'il y a dans $\mathscr{F}_p$ une fonction $f(z)$, avec

$$f'(z_i) = 0 \qquad (i = 1, \ldots, m, \ z_i \subset \mathrm{B}),$$

---

[1] Voir un Mémoire *A calculus of variation* ..., qui paraîtra prochainement dans les *Proc. Lond. Math. Soc.*

il existe aussi dans $\mathscr{F}_p$ une fonction $\varphi(z)$ avec $\varphi'(z_i) = 0$, dont les valeurs couvrent exactement un cercle, éventuellement coupé le long d'arcs de cercle concentriques. C'est cette fonction de $\mathscr{F}_p$ avec les conditions $f'(z_i) = 0$ dont le module maximum est le plus petit possible.

MÉCANIQUE STATISTIQUE. — *Sur une équation générale de la Mécanique statistique.* Note ([1]) de M. Bohuslav Hostinský, transmise par M. Émile Borel.

La configuration instantanée d'un système mécanique S est déterminée par n coordonnées de Lagrange $q_1, q_2, \ldots, q_n$ et par les composantes de vitesse $dq_1/dt, dq_2/dt, \ldots, dq_n/dt$. Considérons le point χ dont les coordonnées sont les $2n$ quantités $q_1, \ldots, q_n, dq_1/dt \ldots dq_n/dt$. Ce point se meut dans un domaine à $2n$ dimensions E; soit F la frontière de E. Le mouvement de χ figure l'évolution du système S. Par exemple, si S est formé par m molécules d'un gaz enfermé dans un vase, nous avons $n = 3m$, les q_i étant les coordonnées cartésiennes des molécules considérées comme des points; les points de la frontière F correspondent soit aux cas où une molécule touche la paroi du vase soit aux cas où une molécule touche une autre.

Nous admettons que, la position P_0 de χ à l'époque t_0 étant connue, il y ait une probabilité $\Phi(P_0, P, t_0, t) d\tau_P$ pour que χ se trouve à l'époque t ($t_0 < t$) à l'intérieur d'un élément infinitésimal $d\tau_P$ de E au voisinage du point P. La probabilité pour que χ arrive de P_0 (époque t_0) à $d\tau_P$ (époque t) sans rencontrer la frontière F sera égale à $j(P_0, P, t_0, t) d\tau_P$ où la fonction $j(P_0, P, t_0, t)$ satisfait aux conditions suivantes :

$$j > 0; \qquad \int_E j \, d\tau_P = 1; \qquad \lim_{t=t_0} j = 0 \quad \text{et} \quad \lim_{t=t_0} \frac{\partial j}{\partial t} = 0 \quad \text{et} \qquad \text{pour } P \neq P_0,$$

$$\int_E j(P_0, Q, t_0, u) j(Q, P, u, t) \, d\tau_Q = j(P_0, P, t_0, t) \qquad (t_0 < u < t).$$

Ces probabilités des déplacements définies au moyen de la fonction j sont analogues de celles que l'on considère dans le cas du mouvement brownien ordinaire. Si χ se trouve à l'époque u dans un élément $d\sigma_M$ de F au voisinage d'un point M de F, et si $d\sigma_N$ est un autre élément de F, la pro-

---

([1]) Séance du 12 septembre 1938.

Commentary on

[4] *Sur un problème d extrémum de la représentation conforme*, Bull. Soc. Math. France **66** (1938), 48 55.

[5] *A method of variation within the family of simple functions*, Proc. London Math. Soc. **44** (1938), 432 449.

[6] *On the coef"cients of simple functions*, Proc. London Math. Soc. **44** (1938), 450 452.

[8] *Sur un théorème de la représentation conforme*, C. R. Acad. Sci. Paris **207** (1938), 520 522.

Schiffer presented his celebrated method of boundary variation in his doctoral thesis of 1939 at the Hebrew University, "Conformal Representation and Univalent Functions" (in Hebrew), under the supervision of Michael Fekete. An announcement had appeared in [3], and a detailed account was first published in [5]. The method was applied in [4] to an extremal problem for transfinite diameter. The papers [6, 8] further demonstrated the power of the variational method with applications to problems in function theory.

Schiffer developed the variational method of [5] with a view to solving the Bieberbach conjecture, which can be stated as follows. A function f analytic and univalent (or "simple") in the unit disk $\mathbb{D} = \{z \in \mathbb{C} : |z| < 1\}$ is said to belong to the class S if $f(0) = 0$ and $f'(0) = 1$. Thus it has a power series development of the form $f(z) = z + \sum_{n=2}^{\infty} a_n z^n$. The Bieberbach conjecture said that for each $f \in S$, the inequality $|a_n| \leq n$ holds for every n, with strict inequality for all n unless f is a rotation of the Koebe function

$$k(z) = \frac{z}{(1-z)^2} = \sum_{n=1}^{\infty} n z^n,$$

which maps $\mathbb{D}$ onto the whole complex plane minus the part of the real axis from $-\frac{1}{4}$ to $-\infty$. Bieberbach made the conjecture in 1916 after giving an elementary proof that $|a_2| \leq 2$. In 1923, Loewner [Lo] developed a new technique, the so-called parametric method, and applied it to show that $|a_3| \leq 3$. Loewner was convinced that because his method effectively represents the class S, it was sure to settle the Bieberbach conjecture,

and he felt frustrated to be stopped at $n = 3$. Since the conjecture amounts to an extremal problem to find $\max_{f \in S} \text{Re}\{a_n\}$, it was natural to mount an attack by a variational method. However, this was more easily said than done.

In the classical calculus of variations, the main difficulty in solving an extremal problem is often to prove the existence of an extremal element. It is then relatively easy to devise a variation, a neighborhood of competitors sufficiently rich to give substantial information about the extremal element. For extremal problems in geometric function theory, on the other hand, one encounters the opposite problem. The existence of an extremal function is assured by Montel's theory of normal families, but the difficulty lies in constructing a nontrivial variation that preserves the class of functions considered. Motivated by Bieberbach's conjecture, Schiffer sought to develop a variation that preserves the class S. Hadamard had introduced a formula for varying Green's function of a smoothly bounded Jordan domain, and Julia had applied Hadamard's method to obtain a suitable variation of a slit domain; but the implementation of those methods required a priori information about the extremal function that was not available. (For further details see the commentary on [13].) Marty [M] had applied a very special variation, precomposition with a Möbius automorphism of $\mathbb{D}$ close to the identity, to show that any function $f \in S$ which maximizes $\text{Re}\{a_n\}$ must satisfy the *Marty relation* $(n+1)a_{n+1} - 2a_2 a_n - (n-1)\overline{a_{n-1}} = 0$. (Observe that the Koebe function satisfies the Marty relation.) In the same paper, Marty also proved that the range of an extremal function is dense in the complex plane: it cannot omit an open set. Both results gave evidence in favor of the Bieberbach conjecture but fell far short of a proof.

Schiffer's achievement in [5] was to devise a more general method of variation that preserves univalence, thereby allowing an extremal function to be compared with a sufficiently large family of neighboring functions to deduce substantial information, often leading to a complete description of the extremal functions. Briefly, Schiffer's method of boundary variation consists of postcomposing a given univalent function with a conformal mapping of the region complementary to a small piece of the boundary of its range. (Details can be found in [5], in Schiffer's expository articles [34, 67], and in [D].) Applied to an extremal problem, the method typically gives (almost automatically) the information that an extremal function maps onto the whole complex plane minus the union Γ of analytic arcs that are trajectories of a certain quadratic differential: $Q(w)dw^2 > 0$, where the function Q is meromorphic on Γ. For the coefficient problem $\max_{f \in S} \mathrm{Re}\{a_n\}$, it turns out that Q is a rational function whose coefficients involve data from the unknown extremal function. Its explicit form is determined in [6], leading to the "Schiffer differential equation" for an extremal function. From this Schiffer concluded in [6] that the omitted set Γ consists of a finite number of unbranched analytic arcs. Much later, he would show [95] that Γ is a single analytic arc with an asymptotic direction at infinity. But in order to verify the Bieberbach conjecture, one needed to prove that Γ is a radial half line, so that the extremal function is the Koebe function.

Schiffer's derivation of the method of boundary variation relied on a topological lemma due to Haslam-Jones [HJ], stating that a continuum whose only limit directions are ± 1 must be a horizontal line segment. The proof of this key lemma was later simplified by several authors (see [S] or [CL]).

In his paper [4], Schiffer applies a primitive version of boundary variation to a problem involving the *n-diameter*

$$d_n(E) = \left\{ \max_{z_1,\ldots,z_n \in E} \prod_{1 \le j < k \le n} |z_j - z_k| \right\}^{\frac{2}{n(n-1)}}$$

of a compact (simply) connected set $E \subset \mathbb{C}$. The problem is couched in a function-theoretic setting, but for each given $n \ge 2$ it reduces to finding configurations of a set E of fixed transfinite diameter $d(E) = \lim_{n \to \infty} d_n(E)$ that will maximize the n-diameter $d_n(E)$. The solution for $n = 2$ was known to be a line segment. Schiffer shows in [4] that for any $n \ge 2$ the extremal configuration is a system of arcs that lie on trajectories of a certain

quadratic differential. For $n = 3$, he concludes that the extremal set E is a "star" consisting of three line segments of equal length meeting at a central point with equal angles $2\pi/3$. As a corollary, he solves a coefficient problem for the class Σ of functions $g(z) = z + b_0 + b_1/z + b_2/z^2 + \cdots$ analytic and univalent in $|z| > 1$. The sharp inequality $|b_1| \le 1$ was known as a trivial consequence of the area theorem. Schiffer proved that $|b_2| \le \frac{2}{3}$ for all $g \in \Sigma$, with equality occurring only for rotations and translations of $g(z) = \{1/k(1/z^3)\}^{1/3} = z - \frac{2}{3}z^{-2} + \ldots$, where k is the Koebe function. Goluzin offered a proof of $|b_2| \le \frac{2}{3}$ around the same time but later acknowledged an error (see[G], p. 279). Schiffer's result gave rise to the conjecture that $|b_n| \le \frac{2}{n+1}$, with equality for $g(z) = \{1/k(1/z^{n+1})\}^{1/(n+1)}$. That conjecture was later disproved for all $n \ge 3$; see the commentary on [59].

In a subsequent paper [11], Schiffer applied his method of boundary variation to a problem closely related to the one in [4], namely, to find the set of smallest transfinite diameter that contains n prescribed points in the plane. In later years, Schiffer [14, 17, 54, 69, 76, 79, 99, 135, 136, 137] returned to this and other extremal problems involving transfinite diameter, a topic pioneered by his thesis adviser Fekete.

The method of boundary variation is a natural device for proving the existence of conformal mappings of multiply connected domains onto canonical domains such as parallel or circular slit domains. This first became apparent in Schiffer's paper [8], where he applied the method to prove that any finitely connected domain (not a punctured sphere) can be mapped conformally onto a disk with concentric circular slits. Contained in the result was a new proof of the Riemann mapping theorem. Given a bounded domain B containing the origin, Schiffer proves that a function f analytic and univalent in B, with local form $f(z) = z + a_2z^2 + \ldots$ and smallest maximum modulus, must map B onto a disk with circular slits. In later works (see for instance [34]), he extended the method to other extremal problems, leading to canonical mappings onto domains of various types. Garabedian and Schiffer [29], and independently Lehto [Le], constructed some of the same canonical mappings by kernel function methods.

Returning to the Bieberbach conjecture in 1955, Garabedian and Schiffer [60] combined variational methods with Loewner's method to prove that $|a_4| \leq 4$. This was a real *tour de force*, but five years later Charzyński and Schiffer [72] found a much simpler proof based only on the Grunsky inequalities, a simple generalization of the area principle. With greater labor, the new "elementary" method then yielded proofs for $n = 6$ and $n = 5$. Finally, de Branges [B] proved a conjecture of I. M. Milin that was known to imply the Bieberbach conjecture. FitzGerald and Pommerenke [FP] then simplified de Branges's proof and translated it into classical terms. When all was said and done, the property of univalence had entered the proof only through Loewner's differential equation, thus validating Loewner's vision of 1923.

Schiffer developed the method of boundary variation [5] and later the method of interior variation [13], in an effort to prove the Bieberbach conjecture. These methods were essential new tools which found important applications to a variety of problems in geometric function theory and to topics such as transfinite diameter, Riemann surface theory, partial differential equations, potential theory, fluid flow, and other areas of mathematical physics. Ironically, however, variational methods played no role in the eventual proof of the Bieberbach conjecture.

References

[B] Louis de Branges, *A proof of the Bieberbach conjecture*, Acta Math. **154** (1985), 137 152.

[CL] Douglas M. Campbell and Jack Lamoreaux, *Continua in the plane with limit directions*, Pacific J. Math. **74** (1978), 37 46.

[D] Peter L. Duren, *Univalent Functions*, Springer-Verlag, 1983.

[FP] Carl H. FitzGerald and Ch. Pommerenke, *The de Branges theorem on univalent functions*, Trans. Amer. Math. Soc. **290** (1985), 683 690.

[G] G. M. Goluzin, *On p-valent functions*, Mat. Sb. **8** (50) (1940), 277 284 (in Russian).

[HJ] U. S. Haslam-Jones, *Tangential properties of a plane set of points*, Quart. J. Math. **7** (1936), 116 123.

[Le] Olli Lehto, *Anwendung orthogonaler Systeme auf gewisse funktionentheoretische Extremal- und Abbildungsprobleme*, Ann. Acad. Sci. Fenn. Ser. A I Math. Phys. 1949, no. 59.

[Lo] Karl Löwner (Charles Loewner), *Untersuchungen über schlichte konforme Abbildungen des Einheitskreises, I*, Math. Ann. **89** (1923), 103 121.

[M] F. Marty, *Sur le module des coef"cients de MacLaurin d une fonction univalente*, C. R. Acad. Sci. Paris **198** (1934), 1569 1571.

[S] Glenn Schober, *Univalent Functions Selected Topics*, Lecture Notes in Math. **478**, Springer-Verlag, 1975.

PETER DUREN

[10] Sur la variation de la fonction de Green de domaines plans quelconques

[10] Sur la variation de la fonction de Green de domaines plans quelconques. *C. R. Acad. Sci. Paris* **209** (1939), 980–982.

D'après (4) et (5), pour $q > p \geq n_1 + n_0$, on a

$$\sum_{n=p}^{q} |a_n e^{-\lambda_n z_0}| = \sum_{n=p}^{q} \frac{a'_n}{\cos\varphi_n} e^{-\lambda_n x_0} \leq \sum_{n=p}^{q} a'_n e^{-\lambda_n x_1} < \varepsilon,$$

c'est-à-dire que la série (1) converge absolument pour $z = z_0$, d'où

(6)
$$C'' \leq C'.$$

Pour toute série de Dirichlet, on a

(7)
$$C' \leq C \leq C''.$$

De (6) et (7) on tire

(8)
$$C' = C = C''.$$

D'autre part, on a

(9)
$$|f^{(m)}(C+1)| \geq |g^{(m)}(C+1)|,$$

en vertu d'un théorème de Landau ([1]); d'après (8) et $a'_n \geq 0$, on voit que le point $z = C$ est singulier pour $g(z)$; de ceci et (9), on déduit : le rayon de convergence de la série

$$\sum_{m=0}^{\infty} (z - C - 1)^m \frac{f^{(m)}(C+1)}{m!}$$

est l'unité. Donc le théorème est prouvé.

THÉORIE DES FONCTIONS. — *Sur la variation de la fonction de Green de domaines plans quelconques.* Note de M. MENAHEM SCHIFFER, présentée par M. Paul Montel.

Soit $\mathscr{E}$ un ensemble borné et fermé dans le plan z avec un diamètre transfini $d(\mathscr{E}) > 0$. La fonction

(1)
$$z^* = z + \varepsilon e^{i\varphi}(z - z_0)^{-1}, \qquad \varepsilon > 0, \qquad 0 \leq \varphi \leq 2\pi, \qquad z_0 \notin \mathscr{E}$$

est univalente et holomorphe dans tout domaine fermé $\mathscr{D}$ contenant $\mathscr{E}$ mais ne contenant pas z_0, pourvu que $\varepsilon = \varepsilon(\mathscr{D}, z_0, \varphi)$ soit suffisamment petit; elle transforme $\mathscr{E}$ en un ensemble $\mathscr{E}^* = \mathscr{E}^*(\varepsilon, \varphi, z_0)$ dans le plan z^* avec les mêmes propriétés. Sur $\mathscr{E}^*$ choisissons les suites $\{z_\nu^{*(n)}\}$ donnant le maximum des produits $\Pi |a_\nu^* - a_\mu^*| (1 \leq \nu < \mu \leq n)$, a_ν^* variant sur $\mathscr{E}^*$. Considérons les

_____
([1]) *Mathematische Annalen*, 61, 1905, p. 1.

polynomes $t_n(z^*) = (z^* - z_1^{*(n)}) \ldots (z^* - z_n^{*(n)})$ et leurs modules maxima μ_n^* sur $\mathscr{E}^*$; les lemniscates $\mathcal{L}_n^*$, définies par $|t_n(z)| = \mu_n^*$, convergent vers $\mathscr{E}^*$, c'est-à-dire (^{1}) que $\widetilde{\mathscr{E}}^*$ est le noyau des $\widetilde{\mathcal{L}}_n^*$ au sens de M. Carathéodory (^{2}). Dans le plan z, on a des courbes analytiques $\mathcal{L}_n$ convergeant vers $\mathscr{E}$ et correspondant aux $\mathcal{L}_n^*$ selon (1).

Considérons les fonctions

$$(2) \qquad g_n^*(z^*; \infty) = \frac{1}{n} \log \left\{ |z^* - z_1^{*(n)}| \ldots |z^* - z_n^{*(n)}| \right\} - \frac{1}{n} \log \mu_n^*,$$

définies pour tout z^* et représentant dans $\widetilde{\mathcal{L}}_n^*$ les fonctions de Green de ces domaines avec le pôle logarithmique à l'infini. Pour $n \to \infty$ ils convergent dans $\widetilde{\mathscr{E}}^*$ vers la fonction de Green $g^*(z^*; \infty)$ de ce domaine (^{3}). De même convergent les fonctions de Green $g_n(z; \zeta)$ de $\widetilde{\mathcal{L}}_n$ avec le pôle logarithmique en $\zeta \subset \widetilde{\mathcal{L}}_n$ vers la fonction de Green $g(z; \zeta)$ de $\widetilde{\mathscr{E}}$ avec le même pôle.

La fonction $g_n^*[z + \varepsilon e^{i\varphi}(z - z_0)^{-1}; \infty] = \gamma_n(z)$ est définie par (2) pour tout point z; elle a des pôles logarithmiques en $z = \infty$, $z = z_0$ et aux $2n$ points $z_\nu^{(n)}$, $z_\nu^{\prime(n)}$ obtenus à l'aide de (1) à partir des n points $z_\nu^{*(n)}$. Aux deux premiers points, $\gamma_n(z)$ devient infini comme $\log|z|$ et $-\log|z - z_0|$; aux autres points, comme $1/n \log|z - z_\nu^{(n)}|$ et $1/n \log|z - z_\nu^{\prime(n)}|$. Un calcul élémentaire montre qu'on a

$$z_\nu^{(n)} = z_\nu^{*(n)} + \varepsilon e^{i\varphi}[z_0 - z_\nu^{*(n)}]^{-1} + O(\varepsilon^2) \qquad \text{et} \qquad z_\nu^{\prime(n)} = z_0 - \varepsilon e^{i\varphi}[z_0 - z_\nu^{*(n)}]^{-1} + O(\varepsilon^2).$$

Les points $z_\nu^{(n)}$ se trouvent près des $z_\nu^{*(n)}$ et leur correspondent par (1), ils se trouvent donc sur $\mathscr{E}$; les $z_\nu^{\prime(n)}$ se trouvent près de z_0. Sur $\mathcal{L}_n$ on a $\gamma_n(z) = 0$, car (1) transforme $\mathcal{L}_n$ en $\mathcal{L}_n^*$. $\gamma_n(z)$ est une fonction harmonique et l'on a l'identité

$$(3) \qquad g_n^*[z + \varepsilon e^{i\varphi}(z - z_0)^{-1}; \infty] = g_n(z; \infty) + g_n(z; z_0) - \frac{1}{n} \sum_{\nu=1}^{n} g_n[z; z_\nu^{\prime(n)}]$$

pour $z \subset \widetilde{\mathcal{L}}_n$, la différence des deux membres étant harmonique et bornée dans ce domaine et s'annulant sur sa frontière.

Introduisons les fonctions analytiques $q_n(z; \zeta)$ et $q_n^*(z^*; \zeta^*)$, déterminées

---

(^{1}) Par $\widetilde{\mathfrak{a}}$ nous désignerons toujours le domaine complémentaire infini de $\mathfrak{a}$ et par $\mathfrak{a} \mathfrak{b}$ le produit (la partie commune) des ensembles $\mathfrak{a}$ et $\mathfrak{b}$ du plan z.

(^{2}) M. Fekete, *Math. Z.*, 37, 1933, p. 638, théorème II.

(^{3}) Ce théorème est dû à M. M. Fekete; la démonstration paraîtra bientôt. Par fonction de Green nous comprenons la notion généralisée; voir Myrberg, *Acta Math.*, 61, 1933, p. 42.

à une constante imaginaire additive près et non nécessairement uniformes dans $\widetilde{\mathcal{L}}_n$ et $\widetilde{\mathcal{L}}_n^*$, dont les parties réelles sont $g_n(z;\zeta)$ et $g_n^*(z^*;\zeta^*)$. On a, grâce à la symétrie des fonctions de Green,

$$(4) \quad \mathcal{R}\{q_n^*[z + \varepsilon e^{i\varphi}(z-z_0)^{-1};\infty]\}$$
$$= \mathcal{R}\left\{ q_n(z;\infty) + q_n(z;z_0) - \frac{1}{n}\sum_{\nu=1}^{n} q_n[z_0 - \varepsilon e^{i\varphi}(z_0 - z_\nu^{*(n)})^{-1} + O(\varepsilon^2);z]\right\},$$

$\mathcal{R}\{x\}$ désignant la partie réelle de x. L'application de la formule de Taylor fournit pour tout z dans $\widetilde{\mathcal{L}}_n$, $\widetilde{\mathcal{L}}_n^*$ différent de z_0

$$(5) \quad \mathcal{R}\left\{ q_n^*(z;\infty) + \frac{\varepsilon e^{i\varphi}}{z-z_0} q_n^{*\prime}(z;\infty) + O(\varepsilon^2)\right\}$$
$$= \mathcal{R}\left\{ q_n(z;\infty) + q'(z_0;z)\varepsilon e^{i\varphi}\frac{1}{n}\sum_{\nu=1}^{n}\frac{1}{z_0 - z_\nu^{*(n)}} + O(\varepsilon^2)\right\},$$

en désignant par $q'(x;y)$ la dérivée $(\partial/\partial x)q(x;y)$.

Dans tout domaine fixe, les fonctions analytiques avec partie réelle non négative forment une famille normale, et l'on peut trouver une borne supérieure pour les $O(\varepsilon^2)$ dans (5), valable pour tout n et pour tout z d'un domaine fermé et fixe dans $\widetilde{\mathcal{S}}$, $\widetilde{\mathcal{S}}^*$ qui ne contient pas z_0. On peut donc passer à la limite $n = \infty$, en se servant encore de l'identité

$$(6) \quad \frac{1}{n}\sum_{\nu=1}^{n}\frac{1}{z_0 - z_\nu^{*(n)}} = q_n^{*\prime}(z_0;\infty)$$

qui découle de (2). Désignons par $q(z;\zeta)$ et $q^*(z;\zeta)$ des fonctions analytiques dans $\widetilde{\mathcal{S}}$, $\widetilde{\mathcal{S}}^*$ avec les parties réelles $g(z;\zeta)$ et $g^*(z;\zeta)$, $\zeta \subset \widetilde{\mathcal{S}}$, $\widetilde{\mathcal{S}}^*$. On a

$$(7) \quad \mathcal{R}\{q^*(z;\infty)\} = \mathcal{R}\{q(z;\infty) + \varepsilon e^{i\varphi}[q'(z_0;z)q^{*\prime}(z_0;\infty)$$
$$- q^{*\prime}(z;\infty)(z-z_0)^{-1}] + O(\varepsilon^2)\}.$$

De (7) découle $q^{*\prime}(z;\infty) = q'(z;\infty) + O(\varepsilon)$ et l'on a *pour tout point z de $\widetilde{\mathcal{S}}$*

$$(8) \quad g^*(z;\infty) = g(z;\infty) + \mathcal{R}\{\varepsilon e^{i\varphi}[q'(z_0;z)q'(z_0;\infty)$$
$$- q'(z;\infty)(z-z_0)^{-1}]\} + O(\varepsilon^2).$$

Cette formule donne la variation de la fonction de Green g du domaine $\widetilde{\mathcal{S}}$, restreint par la seule condition $d(\mathcal{S}) > 0$ (inévitable pour l'existence de g), si $\mathcal{S}$ est soumis à la déformation spéciale (1). Elle permet de calculer aussi cette variation pour toutes les déformations obtenues par composition de transformations de ce type.

[11] Sur la variation du diamètre transfini

[11] Sur la variation du diamètre transfini. *Bull. Soc. Math. France* **68** (1940), 158–176.

SUR LA VARIATION DU DIAMÈTRE TRANSFINI;

Par M. Menahem Schiffer.

(Jérusalem).

1. Introduction. — Considérons dans le plan z tous les ensembles bornés et fermés E. A chacun d'eux adjoignons la suite de nombres suivante :

$$(1) \qquad d_n(E) = \sqrt[\binom{n}{2}]{\max_{a_\nu \subset E} \prod_{\nu < \mu}^{1 \ldots n} |a_\nu - a_\mu|} \qquad (n = 2, 3, \ldots).$$

M. Fekete ([1]) a démontré que ces nombres non négatifs, les « diamètres de l'ordre n de E », convergent pour $n \to \infty$ en décroissant vers une limite $d(E)$, le diamètre transfini de E. Si C est un continu ([2]) borné, on peut représenter son extérieur conformément sur l'extérieur d'un cercle à l'aide d'une fonction univalente $h(z)$ normée à l'infini [à savoir $h(\infty) = \infty$, $h'(\infty) = 1$]; on sait que le rayon de ce cercle sera égal à $d(C)$ [voir ([1]), théorème IX, p. 239]. C'est pourquoi $d(C)$ est appelé, dans le cas d'un continu, le rayon de représentation de C.

Dans une Note précédente ([3]) j'ai traité le problème d'extrémum suivant : Pour tout n fixe trouver parmi tous les continus C avec $d(C) = 1$, dont l'ensemble complémentaire est d'un seul tenant, les continus C_n admettant la valeur maximum Δ_n pour $d_n(C)$. Pour tout n on obtient donc un problème d'extrémum, et

---

([1]) M. Fekete, *Ueber die Verteilung der Wurzeln bei gewissen algebraischen Gleichungen mit ganzzahligen Koeffizienten* (*Math. Zeitschr.*, t. 17, 1923, p. 228-249).

([2]) Dans cette Note nous appellerons continu seulement des *continus propres*, c'est-à-dire des continus contenant deux points au moins.

([3]) M. Schiffer, *Sur un problème d'extrémum de la représentation conforme* (*Bull. Soc. math. France*, t. 66, 1938, p. 48-55).

j'ai développé une méthode de variations infinitésimales ([4]) permettant de traiter le cas général et de caractériser les continus extrémum C_n pour tout n. J'ai démontré que ces continus sont des ensembles sans points intérieurs et composés d'un nombre fini d'arcs analytiques satisfaisant tous à la même équation différentielle du premier ordre.

Nous pouvons en déduire une propriété caractéristique pour les continus C_n. En effet, choisissons sur C_n un système de n points $a_1^{(n)}, \ldots, a_n^{(n)}$ pour lequel le produit

$$(1') \qquad \prod_{\nu < \mu}^{1 \ldots n} | a_\nu - a_\mu | \qquad (a_\nu \subset C_n)$$

devient maximum. Alors *nous pouvons caractériser C_n comme le continu, contenant les points $a_1^{(n)}, \ldots, a_n^{(n)}$ et ayant le diamètre transfini le plus petit possible.* Car, d'une part, supposons qu'il y ait un continu C', contenant les $a_1^{(n)}, \ldots, a_n^{(n)}$ et avec $d(C') < 1$; alors on pourrait construire l'ensemble C'' avec $d(C'') = 1$, en soumettant C' à une homothétie de rapport $d(C')^{-1}$, dont le $d_n(C'')$ deviendrait $d(C')^{-1} d_n(C')$, c'est-à-dire plus grand que $d_n(C_n)$, en contradiction avec la propriété extrémale de C_n. D'autre part, si un continu C' contenant les $a_\nu^{(n)}$ satisfait à $d(C') = 1$, il doit coïncider avec C_n. En effet, on a

$$d_n(C') \geqq d_n(C_n) = \Delta_n,$$

parce que C' contient le système $a_\nu^{(n)}$; mais à cause de $d(C') = 1$, on a par définition $\Delta_n \geqq d_n(C')$. Donc $d_n(C') = \Delta_n$, C' est un continu extrémum pour n et satisfait à la même équation différentielle de premier ordre que C_n. Ces deux continus ayant les points $a_\nu^{(n)}$ communs, ils coïncident entièrement, d'après le théorème d'unicité, pour les équations en question.

M. Pólya posa la question générale de trouver pour n points $a_1, \ldots, a_n$, donnés arbitrairement, un continu borné C qui les contient et possède le diamètre transfini le plus petit possible.

([4]) *Loc. cit.* ([2]). *Voir* aussi : M. SCHIFFER, *A method of variation within the family of simple functions* (*Proc. London Math. Soc.*, 2ᵉ série, t. 44, 1938, p. 432-449).

Après que M. Pólya eut réussi à résoudre ce problème pour certains cas spéciaux ([5]), M. Grötzsch donna la solution la plus générale à l'aide de sa méthode des bandes ([6]).

Dans la présente Note je résous le même problème à l'aide des variations infinitésimales, en me servant d'un théorème de MM. Fekete et Leja (§ 3). On y trouvera une nouvelle méthode d'application de ces variations qui permet aussi la solution d'autres problèmes d'extrémum concernant le diamètre transfini (§§ 4 et 5).

2. Le changement du diamètre transfini par une variation infinitésimale de l'ensemble.

— La solution du problème de Pólya-Grötzsch pour le cas $n = 2$ est connue depuis longtemps. On sait que le continu C contenant deux points donnés et de diamètre transfini $d(\mathrm{C})$ minimum est le segment rectiligne joignant ces deux points. Notre méthode permet de résoudre le problème général; mais pour simplifier les calculs suivants, nous nous bornerons au cas $n = 3$ et nous supposerons $a_1,\ a_2,\ a_3 \equiv 0,\ 1,\ \eta$. De cette recherche spéciale on pourra tirer aisément le procédé général. Soit donc C un continu borné contenant les points 0, 1 et η, dont l'ensemble complémentaire est d'un seul tenant ([7]), et étudions le changement de $d(\mathrm{C})$ pour une déformation infinitésimale qui ne déplace pas ces trois points.

z_0 étant un point à l'extérieur de C, nous formons la fonction

$$(2) \qquad z^* = u_\varepsilon(z) = z + \varepsilon e^{i\varphi}\,\frac{z(z-1)(z-\eta)}{z-z_0} \qquad (\varepsilon > 0,\ 0 \leqq \varphi \leqq 2\pi).$$

Celle-ci est holomorphe et univalente dans tout domaine borné D, contenant C dans son intérieur, mais z_0 dans son extérieur, si $\varepsilon = \varepsilon(\mathrm{D})$ est choisi suffisamment petit. Elle transforme C en un

---

([5]) G. Pólya, *Beitrag zur Verallgemeinerung des Verzerrungssatzes auf mehrfach zusammenhängende Gebiete, III* (*Sitzungsberichte Akad. Berlin*, 1929, p. 55-62).

([6]) H. Grötzsch, *Ueber ein Variationsproblem der konformen Abbildung* (*Berichte*, Leipzig, t. 82, 1930, p. 251-263).

([7]) Cette restriction ne diminue pas la généralité parce qu'on peut adjoindre au continu donné tout son ensemble complémentaire, sauf le domaine contenant le point à l'infini sans changer son diamètre transfini.

continu C_ε^* qui contient aussi les points o, 1, η, parce que $u_\varepsilon(z)$ transforme ces trois points en eux-mêmes. L'ensemble complémentaire de C_ε^* est aussi d'un seul tenant. Pour ε suffisamment petit, C_ε^* est arbitrairement rapproché de C au sens de Fréchet.

Fixons maintenant deux domaines D et D' avec les propriétés mentionnées ci-dessus et tels que D' avec tous ses points frontières se trouve dans D. Soit A le minimum des distances entre les points frontières de D' et de D. Posant

$$(3) \qquad \varepsilon_0 = \frac{A}{2 \, \underset{z \subset D}{\text{borne sup.}} \left| \dfrac{z(z-1)(z-\eta)}{z-z_0} \right|},$$

nous pouvons affirmer, d'après la formule d'inversion de Lagrange, que pour $\varepsilon \leqq \varepsilon_0$ la fonction inverse de (2) $z = u_\varepsilon^{-1}(z^*)$ est régulière et univalente dans $\overline{D}'$, région obtenue de D' en lui adjoignant sa frontière. Cette fonction peut être développée dans cette région sous la forme

$$(4) \qquad u_\varepsilon^{-1}(z^*) = z^* - \frac{\varepsilon}{1!} e^{i\varphi} \frac{z^*(z^*-1)(z^*-\eta)}{z^*-z_0} + \frac{\varepsilon^2}{2!} r(z_1^*\varepsilon),$$

$r(z_1^*\varepsilon)$ représentant une fonction holomorphe de ε et de z^* dans $\overline{D}'$. En outre, nous choisissons $\varepsilon_1 \leqq \varepsilon_0$ tel que tous les ensembles C_ε^* se trouvent dans $\overline{D}'$ si $\varepsilon < \varepsilon_1$.

Après ces préparatifs, calculons et comparons les diamètres transfinis de C et de C_ε^* pour $\varepsilon \leqq \varepsilon(D)$. Nous partons des équations

$$(5) \qquad d(C) = \lim_{n=\infty} d_n(C); \qquad d(C_\varepsilon^*) = \lim_{n=\infty} d_n(C_\varepsilon^*).$$

En outre nous nous servirons de la relation limite suivante, mentionnée à la fin de l'Introduction, qui a été trouvée indépendamment par M. Fekete et par M. Leja ([8]).

---

[8] Dans la forme citée ci-dessus le théorème m'a été indiqué par M. Fekete qui l'a énoncé dans un Mémoire inédit sur l'approximation de fonctions analytiques à l'aide de polynomes, écrit en 1926 et professé dans son cours à l'Université hébraïque dès 1933. Le théorème découle d'un théorème antérieur du même auteur : *Ueber den absoluten Betrag von Polynomen, welche auf einer Punktmenge gleichmässig beschränkt sind*, théorème XIII (*Math. Zeitschr.*, t. 26, 1927. p. 324-344). Le théorème est obtenu aussi en combinant un théorème de M. Kalmàr [*Ueber Interpolation (Matematikai és Physikai Lapok*, 1926, p. 120-149)] et un théorème de M. Fekete [*Ueber Interpolation (Z. f. angew.*

LXVIII.   11

Soit C *un continu borné dont l'ensemble complémentaire est d'un seul tenant. Soit* $z_1^{(n)}, \ldots, z_n^{(n)}$ *l'ensemble des points sur* C, *donnant au produit* ($1'$) *la valeur maximale, les* a_ν *variant sur* C. *Alors on a uniformément dans tout domaine* $\overline{D}$ *borné et fermé qui ne contient pas de points de* C *l'identité*

$$(6) \qquad \lim_{n=\infty} \sqrt[n]{\overline{\prod_{\nu=1}^{n}(z - z_\nu^{(n)})}} = h(z).$$

$h(z)$ *est la fonction univalente et normée à l'infini qui repré- sente l'extérieur de* C *sur le domaine circulaire* $|\zeta| > d(C)$.

Grâce à l'uniformité de la convergence, on déduit de (6) la relation limite pour les dérivés logarithmiques

$$(7) \qquad \lim_{n=\infty} \frac{1}{n} \sum_{\nu=1}^{n} \frac{1}{z - z_\nu^{(n)}} = \frac{h'(z)}{h(z)},$$

valable uniformément dans $\overline{D}$. Pour $|z|$ suffisamment grand, $h(z)$ a le développement $h(z) = z + \alpha + \beta z^{-1} + \ldots$, ce qui entraîne

$$(7') \qquad \frac{h'(z)}{h(z)} = \frac{1}{z} - \frac{\alpha}{z^2} + \ldots.$$

Pour l'usage ultérieur tirons de (7) et de (7'), par comparaison des coefficients du développement de Laurent par rapport à z, l'identité spéciale

$$(8) \qquad \lim_{n=\infty} \frac{1}{n} \sum_{\nu=1}^{n} z_\nu^{(n)} = -\alpha.$$

On obtient des équations analogues par rapport à $C_\varepsilon^\star$, en consi-

---

Math., t. 6, 1926, p. 410-413)]. *Voir aussi :* WALSH, *Interpolation and Approxi- mation*, New York, 1935, théorème III, p. 159, et théorème VI, p. 170.

M. Leja a donné une version équivalente du théorème dans sa Note : *Construction de la fonction analytique effectuant la représentation conforme d'un domaine plan quelconque sur le cercle* (*Math. Ann.*, t. 111, 1935, p. 501-504). *Voir aussi ses Notes antérieures: Une méthode de construction de la fonction de Green appartenant à un domaine plan quelconque* (*C. R. Acad. Sc.*, t. 198, 1934, 231-234); *Sur les suites de polynomes, les ensembles fermés et la fonction de Green* (*Ann. Soc. Polon. math.*, t. 12, 1934, p. 57-71).

dérant les points extrémum $z_\nu^{*(n)}$, la fonction représentant $h_\varepsilon^*(z)$ et le premier coefficient $\alpha^*(\varepsilon)$ de cette dernière.

D'après (1) on a l'inégalité

$$(9) \qquad \log d_n(C_\varepsilon^*) \geqq \frac{1}{\binom{n}{2}} \sum_{\nu<\mu}^{1\ldots n} \log |u_\varepsilon(z_\nu^{(n)}) - u_\varepsilon(z_\mu^{(n)})|,$$

parce que les $u_\varepsilon(z_\nu^{(n)})$ se trouvent sur C_ε^* et leur produit de différences ne peut donc pas surpasser la valeur maximum $d_n(C_\varepsilon^*)^{\binom{n}{2}}$. On a donc

$$(10) \qquad \log d_n(C_\varepsilon^*) \geqq \frac{1}{\binom{n}{2}} \sum_{\nu<\mu}^{1\ldots n} \log |z_\nu^{(n)} - z_\mu^{(n)}|$$

$$+ \frac{1}{\binom{n}{2}} \sum_{\nu<\mu}^{1\ldots n} \log |1 + \varepsilon\, e^{i\varphi} K(z_\nu^{(n)}; z_\mu^{(n)})|,$$

avec

$$(11) \qquad K(z_\nu^{(n)}; z_\mu^{(n)}) = z_\nu^{(n)} + z_\mu^{(n)} - (1+\eta) + z_0 - \frac{z_0(z_0-1)(z_0-\eta)}{(z_\nu^{(n)}-z_0)(z_\mu^{(n)}-z_0)}.$$

Désignons par k le maximum de $|K(z; z')|$ pour tous les couples de points z, z' dans $\overline{D}'$. On a pour $\varepsilon < \frac{1}{2k}$

$$(12) \qquad \log d_n(C_\varepsilon^*) \geqq \log d_n(C) + R\left\{ e^{i\varphi}\, \varepsilon\, \frac{1}{\binom{n}{2}} \sum_{\nu<\mu}^{1\ldots n} K(z_\nu^{(n)}; z_\mu^{(n)}) \right\}$$

$$+ \varepsilon^2 H_n(z_\nu^{(n)}; \varepsilon),$$

avec H_n satisfaisant à l'inégalité

$$(13) \qquad |H_n(z_\nu^{(n)}; \varepsilon)| \leqq \frac{k^2}{2} + \frac{k^3}{3}\varepsilon + \frac{k^4}{4}\varepsilon^2 + \ldots$$

$$= \frac{1}{\varepsilon^2}\left[\log(1-k\varepsilon)^{-1} - k\varepsilon\right] < 4k^2\left[\log 2 - \frac{1}{2}\right].$$

En passant à la limite $n = \infty$, on tire de (11), à l'aide de (7) et (8),

$$(14) \qquad \lim_{n=\infty} \frac{1}{\binom{n}{2}} \sum_{\nu<\mu}^{1\ldots n} K(z_\nu^{(n)}; z_\mu^{(n)}) = -(1+\eta) + z_0 - 2\alpha$$

$$- z_0(z_0-1)(z_0-\eta) \frac{h'(z_0)^2}{h(z_0)^2},$$

et de (5), (12) et (14) suit

$(15)\quad \log d(C_\varepsilon^*) \geqq \log d(C)$

$$+ R\left\{ e^{i\varphi}\varepsilon \left[z_0 - (1 + \tau_i) - 2\alpha \right.\right.$$
$$\left.\left. - z_0(z_0 - 1)(z_0 - \eta)\frac{h'(z_0)^2}{h(z_0)^2} \right] \right\} + \varepsilon^2 H(\varepsilon),$$

$H(\varepsilon)$ satisfaisant également à l'inégalité (13).

Ayant choisi $\varepsilon < \varepsilon_1$, nous pouvons changer dans toutes les considérations précédentes les rôles de C et de C_ε^*. Au lieu de la fonction (2) paraîtra maintenant la fonction (4), différente de la première par le changement de $e^{i\varphi}$ en $-e^{i\varphi}$ et par l'addition du terme $\frac{\varepsilon^2}{2} r(z_1^*\varepsilon)$. $r(z_1^*\varepsilon)$ étant holomorphe dans $\overline{D}'$, on a dans ce domaine

$(16)\qquad\qquad \left| \dfrac{r(z,\,\varepsilon) - r(z',\,\varepsilon)}{z - z'} \right| \leqq b(D').$

On peut donc énoncer une inégalité analogue à (10), dans laquelle il faut remplacer $K(z_\gamma^{(n)};\, z_\mu^{(n)})$ par

$(17)\qquad K^*(z_\nu^{*(n)};\, z_\mu^{*(n)}) = K(z_\nu^{*(n)};\, z_\mu^{*(n)}) + a(z_\nu^{*(n)};\, z_\mu^{*(n)})\, b(D'),$

avec $|a(z_\nu^{*(n)};\, z_\mu^{*(n)})| < \varepsilon$. Maintenant les conclusions précédentes amènent l'inégalité suivante, analogue à (15) :

$(18)\quad \log d(C) \geqq \log d(C_\varepsilon^*)$

$$- R\left\{ e^{i\varphi}\varepsilon \left[z_0 - (1 + \eta) - 2\alpha^*(\varepsilon) \right.\right.$$
$$\left.\left. - z_0(z_0 - 1)(z_0 - \eta)\frac{h_\varepsilon^{*\prime}(z_0)^2}{h_\varepsilon^*(z_0)^2} \right] \right\} + \varepsilon^2 G(\varepsilon).$$

Ici $G(\varepsilon)$ est borné pour ε suffisamment petit. Quand ε converge vers zéro, le continu C_ε^* converge vers C au sens de Fréchet; donc, d'après un théorème classique de M. Carathéodory, on a

$$h_\varepsilon^*(z_0) \to h(z_0), \qquad \alpha^*(\varepsilon) \to \alpha.$$

On déduit de là

$(19)\quad \log d(C) \geqq \log d(C_\varepsilon^*)$

$$- R\left\{ e^{i\varphi}\,\varepsilon \left[z_0 - (1 + \eta) - 2\alpha \right.\right.$$
$$\left.\left. - z_0(z_0 - 1)(z_0 - \tau_i)\frac{h'(z_0)^2}{h(z_0)^2} \right] \right\} + \varepsilon\, o(\varepsilon).$$

$o(\varepsilon)$ désignant un terme satisfaisant à $\lim\limits_{\varepsilon=0} o(\varepsilon) = 0$. De (15) et (19) découle enfin

$$(20) \quad \log d(C_\varepsilon^*) = \log d(C)$$
$$+ R\left\{ e^{i\varphi}\,\varepsilon \left[z_0 - (1+\eta) - 2\alpha \right.\right.$$
$$\left.\left. - z_0(z_0 - 1)(z_0 - \eta)\frac{h'(z_0)^2}{h(z_0)^2} \right]\right\} + \varepsilon\,o(\varepsilon).$$

Le deuxième terme à droite représente la variation de $\log d(C)$ *pour la déformation infinitésimale* (2). Il est évident qu'on peut obtenir de la même manière des formules analogues pour toute autre déformation infinitésimale. En effet, nous nous servirons dans ce mémoire, pour d'autres recherches, des déformations d'un autre type. Mais la transformation spéciale (2) suffit pour résoudre le présent problème.

3. Solution du problème de Pólya-Grötzsch. —

Démontrons d'abord l'existence d'un continu C résolvant le problème de Pólya-Grötzsch. Encore une fois nous nous bornerons au cas 0, 1, η, d'où l'on tire aisément le procédé général. Considérons donc tous les continus C contenant trois points 0, 1, η et dont l'ensemble complémentaire est d'un seul tenant. Parmi eux se trouve le cercle autour de l'origine avec le rayon $\max(1, |\eta|)$. Puisque le diamètre transfini d'un cercle est égal à son rayon, nous en tirons que la limite inférieure δ des $d(C)$ ne surpasse pas la valeur $\max(1, |\eta|)$. D'autre part la solution du problème pour $n = 2$ montre que le diamètre transfini d'un continu contenant deux points n'est jamais inférieur à celui du segment rectiligne qui les joint. Donc $\delta \geq \frac{1}{4}$, car $\frac{1}{4}$ est le diamètre transfini du segment rectiligne $<0, 1>$. La limite inférieure δ de tous les $d(C)$ est donc un nombre positif.

Parmi tous les continus C choisissons une suite partielle C_n telle qu'on ait $\lim\limits_{n=\infty} d(C_n) = \delta$. Considérons la suite correspondante des fonctions univalentes

$$(21) \qquad z = f_n(\xi) = d(C_n)\xi + \alpha_n + \beta_n \xi^{-1} + \dots,$$

qui représentent le domaine $|\xi| > 1$ sur l'extérieur de C_n. Ces fonctions forment une famille normale, comme on voit aisément en passant à la suite des fonctions

$$(21') \qquad f_n\left(\frac{1}{\eta}\right)^{-1} = d(C_n)^{-1}\eta + a_n\eta^2 + \ldots,$$

univalentes et holomorphes dans le cercle $|\eta| < 1$, et en considérant le fait que $\lim_{n=\infty} d(C_n) = \delta > 0$. On peut donc choisir une suite partielle $f_{n'}(\xi)$ convergente uniformément dans tout domaine $|\xi| > r > 1$ vers une fonction univalente et holomorphe

$$(21'') \qquad f(\xi) = \delta\xi + \alpha + \beta\xi^{-1} + \ldots.$$

Dans ce domaine $f(\xi)$ n'atteint pas les valeurs $0, 1, \eta$, parce que les fonctions $f_{n'}(\xi)$ ne le font pas. Donc $f(\xi)$ représente le domaine $|\xi| > 1$ d'une manière biunivoque et conforme sur l'extérieur d'un continu C, qui contient les points $0, 1, \eta$. En outre, on voit que le rayon de représentation de C est δ. On a donc $d(C) = \delta$ et C est un continu résolvant le problème en question.

Soumettons maintenant un continu extrémal C (dont nous venons de démontrer l'existence) aux transformations considérées dans le paragraphe 2. Tous les ensembles C_ε^* obtenus à leur aide sont des ensembles de concurrence à C, parce que tous contiennent aussi les points $0, 1, \eta$. On a donc $\log d(C_\varepsilon^*) \geq \log d(C)$ et à l'aide de (20) on trouve pour $\varepsilon = 0$

$$(22) \qquad \mathrm{R}\left\{e^{i\varphi}\left[z_0 - (1+\eta) - 2\alpha - z_0(z_0-1)(z_0-\eta)\frac{h'(z_0)^2}{h(z_0)^2}\right]\right\} \geq 0.$$

φ étant arbitraire, l'inégalité (22) est possible seulement si l'on a

$$(22') \qquad z_0 - (1+\eta) - 2\alpha = z_0(z_0-1)(z_0-\eta)\frac{h'(z_0)^2}{h(z_0)^2}.$$

Cette égalité doit être valable pour tout z_0 à l'extérieur de C et fournit donc une équation différentielle caractérisant $h(z)$.

Pour trouver C, il vaut mieux considérer la fonction $z = f(\xi)$ inverse de $\xi = h(z)$. $f(\xi) = \xi - \alpha + b\xi^{-1} + \ldots$ représente le domaine $|\xi| > d(C)$ conformément sur l'extérieur de C; cette fonction satisfait à l'équation différentielle

$$(23) \qquad \xi^2 f'(\xi)^2[f(\xi) - (1+\eta+2\alpha)] = f(\xi)[f(\xi)-1][f(\xi)-\eta],$$

ou en écrivant d'une manière un peu plus générale a_1, a_2, a_3 au lieu de 0, 1, η

$(23')$
$$\xi^2 f'(\xi)^2 [f(\xi) - (a_1 + a_2 + a_3 + 2\alpha)]$$
$$= [f(\xi) - a_1][f(\xi) - a_2][f(\xi) - a_3].$$

Dans les équations $(22')$, (23) et $(23')$ paraît toujours le paramètre indéterminé α; celui-ci est fixé par la condition que le continu C contienne les trois points 0, 1, η. Pendant que l'existence d'un tel α résulte de l'existence d'une solution du problème d'extrémum, il n'est nullement trivial qu'il existe un seul α, donc un seul continu résolvant le problème. Ce fait a été démontré, cependant, par M. Grötzsch à l'aide de la méthode de continuité. L'équation $(22')$ se trouve aussi chez M. Grötzsch sous une forme un peu différente [9].

Des propriétés élémentaires du diamètre transfini découle que le continu extrémal C ne possède pas de points intérieurs [10] et que C est donc une coupure du plan z. D'après des théorèmes bien connus de la théorie des équations différentielles, la fonction $f(\xi)$, solution de l'équation différentielle $(23')$, est holomorphe en tout point du cercle $|\xi| = d(C)$, pour lequel $f(\xi)$ est différente de a_1, a_2, a_3 et $a_1 + a_2 + a_3 + 2\alpha$, sauf en un nombre fini de points. Donc C est composé d'arcs analytiques

$$z = z(\varphi) = f\{ e^{i\varphi} d(C) \}$$

satisfaisant tous à l'équation différentielle

(24)
$$\left(\frac{dz}{d\varphi} \right)^2 \frac{a_1 + a_2 + a_3 + 2\alpha - z}{(z - a_1)(z - a_2)(z - a_3)} = 1.$$

De (24) on tire enfin que C est composé de trois arcs analytiques, sortant de a_1, a_2, a_3 respectivement et se rencontrant en

$$a_1 + a_2 + a_3 + 2\alpha$$

sous des angles de $120°$.

Pour le continu $C_3 : z = z(s)$ résolvant le problème d'extrémum

---

[9] *Loc. cit.* [4], p. 262.
[10] Je dois cette remarque à M. Fekete.

traité au paragraphe 1 par rapport à d_3 (C), j'ai trouvé l'équation différentielle ([11])

$$(24') \qquad \left(\frac{dz}{ds}\right)^2 \frac{\frac{1}{3}(a_1 + a_2 + a_3) - z}{(z - a_1)(z - a_2)(z - a_3)} \geqq 0,$$

s désignant un paramètre réel quelconque et a_1, a_2, $a_3 \subset C$ satisfaisant à

$$d_3(C_3) = |a_1 - a_2| \, |a_1 - a_3| \, |a_2 - a_3|.$$

L'inégalité différentielle $(24')$ équivaut à une équation différentielle du type (24); on obtient la dernière en choisissant dans $(24')$ au lieu de s le paramètre $t = \int_0^s r(s)\, ds$, $r^2(s)$ désignant le premier membre de $(24')$. On voit donc que dans le cas spécial du problème de Pólya-Grötzsch, où les points a_1, a_2, a_3, fixant le continu C, sont les points extrémaux du problème que nous venons de mentionner, on sait de plus que le paramètre α est égal à $-\frac{1}{3}(a_1 + a_2 + a_3)$.

4. Problème d'extrémum par rapport à des ensembles composés de deux continus séparés. — J'ai démontré récemment l'inégalité ([12])

$$(25) \qquad d(A + B) \leqq d(A) + d(B),$$

pour tout continu A + B, *composé de deux continus* A *et* B *avec des points communs et situés de manière qu'il existe une courbe de Jordan, séparant les points de* A *et* B *qui ne sont pas communs aux deux ensembles.* L'égalité ne vaut que dans le cas où A + B est un segment rectiligne, composé de deux segments A et B. Ce fait combiné avec l'inégalité (25) donne la solution du problème d'extrémum suivant : On cherche le continu A + B avec $d(A + B)$ donné, qu'on peut décomposer de la façon mentionnée en deux continus A et B de manière

---

([11]) *Loc. cit.* ([3]), p. 53.

([12]) SCHIFFER, *On the subadditivity of the transfinite diameter (Proc. London Math. Soc.,* II. s., sous presse).

que $d(A) + d(B)$ soit le plus petit possible. D'après le précédent on sait que la solution est fournie par un segment rectiligne.

Nous allons proposer maintenant le même problème d'extrémum pour le cas d'un ensemble $A + B$ composé de deux continus A et B, l'ensemble complémentaire de chacun d'eux étant d'un seul tenant. Démontrons d'abord que toute fonction normée à l'infini, univalente et holomorphe dans l'extérieur $\widetilde{A + B}$ de $A + B$ (excepté le pôle à l'infini) représente ce domaine sur l'extérieur $\widetilde{A' + B'}$ d'un ensemble $A' + B'$ (composé évidemment aussi de deux continus séparés A' et B') de diamètre transfini

$$d(A' + B') = d(A + B).$$

En effet, considérons la fonction de Green $g(z; \infty)$ de $A + B$ avec le pôle logarithmique à l'infini et dont l'existence suit de théorèmes classiques. Autour de l'infini elle possède donc le développement

$$g(z; \infty) = \log |z| + \gamma + o\left(\frac{1}{|z|}\right);$$

on sait [13] que γ est égal à $-\log d(A + B)$. Si la fonction

$$z(\xi) = \xi + \alpha + \beta\xi^{-1} + \ldots$$

effectue la représentation de $\widetilde{A' + B'}$ sur $\widetilde{A + B}$, la fonction de Green pour $\widetilde{A' + B'}$ sera $g[z(\xi); \infty]$ et autour de l'infini elle aura le développement

$$g[z(\xi); \infty] = \log |\xi| + \gamma + o\left(\frac{1}{|z|}\right).$$

Donc $A' + B'$ a même deuxième terme γ et par là même diamètre transfini que $A + B$. Nous voyons donc que le diamètre transfini de $A + B$ est invariant par rapport à des représentations conformes biunivoques et normées à l'infini, de son extérieur. Il est évident que le même résultat vaut pour tout ensemble

---

[13] G. Szegö, *Bemerkung zu einer Arbeit von Herrn Fekete : Ueber die Verteilung der Wurzeln bei gewissen algebraischen Gleichungen mit ganzzahligen Kœffizienten* (*Math. Zeitschr*, t. 21, 1924, p. 203-208, voir p. 204).

fermé et borné, pour l'extérieur duquel la fonction de Green existe (au sens classique). Pour le cas que cet extérieur est d'un seul tenant le résultant découle du théorème de M. Fekete [*loc. cit.* (²)].

Après ces préparatifs, cherchons parmi tous les ensembles $A' + B'$ qu'on peut obtenir de $A + B$ à l'aide de telles représentations et pour lesquels on a donc $d(A' + B') = d(A + B)$ celui pour lequel $d(A') + d(B')$ a la valeur la plus petite possible. Si l'on connaît les continus extrémaux A' et B', on peut évaluer la borne inférieure de $d(A) + d(B) — d(A + B)$ pour toute classe d'ensembles, les extérieurs desquels on peut représenter l'un sur l'autre à l'aide d'une fonction univalente et normée à l'infini.

Pour résoudre ce problème nous nous servirons du lemme suivant (¹⁴) :

LEMME. — *a*). *Soit* C *un continu borné dans le plan z contenant plus d'un point et dont l'ensemble complémentaire est d'un seul tenant. Alors il y a pour tout $z_0 \subset C$ et pour toute valeur du paramètre $\rho > 0$ des fonctions $F_\rho(z; z_0, C)$ univalentes et holomorphes dans l'extériëur $\widetilde{C}$ de C (un pôle à l'infini excepté) et possédant autour de l'infini le développement*

(26) $$F_\rho(z; z_0, C) = z — z_0 + \frac{a\rho^2}{z — z_0} + \frac{b\rho^3}{(z — z_0)^2} + \cdots,$$

avec $|a| \leqq 4^2, |b| \leqq 4^3, \ldots$. Ce développement est donc valable dans tout le domaine $|z — z_0| > 4\rho$.

b.) Soit $s(z)$ une fonction analytique sur C non identiquement nulle. Supposons que, pour tout point $z_0 \subset C$ et pour toute suite de fonctions $F_{\rho_\nu}(z; z_0, C)$ pour laquelle on a (¹⁵)

$$\lim_{\rho_\nu \searrow 0} a = \alpha \neq 0,$$

on ait $R\{\alpha s(z_0)\} \geqq 0$. Alors C est une courbe analytique $z(t)$,

---

(¹⁴) Voir la Note *loc. cit.* (⁴), p. 433.

(¹⁵) Il y a telles suites de fonctions, par exemple celle définie par la formule (2″) de la Note citée (⁴), p. 435.

satisfaisant à l'équation différentielle

$$(27) \qquad \left(\frac{dz}{dt}\right)^2 s(z) + 1 = 0.$$

Retournons à notre problème d'extrémum. L'existence d'un ensemble extrémum $A' + B'$ peut être démontrée à l'aide de la conception des familles normales et en se servant du théorème de convergence de Carathéodory-Bieberbach pour la représentation conforme de domaines variables. Sur la partie A' de cet ensemble choisissons un point z_0 et formons la famille $F_\rho(z; z_0, A')$ du lemme. Si ρ est assez petit, on aura dans tout domaine fermé $\overline{D}$ contenant B' mais ne contenant pas z_0

$$(26') \qquad F_\rho(z; z_0, A') = z - z_0 + \frac{a\rho^2}{z - z_0} + \rho^3 h(z; \rho),$$

avec $h(z; \rho)$ holomorphe dans $\overline{D}$. Cette représentation transforme l'extérieur $\widetilde{A}'$ de A' et l'extérieur $\widetilde{A' + B'}$ de $A' + B'$ en ceux de A_ρ^* et de $A_\rho^* + B_\rho^*$. Grâce à l'invariance du diamètre transfini des ensembles par rapport à des représentations conformes de leur extérieur, normées à l'infini, on a

$$d(A_\rho^*) = d(A') \qquad \text{et} \qquad d(A_\rho^* + B_\rho^*) = d(A' + B').$$

Il ne reste donc qu'à étudier combien $d(B')$ est changé par la déformation $(26')$, étant sûr que ce nombre ne peut que croître, vu la propriété minimum de $A' + B'$.

Pour cela nous avons à faire un calcul absolument analogue à celui du paragraphe 2, qui nous fournit immédiatement [16]

$$(28) \qquad \log d(B_\rho^*) = \log d(B') - R\left\{ a\rho^2 \frac{\varphi'(z_0)^2}{\varphi(z_0)^2} \right\} + \rho^2 o(\rho).$$

Ici $\xi = \varphi(z) = z + c_0 + c_1 z^{-1} + \ldots$ représente biunivoquement l'extérieur $\widetilde{B}'$ de B' sur le domaine $|\xi| > d(B')$. De $d(B_\rho^*) \geqq d(B')$

---

[16] Ce résultat découle d'un théorème général permettant d'exprimer la variation de la fonction de Green d'un domaine $\widetilde{B}$ soumis à une déformation infinitésimale du type $(26')$ avec $z_0 \subset \widetilde{B}$. Voir M. SCHIFFER. *Sur la variation de la fonction de Green* (*C. R. Acad. Sc.*, t. 209, 1939, p. 980-982).

découle d'après (28)

$$(29) \qquad R\left\{ a\,\frac{\varphi'(z_0)^2}{\varphi(z_0)^2} \right\} + o(\rho) \lesseqgtr 0.$$

Choisissons maintenant pour z_0 une suite $F_{\rho_\nu}(z\,;\,z_0,\,A')$ pour laquelle $\lim\limits_{\rho_\nu \to 0} a = \alpha \neq 0$ existe. On obtient de (29)

$$(29') \qquad R\left\{ \alpha\left(-\,\frac{\varphi'(z_0)^2}{\varphi(z_0)^2} \right) \right\} \geqq 0.$$

— $\varphi'(z)^2 \varphi(z)^{-2}$ est une fonction analytique sur A' et pas identiquement nulle. Nous pouvons donc appliquer la partie b.) de notre lemme en prenant — $\varphi'(z)^2 \varphi(z)^{-2}$ au lieu de $s(z)$. Nous voyons donc que A' est une courbe analytique $z(t)$, satisfaisant à l'équation différentielle

$$(30) \qquad z'(t)^2\,\frac{\varphi'[z(t)]^2}{\varphi[z(t)]^2} = 1,$$

d'où

$$(30') \qquad \frac{d}{dt}\log\varphi[z(t)] = r = \pm 1,$$

$$(30'') \qquad \varphi[z(t)] = C\,e^{rt}.$$

C désignant une constante. Le résultat obtenu peut être énoncé aussi sous la forme suivante : Si l'on représente $\widehat{B'}$ sur le domaine circulaire $|\xi| > d(B')$, alors A' devient un segment rectiligne T_A dont le prolongement passe par l'origine.

Si T_A a la direction $e^{i\alpha}$, la transformation additionnelle

$$(31) \qquad \eta(\xi) = \xi + \frac{d(B')^2\,e^{2i\alpha}}{\xi}$$

changera le cercle $|\xi| = d(B')$ en un segment rectiligne S_B avec la direction $e^{i\alpha}$, pendant que T_A sera représenté sur un segment rectiligne S_A dans le prolongement de S_B. $\eta[\varphi(z)]$ étant univalente, normée et holomorphe dans $\widetilde{B'}$, on a

$$(32) \qquad d(S_B) = d(B')\,; \qquad d(S_A + S_B) = d(A' + B').$$

Dans toutes nos considérations on peut changer les rôles de A et B'; on obtient donc une fonction $\eta^+[\varphi^+(z)]$ qui représente l'ensemble $A' + B'$ sur deux segments rectilignes S_A^+, S_B^+ situés sur un rayon passant par l'origine avec la direction $e^{i\beta}$. En outre,

$\eta^+[\varphi^+(z)]$ est normée, holomorphe et univalente dans l'extérieur $\widetilde{A'}$ de A'. Donc on a

$$(32') \qquad d(S_A^+) = d(A'); \qquad d(S_A^+ + S_B^+) = d(A' + B').$$

On peut représenter l'extérieur $\widetilde{S_A + S_B}$ de $S_A + S_B$ sur l'extérieur $\widetilde{S_A^+ + S_B^+}$ de $S_A^+ + S_B^+$ d'une manière conforme, univalente et conservant le point à l'infini. On trouve cette représentation en transformant le domaine $\widetilde{S_A + S_B}$ dans le plan η d'abord sur le domaine $\widetilde{A' + B'}$ dans le plan z; après cela on représente ce domaine sur le domaine $\widetilde{S_A^+ + S_B^+}$ dans le plan η^+. La tranformation finale $\eta^+ = \Phi(\eta)$ n'est qu'une homothétie, grâce au théorème suivant :

Soit M + N *un ensemble dans le plan* η, *composé de deux segments rectilignes* M *et* N, *situés sur une même droite; soit* M⁺ + N⁺ *un ensemble du même type dans le plan* η^+. *S'il existe une fonction holomorphe* $\eta^+ = \psi(\eta)$ *représentant biunivoquement l'extérieur* $\widetilde{M + N}$ *de* M + N *sur l'extérieur* $\widetilde{M^+ + N^+}$ *de* M⁺ + N⁺ *en conservant le point à l'infini, alors* $\psi(\eta)$ *n'est qu'une homothétie.*

En démontrant ce théorème, on peut supposer que les deux droites mentionnées coïncident avec les axes réels des plans respectifs. $\psi(\eta)$ est holomorphe dans $\widetilde{M + N}$ (sauf le pôle à l'infini) et peut être prolongé au-dessus de tout point intérieur d'un segment M ou N à l'aide du principe de réflexion. Celle-ci change $\psi(\eta)$ en la fonction $\overline{\psi}(\eta)$, définie à l'aide de l'équation $\overline{\psi}(\eta) = \overline{\psi(\overline{\eta})}$. L'itération d'une réflexion amène toujours à la fonction initiale. On voit donc qu'on obtient toujours l'une des valeurs $\psi(\eta)$ ou $\overline{\psi}(\eta)$, si l'on prolonge la fonction $\psi(\eta)$ le long d'un chemin fermé quelconque en évitant les extrémités de M et N. Un tel prolongement interchange $\psi(\eta)$ et $\overline{\psi}(\eta)$, ou bien ne les change point; les fonctions $\psi(\eta) + \overline{\psi}(\eta)$ et $\psi(\eta) \cdot \overline{\psi}(\eta)$ sont donc uniformes dans tout le plan. D'après un théorème classique de Riemann, ces deux fonctions sont aussi analytiques dans les extrémités de M et de N, et n'ont donc aucune autre singularité que les pôles à l'infini. On a donc

(33) $$\psi(\eta) + \overline{\psi}(\eta) = a_1 \eta + a_2,$$

(34) $$\psi(\eta) \cdot \overline{\psi}(\eta) = a_3 \eta^2 + a_4 \eta + a_5.$$

La fonction $\psi(\eta)$ est une racine de l'équation algébrique

(35) $$\psi(\eta)^2 - (a_1 \eta + a_2)\psi(\eta) + (a_3 \eta^2 + a_4 \eta + a_5) = 0,$$

c'est-à-dire

(35') $$\psi(\eta) = \frac{1}{2}\left\{ a_1 \eta + a_2 + \sqrt{(a_1^2 - 4a_3)\eta^2 + (2a_1 a_2 - 4a_4)\eta + (a_2^2 - 4a_5)} \right\}$$
$$= \frac{1}{2}\left\{ a_1 \eta + a_2 + \sqrt{q(\eta)} \right\}.$$

Les fonctions (33) et (34) étant réelles sur M et N, les coefficients a_i sont réels. Cherchons maintenant les zéros de $q(\eta)$. S'ils sont différents, ils se trouvent nécessairement sur le même segment M ou N, vu l'uniformité de $\psi(\eta)$ dans $\overparen{M + N}$. Supposons qu'ils se trouvent sur M. Alors $q(\eta)$ sera de signe différent sur N et sur le segment de M entre les deux racines de $q(\eta)$. Mais ceci est impossible, $\psi(\eta)$ étant toujours réel sur $M + N$. Donc $q(\eta)$ a une racine double et $\psi(\eta)$ a la forme

(36) $$\psi(\eta) = a\eta + b,$$

comme nous l'avons affirmé.

En particulier, nous avons démontré que $\eta^+ = \Phi(\eta)$ est une homothétie; celle-ci ne change pas le diamètre transfini $d(S_A + S_B)$; on voit donc aisément qu'elle ne change pas $d(S_A)$ et $d(S_B)$. De (32) et de (32') suit grâce à cette invariance

(37) $$d(A') = d(S_A); \qquad d(B') = d(S_B).$$

La valeur minimun pour $d(A) + d(B)$ dans toute classe de pairs de continus séparés A, B, représentables l'un sur l'autre à l'aide d'une fonction univalente et normée à l'infini, est donc atteinte dans le cas où A et B sont deux segments rectilignes sur la même droite. Ainsi le problème d'extrémum est résolu; pour l'évaluation quantitative de $d(A) + d(B) - d(A + B)$, il ne faut connaître que ces nombres pour des segments rectilignes sur une droite.

Nos considérations restent encore valables dans le cas limite que les continus A et B sont rapprochés jusqu'à ce qu'ils aient des points frontières communs. Ce cas contient évidemment le cas particulier décrit au commencement de ce paragraphe où il existe une courbe de Jordan séparant les points non communs de A et B. Soit $A' + B'$ un ensemble du type limite général avec $d(A' + B')$ donné et pour lequel $d(A') + d(B')$ est minimum. L'existence d'un tel ensemble peut être montrée encore une fois par le choix d'une suite d'ensembles $A_n + B_n$ pour laquelle $d(A_n) + d(B_n)$ converge vers la limite inférieure et en se servant des théorèmes de convergence pour les fonctions attachées aux extérieurs de A_n et B_n. D'après les mêmes calculs que dans le cas précédent, on démontre alors : On peut représenter l'extérieur $\overparen{A' + B'}$ de $A' + B'$ à l'aide d'une fonction univalente normée à l'infini sur l'extérieur d'un segment de droite $S_A + S_B$ composé de deux segments S_A et S_B. A tout point frontière de $\overparen{A' + B'}$ appartenant à A' correspond par cette représentation un point sur S_A, et à tout point frontière sur B' un point de S_B. De plus, on montre

$$d(S_A) = d(A') \quad \text{et} \quad d(S_B) = d(B')$$

à l'aide de considérations analogues aux précédentes. Mais nous avons, d'après des théorèmes élémentaires,

$$d(S_A) + d(S_B) = d(S_A + S_B) = d(A' + B'),$$

et par là, grâce à la propriété minimum de l'ensemble $A' + B'$, l'inégalité générale (25). En particulier, la sous-additivité du diamètre transfini est démontrée de nouveau dans le cas mentionné au commencement.

5. Autres applications de la méthode de variation. — Prenons un ensemble $A + B$ composé de deux continus séparés A et B; *il faut représenter l'extérieur $\overparen{A + B}$ de $A + B$ sur l'extérieur $\overparen{A' + B'}$ d'un ensemble $A' + B'$ (du même diamètre transfini) à l'aide d'une fonction univalente, normée à l'infini, de manière que le diamètre transfini de l'un deux* [par exemple $d(B')$] *soit maximun (ou minimun).*

Il est évident que ce problème permet un nombre infini de solutions. Car $A' + B'$ étant déjà un ensemble extrémal (l'existence duquel est évident), on peut faire encore des représentations conformes, holomorphes à l'extérieur de B' et normées à l'infini, sans changer $d(B')$ et $d(A' + B')$. Pour étudier tous les ensembles extrémum, il suffit donc de supposer que B' est un cercle et de chercher la forme de A'. En connaissant cette dernière, dans ce cas spécial, nous la connaîtrons aussi pour tout autre choix de B' en représentant l'extérieur de ce B' sur l'extérieur d'un cercle.

D'abord résolvons le problème de maximum. Nous procédons comme au paragraphe 4. Nous faisons une représentation (26') et continuons le calcul jusqu'à la formule (28) qui nous indique le changement de $\log d(B')$. Puisque $\log d(B')$ est maximal, il faut que

$$(38) \qquad R\left\{ a\, \frac{\varphi'(z_0)^2}{\varphi(z_0)^2} \right\} + o(\rho) \gtrless 0$$

soit satisfait. Si B' est un cercle, nous avons $\varphi(z) \equiv z$. Il suit donc, du lemme du paragraphe 4 que A' est une courbe analytique $z = z(t)$, $z(t)$ satisfaisant à l'équation différentielle

$$(39) \qquad \frac{z'(t)^2}{z(t)^2} = -1,$$

d'où il suit que $z(t) = C e^{it}$ et que A' *est un arc de cercle concentrique à* B'.

Dans le cas du problème minimum, on a le signe $\leqq$ au lieu du $\geqq$ dans (38) et $+1$ au lieu de -1 dans (39). Donc

$$(40) \qquad z(t) = C e^{t},$$

t désignant un paramètre réel. Dans ce cas A' *est un segment rectiligne dirigé vers le centre de* B'.

Il est évident que le même problème se pose pour un ensemble $A + B + \ldots + N$ composé de continus séparés $A, B, \ldots, N$ et qu'il permet une solution absolument analogue.

[13] Variation of the Green function and theory of the p-valued functions

[13] Schiffer, Menaham. "Variation of the Green function and theory of the p-valued functions." *American Journal of Mathematics* **65:2** (1943), 341–360.

VARIATION OF THE GREEN FUNCTION AND THEORY OF THE *p*-VALUED FUNCTIONS.*

By Menahem Schiffer.

1. Introduction. Let Δ be a domain in the ζ-plane, which contains the point $\zeta = 0$ but not $\zeta = \infty$. We consider the family of all functions

$$(1) \qquad f(\zeta) = \zeta + a_2\zeta^2 + a_3\zeta^3 + \cdots + a_n\zeta^n + \cdots,$$

regular and univalent in Δ; this family is known to be compact and therefore we may ask for the extremal functions $f_n(\zeta)$ in the family possessing the largest value of $|a_n|$.

Each extremal function $z = f_n(\zeta)$ maps Δ on a domain D_n in the z-plane, having the same type as Δ and possessing as boundary continua (if there are such at all) analytic curves. By means of the identity

$$(2) \qquad f(\zeta)[1 - xf(\zeta)]^{-1} = \sum_{\nu=1}^{\infty} \{a_\nu + P_\nu(x)\}\zeta^\nu$$

we may attach to each function $f(\zeta)$ a set of polynomials $P_\nu(x)$; then the differential equations, satisfied by the boundary curves of D_n, can be written in the following form. Let $P_n(x)$ be the n-th polynomial, belonging to $f_n(\zeta)$, and let $z(t)$ be a parametric representation of a fixed boundary curve, then we get by appropriate choice of the parameter t [1]

$$(3) \qquad \frac{1}{a_n}\left(\frac{z'(t)}{z(t)}\right)^2 P_n\left(\frac{1}{z(t)}\right) + 1 = 0.$$

In the special case where Δ is simply connected and, therefore, can be supposed to be the unit circle, it is conjectured that $|a_n| \leq n$ holds and that

$$(4) \qquad b_n(\zeta) = \zeta(1 - \zeta)^{-2} = \sum_{\nu=1}^{\infty} \nu\zeta^\nu$$

is, essentially, the only extremal function. Indeed, $b_n(\zeta)$ satisfies the conditions above and there arises the question: Do these determine the extremal function completely. In the equation (3) there appear the coefficients of the

---

* Received June 20, 1941; Revised August 21, 1942.

[1] M. Schiffer, "A method of variation within the family of simple functions," *Proceedings of the London Mathematical Society*, Ser. 2, Vol. 44 (1938), pp. 432-449. M. Schiffer, "On the coefficients of simple functions," *Proceedings of the London Mathematical Society*, Ser. 2, Vol. 44 (1938), pp. 450-452.

341

10

required function; we have, therefore, a difficult functional equation, the solution of which would be decisive for the coefficient problem in the theory of univalent functions.

The above-mentioned results were obtained by means of a method of variations which is based on theorems concerning point sets and conformal representation. In this paper we shall develop an alternative method, using only the most elementary theorems of potential theory, and obtain the same results as those previously obtained in the case of a simply connected domain Δ. In particular, we get in this way a new proof, based on potential theory, of the "$\frac{1}{4}$-theorem" of Koebe-Pick. In the case of multiple connectivity of Δ, the new method differs essentially from the former and is applicable to problems where this failed, but is unapplicable to questions where the old method was useful. The reason for this difference is the following: the old method of variations transforms D_n into a domain of comparison D^*_n of the same conformal type (i. e. D^*_n can be mapped conformally on D_n), while this does not hold for the new method. The latter is, therefore, preferable in problems of pure potential theory, while the former remains valuable in questions of conformal representation of multiply connected domains.

Finally, it will be shown that the new method of variations is also applicable to Riemann surfaces (R. S.) and can be used, therefore, in the theory of p-valued functions.

2. Definition and properties of the variations considered. Let us suppose, for the sake of generality, that we are dealing with a closed R. S. $\Re$, p-sheeted and of genus g. Let $q(z)$ be a uniform function on $\Re$, regular everywhere with the exception of the points $z_0, z_1, \cdots z_m$, which are supposed to be finite and not branch points. At each z_i let $q(z)$ possess a simple pole and the development

$$(5) \qquad q(z) = a_i(z - z_i)^{-1} + b_i + c_i(z - z_i) + \cdots.$$

Then, the function

$$(6) \qquad z^* = z + \rho q(z)$$

will be regular on $\Re$ with the exception of $z = \infty$ and $z = z_i$. Around each point z_i we describe a circle k_i with radius r, and r is chosen so small that no branch point and no point z_k ($\neq z_i$) lies in the interior $\bar{k}_i$ of k_i. Set

$$(7) \qquad m = \text{Max} \, | \, q(z) \, |, \qquad | \, z - z_i \, | = \tfrac{1}{2}r, \qquad i = 0, 1, \cdots m.$$

If ρ satisfies the inequality

$$(8) \qquad | \, \rho \, | < r/2m$$

then $z^*(z)$ is exactly p-valued over the domain $\Re - \sum_{i=0}^{m} \bar{k}_i$. Indeed, let a be a point of the domain; then, according to (7) and (8), we have, for $|z - z_i| = r/2$, $|z - a| > |\rho q(z)|$; it follows, therefore, from the theorem of Rouché that $z^*(z)$ takes on the value a in $\Re - \sum_{i=0}^{m} k_i$ at most p times. On the other hand, each value in the neighborhood of ∞ is taken on exactly p times. Therefore $z^*(z)$ is p-valued in $\Re - \sum_{i=0}^{m} \bar{k}_i$. The representation $z^*(z)$ transforms the circles k_i into analytic curves k^*_i without self-intersections, if only we have

$$(9) \qquad |\rho| < (\operatorname{Max} | \frac{q(x) - q(y)}{x - y} |)^{-1}, \qquad x, y \subset k_i, \qquad i = 0, \cdots m,$$

a condition compatible with (8).

Let us summarize: The representation $z^*(z)$ transforms the domain $\Re - \sum_{i=0}^{m} \bar{k}_i$, contained in the R. S. $\Re$, into a domain bounded by $(m + 1)$ simple curves k^*_i and covering the z^*-plane p times at most. If we add to this domain the interiors $\bar{k}^*_i$ of the curves k^*_i, we get a closed R. S. $\Re^*_{\rho q}$ with p sheets. $\Re^*_{\rho q}$ will be called the R. S. obtained from $\Re$ by means of the variation $V_{\rho q}$.

This variation can be obtained by continuous deformation of $\Re$; it does not alter, therefore, the genus of the surface, since this is a topological invariant. Thus, the variations $V_{\rho q}$ preserve the number of sheets and the genus of the R. S.

Let D be a domain on $\Re$. If no pole z_i of $q(z)$ is situated on the boundary of D, and if r is so small that no point of this boundary is situated in any $\bar{k}_i$, then $V_{\rho q}$ determines in an unambiguous way a variation $D \to D^*_{\rho q}$. Therefore, a domain D on a closed p-sheeted R. S. $\Re$ with genus g is transformed by means of a variation $V_{\rho q}$ into a domain D^* on a R. S. $\Re^*$ of the same type.

It is evident that analogous variations can be defined on non-closed R. S.; for our purposes, however, it will not be necessary to do this.

3. The variation of the Green function. Let us consider a domain D, on a R. S. $\Re$ of the above type, which is bounded by analytic curves. Then the Green functions $g(z; x)$ are known to exist on D, behaving near $z = x_j$ as $-\log|z - x_j|$, harmonic elsewhere in D, and converging to zero when the argument approaches a point of the boundary of D. Let us submit D to a variation $V_{\rho q}$ as defined in 2. For sufficiently small ρ we get a domain D^*

on a R. S. $\Re^*$ with the same properties; we also have new Green functions $g^*(z; x)$ defined for z and x in D^*. Our aim will be to compute $g^*(z; x)$, from $g(z; x)$, neglecting only terms of higher order than ρ.

We consider for this purpose the function

$$(10) \qquad k(z; x) = g^*(z^*(z); x^*(x)) - g(z; x)$$

with z and x in $D - \sum_{i=0}^{m} \bar{k}_i$, and with $z^*(z)$ and $x^*(x)$ obtained from them by means of (6). We suppose further that all z_i are situated in D; otherwise, the z_j exterior to D are to be omitted in the summations of all z_i. $k(z; x)$ is harmonic in $D - \sum_{i=0}^{m} \bar{k}_i$ and is a uniform function of z; for $V_{\rho q}$ transforms this domain into $D^* = \sum_{i=0}^{m} \bar{k}_i$ and there $g^*(z^*; x^*)$ is defined and harmonic. The pole at $z = x$ is cancelled in the difference $g^* - g$. Therefore by Green's identity

$$(11) \qquad k(z; x) = \frac{1}{2\pi} \int_P \{k(t; x) \frac{\partial}{\partial n} g(z; t) - g(z; t) \frac{\partial}{\partial n} k(t; x)\} ds,$$

$P =$ boundary of $D - \sum_{i=0}^{m} \bar{k}_i$, the normal n pointing into the interior of the considered domain.

On the boundary of D, the integrand vanishes, since $k(t; x)$ and $g(z; t)$ do so. Therefore, we have to calculate the integral along the circles k_i only. We introduce (10) into (11) and take into account the identity

$$(12) \qquad \frac{1}{2\pi} \int_{k_i} \{g(z; t) \frac{\partial}{\partial n} g(t; x) - g(t; x) \frac{\partial}{\partial n} g(z; t)\} ds = 0$$

Identity (12) results from the fact that $g(z; t)$ and $g(t; t)$ are harmonic in $\bar{k}_i$ in view of the general identity

$$(13) \qquad \int_{k_i} \{u(t) \frac{\partial}{\partial n} v(t) - v(t) \frac{\partial}{\partial n} u(t)\} ds = 0,$$

which is valid for each couple of functions $u(z)$ and $v(z)$ which are harmonic in $\bar{k}_i$. There remains, therefore, only the calculation of

$$(14) \quad S_i = \frac{1}{2\pi} \int_{k_i} \{g^*(t^*(t); x^*) \frac{\partial}{\partial n} g(z; t) - g(z; t) \frac{\partial}{\partial n} g^*(t^*(t); x^*)\}$$

We put $t = z_i + re^{i\phi}$; then

$$(14') \quad S_i = \frac{1}{2\pi} \int_0^{2\pi} \{g^*[t + \frac{a_i\rho}{re^{i\phi}} + \rho y(re^{i\phi}) ; x^*] \frac{\partial}{\partial r} g(z;t)$$

$$- g(z;t) \frac{\partial}{\partial r} g^*[t + \frac{a_i\rho}{re^{i\phi}} + py(re^{i\phi}) ; x^*] \cdot rd\phi$$

where $y(re^{i\phi})$ denotes an analytic function of its argument $re^{i\phi}$. Let $p^*(u;v)$ and $p(u;v)$ be such analytic functions that $\mathcal{R}\{p^*(u;v)\} = g^*(u;v)$ and $\mathcal{R}\{p(u;v)\} = g(u;v)$. Denoting by $p^{*\prime}$, p' the derivatives of the corresponding functions with respect to the first argument, we get by development into series

$$(15) \quad S_i = \frac{1}{2\pi} \int_0^{2\pi} \{g^*(t;x^*) + \mathcal{R}[p^{*\prime}(t;x^*)(\frac{a_i\rho}{re^{i\phi}} + \rho y(re^{i\phi}))]$$

$$+ O(\rho^2)\} \frac{\partial}{\partial r} g(z;t) \cdot rd\phi$$

$$- \frac{1}{2\pi} \int_0^{2\pi} g(z;t)\{ \frac{\partial}{\partial r}(g^*(t;x^*) + \mathcal{R}[p^{*\prime}(t;x^*)(\frac{a_i\rho}{re^{i\phi}} + py(re^{i\phi}))])$$

$$+ O(\rho^2)\} \cdot rd\phi$$

with $|O(\rho^2)| < C |\rho|^2$. (15) can be simplified by means of the identity (13); we apply it with $v(t) = g(z;t)$, $u(t) = g^*(t;x^*)$, then with $v(t) = g(z;t)$, $u(t) = \mathcal{R}\{p^*(t;x^*)y(t - z_i)\}$ and at last with $v(t) = g(z;t)$, $u(t) = \mathcal{R}\{[p^{*\prime}(t;x^*) - p^{*\prime}(z_i;x^*)]\frac{a_i\rho}{re^{i\phi}}\}$.

Then

$$(16) \quad S_i = \frac{1}{2\pi} \int_0^{2\pi} \mathcal{R}\{p^{*\prime}(z_i;x^*) \frac{a_i\rho}{e^{i\phi}}\} \frac{\partial}{\partial r} g(z;t)d\phi$$

$$+ \frac{1}{2\pi} \int_0^{2\pi} g(z;t) \mathcal{R}\{p^{*\prime}(z_i;x^*)\frac{a_i\rho}{re^{i\phi}}\}d\phi + O(\rho^2).$$

Now, $g(z;t) = g(t;z)$ can be developed into a series of powers of $re^{i\phi}$; thus

$$(17) \quad g(t;z) = g(z_i + re_i{}^{i\phi}z) = g(z_i;z) + \mathcal{R}\{\sum_{\nu=1}^{\infty} \frac{1}{\nu!} p^{(\nu)}(z_i;z)r^\nu e^{i\nu\phi}\},$$

if r is sufficiently small. Therefore

$$(18) \quad S_i = \frac{1}{2\pi} \int_0^{2\pi} \mathcal{R}\{p^{*\prime}(z_i;x^*) \frac{a_i\rho}{e^{i\phi}}\} \mathcal{R}\{\sum_{\nu=1}^{\infty} \frac{1}{(\nu-1)!} p^{(\nu)}(z_i;z)r^{\nu-1}e^{i\nu\phi}\}d\phi$$

$$+ \frac{1}{2\pi} \int_0^{2\pi} [g(z_i;z) + \mathcal{R}\{\sum_{\nu=1}^{\infty} \frac{1}{\nu!} p^{(\nu)}(z_i;z)r^\nu e^{i\nu\phi}\}]$$

$$\times \mathcal{R}\{p^{*\prime}(z_i;x^*) \frac{a_i\rho}{re^{i\phi}}\}d\phi + O(\rho^2).$$

Putting $\rho a_i p^{*\prime}(z_i; x^*) = A e^{i\tau}$ and $p'(z_i; z) = B e^{i\sigma}$, and using the formulas expressing the orthogonality of the trigonometric functions, we get

$$(19) \quad S_i = \frac{1}{2\pi} \int_0^{2\pi} 2AB \cos(\tau - \phi) \cos(\sigma + \phi) d\phi + O(\rho^2)$$

$$= \frac{AB}{2\pi} \int_0^{2\pi} [\cos(\sigma + \tau) + \cos(\tau - \sigma - 2\phi)] d\phi + O(\rho^2)$$

$$= AB \cos(\sigma + \tau) + O(\rho^2) = \mathcal{R}\{\rho a_i p^{*\prime}(z_i; x^*) p'(z_i; z)\} + O(\rho^2).$$

Introduce these values into (11) and note that because of the differentiation in the direction of the inner normal n each integral S_i is to be taken positive. We get

$$(20) \quad g^*(z^*(z); x^*(x)) = g(z; x) + \mathcal{R}\{\rho \sum_{i=0}^m a_i p^{*\prime}(z_i; x^*) \, p'(z_i; z)\} + O(\rho^2).$$

Putting $z^* = z + \rho q(z)$, $x^* = x + \rho q(x)$, we develop $g^*(z, x)$ into a series,

$$(21) \quad g^*(z; x) = g(z; x) + \mathcal{R}\{\rho(\sum_{i=0}^m a_i p^{*\prime}(z_i; x^*) p'(z_i; z)$$

$$- q(z) p^{*\prime}(z; x^*) - q(x) p^{*\prime}(x; z^*))\} + O(\rho^2).$$

It is then obvious that $p^{*\prime}(u; v) = p'(u; v) + O(\rho)$, and therefore we get the final equality

$$(22) \quad g^*(z; x) = g(z; x) + \mathcal{R}\{\sum_{i=0}^m a_i p'(z_i; x) p'(z_i; z)$$

$$- q(z) p'(z; x) - q(x) p'(x; z))\} + O(\rho^2).$$

Thus, we have indeed an expression for $g^*(z; x)$ in terms of $g(z; x)$ with the desired approximation.

It is to be pointed out once more that (22) holds only if all points z_i are situated in D. If some z_i are not in D, the summation is to be applied to the $z_i \subset D$ only.

(22) remains valid, even if D has arbitrary boundaries, if there exists at least one boundary continuum. The variation $V_{\rho q}$ transforms D into an analogous domain D^* and both functions $g(z; x)$ and $g^*(z; x)$ exist. Further, there is easily found a sequence of domains D_n on $\mathfrak{R}$, converging to its kernel D in the Carathéodory sense, and whose boundary is composed of analytic curves. By $V_{\rho q}$ we attach to them domains D^*_n of the same type, which converge to D^*. For the corresponding Green's functions $g_n(z; x) \to g(z; x)$ and $g^*_n(z; x) \to g^*(z; x)$ hold uniformly in each inner partial domain of D and D^*. The functions g^*_n and g_n are connected by formula (22), the residual

member $O(\rho^2)$ of which can be evaluated uniformly for each n, since it contains only $g_n(z;x)$, $g^*_n(z;x)$ and their derivatives, all of which converge to the corresponding values of $g(z;x)$, $g^*(z;x)$ etc. (z and x being situated in the interior of D and D^*). Hence, we get in the limit as $n \to \infty$ the formula (22) for $g^*(z;x)$ and $g(z;x)$.

4. The variation of the representing function belonging to D. We suppose now that D is a simply connected domain on $\mathfrak{R}$ so that there exists, therefore, a function $z = f(\zeta)$, which is analytic in $|\zeta| < 1$ and maps this domain in a one-to-one manner on D. Let $z = 0$ be an interior point of D; then $f(\zeta)$ is fixed unambiguously by the conditions $f(0) = 0$, $f'(0) > 0$. $f(\zeta)$ will be called the representing function belonging to D.

We submit D to a variation $V_{\rho q}$ and get a new domain D^*; by means of (22) we can express the representing function $f^*(\zeta)$ belonging to D^* in terms of $f(\zeta)$, neglecting only terms of a higher order than ρ.

The inverse functions $\phi(z)$ and $\phi^*(z)$ of $f(\zeta)$ and $f^*(\zeta)$ are analytic and univalent in D and D^*, respectively, and map these domains on $|\zeta| < 1$. We have further $\phi(0) = \phi^*(0) = 0$. Hence

$$(23) \qquad g(z;0) = -\log|\phi(z)|; \qquad g^*(z;0) = -\log|\phi^*(z)|.$$

Indeed, the right hand side terms are harmonic in D and D^*, respectively, except at the point $z = 0$ where they become infinite as $-\log|z|$; and they vanish on the boundary of their corresponding domains.

In view of $p(z;0) = -\log\phi(z)$, and because of

$$(23') \qquad\qquad p(z;t) = -\log\frac{\phi(z) - \phi(t)}{1 - \overline{\phi(t)}\phi(z)},$$

we get from (22) and (23)

$$(24) \quad \log|\phi^*(z)| = \log|\phi(z)| - \mathfrak{R}\left\{\rho\left(\sum_{i=0}^{m} a_i \frac{\phi'(z_i)}{\phi(z_i)}\right.\right.$$

$$\times\left[\frac{\phi'(z_i)}{\phi(z_i) - \phi(z)} + \frac{\overline{\phi(z)}\phi'(z_i)}{1 - \overline{\phi(z)}\phi(z_i)}\right] + q(z)\frac{\phi'(z)}{\phi(z)} - q(0)$$

$$\times\left.\left.\left[\frac{\phi'(0)}{\phi(0) - \phi(z)} + \frac{\overline{\phi(z)}\phi'(0)}{1 - \overline{\phi(z)}\phi(0)}\right]\right)\right\} + O(\rho^2).$$

Completing both sides of this equation to analytic functions, we get

$$(25) \quad \log\phi^*(z) = \log\phi(z) - \rho\left(\sum_{i=0}^{m} a_i \frac{\phi'(z_i)^2}{\phi(z_i)[\phi(z_i) - \phi(z)]} + q(z)\frac{\phi'(z)}{\phi(z)}\right.$$

$$+ q(0)\frac{\phi'(0)}{\phi(z)}\Big) - \bar{\rho}\left(\sum_{i=0}^{m}\bar{a}_i\frac{\overline{\phi'(z_i)}^2\phi(z)}{\overline{\phi(z_i)}[1 - \overline{\phi(z_i)}\phi(z)]}\right.$$

$$- \overline{q(0)\phi'(0)}\phi(z)\Big) + i|\rho|C + O(\rho^2),$$

where C denotes a real constant determined by the condition $\phi'(0) > 0$, $\phi^{*\prime}(0) > 0$. We may bring (25) to the form

$$(25') \quad \phi^*(z) = \phi(z) - \rho \left(\sum_{i=0}^{m} a_i \frac{\phi'(z_i)^2 \phi(z)}{\phi(z_i)[\phi(z_i) - \phi(z)]} + q(0)\phi'(0) + q(z)\phi'(z) \right)$$

$$- \bar{\rho} \left(\sum_{i=0}^{m} \bar{a}_i \frac{\overline{\phi'(z_i)^2} \phi(z)^2}{\overline{\phi(z_i)}[1 - \overline{\phi(z_i)}\phi(z)]} - \overline{q(0)\phi'(0)}\phi(z)^2 \right) + i |\rho| C\phi(z) + O(\rho^2)$$

and put $\phi(z) = \zeta$, $\phi'(z) = f'(\zeta)^{-1}$, $\phi(z_i) = \zeta_i$. We have, further,

$$(26) \quad f(\zeta) = z = f^*(\phi^*(z)) = f^*(\phi(z)) + f^{*\prime}(\phi(z))[-\rho(\cdots)$$
$$- \bar{\rho}(\cdots) + i |\rho| C\phi(z)] + O(\rho^2),$$

and since $f^{*\prime}(\zeta) = f'(\zeta) + O(\rho)$, we obtain finally

$$(27) \quad f^*(\zeta) = f(\zeta) + \rho f'(\zeta) \left(\sum_{i=0}^{m} a_i \frac{\zeta}{\zeta_i f'(\zeta_i)^2(\zeta_i - \zeta)} + \frac{q(0)}{f'(0)} + \frac{q(f(\zeta))}{f'(\zeta)} \right)$$

$$+ \bar{\rho} f'(\zeta) \left(\sum_{i=0}^{m} \bar{a}_i \frac{\zeta^2}{\overline{\zeta_i f'(\zeta_i)^2}(1 - \overline{\zeta_i}\zeta)} - \frac{\overline{q(0)}}{\overline{f'(0)}} \zeta^2 \right)$$

$$- i |\rho| C\zeta f'(\zeta) + O(\rho^2).$$

Thus our aim is attained; for the sake of completeness, we compute C explicitly. We have

$$(27') \quad f^{*\prime}(0) = f'(0)\left[1 + \rho \sum_{i=0}^{m} \frac{a_i}{\zeta_i^2 f'(\zeta_i)^2} + \rho q'(0) - i |\rho| C\right] + O(\rho^2).$$

$f^{*\prime}(0)$ being positive, the multiplier of $f'(0)$ must be positive. Hence we get

$$(27'') \qquad\qquad C = \Im \left\{ \frac{\rho}{|\rho|} \left(\sum_{i=0}^{m} \frac{a_i}{\zeta_i^2 f'(\zeta_i)^2} + q'(0) \right) \right\}.$$

5. An application to the coefficient problem for univalent functions.

A function $z = f(\zeta)$ is called univalent in $|\zeta| < 1$, if it maps this domain in a one-to-one manner on a domain D of the z-plane. The z-plane plays here the part of the closed R. S. $\Re$, containing D. According to the discussion of **2** each variation $V_{\rho q}$ transforms the z-plane into itself, since there is only one closed one-sheeted R. S. over the z-plane. For our purposes it is sufficient to choose

$$(28) \qquad\qquad q(z) = \frac{z}{z - z_0}$$

with $z_0 \subset D$.

We seek the functions with the normalization (1) which are regular and univalent in $|\zeta| < 1$, and which possess an n-th coefficient a_n with the largest possible absolute value. Let $f_n(\zeta)$ be such a function and let D_n be the domain, on which it maps $|\zeta| < 1$. The variation $V_{\rho q}$ with $q(z)$ defined as in (28) transforms D_n into a new domain D^*_n with the representing function

$$(29) \quad f^*_n(\zeta) = f_n(\zeta) + \rho\{f'_n(\zeta) \frac{f_n(\zeta_0)}{\zeta_0^2 f'_n(\zeta_0)^2} \frac{\zeta}{1-\zeta/\zeta_0} + \frac{f_n(\zeta)}{f_n(\zeta)-f_n(\zeta_0)}\}$$

$$+ \bar{\rho}f'_n(\zeta) \frac{\overline{f_n(\zeta_0)}}{\overline{\zeta_0 f'_n(\zeta_0)}^2} \frac{\zeta^2}{1-\bar{\zeta_0}\zeta} - i|\rho| C\zeta f'_n(\zeta) + O(\rho^2).$$

$f^*_n(\zeta)$ is also regular and univalent in $|\zeta| < 1$, but it has not the normalization (1). This holds, however, for $f^*_n(\zeta)f^{*\prime}_n(0)^{-1}$, and in view of the extremal property of $f_n(\zeta)$ we obtain

$$(30) \quad |a^*_n f^{*\prime}_n(0)^{-1}| = |a_n + \rho \frac{f_n(\zeta_0)}{\zeta_0^2 f'_n(\zeta_0)^2}((n-1)a_n + (n-1)a_{n-1}\cdot\frac{1}{\zeta_0}$$

$$+ (n-2)a_{n-2}\frac{1}{\zeta_0^2} + \cdots + \frac{1}{\zeta_0^{n-1}}) - \rho \frac{1}{f_n(\zeta_0)} P_n\left(\frac{1}{f_n(\xi_0)}\right)$$

$$+ \bar{\rho} \frac{\overline{f_n(\zeta_0)}}{\overline{\zeta_0^2 f'_n(\zeta_0)}^2}((n-1)a_{n-1}\bar{\zeta_0} + (n-2)a_{n-2}\bar{\zeta_0}^2 + \cdots + \bar{\zeta_0}^{n-1})$$

$$-i|\rho|C(n-1)a_n + O(\rho^2)| \leq |a_n|$$

where $P_n(x)$ is defined by (2). Without loss of generality we may suppose $a_n > 0$; then (30) implies

$$(30') \quad \mathcal{R}\left\{\rho\left[\frac{f_n(\zeta_0)}{\zeta_0^2 f'_n(\zeta_0)^2}\left((n-1)a_n + (n-1)\frac{a_{n-1}}{\zeta_0} + \cdots + \frac{1}{\zeta_0^{n-1}}\right)\right.\right.$$

$$\left.- \frac{1}{f_n(\zeta_0)} P_n\left(\frac{1}{f_n(\zeta_0)}\right)\right] + \bar{\rho} \frac{\overline{f_n(\zeta_0)}}{\overline{\zeta_0^2 f'_n(\zeta_0)}^2}\left((n-1)a_{n-1}\bar{\zeta_0} + \cdots \bar{\zeta_0}^{n-1}\right)\right\}$$

$$+ O(\rho^2) \leq 0.$$

This holds for each value $\rho = |\rho| e^{i\psi}$. If we denote the coefficients of ρ and $\bar{\rho}$ by M and N respectively, and pass to the limit $|\rho| = 0$ in (30'), we find

$$(30'') \qquad \mathcal{R}\{e^{i\psi}M + e^{-i\psi}N\} \leq 0 \quad \text{for} \quad 0 \leq \psi \leq 2\pi.$$

Since obviously $\mathcal{R}\{e^{i\psi}M - e^{-i\psi}\bar{M}\} = 0$, (30'') yields

$$(30''') \qquad \mathcal{R}\{e^{-i\psi}(\bar{M} + N)\} \leq 0$$

for each value of ψ. This is only possible in the case

$$(30^{\text{IV}}) \qquad \bar{M} + N = 0.$$

Introducing the values of M and N into (30^{IV}), we get the equality

(31) $\quad \dfrac{f_n(\zeta_0)}{\zeta_0^2 f'_n(\zeta_0)^2} \left(\dfrac{1}{\zeta_0^{n-1}} + \dfrac{2a_2}{\zeta_0^{n-2}} + \cdots + \dfrac{(n-1)a_{n-1}}{\zeta_0} \right.$

$\left. + (n-1)a_n + (n-1)\bar{a}_{n-1}\zeta_0 + \cdots \zeta_0^{n-1} \right) = \dfrac{1}{f_n(\zeta_0)} \; P_n \left(\dfrac{1}{f_n(\zeta_0)} \right),$

i. e.

(31') $\quad \dfrac{\zeta_0^2 f'_n(\zeta_0)^2}{f_n(\zeta_0)^2} \; P_n \left(\dfrac{1}{f_n(\zeta_0)} \right) = \dfrac{1}{\zeta_0^{n-1}} + \dfrac{2a_2}{\zeta_0^{n-2}} + \cdots$

$\qquad\qquad\qquad\qquad + (n-1)a_n + \cdots + 2\bar{a}_2\zeta_0^{n-2} + \zeta_0^{n-1}.$

(31') holds for each value ζ_0 in $|\zeta| < 1$; hence $f_n(\zeta)$ must satisfy this differential equation everywhere and can be continued with its aid over its whole domain of existence. In particular (31') implies that D_n is bounded by analytic curves satisfying the differential equation (3). For let $z = f(e^{i\tau})$ be the parametric representation of the boundary curve; then (31') yields

(31'') $\qquad\qquad\qquad \dfrac{z'^2}{z^2} \; P_n \left(\dfrac{1}{z} \right) \leq 0$

which is equivalent to (3) in view of the supposition $a_n > 0$.

Our method shows also that there are no points in the z-plane which are exterior to D_n. Suppose, for example, that z_0 were an exterior point of D_n, then use it for the function (28). We transform D_n into D^*_n by $V_{\rho q}$; but (20) has now the form $g^*(z^*; x^*) = g(z^*; x)$, since there is no z_t in D_n. An easy computation shows that the representing function belonging to D^*_n is

(29') $\qquad f^*_n(\zeta) = f_n(\zeta) + \rho \, \dfrac{f_n(\zeta)}{f_n(\zeta) - z_0} - i \, |\rho| \, C\zeta f'_n(\zeta) + O(\rho^2).$

Therefore we have for each sufficiently small ρ

(32) $\qquad |a^*_n f^*_n(0)^{-1}| = |a_n - \rho z_0 P_n \left(\dfrac{1}{z_0} \right) - i \, |\rho| \, C(n-1)a_n + O(\rho^2)| \leq a_n$

and in the same way as above we deduce that $\dfrac{1}{z_0} \, P_n \left(\dfrac{1}{z_0} \right) = 0$. This equation

has a finite number of roots z_0. But were there one point z_0 exterior to D_n, there would exist an infinity of such points. This is impossible, and so we have proved that D_n covers all the plane, analytic slits excepted.

The above derivation of the differential equation (3) gives us the explicit formula (31') for $\dfrac{\zeta^2 f'_n(\zeta)^2}{f_n(\zeta)^2} \; P_n \left(\dfrac{1}{f_n(\zeta)} \right)$. If (3) is proved in an alternative way, Schwarz' principle of reflection implies that the above expression must be a rational function of ζ. By (31′) we expressed the coefficients of this rational function in a simple form by those of $f_n(\zeta)$; in the following paragraph we shall prove this representation in an alternative way.

6. The relation between $P_n(x)$ and the Faber polynomials. We denote by $H_n(\zeta; f)$ the expression $\dfrac{\zeta^2 f'(\zeta)^2}{f(\zeta)^2} P_n\left(\dfrac{1}{f(\zeta)}\right)$. If $z = f_n(\zeta)$ is the function with maximal $a_n > 0$ considered in the last section, we see from (3) that $H_n(\zeta; f_n(\zeta))$ tends to real values, if ζ approaches the unit circle. Hence the principle of reflection yields

$$(33) \qquad H_n((1/\zeta); f_n) = \overline{H(\zeta; f_n)}.$$

This means that $H_n(\zeta; f)$ necessarily has the form $\dfrac{1}{\zeta^{n-1}} + \dfrac{\alpha_2}{\zeta^{n-2}} + \cdots + \dfrac{\alpha_{n-1}}{\zeta}$ $+ a_n + \bar{a}_{n-1}\zeta + \cdots + \zeta^{n-2}$. Hence it suffices to compute the first n coefficients of H_n. Now it can be shown that, if the power series $f(\zeta) = \zeta + a_2\zeta^2$ $+ \cdots + a_n\zeta^n + \cdots$ converges near $\zeta = 0$, the corresponding $H_n(\zeta; f)$ has there the development

$$(34) \quad H_n(\zeta; f) = \frac{1}{\zeta^{n-1}} + \frac{2a_2}{\zeta^{n-2}} + \frac{3a_3}{\zeta^{n-3}} + \cdots + \frac{(n-1)a_{n-1}}{\zeta}$$
$$+ (n-1)a_n + \zeta(\cdots).$$

This identity will be deduced from another, which connects the $P_n(x)$ with the Faber polynomials, whose important rôle in the theory of univalent functions is well known.[2]

If $f(\zeta)$ is a power series of the form (1) converging in a neighborhood of $\zeta = 0$, $F_n(x)$ is called the n-th Faber polynomial belonging to $f(\zeta)$, if it is a polynomial of the n-th degree such that

$$(35) \qquad F_n\left(\frac{1}{f(\zeta)}\right) = \frac{1}{\zeta^n} + A_n\zeta + B_n\zeta^2 + \cdots$$

For each $f(\zeta)$ and for each n there exists exactly one $F_n(x)$. All the polynomials $F_n(x)$ belonging to a fixed function $f(\zeta)$ can be represented by means of a generating function. To show this, consider the function

$$(36) \qquad L(\zeta; \eta) = \log(1 - f(\zeta)/f(\eta)) - \log(1 - \zeta/\eta) - \log\frac{f(\zeta)}{\zeta},$$

---

[2] H. Grunsky, "Koeffizientenbedingungen für schlicht abbildende meromorphe Funktionen," *Math. Zeits.*, vol. 45 (1939), pp. 29-61.

which in the neighborhood of $\zeta = 0$, $\eta = 0$ is an analytic function of its two arguments. In a sufficiently small circle around $\zeta = 0$, $f(\zeta)$ is univalent; thus $(f(\zeta) - f(\eta))(\zeta - \eta)^{-1}$ is regular and non-zero in this circle and $L(\zeta; \eta)$ is regular in it. Therefore $L(\zeta; \eta)$ can be developed in a series of powers of ζ and η

$$(37) \qquad L(\zeta; \eta) = \sum_{\nu,\mu=1}^{\infty} c_{\nu\mu} \zeta^{\nu} \eta^{\mu}.$$

If, on the other hand, we develop $L(\zeta; \eta)$ in powers of ζ for $\eta \neq 0$ fixed, we get

$$(37') \qquad L(\zeta; \eta) = - \sum_{\nu=1}^{\infty} \frac{1}{\nu} F_\nu \left(\frac{1}{f(\eta)} \right) \zeta^\nu + \sum_{\nu=1}^{\infty} \frac{1}{\nu} \frac{\zeta^\nu}{\eta^\nu}.$$

Here $F_\nu(x)$ is a polynomial of the ν-th degree, and a comparison of (37) and (37') yields

$$(37'') \qquad F_n \left(\frac{1}{f(\eta)} \right) = \frac{1}{\eta^n} - n \sum_{\mu=1}^{\infty} c_{n\mu} \zeta^\mu.$$

Hence, $F_n(x)$ is indeed the n-th Faber polynomial belonging to $f(\zeta)$, and all $F_n(x)$ are obtained by means of the generating function

$$(38) \qquad \log(1 - x f(\zeta)) - \log \frac{f(\zeta)}{\zeta} = - \sum_{\nu=1}^{\infty} \frac{1}{\nu} F_\nu(x) \zeta^\nu.$$

From (38) we get by differentiating with respect to x

$$(39) \qquad \frac{f(\zeta)}{1 - x f(\zeta)} = \sum_{\nu=1}^{\infty} \frac{1}{\nu} F'_\nu(x) \zeta^\nu$$

and hence by comparison with (2) [3]

$$(40) \qquad \frac{1}{n} F'_n(x) = a_n + P_n(x).$$

The identity (34) is now easily obtained. Differentiation of (35) with respect to ζ and the application of (40) yield

$$(41) \qquad \frac{f'(\zeta)}{f(\zeta)^2} \left[a_n + P_n \left(\frac{1}{f(\zeta)} \right) \right] = \frac{1}{\zeta^{n+1}} - \frac{A_n}{n} + \cdots$$

and hence

$$(41') \qquad \frac{f'(\zeta)^2}{f(\zeta)^2} P_n \left(\frac{1}{f(\zeta)} \right) = \frac{1}{\zeta^{n+1}} + \frac{2a_2}{\zeta^n} + \cdots + \frac{(n-1)a_{n-1}}{\zeta^3}$$

$$+ \frac{(n-1)a_n}{\zeta^2} + \frac{(n+1)a_{n+1} - 2a_2 a_n}{\zeta} + \cdots$$

From this the desired identity follows immediately.

---

[3] The author found this relation after a most helpful discussion with the late Prof. Schur, whose investigations on Faber polynomials will be published shortly.

In the case of the extremal function $f_n(\zeta)$, we know from (31′) that the coefficient of ζ in the development of $H_n(\zeta; f_n)$ is equal to $(n-1)\bar{a}_{n-1}$; on the other hand, (41′) gives for this coefficient the value $(n+1)a_{n+1} - 2a_2a_n$. Therefore, in this case

$$(42) \qquad (n+1)a_{n+1} = 2a_2a_n + (n-1)\bar{a}_{n-1},$$

a relation established by Marty.

7. A coefficient problem from the theory of p-valued functions. We shall now show, by means of a special problem, how the above method of variation is to be applied in the theory of p-valued functions. Let

$$(1) \qquad f(\zeta) = \zeta + a_2\zeta^2 + a_3\zeta^3 + \cdots + a_n\zeta^n + \cdots$$

be regular in $|\zeta| < 1$, taking on the value zero only at $\zeta = 0$, and taking on each other value at most p times. All these functions $f(\zeta)$ are known to form a compact family, and therefore we may seek those functions $f_n(\zeta)$, for which $|a_n|$ possesses the largest value in the family.

Each function of the family maps $|\zeta| < 1$ on a domain D situated on a p-sheeted R. S. over the z-plane. Let us consider the subclass $\mathfrak{F}_g$ of all functions within the family, the domain D of which can be stretched over a p-sheeted R. S. with genus $\leq g$. $\mathfrak{F}_g$ also is a compact family, and we first consider the coefficent problem for this. Knowing the maximum $\mu_{n,g}$ for $|a_n|$ with respect to $\mathfrak{F}_g$, we get the maximum μ_n for $|a_n|$ with respect to the whole family by the equality

$$(43) \qquad \mu_n = \lim_{g=\infty} \mu_{n,g}.$$

It is sufficient, therefore, to restrict ourselves to the coefficient problem for the family $\mathfrak{F}_g$; our method will enable us to make some assertions concerning the extremal functions.

Let, therefore, g be fixed and let $f_n(\zeta)$ be a function of the family $\mathfrak{F}_g$ with maximal $|a_n|$. This function determines a domain D_n on a R. S. $\mathfrak{R}_n$ which covers the point 0 exactly once, and does not cover the point ∞. $\mathfrak{R}_n$ is p-sheeted and of genus $\gamma \leq g$. We choose a function $q(z)$ uniform on $\mathfrak{R}_n$, regular and finite in each point with coördinate $z = \infty$, and regular and vanishing in each point with coördinate $z = 0$. We apply the variation $V_{\rho q}$ to D_n. The new domain D^*_n belongs to the representing function $f^*_n(\zeta)$ obtained from $f_n(\zeta)$ by means of (27) and possessing the n-th coefficient

$$(44) \quad a^*_n = a_n + \rho \Big(\sum_{i=0}^{m} \frac{a_i}{\zeta_i^2 f'_n(\zeta_i)^2} \big[na_n + (n-1)\frac{a_{n-1}}{\zeta_i} + \cdots + \frac{1}{\zeta_i^{n-1}} \big]$$

$$+ \{ q(f_n(\xi)) \}_{n\text{-th coeff.}} \Big) + \bar{\rho} \Big(\sum_{i=0}^{m} \frac{\bar{a}_i}{\bar{\zeta}_i^2 \bar{f}'_n(\zeta_i)^2} \big[(n-1)a_{n-1}\bar{\zeta}_i + \cdots + \bar{\zeta}_i^{n-1} \big] \Big)$$

$$- i |\rho| Cna_n + O(\rho^2).$$

$V_{\rho q}$ preserves the genus and the number of sheets; hence $f^*_n(\zeta)f^*_n(0)^{-1}$ is a function of the family $\mathfrak{F}_g$ and therefore

$$(45) \quad | a^*_n f^{*\prime}_n(0)^{-1} | = | a_n + \rho \Big(\sum_{i=0}^{m} \frac{a_i}{\zeta_i f'_n(\zeta_i)^2} \big[(n-1)a_n + (n-1)\frac{a_{n-1}}{\zeta_i}$$

$$+ \cdots + \frac{1}{\zeta_i^{n-1}} \big] + \{ q(f_n(\xi)) \} - a_n q'(f_n(0)) \atop {\scriptstyle n\text{-th coeff.}}$$

$$+ \bar{\rho} \sum_{i=0}^{m} \frac{\bar{a}_i}{\bar{\zeta}_i^2 \bar{f}'_n(\zeta_i)^2} \big[(n-1)a_{n-1}\bar{\zeta}_i + \cdots + \bar{\zeta}_i^{n-1} \big] - i |\rho| C(n-1)a_n$$

$$+ O(\rho^2) | \leq | a_n |.$$

Supposing, once more, a_n to be positive, we get finally by the conclusions of **5** the equation

$$(46) \quad \sum_{i=0}^{m} a_i \frac{1}{\zeta_i^2 f'_n(\zeta_i)^2} \Big(\frac{1}{\zeta_i^{n-1}} + \frac{2a_2}{\zeta_i^{n-2}} + \cdots + \frac{(n-1)a_{n-1}}{\zeta_i} + (n-1)a_n$$

$$+ (n-1)\bar{a}_{n-1}\zeta_i + \cdots + \zeta_i^{n-1} \Big)$$

$$= a_n q'(f_n(0)) - \{ q(f_n(\xi)) \}_{n\text{-th coeff.}}$$

There still remains the problem of examining all the functions $q(z)$ of the type considered, which are uniform on $\mathfrak{R}_n$. We get such a function in the easiest way by putting $q(z) = \dfrac{z}{z - z_0}$ supposing only that no boundary point of D_n and no branch point of R_n has the coördinate z_0. Then (46) yields

$$(47) \quad \sum_{i=0}^{m} \frac{f_n(\zeta_i)}{\zeta_i^2 f'_n(\zeta_i)^2} (1/\zeta_i^{n-1} + \cdots + (n-1)a_n + \cdots + \zeta_i^{n-1})$$

$$= \frac{1}{z_0} P_n \Big(\frac{1}{z_0} \Big), \quad f_n(\zeta_i) = z_0.$$

This equality, however, is not the most general assertion concerning $f_n(\zeta)$. To get such an assertion we shall use in the following section a more general type of functions $q(z)$ uniform on $\mathfrak{R}_n$.

8. The variation function $q(z)$, uniform on $\Re_n$. Let $\Re_n$ be the R. S. over which the domain D_n, belonging to the extremal function $f_n(\zeta)$, is extended. γ denoting the genus of $\Re_n$, there exist 2γ loop-cuts $A_1, B_1, \cdots, A_\gamma, B_\gamma$ on the surface, with the following properties: Each loop-cut A_i meets the cut B_i in exactly one point π_i, but meets no other loop-cut; analogously each B_i meets only the corresponding A_i. Choose a fixed point P on $\Re_n$ and connect it with each π_i by a Jordan curve C_i, which has no points in common with the loop-cuts and the other C_k, the points π_i and P excepted. The A_i, B_i, C_i ($i = 1, 2, \cdots \gamma$) yield a canonical resolution of $\Re_n$.

A function $w(z)$ is called an integral of the first kind on $\Re_n$, if it possesses a finite and uniform derivative with respect to the local parameter at each point of the R. S. $w(z)$ itself is of course not uniform on $\Re_n$, if it is not a constant, but it increases by certain constant numbers, the periods, if its argument crosses the cuts A_i, B_i. There exist exactly $\gamma + 1$ integrals of the first kind $1, w_1(z), \cdots, w_\gamma(z)$ on $\Re_n$, which are linearly independent. We can suppose them normalized in such a manner that $w_i(z)$ has the period δ_{ik} with respect to the crossing of A_k, with $\delta_{ik} = 0$ if $i \neq k$ and $= 1$ if $i = k$. These $w_i(z)$ are called the normal integrals of the first kind (transcendent normalization). $\tau(z; x)$ is called an integral of the second kind with pole x, if its derivative with respect to the local parameter is regular and uniform everywhere on $\Re_n$, except at the point x where it has a double pole. Since if $\tau(z; x)$ is, then $\tau(z; x) - \sum\limits_{\nu=0}^{\gamma} c_\gamma w_\gamma(z)$ is also an integral of the second kind with the same pole, we can normalize $\tau(z; x)$ in such a way that all its periods with respect to the A_i vanish. If x is not a branch point of $\Re_n$ we may choose in its neighborhood z itself as a local parameter. Then $\tau(z; x)$ is to be chosen so that the development

$$(48) \qquad \tau(z; x) = \frac{1}{z - x} + a + b(z - x) + \cdots \text{ holds.}$$

All these conditions together fix $\tau(z; x)$ except for an additive constant; $\tau(z; x)$ is called a normal integral of the second kind.

The periods of $\tau(z; x)$ with respect to the A_k are zero by definition; with respect to the B_k, $\tau(z; x)$ has the periods

$$(49) \qquad P_k(x) = -2\pi i \frac{dw_k(x)}{dx} = -2\pi i w'_k(x).$$

$\omega(z; x_1, x_2)$ is called an integral of the third kind on $\Re_n$ if its derivative with respect to the local parameter is uniform and finite everywhere, except for the points x_1 and x_2, where ω' has two simple poles with the sum of residues zero. Hence $\omega(z; x_1, x_2)$ has logarithmic poles at x_1 and x_2. We normalize $\omega(z; x_1, x_2)$ so that its periods with respect to the A_i vanish and so that the residues of its derivative at x_1 and x_2 are -1 and $+1$ respectively. Then $\omega(z; x_1, x_2)$ will be called a normal integral of the third kind.

Between the normal integrals of the second and third kind we have the important relation

$$(50) \qquad \frac{d}{dz} \omega(z; x_1, x_2) = \omega'(z; x_1, x_2) = \tau(x_1; z) - \tau(x_2; z).$$

This equation permits certain assertions concerning the dependence of $\tau(z; x)$ upon its pole x. Examine, for example, $\tau(x_1; z) - \tau(x_2; z)$ in the neighborhood of a branch point z_b of $\Re_n$, which is supposed different from x_1 and x_2. If $t = (z - z_b)^a$ is the local parameter, we have by (50) $\tau(x_1; z) - \tau(x_2; z)$ $= \alpha(z - z_b)^{a-1} \frac{d}{dt} \omega(z; x_1, x_2)$. Now, $\frac{d}{dt} \omega(z; x_1, x_2)$ is an analytic function of t, x_1 and x_2 and can, therefore, be differentiated with respect to x_1 an arbitrary number of times. This shows that each derivative of $\tau(x_1; z)$ with respect to x_1 has at z_b an infinity of order $(z - z_b)^{a-1}$ at most. This fact will be of use to us later.

Having recalled the notions from the theory of Riemann surfaces which are necessary for our investigation, we proceed to the construction of a function $q(z)$, uniform and meromorphic on $\Re_n$, vanishing at $z = 0$ and finite at $z = \infty$. For this purpose, we fix γ points $z_1, z_2, \cdots z_\gamma$ such that the determinant

$$(51) \qquad\qquad |w'_k(z_i)|_{i,k=1,2,\ldots\gamma} \neq 0.$$

This is always possible, the $w_k(z)$ being linearly independent. The z_i can even be chosen in D_n and different from all the branch points of $\Re_n$. z_0 being also a point of this type, we form

$$(52) \qquad\qquad q_1(z) = \sum_{\nu=0}^{\gamma} a^*_\nu \tau(z; z_\nu)$$

and determine the a^*_ν in such a way that $q_1(z)$ is uniform on $\Re_n$. In view of (49), this requirement is equivalent to the equations

$$(53) \qquad\qquad \sum_{\nu=0}^{\gamma} a^*_\nu w'_k(z_\nu) = 0 \qquad\qquad\qquad k = 1, 2, \cdots \gamma,$$

which have, because of (51), one solution, not taking into account an arbitrary common factor. $q_1(z)$ is uniform on $\Re_n$ and regular everywhere, the simple poles z_ν excepted. In the neighborhood of z_ν, $q_1(z)$ has the development (5). We form now

$$(54) \qquad q(z) = \frac{z}{z-a}\, q_1(z),$$

a being a point in the z-plane which is not a branch point of $\Re_n$ but over which are situated p points of D_n. There exists such a point a, $f_n(\zeta)$ being supposed p-valued; denote by $z_{\gamma+1}, \cdots, z_{\gamma+p}$ the p points of D_n with the coördinate a.

The function $q(z)$ satisfies all our requirements and can be used as the variation function. Thus we get from (46)

$$(46') \qquad \sum_{i=0}^{\gamma+p} a_i\chi(\zeta_i) = a_n q'(f_n(0)) - \{q(f_n(\xi))\}_{n\text{-th coeff.}}\ ;$$

if we put here

$$(46'') \quad \chi(\zeta) = \frac{1}{\zeta^2 f'_n(\zeta)^2}\left(\frac{1}{\zeta^{n-1}} + \frac{2a_2}{\zeta^{n-2}} + \cdots + \frac{(n-1)a_{n-1}}{\zeta} + (n-1)a_n\right.$$
$$\left. + (n-1)\bar{a}_{n-1}\zeta + \cdots + \zeta^{n-1}\right),$$

$f(\zeta_i) = z_i$ and

$$(55) \quad a_i = \frac{z_i}{z_i - a}\ \text{for}\ i=0,\cdots\gamma;\quad a_i = aq_1(z_i)\ \text{for}\ i=\gamma+1,\cdots,\gamma+p.$$

By means of (52) we get from (46')

$$(56) \quad \sum_{\nu=0}^{\gamma} a^*_\nu\left[\frac{z_\nu}{z_\nu - a}\chi(\zeta_\nu) + a\sum_{i=\gamma+1}^{\gamma+p}\chi(\zeta_i)\tau(z_i;z_\nu) + \left\{\frac{f_n(\xi)}{f_n(\xi)-a}\tau(f_n(\xi);z_\nu)\right\}_{n\text{-th coeff.}}\right.$$
$$\left. + \frac{a_n}{a}\tau(f_n(0);z_\nu)\right] = 0.$$

The γ sets of $\gamma+1$ numbers $w'_k(z_0),\cdots w'_k(z_\gamma)$ $(k=1,\cdots\gamma)$ being linearly independent and satisfying the equation (53), we deduce from (56)

$$(57) \quad \frac{z_\nu}{z_\nu - a}\chi(\zeta_\nu) + a\sum_{i=\gamma+1}^{\gamma+p}\chi(\zeta_i)\tau(z_i;z_\nu) + \left\{\frac{f_n(\xi)}{f_n(\xi)-a}\tau(f_n(\xi);z_\nu)\right\}_{n\text{-th coeff.}}$$
$$+ \frac{a_n}{a}\tau(f_n(0);z_\nu) = \sum_{k=1}^{\gamma}\lambda_k w'_k(z_\nu).$$

Since ζ_0 and $z_0 = f_n(\zeta_0)$ are arbitrary, the condition $z_0 \subset D_n$ and $z_0 \neq$ branch point of $\Re_n$ excepted, we can drop the index ν in (57). Applying, further, (47) to $a = z_0$ and $\zeta_{\gamma+1},\cdots\zeta_{\gamma+p}$, we get

$$(47') \qquad a \sum_{i=\gamma+1}^{\gamma+p} \chi(\zeta_i) + \left\{ \frac{f_n(\xi)}{f_n(\xi) - a} \right\}_{n\text{-th coeff.}} + \frac{a_n}{a} = 0.$$

Let y be a regular point of $\Re_n$; we multiply $(47')$ by $-\tau(y; z)$ and add the resulting equation to (57). We get

$$(58) \qquad \frac{z}{z - a} \chi(\zeta) + a \sum_{i=\gamma+1}^{\gamma+p} \chi(\zeta_i) \left(\tau(z_i; z) - \tau(y; z) \right) +$$

$$\left\{ \frac{f_n(\xi)}{f_n(\xi) - a} \left[\tau(f_n(\xi); z) - \tau(y; z) \right] \right\}_{n\text{-th coeff.}} + \frac{a_n}{a} \left[\tau(f_n(0); z) - \tau(y; z) \right]$$

$$= \sum_{k=1}^{\gamma} \lambda_k w'_k(z).$$

Now, we apply equation (50), which shows that $\tau(x; z) - \tau(y; z)$ and all its derivatives with respect to x are analytic functions of z, becoming infinite in the neighborhood of a $(k-1)$-fold branch point z_b no faster than $(z - z_b)^{(1-k)/k}$. Since the n-th coefficient of ξ in the expression $f_n(\xi)(f_n(\xi) - a)^{-1}[\tau(f_n(\xi); z) - \tau(y; z)]$ is a linear combination of $\tau(0; z) - \tau(y; z)$, $\tau'(0; z), \cdots$ $\tau^{(n)}(0; z)$, and since, further, all $w'_k(z)$ behave at z_b like these functions, $\chi(\zeta)$ becomes infinite at ζ_b (corresponding to the $(k-1)$-fold branch point z_b) as $(z - z_b)^{(1-k)/k}$ at most. $\chi(\zeta)$ does not depend on the arbitrary values a and y; hence, in view of (58), it has its only singularities at $z = 0$ and at the points ζ_b.

It is easy to find a linear aggregate of uniform normal integrals on $\Re_n$, which behaves at $z = f_n(0)$ as $z \chi(\zeta)$. To do so, we introduce the normal integrals $\tau_\nu(z; f_n(0))$ possessing everywhere on $\Re_n$ a finite and uniform derivative with respect to the local parameter, the point $f_n(0)$ excepted. In the neighborhood of this point,

$$(59) \qquad \tau_\nu(z; f_n(0)) = (1/z^\nu) + \alpha + \beta z + \cdots$$

and the periods with respect to the loop-cuts A_i are zero. Further, let

$$(60) \qquad P_n(x) = \sum_{l=2}^{n} a_{nl} x^{l-1}$$

be the polynomial defined in (2). Then, the expression

$$(61) \qquad L_n(f_n(\zeta)) = - \sum_{l=2}^{n} \frac{a_{nl}}{l - 1} \tau'_{l-1}(f_n(\zeta); f_n(0))$$

has at $\zeta = 0$ the same infinity as

$$(61') \quad \sum_{l=2}^{n} a_{nl} f_n(\zeta)^{-l} = f_n(\zeta)^{-1} P_n\left(\frac{1}{f_n(\zeta)}\right)$$

$$= \frac{f_n(\zeta)}{f'_n(\zeta)^2}\left(\frac{1}{\zeta^{n+1}} + \frac{2a^2}{\zeta^n} + \cdots + \frac{(n-1)a_n}{\zeta^2} + \cdots\right)$$

the last equation resulting from $(41')$. According to $(46'')$, $L_n(f_n(\zeta))$ has at $\zeta = 0$ the same infinity as $z\chi(\zeta)$; at the branch points z_b, $L_n(z)$ becomes infinite as $(z - z_b)^{(1-k)/k}$ at most. Everywhere else $L_n(z)$ is uniform and regular on $\mathfrak{R}_n$. Hence, $z\chi(\zeta) - L_n(z)$ is uniform on $\mathfrak{R}_n$, becomes infinite at the branch points at most as the derivative with respect to z of a function which is regular in the local parameter, and is regular everywhere else with the possible exception of $z = \infty$.

According to (58) $\chi(\zeta)$ is an algebraic function of z which vanishes at $z = \infty$; hence

$$(62) \quad \chi(\zeta) = c_\kappa z^{-\kappa} + c_\lambda z^{-\lambda} + \cdots, \qquad 0 < \kappa < \lambda < \cdots, \quad c_\kappa \neq 0.$$

Let α be a point on the circle $|\zeta| = 1$, for which $f_n(\alpha) = \infty$; we put $z = f_n(\zeta)$ and deduce from $(46'')$ and (62) that

$$(63) \quad C_\mu(\zeta - \alpha)^\mu + C_{\mu+1}(\zeta - \alpha)^{\mu+1} + \cdots$$

$$= \frac{f'_n(\zeta)^2}{f_n(\zeta)^\kappa}\left(c_\kappa + c_\lambda f_n(\zeta)^{-(\lambda-\kappa)} + \cdots\right), \quad \mu \gtreqqless 0,$$

which means

$$(63') \quad D_\mu(\zeta - \alpha)^{\mu/2} + \cdots = \frac{f'_n(\zeta)}{f_n(\zeta)^{\kappa/2}}\left(d_\kappa + d_\lambda f_n(\zeta)^{-(\lambda-\kappa)} + \cdots\right.$$

By integration we get, if $\kappa \neq 2$

$$(64) \quad E_\mu(\zeta - \alpha)^{(\mu+2)/2} + \cdots = f_n(\zeta)^{1-(\kappa/2)}\left(e_\kappa + e_\lambda f_n(\zeta)^{-(\lambda-\kappa)}\right.$$
$$\left. + \cdots\right) + \text{Const},$$

while for $\kappa = 2$ we have

$$(64') \quad E_\mu(\zeta - \alpha)^{(\mu+2)/2} + \cdots = d_\kappa \log f_n(\zeta) + \frac{d_\lambda}{2 - \lambda} f_n(\zeta)^{-(\lambda-\kappa)}$$
$$+ \cdots + \text{Const}.$$

For $\zeta = \alpha$ we have $f_n(\zeta) = \infty$; this is compatible with (64) and $(64')$ only if $\kappa > 2$, since only in this case does the right side remain finite. Hence

$$(62') \quad z\chi(\zeta) = c_\kappa z^{-(\kappa-1)} + c_\lambda z^{-(\lambda-1)} + \cdots \quad \text{with} \quad \kappa - 1 > 1.$$

Thus $\int z\chi(\zeta)\,dz$ remains finite at $z = \infty$ and $\int\{z\chi(\zeta) - L_n(z)\}dz$ is finite everywhere on $\Re_n$ and possesses a uniform derivative. The expression is therefore an integral of the first kind $w(z)$. Introducing the values of $\chi(\zeta)$ and $L_n(z)$, we finally obtain the following differential equation for $f_n(\zeta)$:

$$(65) \quad \frac{f_n(\zeta)}{\zeta^2 f'_n(\zeta)^2}\left(\frac{1}{\zeta^{n-1}} + \frac{2a_2}{\zeta^{n-2}} + \cdots + (n-1)a_n + \cdots + 2\bar{a}_2\zeta^{n-2} + \zeta^{n-1}\right)$$

$$+ \sum_{l=2}^{n} \frac{a_{nl}}{l-1}\, \tau'_{l-1}(f^n(\zeta)\,;f_n(0)) = w'(f_n(\zeta)).$$

From this differential equation it follows that $\Re_n$ has at most $(n-1)$ branch points in the interior of D_n which can only be situated at those $(n-1)$ values ζ in $|\zeta| < 1$, for which $\frac{1}{\zeta^{n-1}} + \frac{2a_2}{\zeta^{n-2}} + \cdots + 2\bar{a}_2\zeta^{n-2} + \zeta^{n-1} = 0$. For suppose that a branch point z_b corresponds to a point ζ_b, which is not one of the mentioned roots; then $\frac{1}{\zeta^2 f'_n(\zeta)^2}\,[\,\cdots]$ becomes infinite as $(\zeta - \zeta_b)^{-2(k-1)}$, while the other terms of (65) become infinite as $(\zeta - \zeta_b)^{-(k-1)}$ at most. Thus, the terms cannot cancel and (65) is impossible.

It is further easy to show that there are no points on $\Re_n$ exterior to D_n and that D_n is, therefore, a slit domain bounded by analytic curves. For let t be a point on $\Re_n$ exterior to D_n; choose $z_0 \cdots z_\gamma$ in the neighborhood of t and exterior to D_n. The above analysis is once more applicable, only $\chi(\zeta_0), \cdots \chi(\zeta_\gamma)$ must be omitted in all formulas. Instead of (58) we get the formula

$$(58') \quad a\sum_{i=\gamma+1}^{\gamma+p}\chi(\zeta_i)\,[\tau(z_i\,;z) - \tau(y\,;z)] + \left\{\frac{f_n(\xi)}{f_n(\xi) - a}\,[\tau(f_n(\xi)\,;z) - \tau(y\,;z)]\right\}_{n\text{-th}}$$

$$+ \frac{a_n}{a}\,[\tau(f_n(0)\,;z) - \tau(y\,;z)] = \sum_{k=1}^{\gamma}\lambda_k\frac{dw_k(z)}{dz}$$

which holds for each z in the neighborhood of t and for arbitrary a. Therefore, according to the principle of permanence, it must hold everywhere. But in the neighborhood of $\zeta = 0$, the right side of the equation remains finite, while the left becomes infinite, which shows the contradiction.

EINSTEIN INSTITUTE OF THE HEBREW UNIVERSITY,
JERUSALEM, PALESTINE.

Commentary on

[10] *Sur la variation de la fonction de Green de domaines plans quelconques*, C. R. Acad. Sci. Paris **209** (1939), 980 982.

[11] *Sur la variation du diamètre trans"ni*, Bull. Soc. Math. France **68** (1940), 158 176.

[13] *Variation of the Green function and theory of the p-valued functions*, Amer. J. Math. **65** (1943), 341 360.

Let E be a compact set in the plane with transfinite diameter $d(E) > 0$, and let Ω denote the unbounded component of its complement. Let $g(z, \zeta)$ be Green's function of Ω with pole at ζ. For any specified point $z_0 \in \Omega$, the perturbation $z^* = z + \varepsilon e^{i\theta}(z - z_0)^{-1}$ is analytic and univalent outside each neighborhood of z_0, provided that $\varepsilon > 0$ is sufficiently small. Let E^* denote the image of E, and let $g^*(z^*, \zeta^*)$ be Green's function of the corresponding domain Ω^*. In the paper [10], Schiffer introduces the important variational formula

$$g^*(z, \infty) = g(z, \infty) + \varepsilon \operatorname{Re} \left\{ e^{i\theta} \left[p'(z_0, z) p'(z_0, \infty) \right. \right.$$

$$\left. \left. -p'(z, \infty)(z - z_0)^{-1} \right] \right\} + O(\varepsilon^2), \ z \in \Omega,$$

where $p(z, \zeta)$ is an analytic function with real part $g(z, \zeta)$. The proof relies on an interesting representation of Green's function $g(z, \infty)$ in terms of the Fekete points of E, points that maximize the product used to define the n-diameter $d_n(E)$. Schiffer attributes this beautiful relation to Fekete and says that "the proof will appear soon," but in fact Fekete's paper was never published. However, the relation is implicit in a result of Leja [L], at least for connected sets E, so that Ω is simply connected and Green's function has the form $g(z, \infty) = \log |\varphi(z)|$, where φ maps Ω conformally onto $|w| > 1$ with $\varphi(\infty) = \infty$ (cf. [11]).

In any event, Schiffer returns to this topic in his paper [13] and gives a more direct derivation of the variational formula for Green's function, free of all mention of Fekete points. There the method of *interior variation*, as it is now called, is generalized and applied to extremal problems in function theory. In particular, an application

to the Bieberbach conjecture yields the same "Schiffer differential equation" for an extremal function that was derived in [6] via the *boundary variation* introduced in [5]. As Schiffer would observe later (see [17]), the method of interior variation generalizes a classical technique of normal displacement of the boundary as developed by Hadamard [H]. However, Hadamard's variation of Green's function works only for domains with smooth boundary, an a priori assumption generally unwarranted for images of extremal functions in problems of function theory. Schiffer's approach makes no such assumption, but typically deduces that an extremal function maps onto a region bounded by analytic arcs. Once that is known, a method for varying slit mappings due to Julia [J] (based on Hadamard's formula) can legitimately be applied to obtain further information.

In [13], the method of interior variation is applied not only to univalent analytic functions, but more generally to problems for p-valent functions, misleadingly called "p-valued" in the title of the paper. Following the publication of [10, 13], G. M. Goluzin produced a series of papers developing and applying another version of interior variation (see [Go] for references and further details). Schiffer's expository paper [67] gives a nice account of interior variation and its various applications.

The method of interior variation is more flexible than boundary variation and is readily adapted to variation in the presence of constraints; but for problems in univalent function theory it is restricted to mappings of a simply connected domain, which are closely related to Green's function. On the other hand, as noted in [17], other domain functionals are expressible in terms of Green's function. For instance, Szego [S] showed that the transfinite diameter of a compact set E (connected or not) has the form $d(E) = e^{-\gamma}$, where γ is the *Robin constant* of Ω, defined by $\gamma = \lim_{z \to \infty} \{ g(z, \infty) - \log |z| \}$. Here $g(z, \infty)$ is Green's function of Ω, the unbounded component of the complement of E. Thus a variation of Green's function gives a variation of the Robin constant, which in turn leads easily to a variation of the transfinite diameter.

Schiffer knew of Szego's result, which he cites on page 169 of [11]; but earlier in that paper he continues to use the variation of n-diameter from [4], deriving a variation of transfinite diameter as $n \to \infty$. With this variation, he obtains a new solution of a problem posed by Pólya [P], to describe the continuum of smallest transfinite diameter that contains n prescribed points in the plane. Pólya had solved the problem in certain special cases, and Grötzsch [Gr] had given a general solution by his method of strips, a forerunner of extremal length. Schiffer applies his variational method and appeals to the above-mentioned theorem of Fekete and Leja [L] to obtain Grötzsch's result, a differential equation for the extremal curves. Although Schiffer carries out the details only for the special case of three prescribed points, he indicates that the same method solves the general problem.

Later in [11], Schiffer gives a new proof of his theorem in [12] on the subadditivity of transfinite diameter. Specifically, that theorem asserts that if A and B are bounded connected sets in the plane with nonempty intersection, and if there exists a Jordan curve C which separates the two sets so that A lies inside C and B outside except for common points on C, then $d(A \cup B) \le d(A) + d(B)$, with strict inequality unless A and B are adjacent segments of the same line. Viewing it as an extremal problem, Schiffer now adapts his variation of Green's function [10] to the method of boundary variation [5] and invokes Szego's relation [S] to obtain the desired result on subadditivity of transfinite diameter.

References

[Go] G. M. Goluzin, *Geometric Theory of Functions of a Complex Variable*, second edition, Izdat. "Nauka", Moscow, 1966; English transl., American Mathematical Society, 1969.

[Gr] Herbert Grötzsch, *Über ein Variationsproblem der konformen Abbildung*, Berichte über die Verhandlungen der Sächsischen Akademie der Wissenschaften zu Leipzig **82** (1930), 251 263.

[H] Jacques Hadamard, *Mémoire sur le problème d analyse relatif à l équilibre des plaques élastiques encastrées*, Mémoires présentés par divers savants à l'Académie de Sciences **33** (1908), N° 4, 1 128.

[J] Gaston Julia, *Sur une équation aux dérivées fonctionnelles liée à la représentation conforme*, Ann. Sci. École Norm. Sup. (3) **39** (1922), 1 28.

[L] M. Leja, *Construction de la fonction analytique effectuant la représentation conforme d un domaine plan quelconque sur le circle*, Math. Ann. **111** (1935), 501 504.

[P] G. Pólya, *Beitrag zur Verallgemeinerung des Verzerrungssatzes auf mehrfach zusammenhängende Gebiete, III*, Sitzungsberichte der Preussischen Akademie der Wissenschaften, 1929, 55 62.

[S] G. Szego, *Bemerkungen zu einer Arbeit von Herrn M. Fekete: Über die Verteilung der Wurzeln bei gewissen algebraischen Gleichungen mit ganzzahligen Koef"zienten*, Math. Z. **21** (1924), 203 208.

PETER DUREN

[14] The span of multiply connected domains

[14] Menahem Max Schiffer, "The span of multiply connected domains," in *Duke Mathematical Journal*, Volume **10** (1943), 209–216.

THE SPAN OF MULTIPLY CONNECTED DOMAINS

By Menahem Schiffer

1. Let D_n be a domain in the z-plane, containing the point $z = \infty$ and bounded by n continua C_ν ($\nu = 1, \cdots, n$). A function $F(z)$ belongs to the family $\phi(D_n)$, if it is univalent and regular in D_n, the point $z = \infty$ excepted, at which it has the development

$$(1) \qquad F(z) = z + \frac{A_2}{z} + \frac{A_3}{z^2} + \cdots .$$

There exists a function $f(z) \subset \phi(D_n)$, mapping D_n on a domain bounded by slits parallel to the real axis. It can be characterized by the following extremal property [5]:

Among all functions $F(z) \subset \phi(D_n)$, $f(z)$ yields the maximal value of $\Re\{A_2\}$.

For the function $g(z) \subset \phi(D_n)$, mapping D_n on a domain bounded by slits parallel to the imaginary axis, an analogous result holds, namely:

Among all functions $F(z) \subset \phi(D_n)$, $g(z)$ yields the minimal value of $\Re\{A_2\}$. The functions

$$(2) \qquad f(z) = z + \frac{a_2}{z} + \frac{a_3}{z^2} + \cdots , \qquad g(z) = z + \frac{b_2}{z} + \frac{b_3}{z^2} + \cdots$$

will be called henceforth the *slit functions* of D_n.

The number

$$(3) \qquad S(D_n) = \Re\{a_2 - b_2\}$$

gives the breadth of the interval in which $\Re\{A_2\}$ varies for all functions of $\phi(D_n)$. $S(D_n)$ is a functional of D_n and will be called *the span of D_n*. The aim of this paper is to discuss some of its properties and to connect this number with other characteristics of the domain.

In this chapter some elementary theorems on $S(D_n)$ will be recalled. Let us remark first:

I. *The span is a non-increasing function of the domain.*

This theorem is obvious; for suppose $D_n \subset D'_m$, then $\phi(D'_m) \subset \phi(D_n)$; hence, the interval of variation of $\Re\{A_2\}$ is not smaller for D_n than for D'_m. This proves the assertion.

II. *The span of all domains which can be mapped on each other with the aid of univalent and normalized functions $p(z) = z + p_1 + p_2/z + \cdots$ is the same.*

Indeed, take $z = z(\zeta) = \zeta + \pi_1 + \pi_2/\zeta + \cdots$ as the univalent function mapping Δ_n in the ζ-plane on D_n in the z-plane. $F(z) \subset \phi(D_n)$ only if $F[z(\zeta)] -$

Received October 8, 1942.

209

$\pi_1 \subset \phi(\Delta_n)$. To the second coefficient A_2 of each function $F(z)$ corresponds the coefficient $A_2 + \pi_2$ of $F[z(\zeta)] - \pi_1$. Hence:

IIa. *A conformal and normalized mapping of D_n involves a translation in the space of the coefficients A_2.*

Therefore the span of these coefficients is not changed by this mapping, as was to be proved.

III. *If the domain D_n^+ is obtained from D_n by a homothetic transformation with the factor d, then*

(4)
$$S(D_n^+) = d^2 S(D_n).$$

For the slit function $f^+(z^+)$ of D_n^+ is obtained from the slit function $f(z)$ for D_n by the identity

(5)
$$f^+(z^+) = df\left(\frac{z^+}{d}\right) = z^+ + \frac{a_2 d^2}{z^+} + \cdots;$$

and in the same way we have

(5′)
$$g^+(z^+) = dg\left(\frac{z^+}{d}\right) = z^+ + \frac{b_2 d^2}{z^+} + \cdots.$$

Hence (4) follows at once.

The unit circle $U : |z| > 1$ has the slit functions

$$f(z) = z + \frac{1}{z}, \qquad g(z) = z - \frac{1}{z}.$$

Hence $S(U) = 2$. A circle C with radius d, $|z| > d$, has according to (4) the span $S(C) = 2d^2$ and an arbitrary simply connected domain with the mapping radius d also has by Theorem II the span $2d^2$.

IV. *The function*

(6)
$$H_\varphi(z) = e^{-i\varphi}[f(z) \cos \varphi + ig(z) \sin \varphi]$$

belongs for each real value of φ to $\phi(D_n)$ and maps D_n on a domain bounded by slits parallel to the vector $e^{-i\varphi}$.

Indeed, the function $h_\varphi(z) = e^{i\varphi}H_\varphi(z)$ has a simple pole at $z = \infty$ and is regular elsewhere in D_n; on the boundaries C_ν it is bounded. The integral

(7)
$$\frac{1}{2\pi i} \oint_{\Sigma C_\nu} \frac{h_\varphi'(z)}{h_\varphi(z) - \alpha} \, dz = N(\alpha) - 1$$

($N(\alpha)$ denoting the number of points in D_n, where $h_\varphi(z) = \alpha$) has the value zero at infinity. Because it is an integer, it can change only when α reaches a value ω, assumed by $h_\varphi(z)$ on C_ν. On the other hand, by definition of the slit functions, $h_\varphi(z)$ has a constant imaginary part k_ν on each boundary continuum C_ν of D_n. Hence, the integral (7) changes its value only if $\Im\{\alpha\}$ passes the value k_ν. Now each finite value in the α-plane can obviously be connected with infinity by a path avoiding all the lines $\Im\{\alpha\} = k_\nu$. If, therefore, we run from infinity along this path to the point α considered, the value of integral (7) remains always equal to zero. Hence $N(\alpha) = 1$. Thus, $h_\varphi(z)$ is univalent in D_n and has on each C_ν a constant imaginary part. Hence, $H_\varphi(z)$ has just the properties asserted in Theorem IV.

The functions $H_\varphi(z)$ can also be characterized by an extremal property: *Among all functions $F(z) \subset \phi(D_n)$, $H_\varphi(z)$ yields the maximal value of $\Re\{e^{2i\varphi}A_2\}$.* The second coefficient of $H_\varphi(z)$ is

(8) $\quad A_2(\varphi) = (a_2 \cos\varphi + ib_2 \sin\varphi)e^{-i\varphi} = \tfrac{1}{2}(a_2 + b_2) + \tfrac{1}{2}(a_2 - b_2)e^{-2i\varphi}.$

Thus, the locus of all the extremal coefficients $A_2(\varphi)$ is a circumference about $\tfrac{1}{2}(a_2 + b_2)$ with radius $\tfrac{1}{2}|a_2 - b_2|$. Since $\Re\{a_2\} = \Re\{A_2(0)\}$ and $\Re\{b_2\} = \Re\{A_2(\tfrac{1}{2}\pi)\}$ are extremal values for $\Re\{A_2(\varphi)\}$, a_2 and b_2 must obviously have equal imaginary parts, whence

(9) $\qquad\qquad\qquad a_2 - b_2 = \Re\{a_2 - b_2\} = S(D_n).$

Combining (8) with the extremal property of $H_\varphi(z)$ yields the result that outside the circumference C there are no possible values for the coefficients A_2 of $F(z) \subset \phi(D_n)$. To show that each point interior to this circumference is a possible coefficient A_2, let us suppose (without loss of generality in view of IIa) that $b_2 = 0$, that is, D_n is a domain with slits parallel to the imaginary axis. Then the coefficients $A_2(\varphi)$ cover a circumference with radius $\tfrac{1}{2}a_2 = \tfrac{1}{2}S(D_n)$, touching the imaginary axis at the origin. By enlarging D_n (preserving its connectivity), $S(D_n)$ will decrease by Theorem I, and it will do so continuously if D_n changes continuously [1]. Since all the domains so obtained remain slit domains, their corresponding circumferences touch the imaginary axis at the origin. If D_n finally becomes the z-plane n times punctured, $S(D_n) = 0$. Therefore, the intermediate circumferences fill all the interior of the original circumference and each point inside this latter is a coefficient of a function $F(z) \subset \phi(D_n)$. Summarizing we get:

V. *The coefficients A_2 of all functions $F(z) \subset \phi(D_n)$ cover exactly a circle with diameter $S(D_n)$* [3].

2. Henceforth we shall suppose that at least one of the boundary continua C_ν of D_n does not reduce to a point. The domain D_n' is called *conformly equivalent* to D_n, if D_n is mapped on D_n' by a function of the family $\phi(D_n)$. Denote the (generalized) Green's function of D_n with the logarithmic pole at infinity by $G(z)$ and let

$$\lim_{z=\infty} (G(z) - \log|z|) = \lambda.$$

Then $e^{-\lambda} = d(D_n)$, the *transfinite diameter* (a concept due to Fekete, who defined it without conformal mapping [2]) of D_n, and we derive easily from our definition

$$(10) \qquad\qquad d(D_n') = d(D_n).$$

Further let $F(D_n)$ be the inner measure of the complementary set $D_n^{\odot}$ of D_n. We have the remarkable inequality [4]

$$(11) \qquad\qquad F(D_n) \leq \pi d(D_n)^2,$$

which shows that $F(D_n)$ has a common upper bound for all domains which are conformally equivalent to a given domain D_n. There arises therefore the problem:

Given a domain D_n, to find the maximum of $F(D_n')$ for all domains D_n', conformally equivalent to D_n.

The existence of a domain D_n' for which $F(D_n')$ attains its maximal value is a consequence of the compactness of the family $\phi(D_n)$.

Suppose D_n to be a domain with $F(D_n) \geq F(D_n')$ for all D_n' conformally equivalent to D_n. Then D_n surely has exterior points. We subdivide the complement $D_n^{\odot}$ into an "areal" part A, consisting of all points exterior to the closure of D_n and of the limit points of this exterior, and into the remaining "linear" part L (if such there be). If $z_0 \subset A$, then the function

$$(12) \qquad\qquad z^* = z + \frac{a\rho^2}{z - z_0}, \qquad \rho > 0, \qquad |a| = 1$$

belongs to $\phi(D_n)$ for an arbitrary sign of a, if ρ is smaller than the distance of z_0 from D_n. It therefore maps D_n on a conformally equivalent domain D_n^*, and we shall now compute $F(D_n^*)$. For this purpose, we describe a circle K with radius $\rho^{\frac{1}{2}}$ around z_0, choosing ρ so small that this circle also lies entirely in A, and find

$$(13) \qquad\qquad F(D_n^*) = \int_{A-K} \left| 1 - \frac{a\rho^2}{(z - z_0)^2} \right|^2 d\tau + F(K^*),$$

where $F(K^*)$ denotes the area of the map of K. This latter is an ellipse with the semi-axes $\rho^{\frac{1}{2}} + \rho^{3/2}$, $\rho^{\frac{1}{2}} - \rho^{3/2}$ and area $\pi(\rho - \rho^3)$, while the area K is $\pi\rho$. Thus we have

$$(14) \qquad\qquad F(D_n^*) = F(D_n) - 2\Re\left\{ a\rho^2 \int_{A-K} \frac{d\tau}{(z - z_0)^2} \right\} + O(\rho^3)$$

with $|O(\rho^3)| < C\rho^3$. Denote

$$(15) \qquad\qquad \lim_{\rho \to 0} \int_{A-K} \frac{d\tau}{(z - z_0)^2} = \int_A \frac{d\tau}{(z - z_0)^2}$$

and get from (14), in virtue of the extremal property of $D_n : F(D_n^*) \leq F(D_n)$, by passing to the limit $\rho = 0$ the condition

(16)
$$\Re\left\{ a \int_A \frac{d\tau}{(z - z_0)^2} \right\} \geq 0.$$

This being true for each sign of a, we get the equation

(17)
$$\int_A \frac{d\tau}{(z - z_0)^2} = 0,$$

valid for each $z_0 \subset A$.

To deal with (17) which is a rather unusual type of functional equation for D_n, we introduce the function

(18)
$$U(x) = \int_A \log \frac{1}{|z - x|} \, d\tau, \qquad d\tau = d\tau_z,$$

$$z = u + iv, \qquad x = \xi + i\eta,$$

representing the potential of the surface distribution with density 1 on A. $U(x)$ is continuous all over the x-plane and so are its first partial derivatives U_ξ and U_η. For the second derivatives

$$U_{\xi\xi} = -\int_A \frac{d\tau}{|z - x|^2} + 2 \int_A \frac{(\xi - u)^2}{|z - x|^4} \, d\tau;$$

(19)
$$U_{\xi\eta} = 2 \int_A \frac{(\xi - u)(\eta - v)}{|z - x|^4} \, d\tau;$$

$$U_{\eta\eta} = -\int_A \frac{d\tau}{|z - x|^2} + 2 \int_A \frac{(\eta - v)^2}{|z - x|^4} \, d\tau$$

we have the Laplace-Poisson equation

(20)
$$\Delta U = -2\pi$$

in A and from (17)

(21)
$$U_{\xi\xi} = U_{\eta\eta}, \qquad U_{\xi\eta} = 0$$

in A. From (20) and (21) we find for $x = \xi + i\eta$ interior to each component of A

(22)
$$U(x) = -\frac{\pi}{2} (\xi^2 + \eta^2) + \alpha\xi + \beta\eta + \gamma,$$

with real constants α, β, γ; these constants, however, may be different for different components of A. The function

(23)
$$V(x) = \frac{1}{\pi} (U_\xi - iU_\eta) = \frac{1}{\pi} \int_A \frac{d\tau}{z - x}$$

has in the l-th component A_l of A the value $-x^* + c_l$ and is analytic outside A. From its continuity in the whole x-plane, we conclude that the functions

$$(24) \qquad \mathfrak{F}(x) = x - V(x), \qquad \mathfrak{G}(x) = x + V(x)$$

are analytic in the exterior of A and approach the values $2\Re\{x_0\} + c_l$, $2i\Im\{x_0\} + c_l$, as x converges to a point x_0 on the boundary of A_l, and have a simple pole at $x = \infty$. Hence, $\mathfrak{F}(x)$ and $\mathfrak{G}(x)$ are the slit functions for the exterior of A. From (24) we have

$$(25) \qquad \frac{\mathfrak{F}'(x) - \mathfrak{G}'(x)}{\mathfrak{F}'(x) + \mathfrak{G}'(x)} = -V'(x) = -\frac{1}{\pi} \int_A \frac{d\tau}{(z - x)^2}.$$

Now the left side member of (25) is invariant with respect to normalized conformal mapping of the exterior of A; we can, therefore, compute its value on the boundary of A, by mapping first the exterior of A on the exterior of analytic curves $\zeta = \zeta_l(t)$ and taking the value of the corresponding function

$$Q(\zeta) = \frac{\phi'(\zeta) - \Gamma'(\zeta)}{\phi'(\zeta) + \Gamma'(\zeta)}$$

at the corresponding point. But, by definition of the slit functions, $\phi'(\zeta)d\zeta/dt$ is real and $\Gamma'(\zeta)d\zeta/dt$ is purely imaginary on the analytic boundary of the domain considered; thus, the quotient $Q(\zeta)$ has the modulus 1, and consequently we have

$$(26) \qquad \left| \frac{\mathfrak{F}'(x) - \mathfrak{G}'(x)}{\mathfrak{F}'(x) + \mathfrak{G}'(x)} \right| = 1 \qquad \text{on the boundary of } A;$$

hence, by the principle of the maximum, we get

$$(26') \qquad \left| \int_A \frac{d\tau}{(z - x)^2} \right| \leq \pi \qquad \text{for all } x \text{ exterior to } A.$$

We now turn to the linear part L of $D_n^\odot$. Let z_0 be a point of L; we take a continuum C_ρ of L containing z_0, such that its exterior can be mapped univalently on the domain $|z^* - z_0| > \rho$ by a mapping function

$$(27) \qquad z^* = z + k\rho + \frac{a\rho^2}{(z - z_0)} + \frac{b\rho^3}{(z - z_0)^2} + \frac{c\rho^4}{(z - z_0)^3} + \frac{d\rho^5}{(z - z_0)^4} + \cdots.$$

It is known [6] that $|a| \leq 1, |b| \leq 4^3, |c| \leq 4^4, |d| \leq 4^5, \cdots$. We choose now another number $0 < \epsilon < 1$ and form the function

$$(28) \qquad z^{**} = z^* + \frac{\epsilon e^{i\varphi}\rho^2}{z^* - z_0}$$

which maps the domain $|z^* - z_0| > \rho$ univalently on the exterior of an ellipse E with semi-axes $\rho(1 + \epsilon)$, $\rho(1 - \epsilon)$ and area $\pi\rho^2(1 - \epsilon^2)$. Comparing (27) and (28) yields

$$(29) \qquad z^{**} = z + k\rho + \frac{(a + \epsilon e^{i\varphi})\rho^2}{z - z_0} + O(\rho^3);$$

this function maps D_n on a domain D_n^{**} with the area (see (14))

$$(30) \quad F(D_n^{**}) = \pi \rho^2 (1 - \epsilon^2) + F(D_n) - 2\Re\left\{(a + \epsilon e^{i\varphi})\rho^2 \int_A \frac{d\tau}{(z - z_0)^2}\right\} + O(\rho^3).$$

The first term on the right side comes from the area of the ellipse E. In view of the extremal property of D_n, we have $F(D_n^{**}) \leq F(D_n)$; hence, from $\rho \to 0$, we obtain

$$(31) \quad \pi(1 - \epsilon^2) \leq 2\Re\left\{(a + \epsilon e^{i\varphi}) \int_A \frac{d\tau}{(z - z_0)^2}\right\} \leq 2 \mid a + \epsilon e^{i\varphi} \mid \cdot \left| \int_A \frac{d\tau}{(z - z_0)^2} \right|.$$

We now choose φ in such a way that $\operatorname{sgn}(-a) = e^{i\varphi}$; since then $\mid a + \epsilon e^{i\varphi} \mid \leq 1 - \epsilon$, we get

$$(32) \quad \pi(1 + \epsilon) \leq 2 \left| \int_A \frac{d\tau}{(z - z_0)^2} \right| \qquad (0 < \epsilon < 1).$$

Hence

$$(32') \quad \pi \leq \left| \int_A \frac{d\tau}{(z - z_0)^2} \right| \qquad (z_0 \subset L).$$

But this combined with (26') (applicable to all points of L) yields

$$\left| \int_A \frac{d\tau}{(z - z_0)^2} \right| = \pi$$

on L and consequently (by the principle of the maximum) everywhere outside A, including the point at infinity. But at this point the integral considered vanishes. Thus the assumption of the existence of L leads to a contradiction; D_n^0 coincides therefore with A, and $\mathfrak{F}(x)$ and $\mathfrak{G}(x)$ are identical with $f(x)$ and $g(x)$.

We summarize:

The extremal domain D_n has a complement consisting of n domains; its slit functions satisfy the equations

$$(33) \quad f'(x) + g'(x) = 2; \qquad f'(x) - g'(x) = -\frac{2}{\pi} \int_{D_n^*} \frac{d\tau}{(z - x)^2}.$$

Comparing the coefficients of x^{-2} on both sides of the last equality, we find

$$(34) \quad S(D_n) = \frac{2}{\pi} F(D_n).$$

Now $S(D_n)$ is, by Theorem II, the same for all conformally equivalent domains; $F(D_n)$, on the other hand, furnishes the maximal value within this family of domains. Hence, we proved for an arbitrary domain D_n

$$(35) \quad F(D_n) \leq \frac{\pi}{2} S(D_n).$$

Thus we have obtained for the span a new definition:

The span of a domain is $2/\pi$ times the maximal area, the complement of which is conformally equivalent to this domain.

By integrating the first equality (33) we get $\frac{1}{2}[f(x) + g(x)] = x$ in the case of an *extremal* domain D_n. If D_n is an arbitrary domain of the z-plane, let $x = x(z)$ be the function of $\phi(D_n)$ mapping it on an extremal domain in the x-plane with the slit functions $f(x)$ and $g(x)$. Then we have $\frac{1}{2}[f(x(z)) + g(x(z))] = x(z)$, but $f(x(z))$ and $g(x(z))$ are the slit functions of D_n. Hence:

In each domain D_n, the arithmetic mean of the slit functions belongs to $\phi(D_n)$ and renders maximum the area of the complements of all maps, obtained by functions of $\phi(D_n)$.

From this theorem it can easily be shown *that an extremal domain is always bounded by convex curves.*

Finally we apply the inequality (11) to an *extremal* domain D_n; using (34), we get in this particular case,

$$(36) \qquad S(D_n) \leq 2d(D_n)^2.$$

Since both members of the inequality are invariants for conformal normalized mapping, the inequality holds for *each* domain D_n. Equality occurs in (36) only if for the corresponding extremal domain $F(D_n) = \pi d(D_n)^2$ holds. But this equality is known to be valid only in the case of a circle; hence, we have equality in (36) only in the case of simply connected domains, as already pointed out in §1. Thus we get for $S(D_n)$ the double estimate

$$(37) \qquad \frac{2}{\pi} F(D_n) \leq S(D_n) \leq 2d(D_n)^2.$$

BIBLIOGRAPHY

1. C. CARATHÉODORY, *Untersuchungen über die konformen Abbildungen, von festen und veränderlichen Gebieten*, Mathematische Annalen, vol. 72(1912), pp. 107–144.
2. M. FEKETE, *Ueber die Verteilung der Wurzeln bei gewissen algebraischen Gleichungen mit ganzzahligen Koeffizienten*, Mathematische Zeitschrift, vol. 17(1923), pp. 228–249.
3. H. GRÖTZSCH, *Ueber das Parallelschlitztheorem der konformen Abbildung schlichter Bereiche*, Berichte Leipzig, vol. 84(1932), pp. 15–36.
4. G. PÓLYA, *Beitrag zur Verallgemeinerung des Verzerrungssatzes auf mehrfach zusammenhängende Gebiete*, 11, Sitzungsberichte Akad. Berlin, 1928, pp. 280–282.
5. R. DE POSSEL, *Zum Parallelschlitztheorem unendlich-vielfach zusammenhängender Gebiete*, Göttinger Nachrichten, 1931, pp. 199–202.
6. M. SCHIFFER, *A method of variation within the family of simple functions*, Proceedings of the London Mathematical Society, (2), vol. 44(1938), pp. 432–449.

HEBREW UNIVERSITY, JERUSALEM.

Commentary on

[14] *The span of multiply connected domains*, Duke Math. J. **10** (1943), 209 216.

In this seminal paper, Schiffer introduced an important conformal invariant for multiply connected plane domains, the span. The definition is in terms of conformal mappings onto canonical slit domains, and there is a complementary characterization (almost literally), in terms of an extremal problem for area under a conformal mapping. While the foundational results on conformal mapping of multiply connected domains concern the existence and uniqueness of mappings onto various canonical domains, Schiffer's paper is an early example of *using* the mappings as a tool in geometric function theory. Moreover, his study of the extremal problem is clearly influenced by his development of variational methods.

Let D be a finitely connected domain containing ∞ in its interior and consider univalent functions F in D with

$$F(z) = z + \frac{A}{z} + \frac{B}{z^2} + \cdots \qquad (1)$$

near ∞. Let $p(z) = z + a/z + \cdots$ and $q(z) = z + b/z + \cdots$ be the unique conformal mappings of D onto a domain bounded by horizontal slits and vertical slits, respectively. Then

$$\operatorname{Re} b \leq \operatorname{Re} A \leq \operatorname{Re} a.$$

Thus $\operatorname{Re} A$ can vary in an interval of length $\operatorname{Re}\{a - b\}$, and Schiffer calls this length the *span* of D. It is the same for any domain that is the image of D under a mapping of the form (1), so in that sense it is a conformal invariant of D. Schiffer proves, among other basic results, that the coefficients A in (1) cover a disk with diameter equal to the span of D.

Now consider conformal mappings of D by functions as in (1) and ask: How large can the area of the complement be? The answer is $\pi/2$ times the span of D. Also, $\Phi(z) = (1/2)(p(z) + q(z))$ is the conformal mapping of D onto the extremal domain, and $\Phi(D)$ is bounded by convex curves. Schiffer's proof is variational, in the course of which he derives some striking identities for the

slit mappings of the extremal domain; see [K] for later work on these. Apparently unbeknownst to Schiffer at the time, the convexity property of Φ had already been discovered by Grunsky in his dissertation [G]. But convexity of the boundary curves does not guarantee univalence, a key property that was shown by Schiffer as part of his analysis. This point is also discussed in [34] as an independent observation and without variational methods.

An alternate approach to the area problem, once one knows that Φ is univalent and turns out to be the extremal mapping, uses an expression for Φ in terms of kernel functions and harmonic measures, no less striking, and may be found in [34]. Between [14] and [34], Schiffer revisited the area problem in [22] as an application of orthonormal families to conformal mappings, then quite new. The correct upper bound for the omitted area emerges easily, but the analysis of Φ as the extremal mapping is troublesome, and the full range of representations via kernel functions was not yet realized. Nevertheless, this was a new take on such problems and was influential in subsequent papers. For example, similar area problems were considered by Garabedian and Schiffer in [26], no doubt motivated by [14] and the ideas in [22]. See [N] as well for a compact exposition. Kühnau [Ku] introduced a notion of the span, together with an associated area problem, using quasiconformal mappings onto (inclined) parallel slit domains. He assumed that the complex dilatation of the mapping is identically zero near ∞, thus allowing for a local expansion analogous to (1).

In [AB], analytic and geometric characterizations of the span were reconsidered by Ahlfors and Beurling as examples of their general approach to defining conformal invariants and associated null-sets. Briefly, they define (relative) conformal invariants by forming $M_{\mathfrak{F}}(z_0, D) = \sup_{\mathfrak{F}} |f'(z_0)|$, where z_0 is a fixed point in D and f varies in a class $\mathfrak{F}(D)$ of analytic functions in D that is invariant under conformal mappings of D.[1]

---

[1] In [AB] most results are given for z_0 a finite point in D, unlike Schiffer's original normalization, but this is not essential (nor was it essential for Schiffer to use $z_0 = \infty$).

If $M_{\mathfrak{F}}(z_0, D) = 0$ (which generally implies that it vanishes identically), then the complement of D is a null-set for the class $\mathfrak{F}(D)$.

For an analytic approach to the span, consider the class $\mathfrak{D}(D)$ of analytic functions with a fixed bound on the Dirichlet integral, specifically

$$\iint_D |f'(z)|^2 \, dxdy \leq \pi.$$

Then $M_{\mathfrak{D}}(z_0, D)^2 = (1/2)\text{span}(D)$, and the extremal is given in terms of the slit mappings as $(p - q)/\sqrt{2\,\text{span}(D)}$. For a geometric characterization, the authors use omitted area to de"ne a class of competing functions. Namely, let $\mathfrak{S}\mathfrak{E}(D)$ be the set of univalent functions in D such that $1/(f(z) - f(z_0))$ omits a set of area at least π. Then one has again

$$M_{\mathfrak{S}\mathfrak{E}}(z_0, D)^2 = (1/2)\text{span}(D),$$

and the extremal is $(p + q)/\sqrt{2\,\text{span}(D)}$; the paper includes a proof that $p + q$ is univalent.

Since $M_{\mathfrak{D}}(z_0, D) = M_{\mathfrak{S}\mathfrak{E}}(z_0, D)$, these results serve to describe the identical classes of null-sets for $\mathfrak{D}(D)$ and $\mathfrak{S}\mathfrak{E}(D)$. This direction of research demonstrates the lasting influence of the span in studying small boundaries and removable point sets for classes of analytic functions on plane domains, with similar applications to Riemann surfaces. The latter is tied up with the classification problem for Riemann surfaces, for example, assessing the size of the boundary for the pur-

pose of supporting nonconstant analytic or harmonic functions with a finite Dirichlet integral, necessary for existence theorems. Schiffer himself contributed to this in [85], 22 years after [14]. For a technical discussion see [AS, RS, SN, SO]. For an informal and personal account, see [A].

References

[A]   Lars V. Ahlfors, *Riemann surfaces and small point sets*, Ann. Acad. Sci. Fenn. Ser. A I Math. **7** (1982), 49 57.

[AB] Lars Ahlfors and Arne Beurling, *Conformal invariants and function theoretic null-sets*, Acta Math. **83** (1950), 101 129.

[AS] L. Ahlfors and L. Sario, *Riemann Surfaces*, Princeton University Press, 1960.

[G]   Helmut Grunsky, *Neue Abschätzungen zur konformen Abbildung ein- und mehrfach zusammenhängender Bereiche*, Schriften Math. Sem. Inst. Angew. Math. Univ. Berlin **1** (1932), 95 140.

[K]   Yukio Kusunoki, *A new proof of Schiffer s identities on planar Riemann surfaces*, Proc. Japan Acad. Ser. A Math. Sci. **60** (1984), 345 348

[Ku] Reiner Kühnau, *Die Spanne von Gebieten bei quasikonformen Abbildungen*, Arch. Rational Mech. Anal. **65** (1977), 299 303.

[N]   Zeev Nehari, *Conformal Mapping*, McGraw-Hill, 1952; Dover edition, 1975.

[RS] Burton Rodin and Leo Sario, *Principal Functions*, D. Van Nostrand Co., 1968

[SN] L. Sario and M. Nakai, *Classi"cation Theory of Riemann Surfaces*, Springer-Verlag, 1970.

[SO] L. Sario and K. Oikawa, *Capacity Functions*, Springer-Verlag, 1969.

BRAD OSGOOD

[16] Sur l'équation différentielle de M. Löwner

[16] Sur l'équation différentielle de M. Löwner. *C. R. Acad. Sci. Paris* **221** (1945), 369–371.

Considérons maintenant une transformation *quelconque* transformant conformément (Γ) en (C). Les points $\alpha_1, \alpha_2, \alpha_3, \alpha_4$ de Γ′ auront pour images A_1, B_1, C_1, D_1. On démontre que $(A_1, B_1, C_1, D_1) = (ABCD)$. Par une projection stéréographique ou passe au lemme de M. Kravtchenko.

Remarques. — $\delta = (1/\rho^2) - (1/r^2)$ est positif pour $0 < r < (1/2)$; dans ces conditions

$$\Lambda^2 \leqq \frac{\pi\sigma}{\log \dfrac{1}{r}}.$$

Par contre $\hat{\delta} = (1/\rho^2) - (1/r^3)$ est toujours négatif.

THÉORIE DES FONCTIONS. — *Sur l'équation différentielle de M. Löwner.*
Note de M. MENAHEM SCHIFFER, présentée par M. Paul Montel.

Considérons dans le plan w une courbe $\mathcal{C}$ admettant la représentation paramétrique $w = w(\tau)$, $0 \leqq \tau \leqq \infty$, $w(\tau) \neq 0$, $w(\infty) = \infty$, où $w(\tau)$ dépend de τ de façon continue. Désignons par $\mathcal{C}(t)$ la courbe ayant la représentation paramétrique $w(\tau)$, $t \leqq \tau \leqq \infty$, et soit

$$f(z; t) = \gamma(t)[z + a_2(t)z^2 + \dots]$$

la fonction représentant le cercle unité $|z| < 1$ sur le plan w, muni de la fente $\mathcal{C}(t)$. D'après K. Löwner ([1]), $f(z, t)$ est dérivable par rapport à t et, pour τ choisi convenablement, on a

$$(1) \qquad \frac{\partial}{\partial t} f(z; t) = z \frac{1 + xz}{1 - xz} \frac{\partial}{\partial z} f(z; t) \qquad |x(t)| = 1,$$

avec une fonction $x = x(t)$ de module 1, continue en t.

$x(t)$ est une fonction de la ligne $\mathcal{C}$; sa détermination pour une courbe $\mathcal{C}$ donnée est un problème difficile. Il nous paraît donc intéressant de pouvoir caractériser cette fonction pour une classe assez étendue de courbes $\mathcal{C}$. Soit $P_n(x)$ un polynome en x de degré n. Considérons une courbe $\mathcal{C}$ avec une représentation paramétrique $w(t)$ satisfaisant à l'équation différentielle

$$(2) \qquad \mathcal{I}\big\{ w'(t)^2 \, w(t)^{-3} P_n[w(t)^{-1}] \big\} = 0.$$

Des courbes de ce type jouent un rôle important dans le problème des coefficients des fonctions univalentes ([2]). Pour cette classe de courbes $\mathcal{C}$, on peut énoncer une propriété importante de $x(t)$.

Soit $f(z; t)$ la fonction univalente adjointe à $\mathcal{C}(t)$; on a évidemment, pour $|z| = 1$,

$$(3) \qquad \mathcal{I}\left\{ \left(z \frac{\partial}{\partial z} f(z; t) \right)^2 f(z; t)^{-3} P_n[f(z; t)^{-1}] \right\} = 0.$$

([1]) *Math. Annalen*, 89, 1923, pp. 103-121.
([2]) M. SCHIFFER, *Proc. London Math. Society*, (2), 44, 1938, pp. 432-449.

25.

La fonction $f(z; t)$ est régulière sur la frontière du cercle-unité à l'exception d'un pôle double, correspondant au point à l'infini sur $\mathcal{C}(t)$. Mais le terme $[(\partial/\partial z) f]^2 . f^{-3}$ est régulier même en ce point; donc, on tire de (3), à l'aide du principe d'inversion de Schwarz, l'identité

$$(4) \qquad \left(z \frac{\partial}{\partial z} f(z; t) \right)^2 f(z; t)^{-3} \mathrm{P}_n [f(z; t)^{-1}] = q_{n+1}(z; t) = \sum_{\rho=-(n+1)}^{(n+1)} \mathrm{A}_\rho(t) z^\rho,$$

avec

$$\mathrm{A}_{-\rho}(t) = \overline{\mathrm{A}_\rho(t)}.$$

Pour t fixe, on a

$$(4') \qquad \int_0 \sqrt{q_{n+1}(z; t)} \frac{dz}{z} = \int_0^{f(z; t)} \sqrt{\mathrm{P}_n\left(\frac{1}{x}\right) \frac{1}{x^2}} \, dx = \mathcal{J}\{ f(z; t) \}.$$

Donc $\mathcal{J}\{ f(z; t) \}$ satisfait à la même équation différentielle partielle (1) que $f(z; t)$ et, en introduisant $q_{n+1}(z; t)$ à l'aide de $(4')$, on obtient après des transformations élémentaires l'équation différentielle

$$(5) \qquad \frac{\partial}{\partial t} q_{n+1}(z; t) = z \frac{1+xz}{1-xz} \frac{\partial}{\partial z} q_{n+1}(z; t) + \frac{4xz}{(1-xz)^2} q_{n+1}(z; t).$$

En comparant le coefficient de z^ν dans les deux membres de l'équation, on obtient pour les $\mathrm{A}_\nu(t)$ l'équation différentielle

$$(6) \quad \mathrm{A}'_\nu(t) = \nu \mathrm{A}_\nu(t) + 2 \sum_{\rho=-(n+1)}^{\nu-1} (2\nu - \rho) \mathrm{A}_\rho(t) x(t)^{\nu-\rho} \qquad [\nu = -(n+1), \ldots, (n+1)].$$

Dans l'équation (5), il n'y a pas de termes en z^{n+2} ou z^{n+3}; donc nécessairement

$$(7) \qquad (2n+4) \sum_{\rho=-(n+1)}^{n+1} \mathrm{A}_\rho x^{n+2-\rho} - \sum_{\rho=-(n+1)}^{n+1} \rho \mathrm{A}_\rho x^{n+2-\rho} = 0,$$

$$(7') \qquad (2n+6) \sum_{\rho=-(n+1)}^{n+1} \mathrm{A}_\rho x^{n+3-\rho} - \sum_{\rho=-(n+1)}^{n+1} \rho \mathrm{A}_\rho x^{n+3-\rho} = 0,$$

ce qui entraîne les équations

$$(8) \qquad \sum_{\rho=-(n+1)}^{n+1} \mathrm{A}_\rho x^{-\rho} = 0, \qquad\qquad (8') \qquad \sum_{\rho=-(n+1)}^{n+1} \rho \mathrm{A}_\rho x^{-\rho} = 0.$$

On voit aisément que les équations (8) et $(8')$ sont nécessaires et suffisantes pour que $q_{n+1}(z; t)$ ait la forme demandée. D'ailleurs, $(8')$ est une conséquence de (8) et des équations différentielles (6) pour les $\mathrm{A}_\nu(t)$.

Les équations (8) et $(8')$ sont homogènes et linéaires par rapport aux $\mathcal{A}_\nu(t)$. En dérivant par rapport à t et en éliminant les $\mathrm{A}'_\nu(t)$ à l'aide des formules (6), on obtient de nouvelles expressions homogènes et linéaires par rapport aux $\mathrm{A}_\nu,$

avec des coefficients dépendant de $\varkappa(t)$ et de $\varkappa'(t)$. Par différentiations successives, on obtient finalement $(2n+3)$ équations linéaires pour les $(2n+3)$ termes A'_ν, avec des coefficient dépendant de $\varkappa(t)$ et de ses $(2n+1)$ premières dérivées. Il faut donc que le déterminant de ce système d'équations linéaires et homogènes soit égal à zéro, ce qui donne une équation différentielle à coefficients constants pour $\varkappa(t)$, d'ordre $(2n+1)$ au plus. Donc, dans le cas d'une fente (2), la fonction $\varkappa(t)$ satisfait à une équation différentielle et le problème de la représentation conforme du plan muni de cette fente sur le cercle-unité se réduit à l'intégration d'un système d'équations différentielles ordinaires.

Pour $n = 1$ par exemple, on établit pour $\varkappa(t)$ l'équation algébrique

$$(9) \qquad 2\mathscr{C} e^{-2t}\varkappa(t)^2 + \mathscr{O} e^{-t}\varkappa(t) - \overline{\mathscr{O}} e^{-t}\varkappa(t)^{-1} - 2\overline{\mathscr{C}} e^{-2t}\varkappa(t)^{-2} = 0,$$

avec les constantes $\mathscr{C}$ et $\mathscr{O}$.

THÉORIE DES FONCTIONS. — *Sur les suites de fractions rationnelles à zéros et pôles réels.* Note de M. **Hubert Delange**, présentée par M. Paul Montel.

Soit une suite de fractions rationnelles $R_1(z)$, $R_2(z)$, ..., $R_n(z)$, ... ayant chacune tous ses zéros et tous ses pôles réels et soit $\varphi(n)$ une fonction positive de l'entier n.

I. Supposons d'abord tous les zéros et tous les pôles de $R_n(z)$ compris entre deux nombres fixes a et b. Appelons p_n et q_n les degrés du numérateur et du dénominateur de $R_n(x)$, A_n et B_n les coefficients de z^{p_n} et z^{q_n} au numérateur et au dénominateur, et $\nu_n(x)$ le nombre de zéros de $R_n(z)$ *au plus égaux à* x diminué du nombre de pôles $\leq x$.

Nous considérerons les deux cas suivants :

a. Le rapport $[\nu_n(x)]/[\varphi(n)]$ reste borné indépendamment de x et de n.

b. On a, quel que soit n et quel que soit x, $\nu_n(x) \geq 0$.

Dans le cas a, pour que la suite des fonctions $1/[\varphi(n)]\log|R_n(z)|$ soit convergente en dehors du segment $[a, b]$, il faut et il suffit que :

1° la suite des fonctions M_n définies par $M_n(t) = \dfrac{1}{\varphi(n)}\displaystyle\int_a^t \nu_n(u)\,du$ soit convergente sur le segment $[a, b]$ vers une fonction limite M;

2° $(p_n - q_n)/[\varphi(n)]$ ait une limite finie h;

3° $1/[\varphi(n)]\log|A_n/B_n|$ ait une limite finie λ.

Pour le cas b, il faut remplacer dans (1) *convergente* par *vaguement convergente* (¹).

Dans les deux cas, la convergence de $1/[\varphi(n)]\log|R_n(z)|$ dans un domaine D

(¹) Étant donnée une suite de fonctions croissantes ψ_n définies sur un certain intervalle, fini ou non, nous disons qu'elle est vaguement convergente sur cet intervalle vers une fonction croissante ψ si $\psi_n(t)$ tend vers $\psi(t)$ pour toutes les valeurs de t pour lesquelles ψ est continue et aux extrémités de l'intervalle lorsque celles-ci sont finies.

Commentary on

[16] *Sur l équation différentielle de M. Löwner,* C. R. Acad. Sci. Paris **221** (1945), 369 371.

Twenty years after its appearance in 1923, Loewner's method was well established as a powerful device for solving extremal problems. For instance, Grunsky had applied it in 1934 to find the radius of starlikeness for univalent functions of class S. Goluzin had applied it in 1936 to obtain the sharp rotation theorem. Loewner [L] had originally used his method to prove $|a_3| \leq 3$. (See [G] or [D] for further information.) In his short note [16] of 1945, Schiffer developed a clever idea for combining Loewner's method with the variational method in order to mount a stronger attack on the Bieberbach conjecture. Schiffer had applied his method of boundary variation [5] to show in [6] that an extremal function for the coefficient problem $\max_{f \in S} \text{Re}\{a_n\}$ maps the unit disk $\mathbb{D}$ conformally onto the whole plane minus a finite number of unbranched analytic arcs that are trajectories of a quadratic differential: $(1/w^3)P_n(1/w) dw^2 < 0$, where P_n is a certain polynomial of degree $n - 2$. (See also [D]. Observe that our present notation differs from that in Schiffer's paper, where P_n denotes a polynomial of degree n.) In particular, if $w = w(s)$ is a sufficiently smooth (local) parametric representation of an omitted arc, then

$$\text{Im}\left\{ w'(s)^2 w(s)^{-3} P_n(1/w(s)) \right\} = 0. \quad (1)$$

On the other hand, any slit mapping can be described by Loewner's theory. Let C be a Jordan arc in the plane with a continuous parametrization $w = w(\tau)$, $0 \leq \tau \leq \infty$, where $w(\tau) \neq 0$ and $w(\infty) = \infty$. Let $C(t)$ be the portion of C corresponding to $t \leq \tau < \infty$, and let $f(z,t)$ denote the conformal mapping of $\mathbb{D}$ onto the complement of $C(t)$ with standard normalization $f(0,t) = 0$ and $\frac{\partial f}{\partial z}(0,t) > 0$. According to Loewner's theory, with suitable choice of parameter, this function $f(z,t)$ satisfies a differential equation

$$\frac{\partial f}{\partial t}(z,t) = z \frac{1 + \kappa(t)z}{1 - \kappa(t)z} \frac{\partial f}{\partial z}(z,t) \quad (2)$$

for some continuous function $\kappa(t)$ of modulus $|\kappa(t)| = 1$. As Schiffer remarks, the determination of Loewner's function $\kappa(t)$ is a difficult problem. His purpose in [16] is to obtain information about this function for a "rather large class of curves," namely those with the property (1) for some polynomial P_n. Noting that $w = w(\theta) = f(e^{i\theta}, t)$ parametrizes the curve $C(t)$, so that

$$w'(\theta) = ie^{i\theta} \frac{\partial f}{\partial z}(e^{i\theta}, t) = iz \frac{\partial f}{\partial z}(z,t) \quad \text{for } |z| = 1,$$

he infers from (1) that

$$\text{Im}\left\{ z^2 \left[\frac{\partial f}{\partial z}(z,t) \right]^2 \frac{P_n(1/f(z,t))}{f(z,t)^3} \right\} = 0 \quad (3)$$

for $|z| = 1$. Since the expression in (3) is real-valued on the unit circle, the Schwarz reflection principle shows that it is a rational function with special structure. Combining this structural formula with (2), Schiffer derives a differential equation for $\kappa(t)$.

Despite the novelty and interest of Schiffer's work, the results in [16] do not appear to shed much light on the coefficient problem. (See, however, the commentary on [28].) In later investigations, for instance, in Schiffer's monumental paper [60] with Garabedian on the Bieberbach conjecture for the fourth coefficient, Loewner's method is combined with the variational method in more effective ways to solve extremal problems for classes of univalent functions.

References

[D] Peter L. Duren, *Univalent Functions*, Springer-Verlag, 1983.

[G] G. M. Goluzin, *Geometric Theory of Functions of a Complex Variable*, second edition, Izdat. "Nauka", Moscow 1966; English transl., American Mathematical Society, 1969.

[L] Karl Löwner (Charles Loewner), *Untersuchungen über schlichte konforme Abbildungen des Einheitskreises, I*, Math. Ann. **89** (1923), 103 121.

PETER DUREN

151

[17] Hadamard's formula and variation of domain-functions

[17] Schiffer, Menahem. "Hadamard's formula and variation of domain-functions." *American Journal of Mathematics* **68:3** (1946), 417–448.

HADAMARD'S FORMULA AND VARIATION OF DOMAIN-FUNCTIONS.*

By Menahem Schiffer.

1. Hadamard's differential equation for Green's function.

1. The present paper deals with applications of functional analysis to the theory of the logarithmic potential and analytic functions. We consider domains D in the complex x-plane bounded by a finite number of proper continua $C_\nu(\nu = 1, \cdots, n)$. If x and y are points in D, Green's function $g(x;y)$ of D is defined in the following way:

a. $g(x;y)$ is a harmonic function of x throughout D, the point $x = y$ excepted; there however $g(x;y) + \log |x - y|$ is harmonic.

b. If x converges to a boundary continuum C_ν, $g(x;y)$ converges to zero.

The existence of $g(x;y)$ is assured for every domain D; $g(x;y)$ is harmonic in y also and satisfies the symmetry condition $g(x;y) = g(y;x)$. Its importance for Dirichlet's problem is well known. If D is simply connected, $g(x;y)$ is related also to the problem of mapping D onto the exterior E of the unit circle. For, let $p(x;y)$ be an analytic function of x in D, such that $g(x;y) = R\{p(x;y)\}$; then $\phi(x;y) = \exp\{p(x;y)\}$ maps D conformally on E so that the point y goes into infinity.

Using the theory of orthogonal functions we may determine the Green's function in the form of an infinite series for any domain whose boundary satisfies certain general conditions. (See Bergman [1] 1, 2.) On the other hand consideration of the rôle of Green's formula in the theory of functions and potential theory indicates the desirability of obtaining various representations of it and in particular those which show the dependence of this formula on the domain in which it is defined.

There are however only a few elementary domains D with an explicitly known Green's function. Nevertheless it is possible to calculate Green's function (approximately) for domains which are sufficiently near such elementary domains. This is done by means of Hadamard's well known variation formula (see Hadamard 2) which is applicable to every domain D bounded by analytic

---

* Received November 15, 1945.

[1] See Bibliography, p. 448.

curves C_ν. Let every point of the boundary $C = \sum_{\nu=1}^{n} C_\nu$ be defined by a parameter s which measures the lengths of the curves successively and runs, therefore, from 0 to l ($l =$ sum of the lengths of all C_ν). Let $\delta n(s) = \epsilon \nu(s)$ be a
continuous function of s which determines the normal displacement of each
boundary point $z(s)$ of the original domain D. $\delta n(s)$ is taken as positive if
the displacement is in the direction of the outer normal with respect to D.
In this way, we define a new domain D^* with a new Green's function $g^*(x;y)$.
According to Hadamard, we have

(1) $$g^*(x;y) = g(x;y) + \frac{\epsilon}{2\pi} \int_C \frac{\partial g(z;x)}{\partial n_z} \frac{\partial g(z;y)}{\partial n_z} \nu(s_z) ds_z + o(\epsilon).$$

Here, $\frac{\partial}{\partial n_z}[g(z;x)]$ denotes the derivative of $g(z;x)$ in the direction of the
outward normal and $o(\epsilon)$, as usual, a term satisfying the condition
$\lim_{\epsilon=0} \frac{1}{\epsilon} o(\epsilon) = 0$. Using the notations of functional calculus (Volterra 1), we
may give to (1) the form

(1') $$\delta g(x;y) = \frac{1}{2\pi} \int_C \frac{\partial g(z;x)}{\partial n} \frac{\partial g(z;y)}{\partial n} \delta n \, ds.$$

Many properties of $g(x;y)$ and the univalent mapping functions connected with it may be derived from (1') (see Lévy, Julia, Biernacki). This
formula, however, loses its meaning if the initial domain D is not bounded
by smooth curves; on the other hand, in this case a well defined Green's
function also exists. Thus (1') is not applicable to extremum problems concerning Green's function; for one can not be sure that the domain D, belonging
to the extremal function, satisfies the suppositions of Hadamard's formula.

2. We shall now transform formula (1') into such a form that it may
be applied to the most general domain D. We use the artifice of specializing
the variation δn and, by means of partial integration, expressing $\delta g(x;y)$
by values of $g(x;y)$ and its derivatives from the *interior* of D. For this
purpose, we choose a fixed point z_0 in D and consider the representation

(2) $$x^* = x + \frac{e^{2i\phi}\rho^2}{x - z_0}, \quad \rho > 0, \qquad\qquad 0 \leq \phi < 2\pi$$

of the x-plane. It transforms the circumference $|x - z_0| = \rho$ into the seg-
ment $< -2\rho e^{i\phi}, + 2\rho e^{i\phi} >$ and is univalent in its exterior $|x - z_0| > \rho$.
For ρ sufficiently small, this representation is univalent on all curves C_ν and
transforms them in a one-to-one manner into neighboring curves C^*_ν which

enclose a new domain D^* of the x-plane. Let $g^*(x; y)$ denote the Green's function of D^*; we shall compute it from $g(x; y)$ by aid of (1').

We remark that for $z \subset C_v$, $y \subset D$

$$(3) \qquad g(z^*; y) = \frac{\partial g(z; y)}{\partial n} \delta n + o(\rho^2);$$

here δn is the normal shift of C_v at the point z, caused by the variation (2). δn is of order ρ^2. Hence (1) has the form

$$(4) \qquad g^*(x; y) = g(x; y) + \frac{1}{2\pi} \int_C \frac{\partial g(z; x)}{\partial n} g(z^*(z); y) ds + o(\rho^2).$$

Let y_1 be the point which is transformed by (2) into y; since $g(z; y_1)$ vanishes on the boundary of D, we may write instead of (4)

$$(4') \qquad \delta g(x; y) = \frac{1}{2\pi} \int_C \frac{\partial g(z; x)}{\partial n} [g(z^*; y) - g(z; y_1)] ds + o(\rho^2).$$

Now, $g(z^*; y) - g(z; y_1)$ is harmonic in the domain D_0, obtained from D by discarding the interior of the circumference $K_\rho \equiv (|x - z_0| = \rho)$. Hence, Green's formula may be appplied to D_0, yielding

$$(5) \qquad \delta g(x; y) = \frac{1}{2\pi} \int_{K_\rho} \left\{ \frac{\partial g(z; x)}{\partial n} [g(z^*; y) - g(z; y_1)] \right.$$
$$\left. - g(z; x) \frac{\partial}{\partial n} [g(z^*; y) - g(z; y_1)] \right\} ds + g(x; y_1) - g(x^*; y) + o(\rho^2)$$

where the normal derivative is to be taken in the direction of increasing ρ. We suppose x, y, y_1 to lie in the exterior of K_ρ and get, therefore, from Green's formula

$$(5') \qquad \frac{1}{2\pi} \int_{K_\rho} \left\{ \frac{\partial g(z; x)}{\partial n} g(z; y_1) - g(z; x) \frac{\partial g(z; y_1)}{\partial n} \right\} ds = 0.$$

Thus, there remains

$$(6) \qquad \delta g(x; y) = \frac{1}{2\pi} \int_{K_\rho} \left\{ \frac{\partial g(z; x)}{\partial n} g(z^*; y) - g(z; x) \frac{\partial g(z^*; y)}{\partial n} \right\} ds$$
$$+ g(x; y_1) - g(x^*; y) + o(\rho^2)$$

which expresses the variation of $g(x; y)$ by means of the values of $g(x; y)$ on the circumference K_ρ, interior to D, only. This formula may be simplified by means of a development in series. We introduce an auxiliary function $p(x; y)$ analytic in x and satisfying

$$(7) \qquad g(x; y) = R\{p(x; y)\}.$$

$p(x; y)$ is not necessarily uniform in D and contains an arbitrary additive imaginary constant depending on y. On K_ρ we have $z = z_0 + \rho e^{i\tau}$ so that, on applying Taylor's theorem to $p(x; y)$ and by (7)

$$(8) \qquad g(z^*; y) = g(z_0; y) + R\{\rho(e^{i\tau} + e^{i(2\phi-\tau)})p'(z_0; y)\} + O(\rho^2),$$

$$(8') \qquad \frac{\partial}{\partial n} g(z^*; y) = \qquad R\{ (e^{i\tau} - e^{i(2\phi-\tau)})p'(z_0; y)\} + O(\rho),$$

$$(8'') \qquad g(z; x) = g(z_0; x) + R\{\rho(e^{i\tau}p'(z_0; x)\} \qquad + O(\rho^2),$$

$$(8''') \qquad \frac{\partial}{\partial n} g(z; x) = \qquad R\{ e^{i\tau}p'(z_0; x)\} \qquad + O(\rho),$$

$$(8^{IV}) \qquad g(x; y_1) = g(x; y) - R\left\{\frac{e^{2i\phi}\rho^2}{y - z_0} p'(y; x)\right\} \qquad + o(\rho^2),$$

$$(8^{V}) \qquad g(x^*; y) = g(x; y) + R\left\{\frac{e^{2i\phi}\rho^2}{x - z_0} p'(x; y)\right\} \qquad + o(\rho^2).$$

The dash denotes differentiation of $p(x; y)$ with respect to its first argument and $O(\epsilon)$ a term satisfying the condition that $(1/\epsilon)O(\epsilon)$ remains bounded if $\epsilon \to 0$.

Introducing formulae (8) into (6) we obtain

$$(9) \qquad \delta g(x; y) = - R\left\{ e^{2i\phi}\rho^2\left[\frac{p'(x; y)}{x - z_0} + \frac{p'(y; x)}{y - z_0}\right]\right\}$$

$$+ \frac{1}{2\pi}\int_0^{2\pi} 2R\{e^{i\tau}p'(z_0; x)\} \cdot R\{e^{i(2\phi-\tau)}p'(z_0; y)\}\rho^2 d\tau + o(\rho^2)$$

and by an easy transformation

$$(10) \qquad g^*(x; y) = g(x; y)$$

$$+ R\left\{ e^{2i\phi}\rho^2\left[p'(z_0; x)p'(z_0; y) - \frac{p'(x; y)}{x - z_0} - \frac{p'(y; x)}{y - z_0}\right]\right\} + o(\rho^2).$$

In virtue of the identity

$$(10') \qquad g^*(x^*; y^*) = g(x; y) + R\left\{ e^{2i\phi}\rho^2\left[\frac{p'(x; y)}{x - z_0} + \frac{p'(y; x)}{y - z_0}\right]\right\} + o(\rho^2),$$

we may put (10) into the simple form (Schiffer 3)

$$(11) \qquad g^*(x^*; y^*) = g(x; y) + R\{e^{2i\phi}\rho^2 p'(z_0; x)p'(z_0; y)\} + o(\rho^2)$$

giving the law of variation of $g(x; y)$ under the particular transformation (2) of D.

(11) was derived from (1) by partial integration. The latter formula is only valid for analytically bounded domains; the same holds, therefore, for (11). But the most general domain D may be approximated arbitrarily

by analytically bounded domains; Green's function of the approximating domain with all its derivatives tends to the corresponding terms of the given domain, uniformly in every interior part. $o(\rho^2)$ depends only upon $g(x;y)$ and its derivatives at interior points of D; hence it can be estimated uniformly. Thus the validity of (11) can be derived for the most general domains D. This fact proves (11) to be superior to (1'), and numerous applications of this formula are possible in extremum problems, concerning the theory of the logarithmic potential and conformal representation, as will be seen below.

Another application of our method is as follows: In the paper (Schiffer 4) the relation between Green's function and the kernal function (see Bergman 1, §VII) is derived. The methods used in the present paper can be employed in investigating the kernal function and lead to new results concerning the behavior of this function in simply- and multiply-connected domains.

2. Further solutions of the variational equation (11).

1. The differential equation (1') for Green's function admits as solution also other domain functions depending on two variables x, y (see Lévy). If the values of such a function are known for an elementary initial domain, say a circle, it can be calculated in principle for *every* other domain by means of (1'). But it will, in general, be difficult to characterize such a function geometrically, independently of the definition by means of a variational equation. On the other hand, we shall define now an important family of domain functions which occur in the theory of conformal representation of multiply connected domains and which satisfy (11).

In the theory of Green's function the boundary condition plays a decisive part. We shall define a more general type of boundary condition which renders the same service and will be called henceforth the type N. It is characterized by the following property:

Let C_ν $(\nu = 1, \cdots, n)$ be a system of smooth curves enclosing a domain D and let $z^*(z)$ be a conformal representation in the neighborhood of all the C_ν, transforming them into a new system of curves C^*_ν which encloses the domain D^*. A boundary condition is said to be of type N, if it assures in this case the following two facts: $\phi(x)$ satisfying this condition with respect to D, $\psi^*(x)$ with respect to D^*, implies that

a. $\phi(x)$ has continuous derivatives on C_ν, and $\psi^*(x)$ on C^*_ν.

b. The following relation is always valid:

$$(12) \qquad \int_C \{\phi(z)\, \frac{\partial}{\partial n_z}\, \psi^*(z^*(z)) - \psi^*(z^*(z))\, \frac{\partial}{\partial n_z}\, \phi(z)\} ds_z = 0.$$

Next, domain functions $\gamma(x;y)$ of D with the following properties will be considered:

a. $\gamma(x;y)$ is, for $y \subset D$ fixed, a harmonic function of $x \subset D$, the point $x = y$ excepted.

b. In the neighborhood of $x = y$, the expression $\gamma(x;y) + \log|x - y|$ is bounded.

c. The function $\gamma(x;y)$ depends continuously on D, uniformly with respect to x.

d. $\gamma(x;y)$ satisfies as function of x a boundary condition of type N.

We assert that on these conditions $\gamma(x;y)$ has a variation formula (11).

First, it is obvious that $\gamma(x;y)$ is symmetric with respect to both its arguments. For, on one hand, Green's identity yields in the case of a smooth boundary C

$$(13) \quad \frac{1}{2\pi} \int_C \{\gamma(z;x) \frac{\partial}{\partial n} \gamma(z;y) - \gamma(z;y) \frac{\partial}{\partial n} \gamma(z;x)\} ds$$
$$= \gamma(x;y) - \gamma(y;x),$$

while, on the other hand, the boundary condition of $\gamma(x;y)$ ensures, by virtue of (12), that this integral vanishes; we have only to put $\phi(z) = \gamma(z;x)$, $z^*(z) = z$, $\psi^*(z) = \gamma(z;y)$. Hence, we have proved the symmetry of $\gamma(x;y)$ for domains D, bounded by smooth curves. But in view of the continuity of $\gamma(x;y)$ in dependence on the domain, this establishes the same property for domains with general boundary.

Let D denote a domain with smooth boundary curves C_ν; by means of the variation (2) it is transformed into D^* with the corresponding domain function $\gamma^*(x;y)$. As in 1, 2, we construct the domain D_0 by discarding from D the interior of the circumference $K_\rho \equiv (|x - z_0| = \rho)$. In D_0, the function $d(x;y) = \gamma^*(x^*(x);y^*(y)) - \gamma(x;y)$ is harmonic in both its arguments, the logarithmic pole $x = y$ being cancelled by subtraction. Hence, Green's identity yields easily

$$(14) \quad \frac{1}{2\pi} \int_{C+K_\rho} \{\gamma(z;x) \frac{\partial}{\partial n} d(z;y) - d(z;y) \frac{\partial}{\partial n} \gamma(z;x)\} ds$$
$$= d(x;y) = \gamma^*(x^*;y^*) - \gamma(x;y).$$

But γ and γ^* satisfy a boundary condition of type N which shows that the integral over C vanishes. Combining this fact with Green's identity

$$(15) \quad \frac{1}{2\pi} \int_{K_\rho} \{\gamma(z;x) \frac{\partial}{\partial n} \gamma(z;y) - \gamma(z;y) \frac{\partial}{\partial n} \gamma(z;x)\} ds = 0$$

which holds for x and y in D_0, we get finally from (14)

$$(15') \quad \gamma^*(x^*;y^*) = \gamma(x;y)$$
$$+ \frac{1}{2\pi} \int_{K_\rho} \{\gamma(z;x) \frac{\partial}{\partial n_z} \gamma^*(z^*;y^*) - \gamma^*(z^*;y^*) \frac{\partial}{\partial n_z} \gamma(z;x)\} ds_z$$

where the normal derivative is to be taken in the outward direction with respect to D_0. If we want to give to n the direction of increasing radius ρ, as is usual, we get

$$(15'') \quad \gamma^*(x^*;y^*) = \gamma(x;y)$$
$$+ \frac{1}{2\pi} \int_{K_\rho} \{\gamma^*(z^*;y^*) \frac{\partial}{\partial n_z} \gamma(z;x) - \gamma(z;x) \frac{\partial}{\partial n_z} \gamma^*(z^*;y^*)\} ds_z.$$

Thus, we have expressed the variation of $\gamma(x;y)$ by an integral taken along a circumference interior to D as we did in (6) with respect to $g(x;y)$. We may perform now the same formal transformations and developments into series as in 1, 2; these hold for $\gamma(x;y)$ too, since we have used there only the symmetry and harmonicity of $g(x;y)$. We introduce a function $\phi(x;y)$, analytic for given $y \subset D$ with respect to its first argument $x \subset D$ and satisfying

$$(16) \qquad\qquad \gamma(x;y) = R\{\phi(x;y)\}.$$

In general, $\phi(x;y)$ will not be uniform in D; it possesses additive (imaginary) moduli with respect to circuits around the C_ν. We may perform on it the same operations as we did above on $p(x;y)$ and we get, in complete analogy to (11),

$$(17) \qquad \gamma^*(x^*;y^*) = \gamma(x;y) + R\{e^{2i\phi}\rho^2\phi'(z_0;x)\phi'(z_0;y)\} + o(\rho^2).$$

This is the law of variation of $\gamma(x;y)$ under the transformation (2) of a domain D with a smooth boundary. But in view of the uniform continuity of $\gamma(x;y)$ this formula remains valid for domains with general boundary.

2.   Our next task is to point out definite examples of domain-functions of the above type. It is well known that every domain D can be mapped in an infinity of ways on the exterior of a circle, cut along concentric circular slits (Koebe 1, Grötzsch 1). This canonical representation is in many respects a natural generalization of the representation of a simply connected domain on the exterior of a circle.

If $f(x;y)$ is a univalent function of x in a simply-connected domain D and maps D on the exterior E of the unit circle such that $y \subset D$ corresponds to infinity, then the Green's function of D is given by

$$(18) \qquad\qquad g(x;y) = \log|f(x;y)|.$$

In the case of a multiply connected domain, however, there exists for every $m = 1, 2, \cdots, n$ a function $f_m(x; y)$, mapping D on the exterior of the unit circle, slit along $n - 1$ concentric circular arcs, such that C_m corresponds to the unit circle and y to infinity. Thus there exist n domain functions

$$(19) \qquad \gamma_m(x; y) = \log |f_m(x; y)|$$

which are a generalization of Green's function for a simply connected domain. We want to show that each $\gamma_m(x; y)$ has a variation formula (17), in analogy to $g(x; y)$. To prove this, it suffices, in view of 1, to show that the boundary conditions for $\gamma_m(x; y)$ are of type N.

Now, there are two characteristic properties of $\gamma_m(x; y)$: (a) $\gamma_m(z; y)$ is zero for z on C_m and constant on each C_ν. (b) $f_m(x; y)$ being uniform in D, the conjugate function of $\gamma_m(x; y)$, i. e., $I\{\log f_m(x; y)\}$, does not change when x describes a circuit around C_ν ($\nu \neq m$) which means, in the case of a smooth C_ν,

$$(20) \qquad \int_{C_\nu} \frac{\partial}{\partial n_z} \gamma_m(z; y) \, ds_z = 0 \qquad\qquad (\nu \neq m).$$

These two facts represent just a boundary condition of type N. For, if any $\phi(z)$ possesses the properties (a) and (b) on the system of smooth curves C_ν and any $\psi^*(z)$ on the corresponding system C^*_ν, then we have for each curve C_ν

$$(12') \qquad \int_{C_\nu} \{\phi(z) \frac{\partial}{\partial n_z} \psi^*(z^*(z)) - \psi^*(z^*(z)) \frac{\partial}{\partial n_z} \phi(z)\} \, ds_z = 0.$$

In fact, $\phi(z)$ and $\psi^*(z^*(z))$ vanish on C_m; on each other C_ν both functions are constant and may be taken out of integration. Thus, there remain periods of type (20) which vanish. Hence, the boundary conditions considered are of type N and $\gamma_m(x; y)$ satisfies, therefore, a variational equation (17).

3. We mention now another type of canonical representation which may again be considered as a generalization of the representation onto the exterior E of the unit circle in the case of a simply connected domain. It is the representation of D by means of a function $h_m(x; y)$ onto E, slit along $n - 1$ radial segments, such that C_m corresponds to the circle and y to infinity (Grötzsch 1, Rengel 1, Koebe 1). Consider now the domain function

$$(21) \qquad \kappa_m(x; y) = \log |h_m(x; y)|.$$

This function too possesses a variation formula (17), since it has boundary conditions of type N. To show this, we remark the following two facts:

(a) $\kappa_m(x;y)$ vanishes on C_m.   (b) On every C_ν ($\nu \neq m$) the conjugate function of $\kappa_m(x;y)$, i. e., $I\{\log h_m(x;y)\}$ is constant.   Hence, using Cauchy-Riemann's differential equations, we have for $z \subset C_\nu$ ($\nu \neq m$)

$$(22) \qquad \frac{\partial}{\partial n_z} \kappa_m(z;y) = -\frac{\partial}{\partial s_z} I\{\log h_m(z;y)\} = 0.$$

These conditions ensure the validity of relations of type (12) for every smooth C_ν; for, $\kappa^*_m(x^*;y^*)$, too, has a constant conjugate function if z remains on C_ν ($\nu \neq m$).   The boundary conditions (a) and (b) are, therefore, of type N and $\kappa_m(x;y)$ satisfies a variational equation (17).

We may consider other canonical representations of D, by means of univalent functions $l_m(x;y)$ which map D onto E slit by concentric circular arcs and radial segments, such that $y \subset D$ corresponds to infinity, C_m to the circle and the other C_ν to either of the slits.   It follows easily that the domain function

$$(23) \qquad \lambda_m(x;y) = \log |l_m(x;y)|$$

is of the type $\gamma(x;y)$ characterized in 1.   Hence it varies too according to (17).

Thus, we have found various domain functions with the same law of variation as Green's function.   The function $p(x;y)$ in (11) can only with difficulty be interpreted geometrically, whereas the functions $\phi(x;y)$, used in (17) for the variation of our domain functions, are the logarithms of univalent functions yielding canonical representations.   The next paragraphs will show the advantages which evolve from this fact.

4.   Next, we transform (17) in such a way that its relation to Hadamard's formula is exposed.   From (17) we obtain easily in virtue of (2)

$$(17') \qquad \gamma^*(x;y) = \gamma(x;y)$$
$$+ R\left\{ e^{2i\phi}\rho^2\left[\phi'(z_0;x)\phi'(z_0;y) - \frac{\phi'(x;y)}{x-z_0} - \frac{\phi'(y;x)}{y-z_0} \right] \right\} + o(\rho^2).$$

Let now C'' be a system of smooth curves forming the boundary of an arbitrary partial domain of D which contains x, y and z_0; then the residue theorem yields

$$(24) \qquad \frac{1}{2\pi i}\int_{C'} \phi'(z;x)\phi'(z;y)\frac{ds}{z-z_0}$$
$$= \phi'(z_0;x)\phi'(z_0;y) - \frac{\phi'(x;y)}{x-z_0} - \frac{\phi'(y;x)}{y-z_0},$$

whence, in view of (17'),

6

(25) $\gamma^*(x;y) = \gamma(x;y)$

$$+ R\left\{ e^{2i\phi}\rho^2 \cdot \frac{1}{2\pi i} \int_{C'} \phi'(z;x)\phi'(z;y) \frac{ds}{z-z_0} \right\} + o(\rho^2).$$

If the boundary C of D is smooth, we may choose, in particular, $C' = C$; in this case we represent C as in **1, 1** by means of the length parameter s which insures $|z'(s)| = 1$ on every C_ν.

We apply (25) to the functions $\gamma_m(x;y)$ defined by (19); we put

(19') $\phi_m(x;y) = \log f_m(x;y)$, i. e., $\gamma_m(x;y) = R\{\phi_m(x;y)\}$.

By definition of $f_m(x;y)$, $\phi_m(x;y)$ has a constant real part for $z \subset C_\nu$. Hence, along every curve C_ν

$$\frac{d}{ds}\phi_m(z(s);y) = \phi'_m(z;y) \cdot z'$$

is imaginary and its value is, according to Cauchy-Riemann's differential equation,

$$\phi'_m(z;y)z' = -i\frac{\partial}{\partial n_z}\gamma_m(z;y).$$

Hence, we may write (25) in the form

(25') $\gamma^*_m(x;y) = \gamma_m(x;y)$

$$+ \frac{1}{2\pi}\int_C \frac{\partial}{\partial n_z}\gamma_m(z;x) \frac{\partial}{\partial n_z}\gamma_m(z;y) R\left\{ \frac{i}{z'}\frac{e^{2i\phi}\rho^2}{z-z_0} \right\} ds + o(\rho^2).$$

On the other hand, we get obviously for the value of the normal shift by variation (2)

(26) $$\delta n = R\left\{ \frac{i}{z'}\frac{e^{2i\phi}\rho^2}{z-z_0} \right\}.$$

Hence, (25') becomes

(27) $$\delta\gamma_m(x;y) = \frac{1}{2\pi}\int_C \frac{\partial}{\partial n_z}\gamma_m(z;x) \cdot \frac{\partial}{\partial n_z}\gamma_m(z;y)\delta n\, ds,$$

which is just Hadamard's formula for our domain functions. It is derived from (17) for all variations obtainable by superposition of elementary variations (2).

Let $\psi_m(x;y)$ be an analytic function of x, connected with the functions (21) by $\kappa_m(x;y) = R\{\psi_m(x;y)\}$; for $z \subset C_\nu(\nu \neq m)$, $\psi_m(z;y)$ has a constant imaginary part. Hence, on all these curves

(28) $$\frac{d}{ds}\psi_m(z(s);y) = \psi'_m(z;y)z' = \frac{\partial}{\partial s_z}\kappa_m(z;y).$$

A transformation of (25), analogous to the above, yields by means of (28)

$$(28')\qquad \delta\kappa_m(x;y)=\frac{1}{2\pi}\int_{C_m}\frac{\partial}{\partial n_z}\kappa_m(z;x)\frac{\partial}{\partial n_z}\kappa_m(z;y)\delta n\,ds$$

$$-\frac{1}{2\pi}\int_{C-C_m}\frac{\partial}{\partial s_z}\kappa_m(z;x)\frac{\partial}{\partial s_z}\kappa_m(z;y)\delta n\,ds.$$

Comparison of (27) and (28') shows that the same formula (17) appears in its integral representation in very different forms according to the particular boundary conditions satisfied by $\gamma(x;y)$. This shows the advantage of the unifying formula (17).

3. Applications of the variation formula to Green's function.

1. Green's function $g(x;y)$ may be considered as a measure of D with respect to the pair of points x,y in D. In fact, Hadamard's formula (1') shows the monotonic behavior of $g(x;y)$ as a domain function; for if δn is always positive, i. e., if D is enlarged, $g(x;y)$ is seen to increase, since along the entire boundary we have $\frac{\partial}{\partial n_z}g(z;x)<0$.

In the neighborhood of y, we have the development

$$(29)\qquad g(x;y)=\log\frac{1}{|x-y|}+\log\frac{1}{d(y)}+O(|x-y|)$$

where $O(\epsilon)\to 0$ with $\epsilon\to 0$. If $\infty\subset D$, we have further

$$(29')\qquad g(x;\infty)=\log|x|+\log\frac{1}{d(\infty)}+O\left(\frac{1}{|x|}\right).$$

The functional $d(\infty)$ has been used frequently in the theory of conformal representation as a measure for D or its boundary C. $d(\infty)$ is called transfinite diameter, Robin's constant or capacity constant of C, or of the domain D whose boundary is C (see Fekete, Szegö, Nevanlinna). All domains which may be mapped upon each other by univalent functions $f(x)$, normalized as

$$(30)\qquad f(\dot{x})=x+a_0+\frac{a_1}{x}+\cdots$$

at infinity, have the same measure $d(\infty)$. Analogously, $d(y)$ is a conformal invariant with respect to representations of D which map y on itself and have there the derivative 1. If $f(x)$ has at infinity (or at y) the derivative a, however, D is mapped on a domain with the measure $ad(\infty)$ (or $ad(y)$). In particular, all $d(y)$ behave like lengths with regard to homotheties.

If D is simply connected, it is well known that $d(y)$ is the radius of the circle on the exterior of which D may be mapped by a function with the normalization

$$(30') \qquad f(x) = \frac{1}{x-y} + a_0 + a_1(x-y) + \cdots$$

(or (30) if $y = \infty$).

Introducing into (1') and (11) the developments (29) and (29') and letting $x \to y$ we get, by an easy comparison of both sides, the general formulas for the variation of $d(y)$:

$$(31) \qquad \delta \log d(y) = -\frac{1}{2\pi} \int_C \left(\frac{\partial g(z;y)}{\partial n_z} \right)^2 \delta n \, ds$$

$$(31') \qquad \log d^*(y^*) = \log d(y)$$
$$- R \left\{ e^{2i\phi} \rho^2 \left[p'(z_0;y)^2 - \frac{1}{(y-z_0)^2} \right] \right\} + o(\rho^2).$$

Both formulas remain valid for $y = \infty$.

(31) shows that $d(y)$ is a functional which decreases as the domain increases. If we consider the domain function

$$(32) \qquad \Gamma(x;y) = 2g(x;y) + \log d(x) + \log d(y)$$

we find, by (1') and (31), its variation formula

$$(32') \qquad \delta\Gamma(x;y) = -\frac{1}{2\pi} \int_C \left[\frac{\partial g(z;x)}{\partial n} - \frac{\partial g(z;y)}{\partial n} \right]^2 \delta n \, ds$$

which proves that $\Gamma(x;y)$ is a decreasing domain function.

2. To demonstrate the applicability of (31'), we deal now with an extremum problem which will become useful later. We choose a boundary continuum C_m of D and a point $y \subset D$; every conformal representation of D of type (30') transforms C_m into a new continuum $\tilde{C}_m$ with a certain transfinite diameter $\delta = \delta(\infty)$. We ask for the extremal values which δ can attain, if all functions (30'), univalent in D, are considered.

To solve this problem of distortion, we start with a domain D with boundary curves $\tilde{C}_\nu$, obtained from D by such a univalent function (30') that δ is maximal. This domain is by no means fixed uniquely; for a representation (30), univalent in the exterior of $\tilde{C}_m$, will not change δ and its superposition on any mapping function (30') will preserve the normalization (30'). This arbitrariness may be avoided by supposing $\tilde{C}_m$ to be a circle centered at the origin. The radius of the circle is necessarily δ, since the transfinite diameter of a circle is equal to its radius. We proceed to determine the shape of the remaining boundary continua $\tilde{C}_\nu$ $(\nu \neq m)$. For this purpose, we choose a continuum $\tilde{C}_l$ $(l \neq m)$, a fixed point z_0 on it and a subcontinuum Γ of $\tilde{C}_l$, containing z_0, of transfinite diameter ρ. All functions (30) which are univalent in the exterior of Γ permit a development

$$(33) \qquad x^+ = \phi(x) = x + k + \frac{a\rho^2}{x - z_0} + \frac{b\rho^3}{(x - z_0)^2} + \cdots$$

valid in every given domain interior to $\tilde{D}$ for Γ sufficiently small, such that $a, b, \cdots$ have bounds independent of Γ. The superposition of a representation (33) on the original mapping of D on $\tilde{D}$ gives a univalent transformation $(30')$ of D on a new domain D^+. This has a boundary continuum C_m^+ arising from $\tilde{C}_m$ by means of (33). Its transfinite diameter $\delta^+ = \delta^+(\infty)$ satisfies, in view of the maximal property of δ, the inequality

$$(34) \qquad\qquad\qquad\qquad \delta^+ \leq \delta.$$

If, on the other hand, $\pi(x; y)$ is an analytic function of x such that $R\{\pi(x; y)\}$ is Green's function for the exterior of $\tilde{C}_m$, we may find δ^+ with the aid of $(31')$. In fact, the function (33) differs on $\tilde{C}_m$ from

$$(2') \qquad\qquad\qquad x^* = x + k + \frac{a\rho^2}{x - z_0}$$

only in terms of order $o(\rho^2)$. Hence, in view of (31), applicable in the case $C = \tilde{C}_m$, they cause a variation of $\log \delta$ of this order only, in addition to that caused by $(2')$ and given by $(31')$. So, we get

$$(35) \qquad \log \delta^+ = \log \delta - R\{a\rho^2\pi'(z_0; \infty)^2\} + o(\rho^2).$$

Since $\tilde{C}_m$ is a circle of radius δ and has as its Green's function $g(x; \infty)$ $= \log \frac{|x|}{\delta}$, we have

$$(36) \qquad\qquad \pi(x; \infty) = \log x - \log \delta$$

and, finally, the following formula for the variation of $\log \delta$:

$$(35') \qquad\qquad \log \delta^+ = \log \delta - R\left\{ a\rho^2 \cdot \frac{1}{z_0^2} \right\} + o(\rho^2).$$

Comparing (34) with $(35')$ we get the inequality

(34')
$$R\left\{ a\rho^2 \cdot \frac{1}{z_0{}^2} \right\} + o(\rho^2) \geq 0,$$

whatever z_0, Γ and the function (33) may have been.

Now, we have to apply the following lemma (Schiffer 1):

LEMMA. *Let C be a continuum in the x-plane; suppose that there exists an analytic function $s(x) \neq 0$ such that for an arbitrary function (33) univalent in the exterior of an arbitrary subcontinuum Γ of transfinite diameter ρ and containing the arbitrary point z_0 we have*

(37)
$$R\{a\rho^2 s(z_0)\} + o(\rho^2) \geq 0.$$

Then C is an analytic curve, expressed by means of a real parameter in the form $x = x(t)$ such that

(37')
$$x'(t)^2 s[x(t)] + 1 = 0.$$

Applying this lemma to (34') we find $\tilde{C}_l$ to be an analytic curve with the differential equation

(37'')
$$x'(t)^2 x(t)^{-2} + 1 = 0$$

whence

(37''')
$$x(t) = k_l e^{it} \qquad\qquad k_l = \text{constant}.$$

Thus we have proved that δ attains its maximum for the representation of D which transforms C_m into a circle and the other $n-1$ continua C_ν into concentric circular arcs.

As the existence of extremal functions is insured in the case of the problem considered, this theorem proves anew the possibility of this particular type of canonical conformal representation.

Had we raised the question of the minimum of the transfinite diameter δ of all possible $\tilde{C}_m$, obtained by representations (30'), our method would have been exactly the same. Only, instead of (34), we would have used the inverse inequality leading to

(34'')
$$R\left\{ a\rho^2 \frac{1}{z_0{}^2} \right\} + o(\rho^2) \leq 0.$$

A new application of our lemma shows that in this case also every $\tilde{C}_l$ is an analytic curve satisfying now the differential equation

(37 IV)
$$x'(t)^2 x(t)^{-2} = 1$$

which yields

(37 V)
$$x(t) = \kappa_l e^t \qquad\qquad \kappa_l = \text{constant}.$$

Hence: The minimum of δ is obtained for the representation of D which transforms C_m into a circle and the remaining C_ν into radial segments.

The extremum problem considered above was so easily solved because the function $\pi(x; \infty)$, appearing in the variation formula (35), is the logarithm of a univalent function. Thus, we could apply an auxiliary mapping and give to $\pi(x; \infty)$ a suitable particular form. Most variational problems, concerning the transfinite diameter of a multiply connected domain, are considerably more difficult because the function $p(x; y)$, occurring in (31'), is not connected with univalent functions. In the variational rule of $\gamma_m(x; y)$ and $\kappa_m(x; y)$, however, there appears always the logarithm of a univalent function, which facilitates obviously the treatment of these domain functions.

3. The method of treating extremum problems, just applied, permits an important numerical application in the theory of mapping simply connected domains. Consider all functions (30) which are univalent in $|x| > 1$. They map the unit circle $|x| = 1$ on continua C and our aim is to estimate the distortion of the frontier caused by the mapping. For this purpose, we fix on $|x| = 1$ an arc of length α, say $(e^{-i(\alpha/2)}, 1, e^{i(\alpha/2)})$; let C_α be its corresponding image on C. C_α being a continuum, it is permissible to ask about its transfinite diameter δ. The unit circle has the transfinite diameter 1 and since the latter is preserved by mapping functions (30), C has the same. The transfinite diameter being a decreasing functional of the domain, $C_\alpha \subset C$ yields obviously $\delta \leq 1$. It is easily seen that there exist for every α representations bringing δ arbitrarily near to 1. But there arises the question: What is the minimum of δ? The similarity of this problem to the above is clear; in fact our method of solution will be exactly as before.

Let us suppose that δ attains its minimum for a representation (30) which transforms the unit circle into a continuum $\tilde{C}$ and the arc into the subcontinuum $\tilde{C}_\alpha$. We may suppose $\tilde{C}_\alpha$ to be a circle around the origin with radius δ, since an auxiliary representation (30), univalent in the exterior of C_α, may be superposed on the original mapping of $|x| > 1$ onto $\tilde{D}$ without changing the normalization (30) or the transfinite diameter δ of $\tilde{C}_\alpha$. If z_0 belongs to $\tilde{C}$, but not to $\tilde{C}_\alpha$, we choose a subcontinuum Γ of $\tilde{C}$ of transfinite diameter ρ which contains z_0, but no point of $\tilde{C}_\alpha$. Again we superpose on the original function, mapping $|x| > 1$ on $\tilde{D}$, arbitrary functions (33) which are univalent in the exterior of Γ. We obtain domains D^+ with boundary continuum C^+, containing C_α^+ as image of the arc. The transfinite diameter δ^+ of C_α^+ satisfies the inequality $\delta^+ \geq \delta$ in view of the minimum property of δ. Introducing the auxiliary function $\pi(x; y)$, connected with Green's function of the exterior of $\tilde{C}_a$, we may use (35) for the calculation of $\log \delta^+$. According to our suppposition with resect to $\tilde{C}_a$ we have here too $\pi(x; \infty) = \log x/\delta$. Therefore we find by exactly the same reasoning which led to (37 ^{V}) that all points of $\tilde{C}$, not belonging to $\tilde{C}_a$, lie on a radius segment in the exterior of $\tilde{C}_a$. By a rotation we may obtain for $\tilde{C}$ finally the following configuration: It consists of the circle $|x| = \delta$ plus the segment $< -\epsilon, -\delta >$, $(\epsilon > \delta)$. The circle corresponds to the arc $(e^{-i(\alpha/2)}, 1, e^{i(\alpha/2)})$ and the segment to the complementary arc.

Thus, the representation is fixed, except for the still unknown values of δ and ϵ. For their determination we consider the representation

$$(38) \qquad \xi = x + \delta^2/x$$

of the exterior of $\tilde{C}$ which has the normalization (30). It transforms $\tilde{C}_a$ into the segment $< -2\delta, 2\delta >$ and the rest of $\tilde{C}$ into the segment $< -\epsilon - \delta^2/\epsilon, -2\delta >$. On the other hand, the function

$$(38') \qquad f(x) = x + 1/x + 2\delta - 2$$

is univalent for $|x| > 1$ and is of type (30). It transforms the unit circle into a segment such that the arc $(e^{-i(\alpha/2)}, 1, e^{i(\alpha/2)})$ corresponds to the interval $< 2\delta - 2(1 - \cos \alpha/2), 2\delta >$ on the real axis and the complementary arc to the interval $< 2\delta - 4, 2\delta - 2(1 - \cos \alpha/2) >$. In view of the unicity of the mapping function which transforms the said arcs into *different* contiguous segments of the real axis, we get by comparison

$$(38'') \qquad 2\delta - 2(1 - \cos \alpha/2) = -2\delta, \quad \epsilon + \delta^2/\epsilon = 4 - 2\delta.$$

In particular, we get from the first equation (38")

$$(39) \qquad \delta = \sin^2 \alpha/4$$

for the minimum of the transfinite diameter. Thus, we have proved the following distortion theorem:

Every function (30), *univalent in* $|x| > 1$, *maps an arc of the unit circle with aperture* α *on a continuum with transfinite diameter* $d \geq \sin^2 \alpha/4$.

4. If we start with the domain $|x| > r$, every function (30), univalent in this domain, will map an arc with aperture α on a continuum with transfinite diameter $d \geq r \sin^2 \alpha/4$. This leads to an important application: Let C be an arbitrary continuum with transfinite diameter r; then it may be mapped on the circle $|x| = r$ by a function (30), univalent in its exterior.

Divide C into two continua A and B such that they correspond to two complementary arcs on $|x| = r$ with apertures α and $\beta = 2\pi - \alpha$. Then the foregoing theorem yields:

$$(40) \qquad d(A) \geq r \sin^2 \alpha/4, \qquad d(B) \geq r \sin^2 \beta/4 = r \cos^2 \alpha/4.$$

Since $r = d(C)$ and $C = A + B$, we get from (40) by addition

$$(41) \qquad\qquad d(A) + d(B) \geq d(A + B)$$

which establishes in a new way the subadditivity of the transfinite diameter for such a composition (see Schiffer 2).

There are numerous other applications of the theorem of **3** which describes the boundary distortion in case of conformal representation of the unit circle; but we shall not treat them here. We have inserted the proof here only in order to show the use which can be made of (31').

4. The conformal radii $d_m(y)$.

1. It is obvious that the extremal representation (30') of D, transforming $C_{\nu m}$ into a circle of radius δ and all other C_ν into concentric circular arcs, is closely related to the domain function (19), considered in **2**, 2. Suppose, indeed, that $f_m(x; y)$ has in the neighborhood of y the development

$$(42) \quad f_m(x; y) = d_m(y)^{-1} \left[\frac{1}{x - y} + a_0 + a_1(x - y) + \cdots \right], \; d_m(y) > 0;$$

then $d_m(y) \cdot f_m(x; y)$ is of type (30') and defines the extremal representation considered. Hence, $d_m(y)$ is the greatest value for the transfinite diameter which $C_{\nu m}$ can attain by a representation (30') of D. The area $F(C_m)$, enclosed by the curve $C_{\nu m}$ may be estimated by means of its transfinite diameter $d(C_{\nu m})$ (Pólya 1)

$$(43) \qquad\qquad F(C_m) \leq \pi d(C_{\nu m})^2.$$

Thus, $\pi d_m(y)^2$ is the greatest area, obtainable for $C_{\nu m}$ by a representation (30') of D. In virtue of (19) and (42), $d_m(y)$ is connected with $\gamma_m(x; y)$ by

$$(44) \qquad \gamma_m(x; y) = \log \frac{1}{|x - y|} + \log \frac{1}{d_m(y)} + O(|x - y|).$$

If $\infty \subset D$, we have

$$(44') \qquad \gamma_m(x; \infty) = \log |x| + \log \frac{1}{d_m(\infty)} + O\left(\frac{1}{|x|}\right).$$

Thus, $d_m(y)$ is related to $\gamma_m(x; y)$ in exactly the same way as is $d(y)$ to

Green's function. In view of the above geometric interpretation, we shall call $d_m(y)$ the conformal radius of C_m with respect to C in y. If we are speaking of $d(C_m; C)$, the conformal radius of C_m with respect to C, we shall refer to $d_m(\infty)$. In order that it be defined, we must have $\infty \subset D$; this will be supposed henceforth in this paragraph.

$d(C_m; C)$ measures the continuum C_m, taking into account all the boundary C. It is invariant with respect to representations (30), univalent in D. It is linearly homogeneous with respect to homotheties, as is easily checked. Introducting (44') into (27), we get by comparison of coefficients

$$(45) \qquad \delta \log d(C_m; C) = -\frac{1}{2\pi} \int_C \left(\frac{\partial}{\partial n} \gamma_m(z; \infty) \right)^2 \delta n \, ds;$$

introducing (44') into the variation formula (17), with $\phi(x; y) = \log f_m(x; y)$ in our case, we get

$$(46) \qquad \log d(C^*_m; C^*) = \log d(C_m; C).$$
$$- R\{e^{2i\phi} \rho^2 f'_m(z_0; \infty)^2 f_m(z_0; \infty)^{-2}\} + o(\rho^2).$$

(45) proves $d(C_m; C)$ to be a decreasing domain function, a fact which follows also easily from its extremal property. Analogously, $d_m(y)$ and even $2\gamma_m(x; y) + \log d_m(x) + \log d_m(y)$ may be shown to be monotonic domain functions; the proof is analogous to that of (32).

2. The following problem requires the application of (46). Divide the boundary C of D into two point sets A and B, each consisting of a finite number of proper continua, and possessing only a finite number of common points. In particular let F be a continuum of A. Consider the conformal radius $d(F; A)$ which is invariant with respect to all representations (30), univalent in the exterior of A. It will change, in general, with respect to mappings (30) of D. We seek the minimum of $d(F; A)$, taking into account all these representations.

In view of the compactness of the family (30) of all functions, univalent in D, there exists at least one function of the family, for which $d(F; A)$ attains its minimum. It maps D, A, F, B on $\bar{D}$, $\bar{A}$, $\bar{F}$, $\bar{B}$ respectively. It is not uniquely determined since the subsequent superposition of a representation (30), univalent in the exterior of $\bar{A}$, does not change univalency or normalization of the total mapping function and preserves also $d(\bar{F}; \bar{A})$. Thus, it may be supposed that $\bar{F}$ is a circle around the origin and the other continua of $\bar{A}$ are concentric circular arcs. In this case, the function $f_F(x; \infty)$ (which corresponds to $f_m(x; \infty)$ in the case $C_m = F$, $C = \bar{A}$), defining the canonical representation of $\tilde{A}$ with distinction of $\tilde{F}$, has the form $xd(\tilde{F};\tilde{d})^{-1}$, since $d(\tilde{F};\tilde{A})$ is just the radius of $\tilde{F}$. Now, to find the sought for minimum reduces to the determination of the shape of $\tilde{B}$.

z_0 being a point of $\tilde{B}$ not belonging to $\tilde{A}$, let us consider once more the functions (33), univalent in the exterior of a subcontinuum Γ of $\tilde{B}$, which contains z_0 but no points of $\tilde{A}$. Superposing them on the mapping $D \to \tilde{D}$, we get a normalized univalent representation $D \to D^+$. In view of (46) and the simple particular form of $f_F(x;\infty)$, we have, by an argument similar to that which led to (35'),

$$(47) \qquad \log d(F^+;A^+) = \log d(\tilde{F};\tilde{A}) - R\left\{a\rho^2 \cdot \frac{1}{z_0^2}\right\} + o(\rho^2).$$

Since the minimal property of $d(\tilde{F};\tilde{A})$ involves

$$(48) \qquad\qquad \log d(F^+;A^+) \geq \log d(\tilde{F};\tilde{A})$$

for an arbitrary choice of the function (33), the lemma of **3**, 2 yields:

$\tilde{B}$ *consists of segments pointing towards the center of* $\tilde{F}$.

We have, then, proved the possibility of an interesting type of conformal representation. The boundary C of a domain D may be divided arbitrarily into a finite number of continua, contiguous in a finite number of points only, such that one continuum is mapped on a circle, a given number of continua onto concentric circular arcs and the remainder on radial slits. Every representation of this type solves an extremum problem.

3. The considerations of 2 contain, in particular, the solution of the following problem. Divide the continuum C_1 into the continua A and B which have only two points in common. Let C_A be the aggregate of all points of A and all C_ν ($\nu > 1$); analogously C_B will consist of B and all C_ν ($\nu > 1$). Then, we ask for a univalent function (30) in D which imparts to $d(\tilde{A};\tilde{C}_A)$ its minimum, $\tilde{A}$ and $\tilde{C}_A$ corresponding to A and C_A by means of the representation. According to 2 we get the solution:

A conformal representation (30) of D which transforms A into a circle $\tilde{A}$, B into a contiguous radial segment and the remaining C_ν ($\nu > 1$) into circular arcs, concentric to the circle, yields the minimum value for $d(\tilde{A};\tilde{C}_A)$. Obviously, $d(\tilde{A};\tilde{C}_A)$ is the radius of $\tilde{A}$.

This result becomes interesting when compared with the solution of the following question. Consider all functions (30), univalent in D which transform C_1 into a circle. If r denotes the radius of the circle, α the aperture of the arc corresponding to A, what is the maximum value of the expression $r \sin^2 \alpha/4$?

Let $\bar{D}$ be an image of D yielding the maximum; then, $\bar{C}_1$ is a circle of radius r, and we may suppose that the arc $(e^{-i(\alpha/2)}, 1, e^{i(\alpha/2)})$ corresponds to A. Again we choose a point z_0 on a fixed $\bar{C}_\nu$ ($\nu > 1$) and a function (33), univalent in the exterior of a small subcontinuum of $\bar{C}_\nu$ containing z_0. In this domain, the function

$$(49) \qquad u = \phi(x) - k + \frac{a\rho^2}{z_0} = x + \frac{a\rho^2 x}{(x-z_0)z_0} + o(\rho^2)$$

is also univalent. This representation deforms $|x| = r$, but for ρ sufficiently small the point $r^2/\bar{z}_0$ remains inside the image of this circumference, i. e., lies in the exterior of the image of D, given by (49). Now, we add the further mapping

$$(49') \quad u^+ = u - \frac{\bar{a}\rho^2 u^2}{(r^2 - \bar{z}_0 u)\bar{z}_0} = x + \frac{a\rho^2 x}{(x-z_0)z_0} - \frac{\bar{a}\rho^2 x^2}{(r^2 - \bar{z}_0 x)\bar{z}_0} + o(\rho^2)$$

which is univalent in $\bar{D}$ for ρ small enough. Consider the expression

$$(49'') \qquad \psi(x) = x\left[1 + \frac{a\rho^2}{(x-z_0)z_0} - \frac{\bar{a}\rho^2}{(r^2/x - \bar{z}_0)\bar{z}_0}\right];$$

its modulus for $|x| = r$ is $|\psi(x)| = r + o(\rho^2)$. Hence, adding an additional correction term of order $o(\rho^2)$, we infer that $u^+(x)$, so corrected, transforms the circle $|x| = r$ into itself. At the same time, the points $re^{i(\alpha/2)}$ and $re^{-i(\alpha/2)}$ on it are mapped on the points

$$(50) \qquad re^{i(\alpha_{1/2})} = re^{\pm i(\alpha/2)}\left[1 + 2i\, I\left\{\frac{a\rho^2}{re^{\pm i(\alpha/2)} - z_0)z_0}\right\}\right] + o(\rho^2).$$

Thus, the new arc, corresponding to A, has the aperture

$$(51) \qquad \alpha^+ = \alpha_1 - \alpha_2 = \alpha - 4r\sin\alpha/2$$

$$\times R\left\{\frac{a\rho^2}{z_0(r^2 + z_0^2 - 2rz_0\cos\alpha/2)}\right\} + o(\rho^2).$$

The function (49'), though univalent in $\bar{D}$, has not yet the normalization (30), but must be divided by $(1 + \frac{\bar{a}\rho^2}{\bar{z}_0^2} + o(\rho^2))$ for this purpose. This operation preserves the aperture α^+, but yields the radius

$$(52) \qquad r^+ = r\left[1 - R\left\{\frac{a\rho^2}{z_0^2}\right\}\right] + o(\rho^2).$$

The expression $r^+ \sin^2 \alpha^+/4$ belongs to a normalized function, univalent in $\tilde{D}$. By (51) and (52) we have

$$(53) \quad r^+ \sin^2 \alpha^+/4 = r \sin^2 \alpha/4 \left[1 - R \left\{ a\rho^2 \frac{(r+z_0)^2}{z_0^2(r^2 + z_0^2 - 2rz_0 \cos \alpha/2)} \right\} \right] + o(\rho^2)$$

Since $\tilde{D}$ was supposed to be an extremal domain, we have for every choice of (33)

$$(54) \qquad R \left\{ a\rho^2 \frac{(r+z_0)^2}{z_0^2(r^2 + z_0^2 - 2rz_0 \cos \alpha/2)} \right\} + o(\rho^2) \geq 0.$$

Applying now the lemma of **3, 2**, we find every $\tilde{C}_v$ $(v > 1)$ to be an analytic curve $x = x(t)$, satisfying the differential equation

$$(55) \qquad \frac{x'(t)^2}{x(t)^2} \frac{[r + x(t)]^2}{[re^{i(\alpha/2)} - x(t)][re^{-i(\alpha/2)} - x(t)]} + 1 = 0.$$

Thus, the extremal domain $\tilde{D}$ is bounded by a circle $\tilde{C}_1$ of radius r and by $n-1$ curves (55). In order to understand better the structure of $\tilde{D}$, consider the function $\zeta(x)$, univalent in the exterior of $\tilde{C}_1$ and of type (30), which maps the arc $\tilde{A}$ of $\tilde{C}_1$, corresponding to A, on a circle around the origin, while its complement $\tilde{B}$ becomes a segment of the negative axis. By these requirements, $\zeta(x)$ is fixed and may be calculated in an elementary way. The radius of the new circle is $r \sin^2 \alpha/4$. The expression $x^2 \zeta'(x)^2 \zeta(x)^{-2}$ is a regular function for $|x| > r$; it is positive on $\tilde{A}$, negative on $\tilde{B}$, has in $re^{i(\alpha/2)}$ and $re^{-i(\alpha/2)}$ simple poles and in $-r$ a double zero point, as is easily seen from geometric considerations. Taking into account, further, the normalization at infinity, we get by means of Schwarz's principle of reflection

$$(56) \qquad \frac{x^2 \zeta'(x)^2}{\zeta(x)^2} = \frac{(x+r)^2}{(x - re^{i(\alpha/2)})(x - re^{-i(\alpha/2)})} .$$

With its aid, (55) may be written in simple form, if we put $\zeta[x(t)] = \xi(t)$

$$(55') \qquad \frac{\xi'(t)^2}{\xi(t)^2} + 1 = 0; \text{ i. e., } \xi(t) = \rho_v e^{it}, \qquad 0 < \rho_v = \text{constant.}$$

The final result may be formulated in the following way:

Among all univalent representations (30) of D which transform A into a circle $\tilde{A}$ and B into a contiguous radial segment, the maximal radius for $\tilde{A}$ is attained, if all other C_v $(v > 1)$ become circular arcs, concentric to $\tilde{A}$.

Thus, the same representation imparts to $d(A; C_A)$ its minimum and to $r \sin^2 \alpha/4$ its maximum which has the value $d(A; C_A)$. Hence, we get the inequality

(57) $d(A;C_A) \geq r \sin^2 \alpha/4$

valid for every représentation (30) which transforms C_1 into a circle. By means of such representations, there corresponds to A an arc of aperture α, and to its complement B on C_1 an arc of aperture $\beta = 2\pi - \alpha$. In view of (57) and the analogous formula for B

(57′) $d(B;C_B) \geq r \sin^2 \beta/4 = r \cos^2 \alpha/4,$

we get by addition

(58) $d(A;C_A) + d(B;C_B) \geq r.$

Since the maximum of r is given by the conformal radius $d(C_1;C)$ (4, 1), we get

(58′) $d(A;C_A) + d(B;C_B) \geq d(A+B;C)$

in complete analogy to the subadditivity formula (41) for the transfinite diameter.

5. The construction of $\gamma_m(x;y)$ by means of Green's function.

1. The functions $\gamma_m(x;y)$ may be constructed explicitly by means of Green's function $g(x;y)$ and other functions derived from it. In this way, we may derive the variation formula (17) for $\gamma_m(x;y)$ directly from (11). For this construction, we have to study the function $p(x;y)$, introduced in 1, 2 by (7), more thoroughly. Though its real part $g(x;y)$ is uniform in D, it has with respect to circuits around C_ν the periods

(59) $2\pi i \omega_\nu(y) = i \int_{C_\nu} \frac{\partial}{\partial n_z} g(y;z)\, ds_z.$

$\omega_\nu(y)$ is a real-valued harmonic function for $y \subset D$; by virtue of its integral representation it attains on C_ν the value 1, on the remaining C_μ the value 0. $\omega_\nu(y)$ is called the harmonic measure of C_ν in y with respect to D (Nevanlinna 1). Obviously, $\omega_\nu(y)$ is the real part of the analytic function

(60) $w_\nu(y) = \frac{1}{2\pi} \int_{C_\nu} \frac{\partial}{\partial n_z} p(y;z)\, ds_z.$

$w_\nu(y)$ is regular in D and possesses periods with respect to circuits around every C_μ; they are

(61) $2\pi i P_{\mu\nu} = i \int_{C_\mu} \frac{\partial \omega_\nu(y)}{\partial n_y}\, ds_y = \frac{i}{2\pi} \int_{C_\mu} \int_{C_\nu} \frac{\partial^2 g(y;z)}{\partial n_y \partial n_z}\, ds_y ds_z = 2\pi i P_{\nu\mu}.$

The matrix $(P_{\mu\nu})$ is thus seen to be symmetric.

Consider, for $x \subset D$, the harmonic function

$$(62) \qquad v(x) = \sum_{\nu=1}^{n} c_{\nu}\omega_{\nu}(x)$$

which has on the curve C_{ν} the boundary value c_{ν}. By Green's theorem we have

$$(63) \qquad \frac{1}{2\pi} \int\int_{D} (\operatorname{grad} v)^2 \, d\tau = -\frac{1}{2\pi} \int_{C} v \frac{\partial v}{\partial n} \, ds = -\sum_{\mu,\,\nu=1}^{n} P_{\mu\nu}c_{\mu}c_{\nu}.$$

This expression is non-negative and vanishes only for $\operatorname{grad} v \equiv 0$, i. e., $v(x)$ $=$ const.   In this case, all c_{ν} are necessarily equal, which leads to the theorem:

The quadratic form $\sum_{\mu,\,\nu=1}^{n} P_{\mu\nu}c_{\mu}c_{\nu}$ is non-positive and vanishes only if all c_{ν} are equal.

The harmonic function $\sum_{\nu=1}^{n} \omega_{\nu}(x)$ has everywhere on the boundary of D the value 1 and coincides, therefore, in D with the constant 1. Hence, the analytic function $\sum_{\nu=1}^{n} w_{\nu}(x)$ is also constant and its period with respect to a circuit around each C_{μ} vanishes:

$$(64) \qquad 2\pi i \sum_{\nu=1}^{n} P_{\mu\nu} = 0 \qquad\qquad (\mu = 1, \cdots, n).$$

Thus, the sum over every row (or column) in the matrix $(P_{\mu\nu})$ is zero and, consequently, the determinant of this matrix vanishes.

Striking out the n-th row and the n-th column of the matrix $(P_{\mu\nu})$, we obtain a matrix with a corresponding negative-definite form, thus with non-vanishing determinant. Let its inverse matrix be $(p_{\mu\nu}^{(n)})$, the upper index indicating that μ and ν do not assume the value n.

2.   By means of the previously defined concepts we may easily construct the function $\gamma_n(x;y)$. We consider the analytic function of $x \subset D$, for $y \subset D$ fixed,

$$(65) \qquad \phi_n(x;y) = p(x;y) - \sum_{\mu,\,\nu=1}^{n-1} p_{\mu\nu}^{(n)} w_{\mu}(x)\omega_{\nu}(y).$$

It has by (59) and (61) the period

$$(65') \qquad \pi_{\sigma} = 2\pi i \left[\omega_{\sigma}(y) - \sum_{\mu,\,\nu=1}^{n-1} p_{\mu\nu}^{(n)} P_{\mu\sigma}\omega_{\nu}(y) \right]$$

with respect to a circuit around C_{σ}. For $\sigma \neq n$, the definition of $(p_{\mu\nu}^{(n)})$

insures $\pi_\sigma = 0$. In order to calculate π_n we apply (64) and express $P_{\mu\nu}$ by means of all $P_{\mu\tau}$ $(\tau \neq n)$. It is then easily seen that

$$(65'') \qquad \pi_n = 2\pi i [\omega_n(y) + \sum_{\tau=1}^{n-1} \omega_\tau(y)] = 2\pi i.$$

The function $\phi_n(x;y)$, regular for $x \subset D$ with exception of the logarithmic pole at y, does not change, therefore, for circuits around C_ν $(\nu \neq n)$ and increases by $2\pi i$ for a circuit around C_n. Its real part is constant on every C_ν; in particular, it vanishes on C_n. Hence, $\phi_n(x;y) = \log f_n(x;y)$, $f_n(x;y)$ mapping D onto the exterior of the unit circle, slit along concentric circular arcs. Thus, in view of (19), (59), (60) and (65)

$$(66) \qquad \gamma_n(x;y) = R\{\phi_n(x;y)\} = g(x;y) - \sum_{\mu,\nu=1}^{n-1} p_{\mu\nu}{}^{(n)} \omega_\mu(x) \omega_\nu(y).$$

Analogously, we may construct all the functions $\gamma_m(x;y)$ $(1 \leq m \leq n-1)$, by defining the matrix $(p_{\mu\nu}{}^{(m)})$, inverse to the matrix obtained by striking out in $(P_{\mu\nu})$ the m-th row and the m-th column.

3. If y and z are two fixed points in D, consider the analytic function of x

$$(67) \qquad \phi_n(x;y,z) = \phi_n(x;y) - \phi_n(x;z).$$

It is uniform for every circuit, since the periods for a circuit around C_n cancel each other exactly. It has logarithmic poles in y and z and a constant real part on every boundary continuum C_ν, vanishing in particular on C_n. Thus, it is the logarithm of the function which maps D on a plane, slit along concentric circular arcs, so that y goes to infinity, z to zero and that C_n corresponds to an arc of the unit circle.

This representation is connected with the following extremum problem: Determine a function (30'), univalent in D, which yields the maximum value for $|f'(z)|$, $z \subset D$ (see de Possel, Rengel).

Let $\tilde{D}$ be the domain belonging to the extremal function, $\tilde{C}_\nu$ its boundary curves. Using a point z_0 on $\tilde{C}_\nu$, we consider the functions (33), univalent in the exterior of a subcontinuum Γ of $\tilde{C}_\nu$ with transfinite diameter ρ which contains z_0. Superposing such a function on the extremal representation $D \to \tilde{D}$, gives a resultant conformal representation of type (30') of D, since

$$(68) \qquad \phi[f(x)] = f^+(x) = f(x) + \frac{a\rho^2}{f(x) - z_0} + o(\rho^2)$$

has at $x = y$ the principal part $\dfrac{1}{x-y}$. Further, we have

$$(69) \qquad f^{\nu}(z) = f'(z)\left[1 - \frac{a\rho^2}{(f(z) - z_0)^2} + o(\rho^2)\right].$$

In view of the maximum property of $|f'(z)|$ there holds for every function (33)

$$(70) \qquad R\left\{\frac{a\rho^2}{(f(z) - z_0)^2}\right\} + o(\rho^2) \leq 0.$$

Hence, in virtue of Lemma **3**, 2, every C_ν is an analytic curve $x = x(t)$ with the differential equation

$$(71) \qquad x'(t)^2[f(z) - x(t)]^{-2} + 1 = 0.$$

Without restriction of generality we may suppose $f(z) = 0$, since addition of a constant to $f(x)$ changes neither its derivative nor its normalization (30′). Thus, all $\tilde{C}_\nu$ have the form

$$(70') \qquad x(t) = c_\nu e^{it}, \qquad\qquad 0 < c_\nu = \text{constant},$$

i. e., they are all circular arcs around the common center $f(z) = 0$.

Thus, the extremal function $f(x)$, yielding this representation, coincides, by the unicity theorem for this type of conformal representation, with a constant multiple of $\exp\{\phi_n(x; y, z)\}$. Hence,

$$(72) \qquad \log|f(x)| = k + g(x; y) - g(x; z)$$
$$- \sum_{\mu,\nu=1}^{n-1} p_{\mu\nu}{}^{(n)}\omega_\mu(x)[\omega_\nu(y) - \omega_\nu(z)].$$

In order to eliminate the constant k, we combine (72) with (29) and (30′); for $x \to y$, we get

$$(72') \qquad 0 = k + \log\frac{1}{d(y)} - g(y; z)$$
$$- \sum_{\mu,\nu=1}^{n-1} p_{\mu\nu}{}^{(n)}\omega_\mu(y)[\omega_\nu(y) - \omega_\nu(z)].$$

On the other hand, for $x \to z$, we get in the same way by means of the development $f(x) = f'(z)(x - z) + \cdots$, valid in the neighborhood of the point z,

$$(72'') \qquad \log|f'(z)| = k + g(z; y)$$
$$- \log\frac{1}{d(z)} - \sum_{\mu,\nu=1}^{n-1} p_{\mu\nu}{}^{(n)}\omega_\mu(z)[\omega_\nu(y) - \omega_\nu(z)].$$

By subtracting (72′) from (72″) we get

$$(73) \qquad \log|f'(z)| = 2g(z; y) + \log d(z) + \log d(y)$$
$$+ \sum_{\mu,\nu=1}^{n-1} p_{\mu\nu}{}^{(n)}[\omega_\mu(y) - \omega_\mu(z)][\omega_\nu(y) - \omega_\nu(z)].$$

7

The last term is non-positive, since $(p_{\mu\nu}{}^{(n)})$ is the inverse of a matrix with a non-positive quadratic form. The maximum for $|f'(z)|$ is $\geq \dfrac{1}{|z-y|^2}$, since this value is always attained in the particular case $f(x) = \dfrac{1}{x-y} - \dfrac{1}{z-y}$. Thus we have proved for the function (32) the interesting inequality

$$(74) \qquad \Gamma(x;y) = 2g(x;y) + \log d(x) + \log d(y) \geq \log \frac{1}{|x-y|^2}.$$

From (66) we obtain, by comparing coefficients

$$(75) \qquad \log \frac{1}{d_n(y)} = \log \frac{1}{d(y)} - \sum_{\mu,\nu=1}^{n-1} p_{\mu\nu}{}^{(n)} \omega_\mu(y) \omega_\nu(y).$$

Applying this identity and (66) to (73), we get

$$(76) \qquad \log|f'(z)| = 2\gamma_n(z;y) + \log d_n(z) + \log d_n(y).$$

Thus, we have established a simple relation between the domain function $\gamma_n(z;y)$ and the maximal derivative of all functions (30'), univalent in D. It makes obvious the fact, established in **4, 1**, that the right hand side of (76) is a monotone decreasing function of the domain (diminution of the family of competing functions decreases the maximum).

4. After having shown the significance of the expressions $w_\nu(x)$ and $P_{\mu\nu}$, we ask now about their variational formulae. We shall derive them from (11) and get in this way anew the variation formula for $\gamma_n(x;y)$. The variation (2) transforms the domain D with boundary curves C_ν and Green's function $g(x;y)$ into a domain D^* with boundary curves C^*_ν and Green's function $g^*(x;y)$. The connection between Green's functions of both domains is established by (11). If the point x describes a circuit around C_ν, its image point x^* turns around the corresponding C^*_ν, and, by (11) and (59), the period of $p^*(x^*;y^*)$ with respect to this circuit is

$$(77) \qquad 2\pi i\omega^*_\nu(y^*) = i\int_{C_\nu} \frac{\partial}{\partial n_x} g^*(y^*;x^*)\,ds_x$$

$$= i\int_{C_\nu} \frac{\partial g(y;x)}{\partial n_x}\,ds_x + iR\{e^{2i\phi}\rho^2 p'(z_0;y)\int_{C_\nu} \frac{\partial p'(z_0;x)}{\partial n_x}\,ds_x\} + o(\rho^2)$$

whence, by virtue of (59) and (60),

$$(78) \qquad \omega^*_\nu(y^*) = \omega_\nu(y) + R\{e^{2i\phi}\rho^2 p'(z_0;y)w'_\nu(z_0)\} + o(\rho^2).$$

This formula yields the variation of the harmonic measure $\omega_\nu(y)$. Using (61), we derive herefrom the variation formula for the $P_{\mu\nu}$. In fact, we have from (78):

$$(79) \quad P^*{}_{\mu\nu} = \frac{1}{2\pi} \int_{C\mu} \frac{\partial}{\partial n_y} \omega^*{}_\nu(y^*) \, ds_y$$

$$= P_{\mu\nu} + R\{e^{2i\phi}\rho^2 w'_\nu(z_0) \cdot \frac{1}{2\pi} \int_{C\mu} \frac{\partial p'(z_0; y)}{\partial n_y} \, ds_\nu\} + o(\rho^2),$$

whence, in view of (60),

$$(80) \quad P^*{}_{\mu\nu} = P_{\mu\nu} + R\{e^{2i\phi}\rho^2 w'_\mu(z_0) w'_\nu(z_0)\} + o(\rho^2).$$

The expressions $p_{\mu\nu}{}^{(n)}$ are defined by the system of equations

$$(81) \quad \sum_{\lambda=1}^{n-1} p_{\mu\lambda}{}^{(n)} P_{\lambda\sigma} = \delta_{\mu\sigma} = \begin{cases} 1 & \mu = \sigma \\ 0 & \mu \neq \sigma \end{cases} \qquad (\mu, \sigma = 1, \cdots, n-1).$$

We derive from (80) and (81) the following system of equations for $p_{\mu\nu}{}^{(n)*}$:

$$(82) \quad \delta_{\mu\sigma} = \sum_{\lambda=1}^{n-1} p_{\mu\lambda}{}^{(n)*} P^*{}_{\lambda\sigma} = \sum_{\lambda=1}^{n-1} p_{\mu\lambda}{}^{(n)*} P_{\lambda\sigma}$$

$$+ R\{e^{2i\phi}\rho^2 w'_\sigma(z_0) \sum_{\lambda=1}^{n-1} p_{\mu\lambda}{}^{(n)} w'_\lambda(z_0)\} + o(\rho^2).$$

Multiplying both sides with $p_{\sigma\nu}{}^{(n)}$ and summing up with respect to σ, we get from (81), in view of the symmetry property $p_{\mu\nu}{}^{(n)} = p_{\nu\mu}{}^{(n)}$

$$(83) \quad p_{\mu\nu}{}^{(n)*} + R\{e^{i2\phi}\rho^2 \sum_{\sigma=1}^{n-1} p_{\nu\sigma}{}^{(n)} w'_\sigma(z_0) \cdot \sum_{\lambda=1}^{n-1} p_{\mu\lambda}{}^{(n)} w'_\lambda(z_0)\} + o(\rho^2) = p_{\mu\nu}{}^{(n)},$$

i. e.,

$$(83') \quad p_{\mu\nu}{}^{(n)*} = p_{\nu\mu}{}^{(n)} - R\{e^{2i\phi}\rho^2 \sum_{\kappa=1}^{n-1} p_{\mu\kappa}{}^{(n)} w'_\kappa(z_0) \cdot \sum_{\lambda=1}^{n-1} p_{\nu\lambda}{}^{(n)} w'_\lambda(z_0)\} + o(\rho^2).$$

From (66), (11), (78) and (83') we get finally

$$(84) \quad \gamma^*{}_n(x^*; y^*) = \gamma_n(x; y)$$

$$+ R\{e^{2i\phi}\rho^2 [p'(z_0; x) - \sum_{\mu,\nu=1}^{n-1} p_{\mu\nu}{}^{(n)} w'_\mu(z_0)\omega_\nu(x)][p'(z_0; y)$$

$$- \sum_{\mu,\nu=1}^{n-1} p_{\mu\nu}{}^{(n)} w'_\mu(z_0)\omega_\nu(y)] + o(\rho^2),$$

and using definition (65)

$$(84') \quad \gamma^*{}_n(x^*; y^*) = \gamma_n(x; y) + R\{e^{2i\phi}\rho^2 \phi'_n(z_0; x)\phi'_n(z_0; y)\} + o(\rho^2).$$

This coincides exactly with the variation formula (17) for $\gamma_n(x; y)$.

6. The functionals of a Riemann surface and their variation.

1. The above formai relations between the functionals of a domain D: $g(x;y)$, $\omega_\nu(y)$ and $P_{\mu\nu}$ recall similar connections between the elementary integrals on a Riemann surface and their periods. These analogies are not accidental; they result from the close relations between the theory of conformal representation of multiply connected domains and the theory of Riemann surfaces (see Schottky).

The elementary integrals are functionals of the Riemann surface and we want to study how their variation depends upon that of the surface. To this purpose, we have first to recall certain concepts from the theory of Riemann surfaces.

If the Riemann surface P has the genus $p \geq 1$, there exists on it a canonical system of $2p$ closed rectifiable curves a_j, b_j $(j = 1, \cdots, p)$, such that every pair a_j, b_j has exactly one point of intersection, while no two curves of this system have further common points. Let $v_j(x)$ be analytic on P (i. e., it has a convergent development into power series at every point of P), with the period 1 with respect to a circuit on a_j and the period 0 with respect to circuits on every other a_k. $v_j(x)$ is fixed by this condition, up to an additive constant and is called the j-th elementary integral of first kind on P. The periods of all $v_j(x)$ with respect to circuits on the curves b_k form the matrix (π_{jk}).

Let $t(x;y)$ be an analytic function of x on P, the point y excepted where it has a simple pole with residue 1. Let $t(x;y)$ remain unchanged if x describes any curve a_j. These conditions fix $t(x;y)$ up to an additive constant; $t(x;y)$ is called an elementary integral of the second kind on P.

Let $w(x;y,z)$ be an analytic function of x on P, the points y and z excepted where it has logarithmic poles with the principal part $\log \dfrac{x-z}{x-y}$. Let $w(x;y,z)$ not change if x describes a curve a_j. These conditions fix $w(x;y,z)$ up to an additive constant. $w(x;y,z)$ is called an elementary integral of the third kind on P.

The elementary integrals of the first and second kind are uniform functions of x on the surface $\tilde{P}$, obtained from P by cutting it along all the curves of the system a_j, b_j. The boundary Σ of P consists of the curves a_j, b_j; but every side of each curve being counted separately, every curve appears twice in Σ. Let $f(x)$ be uniform and meromorphic on $\tilde{P}$; then

(85)
$$S\{f(x)\} = \frac{1}{2\pi i} \int_\Sigma f(x)\, dx$$

represents the sum of all residues of $f(x)$ on $\bar{\mathbf{P}}$. If, in particular, $f(x)$ is a linear aggregate of a finite number of elementary integrals with coefficients uniform on $\mathbf{P}$, we may, on the other hand, evaluate the expression (85) by using the characteristic periodicity relations of the elementary integrals, mentioned above. Riemann's method of boundary integration compares both results and obtains in this way relations between elementary integrals and their periods.

Since we shall use the result later on, we apply Riemann's method to

$$(86) \qquad S\{t(x;u)w'(x;y,z)\} = \frac{1}{2\pi i}\int_{\Sigma} t(x;u)w'(x;y,z)\,dx, \qquad u \subset \bar{\mathbf{P}}$$

(the dash denotes as usual the derivative with respect to the first variable). Integrating along Σ, we have to pass every curve a_j and b_j twice, but in opposite directions. We get from one bank of the curve a_j to the other by following the curve b_j, and from one bank of b_j to the other by describing a_j (starting always in their common point). $t(x;u)$ having the period zero with respect to every circuit a_j, it has the same value on both banks of every b_j; hence, the integrations along the two banks of b_j in (86) cancel each other. The values of $t(x;u)$ on both banks of a_j differ by a constant period, depending, however, upon u, say $p_j(u)$. Thus

$$(86') \qquad S\{t(x;u)w'(x;y,z)\} = -\sum_{k=1}^{p}\frac{1}{2\pi i}\int_{a_k} w'(x;y,z)\,dx \cdot p_k(u).$$

Now, $w(x;y,z)$ does not change after describing a full circuit a_k; hence, each integral in (86') vanishes, and their sum also.

On the other hand, we have, by virtue of the residue theorem,

$$(87) \qquad S\{t(x;u)w'(x;y,z)\} = w^t(u;y,z) - [t(y;u) - t(z;u)].$$

This being zero by the foregoing argument, we have established the following relation between elementary integrals of the second and the third kind:

$$(88) \qquad w'(x;y,z) = t(y;x) - t(z;x).$$

In the same way the following relations are proved: The equalities

$$(89) \qquad \pi_{jk} = \pi_{kj};$$

the periodicity relations

$$(90) \qquad \frac{1}{2\pi i}\int_{b_j} t'(x;y)\,dx = -v'_j(y),$$

$$(91) \qquad \frac{1}{2\pi i}\int_{b_j} w'(x;y,z)\,dx = -[v_j(y) - v_j(z)],$$

and the symmetry relations

$$(92) \qquad\qquad t'(x;y) = t'(y;x)$$

$$(93) \qquad w(x;y,z) - w(u;y,z) = w(y;x,u) - w(z;x,u).$$

2. After these preparations let us examine the change of the elementary integrals due to a variation of **P**. It is known that they depend continuously upon **P** (see Ritter, Koebe 2). But no explicit formula, describing the dependency, has been given as yet. We consider the following particular variations of **P**.

Let $f(x)$ be uniform and meromorphic on **P** with simple poles z_ν $(\nu = 1, \cdots, N)$ and residues r_ν at z_ν. For sake of simplicity we suppose that no z_ν coincides with a branch point of **P** or lies on a curve of our canonical system. Consider the function

$$(94) \qquad\qquad x^* = x + \rho^2 f(x).$$

Fix around each z_ν a small circumference k_ν, not containing any branch point of **P** or a point of the canonical curve system, such that no two circles overlap. Discarding the interiors of k_ν from **P**, we obtain a surface $\mathbf{P}_0$ with N holes. If ρ is sufficiently small, $\mathbf{P}_0$ is mapped by (94) in a one-to-one manner on a surface $\mathbf{P}^*_0$ with N holes, which can be completed to a closed Riemann surface $\mathbf{P}^*$ (see Schiffer 3).

Our purpose is to calculate for this variation $\mathbf{P} \to \mathbf{P}^*$ the corresponding variation formulae for the elementary integrals and their periods. We remark that **P** and $\mathbf{P}^*$ are of the same genus and that the variation (94) transforms the system a_j, b_j again into a canonical system a^*_j, b^*_j on $\mathbf{P}^*$. The functions $v^*_j(x)$, $t^*(x;y)$, $w^*(x;y,z)$ being the elementary integrals on $\mathbf{P}^*$, the functions $v^*_j(x^*(x))$, $t^*(x^*(x);y^*(y))$ and $w^*(x^*(x);y^*(y),z^*(z))$ are given on $\mathbf{P}_0$ by means of (94). If x describes the system a_j, b_j, its image x^* does the same with respect to a^*_j, b^*_j. We establish now the equality

$$(95) \quad \frac{1}{2\pi i} \int_\Sigma t(x;u) \frac{d}{dx} w^*(x^*;y^*,z^*) dx = \sum_{j=1}^p v'_j(u) \int_{a_j} \frac{d}{dx} w^*(x^*;y^*,z^*) dx,$$

by means of the periodicity properties of $t(x;u)$. In fact, $t(x;u)$ has the periods zero with respect to the circuits a_j and, by (90), the periods $-2\pi i\, v'_j(u)$ regarding the b_j. Further, $w^*(x^*;j^*,z^*)$ does not change, if x^* makes a full circuit around a^*_j, i. e., if x describes a_j. Thus,

$$(96) \qquad\qquad \frac{1}{2\pi i} \int_\Sigma t(x;u) \frac{d}{dx} w^*(x^*;y^*,z^*) dx = 0.$$

If, on the other hand, u, y, z lie in P_0, we may evaluate the last integral by Cauchy's theorem and so derive, from (96), the equality

$$(97) \qquad 0 = \frac{d}{du} w^*(u^*; y^*, z^*) - [t(y; u) - t(z; u)]$$
$$+ \sum_{\nu=1}^{N} \frac{1}{2\pi i} \int_{k\nu} t(x; u) \frac{d}{dx} w^*(x^*; y^*, z^*) dx.$$

By elementary development into series, using (94) and the residue theorem, we find

$$(98) \qquad \frac{1}{2\pi i} \int_{k\nu} t(x; u) \frac{d}{dx} w^*(x^*; y^*, z^*) dx = - r_\nu \rho^2 t'(z_\nu; u) w'(z_\nu; y, z) + o(\rho^2).$$

In view of (88) and (98) we may write, instead of (97),

$$(99) \qquad \frac{d}{du} w^*(u^*; y^*, z^*) = \frac{d}{du} w(u; y, z) + \sum_{\nu=1}^{N} r_\nu \rho^2 t'(z_\nu; u) w'(z_\nu; y, z) + o(\rho^2).$$

By virtue of (92), (99) may be integrated with respect to u from x_0 to x. Applying once more (88), we find

$$(100) \qquad w^*(x^*; y^*, z^*) - w^*(x_0^*; y^*, z^*) = w(x; y, z) - w(x_0; y, z)$$
$$+ \sum_{\nu=1}^{N} r_\nu \rho^2 w'(z_\nu; x, x_0) w'(z_\nu; y, z) + o(\rho^2).$$

Now, let x describe the curve b_j; then x^* will describe b^*_j and, in view of (91), (88) and (90), we get by comparison of the periods on both sides of (100):

$$(101) \qquad v^*_j(y^*) - v^*_j(z^*) = v_j(y) - v_j(z)$$
$$+ \sum_{\nu=1}^{N} r_\nu \rho^2 v'_j(z_\nu) w'(z_\nu; y, z) + o(\rho^2).$$

Next, let y describe the curve b_k and compare once more the increments on both sides:

$$(102) \qquad \pi^*_{jk} = \pi_{jk} - \sum_{\nu=1}^{N} r_\nu \rho^2 v'_j(z_\nu) v'_k(z_\nu) + o(\rho^2).$$

The great similarity between (100), (101) and (102), on the one hand, and (11), (72) and (80), on the other, is obvious. The above formulas determine explicitly the variation of the elementary integrals and their periods on Riemann surfaces. To their applications on extremum problems concerning Riemann surfaces we hope to return elsewhere.

EINSTEIN INSTITUTE,
THE HEBREW UNIVERSITY,
JERUSALEM, PALESTINE.

BIBLIOGRAPHY.

S. Bergman. (1) *Partial differential equations, Advanced topics (Conformal mapping of multiply connected domains)*, Publication of Brown University, Providence, R. I., 1941.

———. (2) A remark on the mapping of multiply connected domains, *American Journal of Mathematics*, vol. 68 (1946), pp. 20-28.

M. Biernacki. "Sur quelques majorantes de la théorie des fonctions univalentes, Comptes Rendus, vol. 201 (1935), pp. 256-258.

M. Fekete. "Über die Verteilung der Wurzeln bei gewissen algebraischen Gleichungen mit ganzzahligen Koeffizienten," *Math. Zeitschrift*, vol. 17 (1923), pp. 228-249.

H. Grötzsch. "Über die Verzerrung bei schlichter konformer Abbildung mehrfach-zusammenhängender schlichter Bereiche II," *Berichte Leipzig*, vol. 81 (1929), pp. 217-221.

J. Hadamard. "Mémoire sur le problème d'analyse relatif à l'équilibre des plaques élastiques encastrées," *Mémoires présentés par divers savants à l'Académie des Sciences*, vol. 33 (1908).

G. Julia. "Sur une équation aux dérivées fonctionelles liée à la représentation conforme," *Annales de l'Ecole Normale* (3), vol. 39 (1922), pp. 1-28.

P. Koebe. (1) Abhandlungen zur Theorie der konformen Abbildung IV," *Acta Math.*, vol. 41 (1918), pp. 305-344.

———. (2) "Über die Uniformisierung algebraischer Kurven IV," *Math. Annalen*, vol. 75 (1914), pp. 42-129.

P. Lévy. *Leçons d'Analyse fonctionelle*, Paris, 1922.

R. Nevanlinna. *Eindeutige analytische Funktionen*, Berlin, 1936.

G. Pólya. "Beitrag zur Verallgemeinerung des Verzerrungssatzes auf mehrfach-zusammenhängende Gebiete II," *Sitzungsberichte Akad. Berlin* (1928), pp. 280-282.

R. de Possel. "Sur quelques problèmes de représentation conforme," *Comptes Rendus*, vol. 194 (1932), pp. 42-44.

E. Rengel. "Existenzbeweise für schlichte Abbildungen mehrfach zusammenhängender Bereiche auf gewisse Normalbereiche," *Jahresb. D. Math. Ver.*, vol. 44 (1934), pp. 51-55.

E. Ritter. "Die Stetigkeit der automorphen Funktionen bei stetiger Abänderung des Fundamentalbereiches," *Math. Annalen*, vol. 45 (1894), pp. 473-544 and vol. 46 (1895), pp. 200-248.

M. Schiffer. "A method of variation within the family of simple functions," *Proceedings of the London Mathematical Society* (2), vol. 44 (1038), pp. 432-449.

———. (2) "On the subadditivity of the transfinite diameter," *Proceedings of the Cambridge Philosophical Society*, vol. 37 (1941), pp. 373-383.

———. (3) "Variation of the Green function and theory of the p-valued functions," *American Journal of Mathematics*, vol. 65 (1943), pp. 341-360.

———. (4) "On the kernal function of a system of ortho-normal functions," to appear later.

E. Schottky. "Über die konforme Abbildung mehrfach zusammenhängender ebener Flächen," *Journ. f. reine u. angew. Math.*, vol. 83 (1877), pp. 300-351.

G. Szegö. "Bemerkung zu einer Arbeit von Herrn M. Fekete," *Math. Zeitschr.*, vol. 21 (1924), pp. 203-208.

V. Volterra. *Theory of Functionals*, London, 1931.

Commentary on

[17] *Hadamard s formula and variation of domain-functions*, Amer. J. Math. **68** (1946), 417 448.

In this important paper, Schiffer opens new pathways for the variational method. He had previously introduced a generalization of the classical formula of Hadamard [H] for variation of Green's function. Hadamard's analysis relied on normal displacement of the boundary and required that the domain be smoothly bounded. Recognizing the need for greater generality for application to extremal problems, Schiffer [13] developed the method of interior variation and applied it to the coefficient problem for univalent functions. In [17], he turns attention to a variety of extremal problems involving quantities expressible in terms of Green's function: transfinite diameter, harmonic measures, canonical mappings of multiply connected domains, the Bergman kernel function, etc. Here the insight is that a variation of Green's function induces a variational formula for any such quantity.

The paper begins, rather surprisingly, with an application of Hadamard's formula to derive the variational formula for Green's function given previously in [13]. Assuming first that D is a domain bounded by analytic curves $C_1, C_2, \ldots, C_n$, Schiffer chooses an arbitrary point $z_0 \in D$ and introduces the mapping $z^* = z + e^{2i\varphi}\rho^2/(z - z_0)$, which carries the circle $\{z : |z - z_0| = \rho\}$ to a segment of length 4ρ and is univalent in the region $\{z : |z - z_0| > \rho\}$. For small ρ, the curves C_k are mapped to nearby analytic curves C_k^* bounding a domain D^*. Letting $g(z, \zeta)$ denote Green's function of D and $g^*(z, \zeta)$ that of D^* and choosing an analytic completion $p(z, \zeta)$ of $g(z, \zeta)$, he derives his familiar variational formula

$$g^*(z^*, \zeta^*) = g(z, \zeta)$$
$$+ \operatorname{Re}\{e^{2i\varphi}\rho^2 p'(z_0, z)p'(z_0, \zeta)\} + o(\rho^2)$$

from Hadamard's formula. An approximation argument then removes the restriction to domains with analytic boundary.

Schiffer now observes that the same method of derivation gives a variation for a more general class of domain-functions. One such function is $\log|f_m(z, \zeta)|$, where D is a domain bounded by n proper continua C_k and $f_m(z, \zeta)$ maps D conformally onto the exterior of the unit disk with $n - 1$ concentric circular slits, sending C_m to the unit circle and ζ to ∞. However, Schiffer points out that its variational formula also comes directly from the variation of Green's function, since $f_m(z, \zeta)$ can be constructed from the harmonic measures of boundary components (cf. Nehari [N]), which are represented via Green's function.

Similarly, the variation of Green's function of a domain D containing ∞ leads to a variation of the transfinite diameter $d = d(\partial D)$, since (as Schiffer [11] had noted), a theorem of Szego [S] says that $d = e^{-\gamma}$, where $\gamma = \lim_{n \to \infty}\{g(z, \infty) - \log|z|\}$. The variational formula for transfinite diameter is then applied to some specific extremal problems. A conformal mapping f of D is said to be *admissible* if it has the form $f(z) = z + b_0 + b_1/z + \ldots$ near ∞. It follows from Szego's theorem that $d(\partial D)$ is invariant under admissible mappings, but the transfinite diameter of a subset of ∂D may be distorted. Schiffer raises the question of extremal distortion of $d(C_m)$ for some particular boundary component C_m. He finds that among all admissible mappings, $d(f(C_m))$ is smallest when f maps C_m to a circle and sends the remaining $n - 1$ boundary components to radial slits. The maximum is attained when f maps C_m to a circle and the other components to concentric circular slits. Schiffer [99, 136, 137] later returned to the minimum problem, which led ultimately to the concept of Robin capacity; and Thurman [T] solved the maximum problem in greater generality (see commentary on [137]). Taking D to be the exterior of the unit disk, Schiffer [17] uses a variational method to obtain the sharp inequality $d(f(A)) \geq \sin^2(\alpha/4)$ for all admissible mappings f, where A is a closed arc of length α on the unit circle. A calculation based on Szego's theorem shows that $d(A) = \sin(\alpha/4)$. Some years later, Pommerenke [P1] applied the Goluzin inequalities (see [P2]) to show that $d(f(A)) \geq d(A)^2$ for every closed subset A of the unit circle. A variational proof, showing further that the inequality is sharp, appears in [137].

Schiffer's paper [19] can be viewed as a sequel to [17]. There he proved the relation $K(z, \zeta) = -(2/\pi)\partial^2 g(z, \zeta)/\partial z \partial \overline{\zeta}$ between Green's function and the Bergman kernel function, using it to obtain a variation of the kernel function which led to useful information. In fact, this line of investigation gradually blossomed into a major application of the variational method (see [51]). Also in [17] are early steps toward a variational treatment of functionals of Riemann surfaces. Eight years later, a series of joint papers with Spencer would culminate in a book [55] on this topic.

References

[H] Jacques Hadamard, *Mémoire sur le problème d analyse relatif à l équilibre des plaques élastiques encastrées*, Mémoires présentés par divers savants à l'Académie de Sciences **33** (1908), N° 4, 1 128.

[N] Zeev Nehari, *Conformal Mapping*, McGraw-Hill, 1952; Dover edition, 1975.

[P1] Ch. Pommerenke, *On the logarithmic capacity and conformal mapping*, Duke Math. J. **35** (1968), 321 325.

[P2] Ch. Pommerenke, *Univalent Functions*, Vandenhoeck & Ruprecht, 1975.

[S] G. Szego, *Bemerkungen zu einer Arbeit von Herrn M. Fekete: Über die Verteilung der Wurzeln bei gewissen algebraischen Gleichungen mit ganzzahligen Koef"zienten*, Math. Z. **21** (1924), 203 208.

[T] Robert E. Thurman, *Upper bound for distortion of capacity under conformal mapping*, Trans. Amer. Math. Soc. **346** (1994), 605 616.

PETER DUREN

[19] The kernel function of an orthonormal system

[19] Menahem Max Schiffer. "The kernel function of an orthonormal system," in *Duke Mathematical Journal*, Volume **13** (1946), 529–540.

THE KERNEL FUNCTION OF AN ORTHONORMAL SYSTEM

By Menahem Schiffer

1. Let B be a finite domain in the complex z-plane bounded by a finite number of smooth curves C_1, C_2, $\cdots$, C_n. A set of functions $\varphi_\nu(z)$ ($\nu = 1, 2, \cdots$), analytic in B and satisfying the condition

$$(1) \qquad \iint_B \varphi_\mu(z)\overline{\varphi_\nu(z)} \, d\omega_z = \delta_{\mu\nu}, \qquad z = x + iy, \qquad d\omega_z = dx \, dy,$$

is called orthonormal with respect to B.

The family of all functions $f(z)$, analytic in B and with $\iint_B |f(z)|^2 \, d\omega_z < \infty$, will be denoted by $L^2(B)$. If every $f(z) \subset L^2(B)$ may be expanded into a series

$$(2) \qquad f(z) = \sum_{\nu=1}^{\infty} c_\nu \varphi_\nu(z), \qquad c_\nu = \iint_B f(\zeta)\overline{\varphi_\nu(\zeta)} \, d\omega_\zeta$$

which converges uniformly in every interior sub-domain of B, the set $\varphi_\nu(z)$ is called a closed system.

Bergman [1, 2] proved for every domain B the existence of closed orthonormal systems and indicated a method for their construction. For every given domain B there exist an infinity of closed orthonormal sets. Bergman further defines

$$(3) \qquad K(z; \bar{\zeta}) = \sum_{\nu=1}^{\infty} \varphi_\nu(z)\overline{\varphi_\nu(\zeta)}$$

to be the kernel function of B. He shows that the infinite series in (3) converges uniformly if z and ζ are restricted to an interior partial domain of B and that, therefore, $K(z; \bar{\zeta})$ is an analytic function of z and $\bar{\zeta}$. $K(z; \bar{\zeta})$ is independent of the particular closed orthonormal set $\varphi_\nu(z)$, and is uniquely determined by the domain B. The kernel function plays an important role in the theory of conformal representation [2, 3] and is of considerable practical importance, especially as its numerical computation may be performed with ease.

There arises, however, the question of how $K(z; \bar{\zeta})$ is related to the other domain functions which appear in the theory of conformal representation, in particular to Green's function. In this paper, this relation will be established and thereby a twofold purpose will be attained. On the one hand, all theoretical results obtained with respect to Green's function become at once applicable to the kernel function; on the other hand, convenient methods for computing the kernel function are thus made available for practical applications of Green's function. It will be shown below that certain results concerning Green's function

Received January 14, 1946.

529

become much simpler if expressed in terms of the kernel function, which demonstrates the theoretical importance of the latter.

2. By definition (3) of the kernel function and in view of (2), every function $f(z) \subset L^2$ satisfies the integral equation

$$(4) \qquad f(z) = \iint_B K(z; \bar{\zeta}) f(\zeta) \, d\omega_\zeta \ .$$

This property to reproduce by integration every function $f(z) \subset L^2$ characterizes the kernel function uniquely. For, if there exists another function $k(z; \bar{\zeta})$ of this kind, analytic in z and ζ and belonging to L^2 with respect to both arguments, consider the integral

$$(5) \qquad \iint_B K(z; \bar{\zeta}) \overline{k(v; \bar{\zeta})} \, d\omega_\zeta = \overline{\iint_B k(v; \bar{\zeta}) K(\zeta; \bar{z}) \, d\omega_\zeta} \ .$$

In view of (4), the left-hand side has the value $\overline{k(v; \bar{z})}$; the right-hand integral may be evaluated by using the analogous property of $k(v; \bar{\zeta})$ and has, therefore, the value $\overline{K(v; \bar{z})}$. Hence, the identity of $k(v; \bar{z})$ with $K(v; \bar{z})$ is established.

Identity (4) leads immediately to a remarkable property of the kernel function. By Schwarz's inequality we obtain from (4)

$$(6) \qquad \begin{aligned} | f(z) |^2 &\leq \iint_B | f(\zeta) |^2 \, d\omega_\zeta \cdot \iint_B K(z; \bar{\zeta}) \overline{K(z; \bar{\zeta})} \, d\omega_\zeta \\ &\leq K(z; \bar{z}) \cdot \iint_B | f(\zeta) |^2 \, d\omega_\zeta \ , \end{aligned}$$

whence, for every function $f(z) \subset L^2$,

$$(6a) \qquad (K(z; \bar{z}))^{-1} \leq \iint_B \left| \frac{f(\zeta)}{f(z)} \right|^2 d\omega_\zeta$$

and equality holds only for $f(\zeta) = K(\zeta; \bar{z})$. Thus, we may define $K(z; \bar{\zeta})$ by an extremum property within the family of all $f(z) \subset L^2$. This definition was, in fact, used by Bergman [2] who, however, derived it from the original definition (3) of the kernel function and the orthonormality of the $\varphi_\nu(z)$.

3. Green's function of the domain B

$$(7) \qquad g(z; \zeta) = \log \frac{1}{|z - \zeta|} + h(z; \zeta)$$

is, for fixed $\zeta \subset B$, harmonic and regular in $z \subset B$ with the exception of $z = \zeta$ where $g(z; \zeta)$ has a logarithmic infinity; it vanishes if z is a point of the boundary of B. As is well known, $g(z; \zeta) \equiv g(\zeta; z)$; consequently, $g(z; \zeta)$ is also harmonic in ζ. We shall now obtain, by suitable differentiations of $g(z; \zeta)$, an analytic function of z and $\bar\zeta$ which belongs to L^2 with respect to both arguments and has the characteristic property (4) of the kernel function. We shall thus have effected the construction of $K(z; \bar\zeta)$ from the real-valued harmonic function $g(z; \zeta)$.

For the actual carrying out of this construction, we require the differential operators

$$(8) \qquad \frac{\partial}{\partial z} = \frac{1}{2}\left(\frac{\partial}{\partial x} - i\frac{\partial}{\partial y}\right), \qquad \frac{\partial}{\partial \bar z} = \frac{1}{2}\left(\frac{\partial}{\partial x} + i\frac{\partial}{\partial y}\right)$$

and the function

$$(9) \qquad M(z; \bar\zeta) = \frac{\partial^2 g(z; \zeta)}{\partial z\partial\bar\zeta}$$

which is obviously analytic in z and $\bar\zeta$, $g(z; \zeta)$ being harmonic in both its arguments. It is easily verified that the logarithmic pole of $g(z; \zeta)$ is suppressed by this particular process of differentiation and that $M(z; \bar\zeta)$ is regular everywhere in B.

Consider now the integral

$$(10) \qquad \mathfrak{F}(z) = \iint_B f(\zeta)M(z; \bar\zeta)\,d\omega_\zeta = \iint_B f(\zeta)\frac{\partial^2 g(z; \zeta)}{\partial z\partial\bar\zeta}\,d\omega_\zeta$$

with arbitrary $f(z) \subset L^2$. We evaluate $\mathfrak{F}(z)$ by integrating by parts; but by this process a pole at $\zeta = z$ reappears in Green's function. In order to avoid difficulties, we surround this point by a circle $|\zeta - z| \le \epsilon$ which is removed from B; the remaining domain will be denoted by B_ϵ. If, in (10), we integrate only over B_ϵ instead of over B, we commit an error $O(\epsilon)$, where $O(\epsilon)$ denotes a term with $\lim_{\epsilon \to 0} (O(\epsilon)/\epsilon) < \infty$. On the other hand, we know that $\partial g(z; \zeta)/\partial z$ is harmonic and regular for $\zeta \subset B_\epsilon$ and so we may now integrate in (10) by parts. This yields

$$(11) \qquad \mathfrak{F}(z) = \frac{1}{2i}\oint f(\zeta)\frac{\partial g(z; \zeta)}{\partial z}\,d\zeta - \iint_{B_\epsilon} \frac{\partial f(\zeta)}{\partial\bar\zeta}\frac{\partial g(z; \zeta)}{\partial z}\,d\omega_\zeta + O(\epsilon).$$

Here the line integral is to be taken over all the boundary curves of B_ϵ in the positive sense with respect to B_ϵ.

In view of definition (8) and of Cauchy-Riemann's differential equations, we have for every analytic function of ζ

$$(12) \qquad \frac{\partial f(\zeta)}{\partial\bar\zeta} \equiv 0.$$

Hence, the integral over B_ϵ in (11) vanishes.

We further remark that, for ζ on the boundary of B, we have, identically in

z, $g(z; \zeta) \equiv 0$. Hence, the derivative $\partial g(z; \zeta)/\partial z$ also vanishes identically in z if ζ is a point of the boundary of B. Thus, (11) may be rewritten in the form

$$(11a) \quad \mathfrak{F}(z) = -\frac{1}{2i} \oint_{|\zeta - z| = \epsilon} f(\zeta)\left[\frac{1}{2(\zeta - z)} + \text{bounded function}\right] d\zeta + O(\epsilon),$$

where the integration is now performed over the circumference in the positive sense. By Cauchy's residue theorem, we finally obtain

$$(11b) \qquad\qquad \mathfrak{F}(z) = -\frac{\pi}{2} f(z) + O(\epsilon).$$

We now let ϵ tend to zero and, by comparing the result of (11b) with the definition (10) of $\mathfrak{F}(z)$, we arrive at the identity

$$(13) \qquad\qquad f(z) = \iint_B f(\zeta)\left[-\frac{2}{\pi}\frac{\partial^2 g(z; \zeta)}{\partial z \partial \bar{\zeta}}\right] d\omega_\zeta .$$

This shows that the kernel of the integral equation (13) is equal to $K(z; \bar{\zeta})$, i.e.,

$$(14) \qquad\qquad K(z; \bar{\zeta}) = -\frac{2}{\pi}\frac{\partial^2 g(z; \zeta)}{\partial z \partial \bar{\zeta}}.$$

This identity was obtained for simply-connected domains by Greenstone in another way [4]. Equation (13) which is fundamental for our proof is due to Wirtinger [13].

4. Very similar transformations may be effected in the case of the integral

$$(15) \qquad\qquad y(\bar{z}) = \iint_B f(\zeta)\frac{\partial^2 g(z; \zeta)}{\partial \bar{z} \partial \bar{\zeta}} d\omega_\zeta$$

which represents an analytic function of $\bar{z}$. The kernel $\partial^2 g(z; \zeta)/\partial \bar{z}\partial\bar{\zeta}$ has for $\bar{z} = \zeta$ a double pole, but nevertheless $y(\bar{z})$ may be defined as the limit of corresponding integrals $y_\epsilon(\bar{z})$ with respect to domains B_ϵ. As before, we find by integration by parts that

$$(16) \qquad y_\epsilon(\bar{z}) = -\frac{1}{2i}\oint_{|\zeta - z| = \epsilon} f(\zeta)\left[\frac{1}{2(\bar{\zeta} - \bar{z})} + \text{bounded function}\right] d\zeta.$$

Putting $\zeta - z = \epsilon e^{i\varphi}$, $f(\zeta) = f(z) + \epsilon e^{i\varphi} f'(z) + \cdots$, we obtain

$$y_\epsilon(\bar{z}) = -\frac{1}{2i}\int_0^{2\pi} [f(z) + \epsilon e^{i\varphi} f'(z) + \cdots]$$

$$(16a)$$

$$\cdot \left[\frac{1}{2\epsilon}\cdot e^{i\varphi} + \text{bounded function}\right] i\epsilon e^{i\varphi}\, d\varphi$$

whence, in view of the orthogonality relations between the $e^{i\nu\varphi}$,

(16b)
$$y_\epsilon(\bar z) = O(\epsilon).$$

Taking again the limit for $\epsilon \to 0$, we have established the identity

(17)
$$\iint_B f(\zeta) \frac{\partial^2 g(z; \zeta)}{\partial \bar z \partial \bar\zeta}\, d\omega_\zeta \equiv 0$$

for every $f(z) \subset L^2$.

Let us now define the function

(18)
$$L(z; \zeta) = \frac{1}{2(z - \zeta)^2} + \frac{\partial^2 g(z; \zeta)}{\partial z \partial \zeta}$$

which is analytic and regular for z and ζ in B. According to (4), we have

(19)
$$L(z; v) = \iint_B K(z; \bar\zeta) L(\zeta; v)\, d\omega_\zeta$$

$$= \iint_B \frac{K(z; \bar\zeta)}{2(\zeta - v)^2}\, d\omega_\zeta + \iint_B K(z; \bar\zeta) \frac{\partial^2 g(v; \zeta)}{\partial v \partial \zeta}\, d\omega_\zeta\,.$$

In view of (17), the last integral on the right-hand side vanishes, whence

(20)
$$\frac{\partial^2 g(z; \zeta)}{\partial z \partial \zeta} = -\frac{1}{2(z - \zeta)^2} + \iint_B \frac{K(z; \bar t)}{2(t - \zeta)^2}\, d\omega_t\,.$$

Formulae (14) and (20) yield expressions for the mixed derivatives of Green's function in terms of the kernel function.

5. As an application of our identification of the kernel function with a particular derivative of Green's function, we shall develop the theory of variation of the *kernel function* in dependence of the domain B. Suppose that every point on the boundary of B is shifted in the direction of the normal by a distance δn, where δn is a continuous single-valued function on the boundary and is counted positive for inward displacements. As shown by Hadamard [6], Green's function will vary with B according to the formula

(21)
$$\delta g(z; \zeta) = -\frac{1}{2\pi} \oint \frac{\partial g(z; t)}{\partial n_t} \frac{\partial g(t; \zeta)}{\partial n_t} \delta n_t\, ds_t\,,$$

where the integration with respect to t is to be performed over the whole boundary of B.

Now we have, for $t = \sigma + i\tau$ on the boundary of B,

(22)
$$\frac{\partial g(z; t)}{\partial \sigma} = \frac{\partial g(z; t)}{\partial n_t} \cos (n, \sigma); \qquad \frac{\partial g(z; t)}{\partial \tau} = \frac{\partial g(z; t)}{\partial n_t} \cos (n, \tau);$$

whence

(22a) $\dfrac{\partial g(z;\,t)}{\partial t} = \dfrac{1}{2}\dfrac{\partial g(z;\,t)}{\partial n_t}\,e^{i\,(n,\sigma)};\qquad \dfrac{\partial g(t;\,\zeta)}{\partial t} = \dfrac{1}{2}\dfrac{\partial g(t;\,\zeta)}{\partial n_t}\,e^{-i\,(n,\sigma)};$

and

(22b) $\dfrac{\partial g(z;\,t)}{\partial n_t}\,\dfrac{\partial g(t;\,\zeta)}{\partial n_t} = 4\,\dfrac{\partial g(z;\,t)}{\partial t}\,\dfrac{\partial g(t;\,\zeta)}{\partial t}.$

Differentiating (21) with respect to z and $\bar\zeta$ we obtain, in view of (14) and (22b),

(23) $\delta K(z;\,\bar\zeta) = \oint K(z;\,\bar t)K(t;\,\bar\zeta)\,\delta n_t\,ds_t$

which is a particularly simple variation formula for $K(z;\,\bar\zeta)$. This type of variational equation was considered first by Hadamard [7] (see also Lévy [8]).

Putting $\zeta = z$ in (23), we obtain

(24) $\delta K(z;\,\bar z) = \oint |\,K(z;\,\bar t)\,|^2\,\delta n_t\,ds_t\,.$

If we choose δn_t always negative, i.e. if we expand the domain B everywhere, we obviously have $\delta K(z;\,\bar z) < 0$. The function $K(z;\,\bar z)$, which by definition is non-negative, decreases, therefore, when its domain of definition increases. This result is also due to Bergman [2] who applied it in connection with certain generalizations of the lemma of Schwarz-Pick in the theory of conformal representation.

In the same way, a somewhat more general result may be established. In virtue of (23), we have

(25) $\delta\{K(z;\,\bar z) + K(z;\,\bar\zeta) + K(\zeta;\,\bar z) + K(\zeta;\,\bar\zeta)\} = \oint |\,K(z;\,\bar t) + K(\zeta,\,\bar t)\,|^2\,\delta n_t\,ds_t$

which proves that $K(z;\,\bar z) + K(z;\,\bar\zeta) + K(\zeta;\,\bar z) + K(\zeta;\,\bar\zeta)$ is a decreasing domain function.

6. So far we have only considered domains B of finite connectivity and with smooth boundary, so as to be able to carry out all the integrations required. But it is well known that Green's function may also be defined for much more general domains B. By the usual method of approximating a general domain by smoothly bounded domains of finite connectivity, we may also establish the identities (14) and (20) for the general case. The theory of variation of Green's function is however restricted to the case of smooth boundaries, since only in this case the existence of a normal in every point of the boundary is assured. On the other hand, in dealing with extremum problems an unknown domain has often to be characterized by the property that a certain variation vanishes, even if it is not certain at all that the domain has a smooth boundary. For this purpose, another kind of variation has proved useful and shall be discussed now.

Let z_0 be an interior point of the domain B in question. The transformation

(26) $z^* = z + \dfrac{\rho^2 e^{i\varphi}}{z - z_0},\qquad \rho > 0,\qquad 0 \le \varphi \le 2\pi$

yields a univalent conformal representation of the domain $|z - z_0| > \rho$. For ρ small enough, the entire boundary of B is situated inside this domain and is therefore mapped in a one-to-one manner upon a new boundary set which defines a new domain B^*. The variation of Green's function $g(z; \zeta)$ with respect to the variation $B \to B^*$ is given by the formula

$$
g^*(z; \zeta) = g(z; \zeta) + \Re\left\{\rho^2 e^{i\varphi}\right.
$$

(27)

$$
\cdot\left[4\frac{\partial g(z_0\ ; z)}{\partial z_0}\frac{\partial g(z_0\ ; \zeta)}{\partial z_0} - \frac{2}{z - z_0}\frac{\partial g(z; \zeta)}{\partial z} - \frac{2}{\zeta - z_0}\frac{\partial g(z; \zeta)}{\partial \zeta}\right]\right\} + o(\rho^2),
$$

where $\Re$ = real part and $g^*(z; \zeta)$ is Green's function of B^* and $o(\rho^2)$ denotes a term satisfying for $\rho \to 0$ the limit relation $\lim o(\rho^2)/\rho^2 = 0$. (See [10].)

Differentiating the variation formula (27) with respect to z and ζ and using the identities (14) and (20), we easily obtain the following variation formula for the kernel function:

$$
K^*(z; \bar{\zeta}) = K(z; \bar{\zeta}) + \rho^2 e^{i\varphi}\left(2K(z_0\ ; \bar{\zeta})L(z_0\ ; z) - \frac{\partial}{\partial z}\left[\frac{K(z; \bar{\zeta}) - K(z_0\ ; \bar{\zeta})}{z - z_0}\right]\right)
$$

(28)

$$
+ \rho^2 e^{-i\varphi}\left(2K(z; \bar{z}_0)\overline{L(z_0\ ; \zeta)} - \frac{\partial}{\partial\bar{\zeta}}\left[\frac{K(z; \bar{\zeta}) - K(z; \bar{z}_0)}{\bar{\zeta} - \bar{z}_0}\right]\right) + o(\rho^2).
$$

In the same way we may establish the variation formula for $L(z; \zeta)$ by differentiation of (27) with respect to z and ζ. An easy calculation yields

$$
L^*(z; \zeta) = L(z; \zeta) + \rho^2 e^{i\varphi}(2L(z_0; z)L(z_0; \zeta)
$$

$$
- \frac{\partial}{\partial z}\left[\frac{L(z; \zeta) - L(z_0\ ; \zeta)}{z - z_0}\right] - \frac{\partial}{\partial\zeta}\left[\frac{L(z; \zeta) - L(z; z_0)}{\zeta - z_0}\right]\right)
$$

(29)

$$
+ \rho^2 e^{-i\varphi}\cdot\frac{\pi^2}{2}\,K(z; \bar{z}_0)K(\zeta; \bar{z}_0) + o(\rho^2).
$$

Formulae (28) and (29) show the importance of developing the theory of $L(z; \zeta)$ together with that of $K(z; \bar{\zeta})$. The variations of both functions are expressed by a system of two variational equations. In contradistinction to (27), the formulae (28) and (29) do not contain derivatives with respect to z_0, a fact which is of importance in applications to extremum problems.

7. In §3 we obtained an integral equation by a method of integration by parts which utilizes the vanishing of Green's function on the boundary of B. This suggests applying the same method to functions harmonic in B which have constant values on each boundary continuum of the domain and possess a logarithmic singularity.

To this end, let us consider the function

$$(30) \qquad \varphi_n(z; \zeta) = d_1(\zeta)(z - \zeta) + d_2(\zeta)(z - \zeta)^2 + \cdots ,$$

which, for fixed $\zeta \subset B$, is analytic in $z \subset B$ and maps B on the interior of the unit circle slit along concentric circular arcs, such that ζ corresponds to the center of the circle and the boundary continuum C_n to the circumference of the unit circle. For fixed $\zeta \subset B$,

$$(31) \qquad g_n(z; \zeta) = - \log | \phi_n(z; \zeta) |$$

is harmonic in z. $g_n(z; \zeta)$ has a logarithmic pole at $z = \zeta$, in the same way as Green's function, and if z varies on the m-th boundary continuum C_m, we have

$$(32) \qquad g_n(z; \zeta) = k_{m,n}(\zeta) \qquad\qquad \text{for } z \subset C_m .$$

In particular we have $k_{n,n}(\zeta) = 0$, by definition. It is known [9 and 12] that $g_n(z; \zeta)$ is symmetric in both arguments, i.e. $g_n(z; \zeta)$ is also harmonic in ζ. The function

$$(33) \qquad M_n(z; \bar{\zeta}) = \frac{\partial^2 g_n(z; \zeta)}{\partial z \, \partial \bar{\zeta}}$$

is therefore analytic in z and $\bar{\zeta}$ and it is easily seen that $M_n(z; \bar{\zeta})$ is regular everywhere in B. As in §3, we consider now the integral

$$(34) \qquad \mathfrak{F}_n(z) = \iint_B f(\zeta) \, M_n(z; \bar{\zeta}) \, d\omega_\zeta$$

for an arbitrary function $f(z) \subset L^2$. In exactly the same way as before, we find by integration by parts

$$(35) \qquad \mathfrak{F}_n(z) = \frac{1}{2i} \oint f(\zeta) \, \frac{\partial g_n(z; \zeta)}{\partial z} \, d\zeta + O(\epsilon),$$

where the integral is to be taken over the boundary of B_ϵ in the positive sense with respect to the domain. In view of (32), (35) may be written

$$
\mathfrak{F}_n(z) = \sum_{\nu=1}^n \frac{1}{2i} \oint_{C_\nu} f(\zeta) \, \frac{\partial}{\partial z} (k_{\nu,n}(z)) \, d\zeta
$$

$$(35a)$$

$$
- \frac{1}{2i} \oint_{|\zeta - z| = \epsilon} f(\zeta) \, \frac{\partial g_n(z; \zeta)}{\partial z} \, d\zeta + O(\epsilon).
$$

Let us now make an additional assumption regarding $f(z)$. We shall suppose that the integral of $f(z)$ is a single-valued function in B, i.e. there exists in B a single-valued function $F(z)$ such that $f(z) = \partial F(z)/\partial z$. With this assumption we may simplify (35a) to

$$(35b) \qquad \mathfrak{F}_n(z) = -\frac{1}{2i} \oint_{|\zeta - z| = \epsilon} f(\zeta) \, \frac{\partial g_n(z; \zeta)}{\partial z} \, d\zeta + O(\epsilon),$$

and, making the same transformations as in §3, we finally obtain

(36) $$\mathfrak{F}_n(z) = -\frac{\pi}{2}\, f(z),$$

i.e.

(36a) $$f(z) = \iint_B f(\zeta)\left[-\frac{2}{\pi}\frac{\partial^2 g_n(z;\,\zeta)}{\partial z\,\partial\bar\zeta}\right] d\omega_\zeta .$$

We have thus proved: The process of integration (36a), involving the function

(37) $$K^\star(z;\,\bar\zeta) = -\frac{2}{\pi}\frac{\partial^2 g_n(z;\,\zeta)}{\partial z\,\partial\bar\zeta}$$

reproduces every function $f(z)$ of the subclass l^2 of L^2 composed of all functions possessing a single-valued integral in B.

Bergman [2] considered systems of orthonormal functions of the class l^2. He showed that there always exist orthonormal systems which are closed with respect to l^2 and he defined for them the kernel function which, of course, satisfies also the integral equation (36a). As in §2, it is easily seen that the kernel function of the systems considered coincides with our function $K^\star(z;\,\bar\zeta)$ defined in (37).

Obviously, we might in exactly the same way prove

(37a) $$K^\star(z;\,\bar\zeta) = -\frac{2}{\pi}\frac{\partial^2 g_\nu(z;\,\zeta)}{\partial z\,\partial\bar\zeta} \qquad (\nu = 1, 2, \cdots, n)$$

where $g_\nu(z;\,\zeta)$ is defined similarly as $g_n(z;\,\zeta)$, differing from it only by the fact that C_ν now plays the role of C_n. It is interesting to note that the particular type of differentiation used in (37a) leads to a result which is independent of ν.

8. The function

(38) $$\log\phi_n(z;\,\zeta) = \log(z - \zeta) + \log d_1(\zeta) + \cdots ,$$

is, for fixed $\zeta \subset B$, analytic and regular in z, apart from the logarithmic singularity $z = \zeta$. It has a constant real part if z varies on a fixed boundary continuum C_m ; in fact, we conclude from (31) and (32)

(39) $$\mathfrak{R}\{\log\phi_n(z;\,\zeta)\} = -k_{m,n}(\zeta) \qquad \text{for } z \subset C_m .$$

Hence, if we define for $\zeta = \xi + i\eta$ the functions

(40) $$-\frac{\partial}{\partial\xi}[\log\phi_n(z;\,\zeta)] = \frac{1}{z - \zeta} + \alpha_0(\zeta) + \alpha_1(\zeta)(z - \zeta) + \cdots ,$$

(40a) $$-\frac{\partial}{\partial\eta}[\log\phi_n(z;\,\zeta)] = i\left[\frac{1}{z - \zeta} + \beta_0(\zeta) + \beta_1(\zeta)(z - \zeta) + \cdots\right],$$

each of them is, for fixed $\zeta \subset B$, analytic and regular in z with the exception of the pole $z = \zeta$, single-valued, and of constant real part if z is situated on a fixed C_m. Since each of these functions has only one simple pole and all the values it attains on the boundary of B are situated on linear segments parallel to the imaginary axis it follows, by a well-known argument, that both are univalent. If we put

$$(41) \quad f(z; \zeta) = i \frac{\partial}{\partial \eta}[\log \phi_n(z; \zeta)] = \frac{1}{z - \zeta} + \beta_0(\zeta) + \beta_1(\zeta)(z - \zeta) + \cdots,$$

$$(41a) \quad g(z; \zeta) = -\frac{\partial}{\partial \xi}[\log \phi_n(z; \zeta)] = \frac{1}{z - \zeta} + \alpha_0(\zeta) + \alpha_1(\zeta)(z - \zeta) + \cdots,$$

$f(z; \zeta)$ maps B in a one-to-one manner on a plane slit along linear segments parallel to the real axis, the point $\zeta \subset B$ corresponding to infinity; in the same way, $g(z; \zeta)$ maps B on a plane slit along linear segments parallel to the imaginary axis.

The pair of univalent functions just constructed is well-known in the theory of conformal representation [5]. In virtue of definitions (31) and (37), we may express $K^\star(z; \bar{\zeta})$ in terms of these functions as follows:

$$(42)$$
$$K^\star(z; \bar{\zeta}) = \frac{1}{2\pi}\left[\frac{\partial f(z; \zeta)}{\partial z} - \frac{\partial g(z; \zeta)}{\partial z}\right]$$
$$= \frac{\beta_1(\zeta) - \alpha_1(\zeta)}{2\pi} + \frac{2(\beta_2(\zeta) - \alpha_2(\zeta))}{2\pi}(z - \zeta) + \cdots.$$

In particular, one has for $z = \zeta$

$$(42a) \qquad K^\star(\zeta; \bar{\zeta}) = \frac{1}{2\pi}(\beta_1(\zeta) - \alpha_1(\zeta)).$$

Now, the difference

$$(43) \qquad S(\zeta) = \beta_1(\zeta) - \alpha_1(\zeta),$$

which is, in view of (42a), always positive, is a well-known functional of the domain B. It is closely related to the family $\mathfrak{F}_\zeta$ of all functions which are univalent in B and have at $z = \zeta$ the normalization

$$(44) \qquad f(z) = \frac{1}{z - \zeta} + a_0 + a_1(z - \zeta) + \cdots.$$

The coefficients a_1 corresponding to all the functions of $\mathfrak{F}_\zeta$ cover exactly a circle of the complex plane with the diameter $S(\zeta)$.

Each function $u = f(z) \subset \mathfrak{F}_\zeta$ maps B onto an infinite domain D of the u-plane. Consider the complement D_1 of D in the u-plane which consists of n continua. It can be shown [11] that the area of D_1 is always less than $\frac{1}{2}\pi S(\zeta)$

and that there always exists an extremal function $f(z) \subset \mathfrak{F}_\zeta$ for which this maximal value is attained in fact. We call, therefore, the functional

$$(45) \qquad A_e(\zeta) = \frac{\pi}{2} S(\zeta) = \pi^2 K^\star(\zeta; \bar{\zeta})$$

the maximal external area of B with respect to ζ.

Further let G_ζ be the family of all functions $f(z)$ which are analytic and single-valued in B and have at $z = \zeta$ the normalization

$$(46) \qquad f(z) = (z - \zeta) + b_2(z - \zeta)^2 + \cdots .$$

Every $f(z) \subset G_\zeta$ maps B upon a domain which is not necessarily simply covered and which has the area

$$(47) \qquad A = \iint_B | f'(t) |^2 d\omega_t .$$

Applying the Schwarz inequality to the integral equation (36a) with $f'(z)$ instead of $f(z)$, one obtains immediately

$$(48) \qquad | f'(z) |^2 \leq \iint_B | f'(t) |^2 d\omega_t \cdot K^\star(z; \bar{z}).$$

Putting here $z = \zeta$ and remarking that, in virtue of (46), $f'(\zeta) = 1$, one obtains the inequality

$$(49) \qquad A \geq (K^\star(\zeta; \bar{\zeta}))^{-1}.$$

The minimal area is attained in the case $f(z) = \int_\zeta^z [K^\star(t; \bar{\zeta})/K^\star(\zeta; \bar{\zeta})] \, dt$ which, as is easily seen, belongs to G_ζ . In virtue of this minimum property, we call

$$(50) \qquad A_i(\zeta) = (K^\star(\zeta; \bar{\zeta}))^{-1}$$

the minimal interior area of B with respect to ζ.

From (45) and (50) we deduce the interesting identity

$$(51) \qquad A_i(\zeta) \cdot A_e(\zeta) = \pi^2$$

which connects the maximal exterior area of B with its minimal interior area with respect to an arbitrary point ζ of B.

BIBLIOGRAPHY

1. S. BERGMAN, *Über die Entwicklung der harmonischen Funktionen der Ebene und des Raumes nach Orthogonalfunktionen*, Mathematische Annalen, vol. 86(1922), pp. 236–271.
2. S. BERGMAN, *Partial Differential Equations*, Advanced Topics, Brown University, Providence, 1941.
3. S. BERGMAN, *A Remark on the Mapping of Multiply-Connected Domains*, American Journal of Mathematics, vol. 68(1946), pp. 20–28.
4. L. GREENSTONE, *Mapping of Multiply Connected Domains by Analytic Functions*, Transactions of the American Mathematical Society, vol. 61(1947).

5. H. Grötzsch, *Über das Parallelschlitztheorem der konformen Abbildung schlichter Bereiche*, Berichte Verhandlungen sächsischen Akademie Leipzig, vol. 84(1932), pp. 15–36.

6. J. Hadamard, *Sur les opérations fonctionnelles*, Comptes Rendus, Paris, vol. 136(1903), pp. 351–354.

7. J. Hadamard, *Mémoire sur le problème d'analyse relatif à l'équilibre des plaques élastiques encastrées*, Mémoires des Savants étrangers, vol. 33(1908), pp. 1–128.

8. P. Lévy, *Leçons d'Analyse Fonctionnelle*, Paris, 1922.

9. C. C. Lin, *On the motion of vortices in two dimensions*, University of Toronto Studies, Applied Mathematics, No. 5, 1943.

10. M. Schiffer, *Variation of the Green Function and Theory of the p-valued Functions*, American Journal of Mathematics, vol. 65(1943), pp. 341–360.

11. M. Schiffer, *The Span of Multiply Connected Domains*, this Journal, vol. 10(1943), pp. 209–216.

12. M. Schiffer, *Hadamard's Formula and Variation of Domain-Functions*, American Journal of Mathematics, vol. 68(1946), pp. 417–448.

13. W. Wirtinger, *Über eine Minimalaufgabe im Gebiet der analytischen Funktionen*, Monatshefte für Mathematik und Physik, vol. 39(1932), pp. 377–384.

The Hebrew University, Jerusalem.

Commentary on

[19] *The kernel function of an orthonormal system*, Duke Math. J. **13** (1946), 529 540.

This is the first in a series of well-known papers (the remainder written with Stefan Bergman) dealing with the reproducing kernel in a Hilbert space of solutions of an elliptic partial differential equation. Here the elliptic operator is simply $\frac{\partial}{\partial \bar{z}}$, and so the kernel function $K(z, \bar{\zeta})$ is the classical Bergman kernel. Schiffer offers a new proof of the relation

$$K(z, \bar{\zeta}) = -\frac{2}{\pi} \frac{\partial^2 g(z, \zeta)}{\partial z \partial \bar{\zeta}}$$

between $K(z, \bar{\zeta})$ and Green's function $g(z, \zeta)$ of the domain, and then applies it to study the variation of the kernel function with respect to that of the domain. A similar investigation is also carried out for the reproducing kernel in the space of analytic functions that possess single-valued antiderivatives. "Green's functions" associated with the latter kernel were later used by V. A. Zmorovich [Z] and P. M. Tamrazov [T] in a geometric construction of analogues of Blaschke products in finitely connected domains. (For a more streamlined exposition of the results in [Z, T], see [CW, Kh].)

The techniques developed in this paper have been used and extended over the years; see the references in the commentaries on [20, 23].

References

[CW] R. Coifman and Guido Weiss, *A kernel associated with certain multiply connected domains and its applications to factorization theorems*, Studia Math. **28** (1966/1967), 31 68.

[Kh] D. Khavinson, *Factorization theorems for different classes of analytic functions in multiply connected domains*, Pacific J. Math. **108** (1983), 295 318.

[T] P. M. Tamrazov, *A generalized Blaschke product in domains of arbitrary connectivity*, Dopovidi Akad. Nauk Ukraïn. RSR **1962** (1962), 853 856 (Ukrainian).

[Z] V. A. Zmorovic, *On the generalisation of Schwarz s integral formula on n-connected circular domains*, Dopovidi Akad. Nauk Ukraïn. RSR **1958** (1958), 489 492 (Ukrainian).

DMITRY KHAVINSON

[20] (with S. Bergman) A representation of Green's and Neumann's functions in the theory of partial differential equations of second order

[20] Menahem Max Schiffer, S. Bergman. "A representative of Green's and Neumann's functions in the theory of partial differential equations of second order," in *Duke Mathematical Journal*, Volume **14** (1947), 609–638.

A REPRESENTATION OF GREEN'S AND NEUMANN'S FUNCTIONS IN THE THEORY OF PARTIAL DIFFERENTIAL EQUATIONS OF SECOND ORDER

By S. Bergman and M. Schiffer

Introduction. Despite the fact that Laplace's differential equation in two variables is a particular case of a partial differential equation of elliptic type, the investigations carried out in this special theory are to a large extent methodologically isolated. The reason is that the powerful tools of complex variables and the theory of analytic functions are not applicable to more general equations.

Recently, new methods have been introduced into the theory of analytic and harmonic functions which are based upon the concept of orthogonal systems, and are of much greater generality than the older ones. They were, in fact, first developed in the theory of analytic functions of two complex variables and led to important results in this difficult branch of analysis [1].

Not long ago it was found possible to make a further advance in this approach and to show that various important functions connected with a plane domain, such as harmonic measure and Green's function, can be expressed in a particularly simple form in terms of orthogonal functions [3] and [10].

In this paper these methods will be developed and applied to the theory of an extended class of partial differential equations of elliptic type. We shall obtain a unified theory and, in particular, a construction for Green's and Neumann's functions for a given domain in terms of orthogonal functions. This construction seems to be of real value for actual computations. Our theory permits further an easy investigation of the variation of the above fundamental solutions with varying domain of definition.

In the last section, we shall discuss in more detail the extent and character of our methods.

1. **Generalities and definitions.** In order to show the ideas in full clarity, without being hampered by too many formalisms, we shall develop the theory at first in the case of a certain simple differential equation. The methods are, however, quite general and in §7 we shall indicate the possible extensions of the theory. Let us consider, therefore, the differential equation

$$(1) \qquad \Delta\varphi - P\varphi = \frac{\partial^2 \varphi}{\partial x^2} + \frac{\partial^2 \varphi}{\partial y^2} - P\varphi = 0, \qquad P(x, y) > 0,$$

Received February 7, 1947. Research paper done under Navy Contract NOrd 8555-Task F, at Harvard University. The ideas expressed in this paper represent the personal views of the authors, and are not necessarily those of the Bureau of Ordnance.

609

in a given domain B of the (x, y)-plane. We suppose $P(x, y)$ to be a continuous positive function of (x, y).

We introduce now a metric in function-space defined by the scalar product.

$$(2) \qquad\qquad D\{\varphi, \psi\} = \iint_B [\varphi, \psi]\, dx\, dy,$$

where the integral over B is understood in the Lebesgue sense and where

$$(3) \qquad\qquad [\varphi, \psi] = \frac{\partial \varphi}{\partial x}\frac{\partial \psi}{\partial x} + \frac{\partial \varphi}{\partial y}\frac{\partial \psi}{\partial y} + P\varphi\psi.$$

This product obviously satisfies the Schwarz inequality

$$(4) \qquad\qquad D\{\varphi, \psi\}^2 \leq D\{\varphi, \varphi\} \cdot D\{\psi, \psi\}$$

and hence the triangle inequality

$$(5) \qquad\qquad (D\{\varphi - \psi, \varphi - \psi\})^{\frac{1}{2}} \leq (D\{\varphi, \varphi\})^{\frac{1}{2}} + (D\{\psi, \psi\})^{\frac{1}{2}}.$$

Let us note the following identity. If φ and ψ are both solutions of (1), we have

$$(6) \qquad\qquad \Delta(\varphi\psi) = 2[\varphi, \psi],$$

a relation which will be used later on.

We define *the class* Λ of solutions φ of (1) for which $D\{\varphi, \varphi\} < \infty$ and which have second derivatives in B. It is well known [4] that one can construct a complete set of solutions $\{\varphi_\nu(x, y)\}$ of (1), of class Λ, such that every solution of Λ may be represented as a linear combination of this set. Writing, instead of the pair of coordinates (x, y), the letter Z which denotes the point (x, y), we may express

$$(7) \qquad\qquad \varphi(Z) = \lim_{n=\infty} \sum_{\nu=1}^{n} A_\nu^{(n)}\varphi_\nu(Z),$$

and the convergence is uniform in every given interior partial domain of B.

We may suppose without loss of generality that the complete system of functions $\{\varphi_\nu\}$ is orthonormal with respect to the above metric, that is

$$(8) \qquad\qquad D\{\varphi_\nu, \varphi_\mu\} = \delta_{\nu\mu}, \qquad \delta_{\nu\mu} = \begin{cases} 0 & (\nu \neq \mu) \\ 1 & (\nu = \mu). \end{cases}$$

In this case, (7) may be put into the simpler form

$$(7') \qquad\qquad \varphi(Z) = \sum_{\nu=1}^{\infty} A_\nu\varphi_\nu(Z),$$

and this series converges uniformly in every interior partial domain of B. The coefficients A_ν are, in view of (8), given by the Fourier formula

$$(7'') \qquad\qquad A_\nu = D\{\varphi, \varphi_\nu\}.$$

Bergman [4] has shown that the bilinear expression

$$(9) \qquad K(Z, W) = \sum_{\nu=1}^{\infty} \varphi_\nu(Z)\varphi_\nu(W) \qquad\qquad (Z \equiv (x, y) \,\varepsilon\, B, W \equiv (u, v) \,\varepsilon\, B),$$

converges uniformly in every closed subdomain of B, and is independent of the particular choice of the complete orthonormal system $\{\varphi_\nu(Z)\}$. This function is called the kernel function of the domain B with respect to the differential equation (1), and is one of the most important domain functions connected with it and a given domain B.

We mention at first the following obvious but fundamental identity. Let $\varphi(Z)$ be a solution of (1) of class Λ. Then from (7'), (8) and (9) one has immediately

$$(10) \qquad\qquad \varphi(Z) = D_W\{K(Z, W), \varphi(W)\} \qquad\qquad (Z \,\varepsilon\, B),$$

where the subscript W of D indicates the variables over which the D-process is to be carried out. This integro-differential equation which is satisfied by every solution $\varphi(Z) \,\varepsilon\, \Lambda$ may be used to derive two important properties of the kernel function which define it independently of any orthonormal system:

(a) Combining (10) with (4) yields

$$(11) \qquad\qquad \varphi(T)^2 \leq D_W\{K(T, W), K(W, T)\}D\{\varphi, \varphi\},$$

and equality holds only for $\varphi_T(Z) = K(Z, T)$. Since $\varphi_T(W) = K(W, T)$ is a solution of (1), we have further by virtue of (10)

$$(12) \qquad\qquad D_W\{K(T, W), K(W, T)\} = K(T, T),$$

whence

$$(13) \qquad\qquad \varphi(T)^2 \leq K(T, T)D\{\varphi, \varphi\}.$$

Thus we have proved the following extremum theorem:

Consider all solutions $\varphi(Z)$ of class Λ which satisfy at a fixed point $T \,\varepsilon\, B$ the equation $\varphi(T) = 1$. That function $\varphi_T(Z)$ of this class which yields the minimum norm $D\{\varphi, \varphi\}$ is just the function $K(Z, T) \cdot [K(T, T)]^{-1}$ and the minimum norm is $[K(T, T)]^{-1}$.

It is obvious that this theorem provides an independent definition of the kernel function.

(b) The following uniqueness theorem will be of great use in this paper.

Let $L(Z, W)$ be a symmetric function of Z and W which belongs as a function of each variable to the class Λ. If for every solution $\varphi(Z)$ of the class Λ the identity

$$(10') \qquad\qquad \varphi(Z) = D_W\{L(Z, W), \varphi(W)\} \qquad\qquad (Z \,\varepsilon\, B)$$

holds, then one has necessarily

(14) $$L(Z, W) \equiv K(Z, W) \qquad (Z \, \varepsilon \, B, \, W \, \varepsilon \, B).$$

Proof. Apply (10') to the particular function $\varphi_T(Z) = K(Z, T)$. One has

(15) $$K(Z, T) = D_W\{L(Z, W), K(W, T)\} \qquad (Z \, \varepsilon \, B, \, T \, \varepsilon \, B).$$

On the other hand, from the symmetry of K and L, the identity (10) and the fact that $L(W, Z)$ belongs as a function of W to class Λ, one has

(15') $$K(Z, T) = D_W\{K(T, W), L(W, Z)\} = L(T, Z) = L(Z, T)$$

which proves our assertion.

The kernel function is closely connected with the boundary value theory of (1) with respect to B. In order to make this clear, let us put Green's identity into the following form:

Let us suppose that each boundary continuum of B is a smooth curve and that $P(Z)$ is still continuous in the smooth boundary C of B. Let $f(Z)$ be continuously differentiable in B and continuous on C, while $\varphi(Z)$ is a function of class Λ with continuous derivatives on C. Then one has the identity

(16) $$D\{f, \varphi\} = - \oint_C f(W) \frac{\partial \varphi(W)}{\partial n_W} ds_W$$

where the integration is performed in positive sense over the boundary C of B, and where n_W denotes the interior normal of C at the point W. Green's identity (16) will, as usual, be our main tool in the following considerations. We shall, therefore, in the beginning have to suppose that B is smoothly bounded and derive our results in this particular case. Later, we shall be able to drop this restrictive assumption and to obtain identities for the most general case. But always B will be supposed to be of the aforementioned special type, if the contrary is not explicitly stated.

We shall prove in the next paragraph that $K(Z, W)$ is continuous and has a continuous normal derivative on C if C is composed of smooth curves. Hence we may apply (16) in the following form. Let $\varphi(Z)$ be of class Λ and continuous on C. Then in view of (10) and (16)

(17) $$\varphi(Z) = D_W\{K(Z, W), \varphi(W)\} = - \oint_C \varphi(W) \frac{\partial}{\partial n_W} K(Z, W) \, ds_W \qquad (Z \, \varepsilon \, B),$$

which gives a representation of such a solution $\varphi(Z)$ of class Λ in terms of its boundary values $\varphi(W)$. If, on the other hand, $\varphi(Z)$ has continuous normal derivatives, we might apply (16) the other way and obtain

(18) $$\varphi(Z) = D_W\{K(Z, W), \varphi(W)\} = - \oint_C K(Z, W) \frac{\partial}{\partial n_W} \varphi(W) \, ds_W \qquad (Z \, \varepsilon \, B),$$

which yields a representation of the solution in terms of its normal derivatives on the boundary. This shows obviously that the kernel function permits a solution of Dirichlet's and Neumann's problem and must therefore be closely related to Green's and Neumann's functions.

The connection between the kernel function, Green's function and Neumann's function is the main result of this paper and will be established in the next paragraph.

2. The kernel function in terms of Green's and Neumann's functions. One of the most important concepts in the theory of differential equations is that of the fundamental solution. We call $S(Z, W)$ a fundamental solution of (1) with respect to the parameter point W, if it is a solution of (1) everywhere in B with the exception of the point $Z = W$, and if $S(Z, W) + \log |z - w|$ is continuous at $Z = W$. (In this notation, we coordinate to every point $Z \equiv (x, y)$ the complex number $z = x + iy$. This notation will be used throughout the whole paper.) One shows easily that $S(Z, W)$ may be written in the form

$$(19) \qquad S(Z, W) = A(Z, W) \log \frac{1}{|z - w|} + a(Z, W),$$

where A is symmetric in Z and W, of class Λ, a solution of (1) in each of its arguments, and satisfying identically in Z

$$(19') \qquad\qquad\qquad A(Z, Z) \equiv 1.$$

$a(Z, W)$ is not defined in a unique way as there exist infinitely many fundamental solutions in a given domain B.

We shall now introduce the two important fundamental solutions which can be defined with respect to a given domain B. We still assume that the domain B has a smooth boundary C. We define Green's function $G(Z, W)$ of B with respect to the differential equation (1) by the two requirements:

(a) $G(Z, W)$ is a fundamental solution of (1) for the domain B.

(b) For fixed $W \, \varepsilon \, B$ and $Z \, \varepsilon \, C$ we have $G(Z, W) \equiv 0$.

These two conditions determine $G(Z, W)$ uniquely and from them there can also be derived the symmetry property:

$$(20) \qquad\qquad\qquad G(Z, W) = G(W, Z).$$

It is known that $G(Z, W)$ has on C continuous normal derivatives in Z for fixed $W \, \varepsilon \, B$, and by means of it every solution of (1) of class Λ which is continuous on C may be represented by its boundary values in the form

$$(21) \qquad\qquad \varphi(Z) = \frac{1}{2\pi} \oint_C \varphi(W) \frac{\partial}{\partial n_W} G(Z, W) \, ds_W \qquad (Z \, \varepsilon \, B).$$

Neumann's function $N(Z, W)$ of B with respect to (1) is defined by the properties:

(a) $N(Z, W)$ is a fundamental solution of (1) in B.

(b) $\partial N(Z, W)/\partial n_z \equiv 0$ for $W \, \varepsilon \, B$ fixed and $Z \, \varepsilon \, C$.
This definition is unambiguous; we derive from it the symmetry relation

$$(22) \qquad\qquad\qquad N(Z, W) = N(W, Z).$$

It is known that $N(Z, W)$ is for fixed $W \, \varepsilon \, B$ continuous on C in Z, and if $\varphi(Z)$ is of class Λ and has continuous normal derivatives on C, we have

$$(23) \qquad\qquad \varphi(Z) = -\frac{1}{2\pi} \oint_C N(Z, W) \frac{\partial}{\partial n_W} \varphi(W) \, ds_W \qquad\qquad (Z \, \varepsilon \, B),$$

which represents $\varphi(Z)$ by the values of its normal derivative on the boundary C.

In this connection it should be remarked that the existence and uniqueness of Green's and Neumann's functions are guaranteed by the condition $P > 0$ and that at this point we have made the most important use of this assumption.

One might now try to replace the integrations (21) and (23) over the boundary of B by certain integrations over the domain B itself in order to drop the very restrictive assumptions on the type of the boundary C. However, Green's identity (16) holds only for functions f and φ with continuous derivatives in B, and is therefore not applicable to G or N. Now we note that the difference

$$(24) \qquad\qquad\qquad R(Z, W) = N(Z, W) - G(Z, W)$$

is continuous in B since the logarithmic infinity has been removed by subtraction. R is a continuous and symmetric function of Z and W and is of class Λ in each argument. Now let $\varphi(Z)$ be of class Λ and still continuous on C. Then we get from (16), (24) and (21)

$$(25) \qquad \begin{aligned} D_W\{R(Z, W), \varphi(W)\} &= -\oint_C \varphi(W) \frac{\partial}{\partial n_W} (N(Z, W) - G(Z, W)) \, ds_W \\ &= \oint_C \varphi(W) \frac{\partial}{\partial n_W} G(Z, W) \, ds_W = 2\pi\varphi(Z), \end{aligned}$$

which shows that

$$(26) \qquad\qquad \varphi(Z) = D_W\left\{\frac{1}{2\pi} (N(Z, W) - G(Z, W)), \varphi(W)\right\} \qquad\qquad (Z \, \varepsilon \, B),$$

if $\varphi(Z)$ is still continuous on C. From this, the validity of (26) for every function of class Λ is easily derived. For each $\varphi(Z)$ of class Λ is continuous on the smooth boundary C^* of an interior domain B^*. Thus we get a representation

$$(26') \qquad\qquad \varphi(Z) = D_W^*\left\{\frac{1}{2\pi} (N^*(Z, W) - G^*(Z, W)), \varphi(W)\right\} \qquad\qquad (Z \, \varepsilon \, B^*)$$

with respect to B^*. If B^* converges to B, we get in view of the continuity of G and N (26) in whole generality, *i.e.*, for every function $\varphi(Z)$ of class Λ.

Hence in view of the uniqueness theorem of §1

$$(27) \qquad K(Z, W) = \frac{1}{2\pi} (N(Z, W) - G(Z, W)) \qquad (Z \, \varepsilon \, B, W \, \varepsilon \, B).$$

As a first application of this identity we draw from (27) the conclusion that $K(Z, W)$ is continuous and has continuous normal derivatives on the smooth boundary C, as was already stated in §1.

After having derived the identity (27) under our restrictive assumptions on the boundary C, we may now drop this assumption, since K and G are continuous functionals of the domain and are defined for the most general type of boundary curves. N is of course originally defined only for domains with smooth boundary curves, since only in this case can one speak of normal derivatives. By the relation (27) we may extend the definition of Neumann to the most general domains, too; this fact will prove of the greatest importance in extremum problems. Thus we obtain the following fundamental theorem:

For every domain B the kernel function is connected with Neumann's and Green's functions by the identity (27).

Obviously, our general definition for Neumann's function is closely connected with the generalization of the second boundary value problem given by Courant [5].

In view of (27) we can construct the kernel function if Green's and Neumann's functions are known. In general, however, it is very difficult to determine these fundamental solutions for a given domain, while the computation of the kernel function by orthogonalization of a complete system is much easier. Thus we are led to the problem of expressing Green's and Neumann's functions separately in terms of the kernel function. This is in fact possible, as may be seen from the following considerations.

We note that if $\varphi(Z)$ is of class Λ, the integral expression $D_W\{G(Z, W),$ $\varphi(W)\}$ still converges in spite of the logarithmic pole at $Z = W$. We consider the domain B_ρ, obtained from B by eliminating a circle of radius ρ around the point Z. Applying Green's identity (16) on this domain, we get in view of the vanishing of $G(Z, W)$ for $W \equiv (u, v) \, \varepsilon \, C$

$$(28) \qquad \iint_{B_\rho} [G(Z, W), \varphi(W)] \, du \, dv = - \oint_{|w-z|=\rho} G(Z, W) \frac{\partial}{\partial n_W} \varphi(W) \, ds_W \qquad (Z \, \varepsilon \, B),$$

and in the limit $\rho = 0$, the identity

$$(29) \qquad D_W\{G(Z, W), \varphi(W)\} \equiv 0.$$

In the same way we calculate by means of Green's identity

$$\iint_{B_\rho} [N(Z, W), \varphi(W)] \, du \, dv = - \oint_C N(Z, W) \frac{\partial}{\partial n_W} \varphi(W) \, ds_W$$

(30)

$$- \oint_{|w-z|=\rho} N(Z, W) \frac{\partial}{\partial n_W} \varphi(W) \, ds_W.$$

In the limit $\rho = 0$, we get in view of (23)

(31) $D_W\{N(Z, W), \varphi(W)\} = 2\pi\varphi(Z).$

Again (29) and (31) have been derived for functions $\varphi(z)$ of class Λ with continuous normal derivatives on C only. But as before this result is extended to the whole class Λ by using analogous identities for interior domains and using properties of continuity of Green's and Neumann's functions in their dependence of the domain. Thus, we may enounce the following theorem:

Green's function is orthogonal to every function of class Λ. The scalar product of Neumann's function with every function of class Λ reproduces the latter multiplied by 2π.

Combining (28) with (31) leads us again to (26), but we are more interested in the following application:

Apply identity (29) in the particular case $\varphi_T(W) = K(W, T)$. We have

(29') $D_W\{G(Z, W), K(W, T)\} \equiv 0$

for every value of $Z \, \varepsilon \, B$ and $T \, \varepsilon \, B$.

Suppose now that a particular symmetric fundamental solution $S(Z, W)$ is known. Then we may write

(32) $G(Z, W) = S(Z, W) + H(Z, W),$

where $H(Z, W)$ is of class Λ with respect to both arguments. Introducing this representation into (29') and noting that $H(Z, W)$ satisfies the identity (10), we get

(33) $D_W\{S(Z, W), K(W, T)\} + H(Z, T) = 0$

whence in view of (32)

(34) $-D_W\{S(Z, W), K(W, T)\} + S(Z, T) = G(Z, T).$

Thus *we may construct Green's function by a simple process of integration if we know any symmetric fundamental solution $S(Z, W)$ and the kernel function $K(Z, W)$.* There exist well-known procedures for constructing fundamental solutions. Thus we solve the problem of finding Green's function completely if we can determine the kernel function $K(Z, W)$. With known kernel and Green's functions, one obtains Neumann's function immediately from (27).

On the other hand, it is sufficient to know the kernel function alone in order to solve Dirichlet's as well as Neumann's problem. In fact, one has in view of (27)

$$(35) \qquad K(Z, W) = \frac{1}{2\pi} N(Z, W)$$

for $Z \, \varepsilon \, C$ and $W \, \varepsilon \, B$ and

$$(36) \qquad \frac{\partial}{\partial n_z} K(Z, W) = - \frac{1}{2\pi} \frac{\partial}{\partial n_z} G(Z, W)$$

for $Z \, \varepsilon \, C$ and $W \, \varepsilon \, B$.

For solving Dirichlet's and Neumann's problem this information is already sufficient, as is easily seen from (21) and (23). The equations (35) and (36) together with equations (21) and (23) lead at once to the formulas (17) and (18).

3. **The kernel function as operator in function space.** The above considerations can be interpreted in a very intuitive way if we use the geometric concept of functional space. In fact, let us consider the linear space Ω of all functions $f(Z)$ possessing continuous derivatives in the given domain B and with $D\{f, f\} < \infty$. In this space we may use the metric defined by the scalar product (3) and the linear operation

$$(37) \qquad \psi(Z) = D_W\{K(Z, W), f(W)\}$$

which makes correspond to every element $f(Z)$ in our space a solution $\psi(Z)$ of class Λ. Considering the fact that all functions of class Λ form a linear subspace Ω_1 of Ω, we may conceive of this operation as a projection of $f(Z)$ into Ω_1. We wish to study the geometric meaning of this type of projection.

We again assume B to be bounded by smooth curves and apply Green's identity (16) on (37). We get, if $f(Z)$ is still continuous on C,

$$(38) \qquad \psi(Z) = - \oint_C f(W) \frac{\partial}{\partial n_W} K(Z, W) \, ds_W$$

and in view of (36) and (21)

$$(39) \qquad \psi(Z) = \frac{1}{2\pi} \oint_C f(W) \frac{\partial}{\partial n_W} G(Z, W) \, ds_W$$

is just that function of class Λ which has on the boundary C of B the same values as the given function $f(Z)$. This fact characterizes the projection unambiguously.

In particular, we know now that every function $f(Z)$ which vanishes on C is projected by the operation (37) into the solution 0, since this is the only solution of class Λ which vanishes on C.

If we express the function $f(Z) \, \varepsilon \, \Omega$ in the form

$$(40) \qquad f(Z) = \psi(Z) + r(Z),$$

where $\psi(Z)$ is its projection into Ω_1, the remainder $r(Z)$ will be orthogonal to every function $\varphi(Z)$ of Ω_1. In fact, we have

$$
\begin{aligned}
(41) \qquad D\{r, \varphi\} &= D_Z\{r(Z), D_W\{K(Z, W), \varphi(W)\}\} \\
&= D_W\{D_Z\{r(Z), K(Z, W)\}, \varphi(W)\} = 0
\end{aligned}
$$

since $r(Z) = 0$ on the boundary C and hence $D_Z\{r(Z), K(Z, W)\} \equiv 0$. Thus $K(Z, W)$ establishes exactly an orthogonal projection of $f(Z)$ into Ω_1.

THEOREM. *The process of scalar multiplication with the kernel function associates with every function $f(Z)$ ε Ω a function $\psi(Z)$ ε Ω_1. This association has the character of an orthogonal projection. Both associated functions have the same boundary values on C.*

We may now generalize a well-known problem of Wirtinger [12] (see also Martin [8]) and give its complete solution. We consider a given function $f(Z)$ in Ω and ask for the nearest function $\varphi(Z)$ in Ω_1 with respect to our metric. In order to solve this problem we use for $f(Z)$ the representation (40), and if $\varphi(Z)$ is an arbitrary function of Ω_1, we have

$$
\begin{aligned}
(42) \qquad D\{f - \varphi, f - \varphi\} &= D\{\psi - \varphi + r, \psi - \varphi + r\} \\
&= D\{\psi - \varphi, \psi - \varphi\} + D\{r, r\} + 2D\{\psi - \varphi, r\}.
\end{aligned}
$$

Now $D\{\psi - \varphi, r\} = 0$ since $\psi - \varphi \varepsilon \Omega_1$ and r is orthogonal to every function of this space. Thus

$$
(42') \qquad\qquad D\{f - \varphi, f - \varphi\} \geq D\{r, r\}
$$

and equality holds only for $\varphi \equiv \psi$. Hence $\psi(Z)$ is the nearest function in Ω_1 to $f(z)$.

THEOREM. *To every function $f(Z)$ ε Ω there exists exactly one function $\psi(Z)$ ε Ω_1 which is nearest to it. $\psi(Z)$ is obtained from $f(Z)$ by orthogonal projection.*

It proves of great advantage to enlarge the space of functions Ω by considering all functions $f(Z)$ with continuous derivatives in B except at a finite number of points W where $f(Z)$ may have a logarithmic infinity. This space will be denoted as T; its subspace containing the solutions of (1) will be denoted correspondingly as T_1. Let $h(Z, W)$ be an arbitrary function of T with logarithmic pole at W. Its projection into T_1 is

$$
(43) \qquad\qquad \chi(Z, W) = D_T\{K(Z, T), h(T, W)\}
$$

and this function of Z lies even in Ω_1 since it is of class Λ.

Let us suppose $h(Z, W)$ still continuous for $Z \varepsilon C$. Applying on (43) Green's identity (16), one obtains (after first cutting out the point $T = W$, then going to the limit as usual)

$$\chi(Z, W) = - \oint_C h(T, W) \frac{\partial}{\partial n_T} K(T, Z) \, ds_T$$

(44)

$$= \frac{1}{2\pi} \oint_C h(T, W) \frac{\partial}{\partial n_T} G(T, Z) \, ds_T .$$

Hence $\chi(Z, W)$ is a continuous solution of (1) which has the same boundary values as the projected function $h(Z, W)$.

This result clarifies numerous facts obtained in the preceding paragraph. Obviously

(29') $$D_W\{G(Z, W), K(W, T)\} \equiv 0 \qquad (Z \varepsilon B, T \varepsilon B),$$

since $G(Z, W)$ is of class T and has the boundary value 0. This follows from (29) at once, but conversely (29) may easily be derived from (29'). Further, every fundamental solution $S(Z, W)$ lies in T. Hence

(45) $$\sigma(Z, W) = D_T\{K(Z, T), S(T, W)\}$$

is of class Λ, with the same boundary values on C as S. Hence $S(Z, W) - \sigma(Z, W) = G(Z, W)$ as indicated in (34), since this function is a fundamental solution which vanishes on the boundary C of B.

We may further define $G(Z, W)$ unambiguously as that function of class T_1 which behaves at $Z = W$ like $\log(1/|z - w|)$ and is orthogonal to the whole subspace Ω_1. This follows at once from (29).

We close this section by calculating the scalar products of the fundamental solutions $G(Z, W)$ and $N(Z, W)$ which are of great importance in the geometry of T_1. We exclude for this purpose from B two circles of radius ρ around the points Z and W, and denote the remaining domain $B_{\rho, z, w}$. By Green's identity we have (for $T \equiv (\xi, \eta)$),

$$\iint_{B_{\rho, z, w}} [G(Z, T), G(T, W)] \, d\xi \, d\eta$$

(46)

$$= - \oint_{|t-z|=\rho} G(Z, T) \frac{\partial}{\partial n_T} G(T, W) \, ds_T - \oint_{|t-w|=\rho} G(Z, T) \frac{\partial}{\partial n_T} G(T, W) \, ds_T .$$

In the limit $\rho = 0$, (46) yields the result:

(47) $$D_T\{G(Z, T), G(T, W)\} = 2\pi G(Z, W).$$

Analogously, one finds

(48) $$D_T\{N(Z, T), N(T, W)\} = 2\pi N(Z, W)$$

and

(49) $$D_T\{G(Z, T), N(T, W)\} = 2\pi G(Z, W).$$

Subtraction of (49) from (47) yields again (29'), while subtraction of (48) from (49) leads to an easy consequence of (31). Formulas (47), (48) and (49) are very useful for claculations in our metric. They are special cases of the following two identities which can be proved in just the same way:

Let $S(Z, W)$ be any fundamental solution of (1). Then one has always

$$(50) \qquad D_T\{G(Z, T), S(T, W)\} = 2\pi G(Z, W)$$

and

$$(51) \qquad D_T\{N(Z, T), S(T, W)\} = 2\pi S(Z, W).$$

Subtracting (51) from (50) leads us again to (34).

The most general case arises if $H(Z, W)$ is an arbitrary function of class T which behaves at W as $\log(1/|z - w|)$. Then Green's identity still yields

$$(52) \qquad D_T\{N(Z, T), H(T, W)\} = 2\pi H(Z, W),$$

while

$$(53) \qquad D_T\{G(Z, T), H(T, W)\} = 2\pi(H(Z, W) - \chi(Z, W)),$$

where $\chi(Z, W)$ is of class Λ and has the same boundary values on C as $H(Z, W)$.

It is very interesting that *scalar multiplication with Neumann's function is the identity operation in the* T-*space.*

4. Digression on harmonic functions.

In the above theory the requirement $P(Z) > 0$ plays an essential role. In order to show its importance clearly, we develop now the theory of the Laplace equation

$$(54) \qquad \Delta\varphi = 0$$

along parallel lines and we shall soon recognize the difference between the two cases. Incidentally, the theory of orthonormal systems and their kernel function was initially developed just for this particular differential equation (Bergman [2]) and this fact makes the following investigation still more interesting.

In the case of the Laplace equation we have the orthonormalization conditions

$$(55) \quad D\{\varphi_\nu, \varphi_\mu\} = \iint_B [\varphi_\nu, \varphi_\mu]\, dx\, dy = \iint_B \left(\frac{\partial\varphi_\nu}{\partial x}\frac{\partial\varphi_\mu}{\partial x} + \frac{\partial\varphi_\nu}{\partial y}\frac{\partial\varphi_\mu}{\partial y}\right) dx\, dy = \delta_{\nu\mu},$$

where the $\{\varphi_\nu(Z)\}$ represent a system of functions harmonic in B. Now it is at once obvious that no complete system can be found, for the constant $\varphi(Z) \equiv 1$ is on the one hand a solution of (54), and on the other hand orthogonal to each harmonic function, even to itself.

Thus one has to restrict somehow the family of all functions $\varphi(Z)$, harmonic in B, by an additional condition which excludes all constants. For example, we might require the homogeneous linear condition

$$(56) \qquad \oint_C \varphi(Z) \, ds_Z = 0$$

for all harmonic functions considered.

Further, it is well known that in the case of the Laplace equation the definition of Neumann's function has to be changed. One can no more require a fundamental solution with vanishing normal derivatives on the boundary C, but can only prescribe a constant normal derivative on C. Thus, in this case Neumann's function is defined by the two conditions:

(a) $N(Z, W)$ is a fundamental solution of (54).

(b) For fixed $W \, \varepsilon \, B$ and $Z \, \varepsilon \, C$, one has $\partial N(Z, W)/\partial n_Z = c$. The value of the constant c follows immediately from Gauss' theorem valid for every fundamental solution $S(Z, W)$ of (54):

$$(57) \qquad \oint_C \frac{\partial}{\partial n_Z} S(Z, W) \, ds_Z = 2\pi.$$

From (57) one derives at once

$$(58) \qquad c = \frac{2\pi}{l},$$

where l is the total length of the boundary C. Even now, $N(Z, W)$ is defined only up to an additive constant; we determine the latter by the requirement

$$(56') \qquad \oint_C N(Z, W) \, ds_Z \equiv 0 \qquad\qquad (W \, \varepsilon \, B).$$

For every harmonic function φ continuous on C and for which $D\{\varphi, \varphi\} < \infty$ we can calculate by means of Green's identity (16)

$$(59) \qquad \begin{aligned} D_W\{N(Z, W), \varphi(W)\} &= -\oint_C \varphi(W) \frac{\partial}{\partial n_W} N(Z, W) \, ds_W + 2\pi\varphi(Z) \\[2mm] &= -\frac{2\pi}{l} \oint_C \varphi(W) \, ds_W + 2\pi\varphi(Z). \end{aligned}$$

We define the subclass $\mathfrak{F}$ of harmonic functions φ in B for which

$$(59') \qquad D\{\varphi, \varphi\} < \infty$$

$$(59'') \qquad D_W\{N(Z, W), \varphi(W)\} = 2\pi\varphi(Z).$$

If $\varphi(Z)$ is still continuous on C, condition (59″) is equivalent to (56).

Within the family $\mathfrak{F}$ equations (29) and (31) of §2 are still valid. Thus identity (27) is valid again with respect to a kernel function of orthonormal functions of $\mathfrak{F}$.

We remark further that every harmonic function $\varphi_\nu(Z)$ possesses a conjugate function $\psi_\nu(Z)$ related to it by the Cauchy-Riemann equations

$$(60) \qquad \frac{\partial \varphi_\nu}{\partial x} = \frac{\partial \psi_\nu}{\partial y}, \qquad \frac{\partial \varphi_\nu}{\partial y} = - \frac{\partial \psi_\nu}{\partial x}.$$

Obviously one has $[\varphi_\nu, \varphi_\nu] = [\psi_\nu, \psi_\nu]$ and $[\varphi_\nu, \psi_\nu] = 0$, whence *a fortiori*

$$(61) \qquad D\{\varphi_\nu, \varphi_\nu\} = D\{\psi_\nu, \psi_\nu\}, \qquad D\{\varphi_\nu, \psi_\nu\} = 0.$$

From (61) one recognizes that the concept of conjugate functions will play an important role in the construction of an orthonormal system and this will be in fact shown below.

However, there is one difficulty in considering the conjugate functions to a complete system of $\mathfrak{F}$. If B is not simply connected, but bounded by $n > 1$ continua C_ν ($\nu = 1, 2, \cdots, n$), the conjugates of uniform harmonic functions are not necessarily uniform and do not belong therefore to $\mathfrak{F}$. In order to overcome this difficulty, we separate the linear space of all functions harmonic in B in the following way:

We introduce the harmonic measures $\omega_\nu(Z)$ of the boundary continua C_ν, i.e., those functions $\omega_\nu(Z)$ which are harmonic in B and vanish on every boundary continuum C_μ ($\mu \neq \nu$) and have on C_ν the boundary value 1. We state the following theorem:

Every function $\omega_\nu(Z)$ is orthogonal to every harmonic function $\phi(Z)$ with uniform conjugate $\psi(Z)$.

For the proof we simply apply to the scalar product $D\{\omega_\nu, \phi\}$ Green's identity (16),

$$(62) \qquad D\{\omega_\nu, \phi\} = - \oint_C \omega_\nu \frac{\partial \phi}{\partial n} ds = - \oint_{C_\nu} \frac{\partial \phi}{\partial n} ds = - \oint_{C_\nu} \frac{\partial \psi}{\partial s} ds = 0.$$

Hence the space of all harmonic functions is split up into two complementary orthogonal spaces: the one of the $\omega_\nu(Z)$ and the other of functions ϕ with uniform conjugates.

In computing the kernel function $K(Z, W)$ we take at first the $\omega_\nu(Z)$ and orthonormalize them; then we add to these orthonormal functions a sequence of pairs of conjugate functions $\varphi_\nu(Z)$ and $\psi_\nu(Z)$. The conditions of orthonormalization in this set are

$$(63) \qquad \iint_B \left(\frac{\partial \varphi_\nu}{\partial x} \frac{\partial \varphi_\mu}{\partial x} + \frac{\partial \varphi_\nu}{\partial y} \frac{\partial \varphi_\mu}{\partial y} \right) dx \, dy = \delta_{\nu\mu},$$

$$\iint_B \left(\frac{\partial \varphi_\nu}{\partial x} \frac{\partial \psi_\mu}{\partial x} + \frac{\partial \varphi_\nu}{\partial y} \frac{\partial \psi_\mu}{\partial y} \right) dx \, dy = \iint_B \left(- \frac{\partial \varphi_\nu}{\partial x} \frac{\partial \varphi_\mu}{\partial y} + \frac{\partial \varphi_\mu}{\partial x} \frac{\partial \varphi_\nu}{\partial y} \right) dx \, dy = 0.$$

Introducing the analytic functions of complex argument $z = x + iy$

(64) $$F_\nu(z) = \varphi_\nu(Z) + i\psi_\nu(Z),$$

we may condense these pairs of conditions into the simpler form

(65) $$\iint\limits_B \overline{F'_\nu(z)} F'_\mu(z) \, dx \, dy = \delta_{\nu\mu} \, ,$$

where $F'(z) = dF(z)/dz$. Thus we are led to a new type of orthonormalization. First let us compute the kernel function $K(Z, W)$. One has obviously

(66)
$$K(Z, W) = \sum_{\mu,\nu=1}^n c_{\mu\nu}\omega_\mu(Z)\omega_\nu(W) + \sum_{\nu=1}^\infty (\varphi_\nu(Z)\varphi_\nu(W) + \psi_\nu(Z)\psi_\nu(W))$$
$$= \sum_{\mu,\nu=1}^n c_{\mu\nu}\omega_\mu(Z)\omega_\nu(W) + \mathrm{Re}\left\{ \sum_{\nu=1}^\infty F_\nu(z)\overline{F_\nu(w)}\right\},$$

where the first sum on the right side originates from the orthonormalization of the $\omega_\nu(Z)$. In view of (27) we have

(66′)
$$K(Z, W) = \sum_{\mu,\nu=1}^n c_{\mu\nu}\omega_\mu(Z)\omega_\nu(W) + \frac{1}{2}\sum_{\nu=1}^\infty (F_\nu(z)\overline{F_\nu(w)} + \overline{F_\nu(z)}F_\nu(w))$$
$$= \frac{1}{2\pi}(N(Z, W) - G(Z, W)).$$

We introduce the differential operators

(67) $$\frac{\partial}{\partial z} = \frac{1}{2}\left(\frac{\partial}{\partial x} - i\frac{\partial}{\partial y}\right), \qquad \frac{\partial}{\partial \bar{z}} = \frac{1}{2}\left(\frac{\partial}{\partial x} + i\frac{\partial}{\partial y}\right),$$

and define a set of uniform analytic functions in B by

(68) $$2\frac{\partial}{\partial z}\omega_\nu(Z) = U'_\nu(z).$$

Then we derive from (66′)

(69) $$\frac{\partial^2 K(Z, W)}{\partial z \, \partial w} = \frac{1}{4}\sum_{\mu,\nu=1}^n c_{\mu\nu}U'_\mu(z)U'_\nu(w) = \frac{1}{2\pi}\left(\frac{\partial^2 N}{\partial z \, \partial w} - \frac{\partial^2 G}{\partial z \, \partial w}\right).$$

This shows an interesting relation between derivatives of Green's and Neumann's functions:

(69′) $$\frac{\partial^2 N}{\partial z \, \partial w} - \frac{\partial^2 G}{\partial z \, \partial w} = \frac{\pi}{2}\sum_{\mu,\nu=1}^n c_{\mu\nu}U'_\mu(z)U'_\nu(w).$$

Analogously, we have by differentiation of (66') with respect to z and $\overline{w}$

(70)
$$\frac{\partial^2 K(Z, W)}{\partial z \, \partial \overline{w}} = \frac{1}{4} \sum_{\mu, \nu=1}^{n} c_{\mu\nu} U'_\mu(z) \overline{U'_\nu(w)} + \frac{1}{2} \sum_{\nu=1}^{\infty} F'_\nu(z) \overline{F'_\nu(w)}$$

$$= \frac{1}{2\pi} \left(\frac{\partial^2 N}{\partial z \, \partial \overline{w}} - \frac{\partial^2 G}{\partial z \, \partial \overline{w}} \right).$$

In the beginning of the theory of orthonormal systems one introduced directly a complete system of analytic functions $\{f_\nu(z)\}$ orthonormalized by the condition

(71)
$$\iint_B \overline{f_\nu(z)} f_\mu(z) \, dx \, dy = \delta_{\nu\mu}$$

which is of the type (65). Then, one defined the kernel function

(72)
$$k(z, \overline{w}) = \sum_{\nu=1}^{\infty} f_\nu(z) \overline{f_\nu(w)}.$$

Comparing these definitions with our above representation, one recognizes that the $f_\nu(z)$ are the derivatives of a complete system of harmonic functions of the family $\mathfrak{F}$. It is obviously preferable to work with these derivatives since we may now drop the additional condition (59'). For, to every analytic function $f_\nu(z)$ one can find an integral $F_\nu(z)$ such that its real part $\varphi_\nu(Z)$ satisfies (59'). This requirement determines the arbitrary constant of integration of $\varphi_\nu(Z)$. But only by identifying the orthonormalization (71) with (55), is it possible to imbed the special theory of the Laplace equation into our much more general investigations.

One recognizes easily that, in view of (66'),

(73)
$$\frac{\partial^2 K(Z, W)}{\partial z \, \partial \overline{w}} = \frac{1}{2\pi} \left(\frac{\partial^2 N}{\partial z \, \partial \overline{w}} - \frac{\partial^2 G}{\partial z \, \partial \overline{w}} \right) = \frac{1}{4} \sum_{\mu, \nu=1}^{n} c_{\mu\nu} U'_\mu(z) \overline{U'_\nu(w)}$$

$$+ \frac{1}{2} \sum_{\nu=1}^{\infty} F'_\nu(z) \overline{F'_\nu(w)} = \frac{1}{2} k(z, \overline{w}) - \frac{1}{4} \sum_{\mu, \nu=1}^{n} c_{\mu\nu} U'_\mu(z) \overline{U'_\nu(w)}$$

since $k(z, \overline{w})$ is just composed of the derivatives of those functions which build up $K(Z, W)$, but without distinguishing between the $U'_\nu(z)$ and $F'_\nu(z)$.

Schiffer [10] proved the identity

(74)
$$k(z, \overline{w}) = -\frac{2}{\pi} \frac{\partial^2 G}{\partial z \, \partial \overline{w}}.$$

Comparing this with (73) leads to the identity

(74')
$$\frac{\partial^2 N}{\partial z \, \partial \overline{w}} + \frac{\partial^2 G}{\partial z \, \partial \overline{w}} = -\frac{\pi}{2} \sum_{\mu, \nu=1}^{n} c_{\mu\nu} U'_\mu(z) \overline{U'_\nu(w)},$$

an equation which may also be derived directly without great difficulty.

The surprisingly simple relations (69') and (74') between the second derivatives of Green's and Neumann's functions arise of course from the well-known fact that for Laplace's equation in the plane, the Neumann problem for harmonic functions coincides with the Dirichlet problem for its conjugate.

Finally, we remark that the derivatives $U'_\nu(z)$ of the harmonic measures $W_\nu(Z)$, defined by (68), may easily be expressed in terms of the kernel function $k(z, \bar{w})$. By definition of the harmonic measure and in view of (21), we have

$$(75) \qquad \omega_\nu(Z) = \frac{1}{2\pi} \oint_{C_\nu} \frac{\partial}{\partial n_W} G(Z, W)\, ds_W = \frac{1}{\pi} \oint_{C_\nu} \frac{\partial}{\partial \bar{w}} G(Z, W)\, d\bar{w}.$$

Hence, differentiating with respect to z and using (68) and (74), we get (see [3])

$$(75') \qquad U'_\nu(Z) = - \oint_{C_\nu} k(z, \bar{w})\, d\bar{w}.$$

5. The dependence of the kernel function on the domain.

Up to this point, we have studied the kernel function for a given fixed domain B. We now wish to show how the kernel function varies with its domain of definition. For this purpose we have to suppose that B is imbedded in the interior of a larger domain B° in which $P(Z)$ is still continuous and positive.

We suppose B to be bounded by N smooth closed curves C_ν, $\nu = 1, 2, \cdots$, N. Let s be the length parameter of all the boundary curves such that for s running from zero to the total length l of the boundary C, the point $Z(s)$ describes the whole boundary. We define now a differentiable function $\rho(s)$ for $0 \leq s \leq l$, such that $\rho(s) = \rho(s')$ if $Z(s) = Z(s')$, and we choose a positive number ϵ. We shift every boundary point $Z(s)$ along the normal of C at Z by an amount $\epsilon\rho(s)$; this shift is considered positive if it has the direction of the inward normal, negative if it goes outwards. We denote this shift by

$$(76) \qquad \delta n = \epsilon\rho(s).$$

For ϵ small enough, C is transformed into a new set C^* of smooth non-intersecting curves C_ν^*, $\nu = 1, 2, \cdots, N$, which enclose a new domain B^*. We may choose ϵ so small that the domain B^* is still contained in B°.

We may therefore ask for the kernel function $K^*(Z, W)$ of B^* with respect to the differential equation (1). We shall express it asymptotically in terms of $K(Z, W)$ and δn.

At first, let us suppose $\delta n \geq 0$, which means that $B^* \subset B$. Every $\varphi(Z)$ of class Λ belongs therefore to the corresponding class Λ^* with respect to B^*. Hence we may represent $\varphi(Z)$ in a twofold way:

$$(10) \qquad \varphi(Z) = D_W\{K(Z, W), \varphi(W)\} \qquad\qquad (Z \,\varepsilon\, B),$$

or

$$(10') \qquad \varphi(Z) = D_W^*\{K^*(Z, W), \varphi(W)\} \qquad\qquad (Z \,\varepsilon\, B^*),$$

where D^* denotes the D-process with respect to B^* and Z is supposed to lie in B^*. Since it is well known that Green's and Neumann's functions vary continuously with the domain of definition, the same holds for $K(Z, W)$ in view of (27). Hence we may write

$$(77) \qquad K^*(Z, W) = K(Z, W) + \delta K(Z, W) \qquad (Z \, \varepsilon \, B^*;\, W \, \varepsilon \, B^*),$$

where $\delta K(Z, W) \to 0$ for $\epsilon \to 0$. If we suppose further that $\varphi(W)$ has continuous derivatives on C, we may, in view of definition (2), write

$$(78) \qquad \begin{aligned} D_W\{\dot{K}(Z, W), \varphi(W)\} &= D_{\!\!\!\not{W}}^{*}\{K(Z, W), \varphi(W)\} \\ &\quad + \oint_C [K(Z, W), \varphi(W)]\, \delta n_W\, ds_W + o(\epsilon), \end{aligned}$$

where $o(\epsilon)$ denotes a quantity such that $\lim\, (o(\epsilon) \cdot \epsilon^{-1}) = 0$ for $\epsilon \to 0$. Comparing now (10) with (10') and using (77) and (78), we get

$$(79) \qquad D_{\!\!\!\not{W}}^{*}\{\delta K(Z, W), \varphi(W)\} = \oint_C [K(Z, W), \varphi(W)]\, \delta n_W\, ds_W + o(\epsilon).$$

Let us choose now in particular $\varphi_T(W) = K(W, T)$ which satisfies all our assumptions. We obtain

$$(80) \qquad \begin{aligned} D_{\!\!\!\not{W}}^{*}\{\delta K(Z, W), K(W, T)\} &= D_{\!\!\!\not{W}}^{*}\{\delta K(Z, W), K^*(W, T)\} + o(\epsilon) \\ &= \oint_C [K(Z, W), K(W, T)]\, \delta n_W\, ds_W + o(\epsilon). \end{aligned}$$

In view of (10') and the fact that $\delta K(Z, W)$ is of class Λ^*, we find finally

$$(81) \qquad \delta K(Z, W) = \oint_C [K(Z, T), K(T, W)]\, \delta n_T\, ds_T + o(\epsilon).$$

This is the required variation formula, derived now only in the case $\delta n \geq 0$.

Even in this particular case, an interesting application of the variation formula may be made. Putting $Z = W$, we get

$$(81') \qquad \delta K(Z, Z) = \oint_C [K(Z, T), K(T, Z)]\, \delta n_T\, ds_T + o(\epsilon),$$

which shows that $K(Z, Z)$ increases monotonically with decreasing domain. This fact is also obvious from the minimum theorem of §1, but it is important that the variation method give an independent proof which may easily be generalized. The dependence of the kernel function and associate function upon the domain has been studied formerly by the method of certain minimum principles [2]. Our new approach leads to a powerful and very useful tool for investigations of this type.

Let us consider now the general case where the sign of δn is arbitrary. Using a device of Hadamard, we introduce a third domain of comparison: B^+ which is smoothly bounded, includes B as well as B^*, and lies in an ϵ-neighborhood of both. We may calculate by the previous method the variation of the kernel function $K^+(Z, W)$ if we change B^+ into B^* or B. Eliminating then $K^+(Z, W)$ from the resulting two formulas, we get (81) anew, without any restriction on δn.

At last, let us change our notations so that $\delta K(Z, W)$ shall now denote the first variation of the kernel function in the sense of functional analysis (see [11] and [7]). Then, we have the theorem:

The variation $\delta K(Z, W)$ of the kernel function is determined by the formula

$$(81'') \qquad \delta K(Z, W) = \oint_C [K(Z, T), K(T, W)] \, \delta n_T \, ds_T .$$

If we review our method of proof, we see that the main tools were the identities (10) and (10'). Now there holds exactly the same identity for Neumann's function, namely, (31). Hence we may find in the same way the variation of Neumann's function. We obtain for every function $\varphi(Z)$ of class Λ with continuous derivatives on C

$$(82) \qquad D_W^*\{\delta N(Z, W), \varphi(W)\} = \oint_C [N(Z, W), \varphi(W)] \, \delta n_W \, ds_W$$

if $\delta n \geq 0$ along C and $Z \, \varepsilon \, B^*$. Although $N(Z, W)$ has a logarithmic pole at $Z = W$, $\delta N(Z, W)$ is already of class Λ^*. Hence, putting $\varphi_T(W) = K(W, T)$, we may proceed as before and obtain in the case considered the theorem:

The variation of Neumann's function is given by the formula

$$(83) \qquad \delta N(Z, W) = \oint_C [N(Z, T), K(T, W)] \, \delta n_T \, ds_T .$$

Finally, we free ourselves by the former device from the assumption $\delta n \geq 0$ and obtain (83) in all generality for arbitrary sign of δn.

The same reasoning holds also for Green's function in view of identity (29). We obtain in exactly the same way the theorem:

The variation of Green's function is given by the formula

$$(84) \qquad \delta G(Z, W) = \oint_C [G(Z, T), K(T, W)] \, \delta n_T \, ds_T .$$

We may bring the equations (83) and (84) into a more symmetric form by use of (27). We recall the definition (3) of our bracket symbol, which may also be put into the form

$$(3') \qquad [f, g] = (\operatorname{grad} f, \operatorname{grad} g) + P \cdot fg \qquad \left(\operatorname{grad} f \equiv \left(\frac{\partial f}{\partial x}, \frac{\partial f}{\partial y}\right)\right).$$

We know now by definition that grad G is normal to C, while grad N has always the tangential direction; since, further, G vanishes on C, we have identically for $T \varepsilon C$

$$(85) \qquad\qquad [G(Z, T), N(T, W)] \equiv 0 \qquad\qquad (Z \varepsilon B; W \varepsilon B).$$

Thus we get from (83) and (84) by use of (27) and (85),

$$(86) \qquad\qquad \delta N(Z, W) = \frac{1}{2\pi} \oint_C [N(Z, T), N(T, W)] \, \delta n_T \, ds_T$$

and

$$(87) \qquad\qquad \delta G(Z, W) = - \frac{1}{2\pi} \oint_C [G(Z, T), G(T, W)] \, \delta n_T \, ds_T .$$

These two formulas are very analogous to the variation formula for $K(Z, W)$ and have the further advantage that the variation formula of a given function contains only the functions considered.

In view of (3'), formula (87) can also be given the form

$$(87') \qquad\qquad \delta G(Z, W) = - \frac{1}{2\pi} \oint_C \frac{\partial G(Z, T)}{\partial n_T} \frac{\partial G(T, W)}{\partial n_T} \, \delta n_T \, ds_T .$$

This formula was discovered in the case of Laplace's equation by Hadamard (see [6] and [7]) and numerous applications to the theory of functions and conformal mapping were derived from it (see [9]).

Hadamard gave a variation formula for Neumann's function of Laplace's equation too; but it appears that the above proof is shorter and more natural than the older one. In particular, our proof shows that all three variation formulas are derived from the same source.

As a particular application of the variation formulas, we consider certain functionals of B with respect to the differential equation (1) which have an interesting behavior under variation of their domain. Define two functionals by the following formulas:

$$(88) \qquad\qquad \gamma(Z) = \lim_{W \to Z} [G(Z, W) + \log |z - w|]$$

and

$$(88') \qquad\qquad \nu(Z) = \lim_{W \to Z} [N(Z, W) + \log |z - w|].$$

In the case of Laplace's equation, $\gamma(Z)$ permits a physical interpretation as the capacity constant of the domain and is also connected with problems of conformal mapping. It becomes obvious that $\gamma(Z)$ and $\nu(Z)$ are of interest in our general case, too, if one considers their variation formulas, derived from (86) and (87)

$$(89) \qquad\qquad \delta\gamma(Z) = - \frac{1}{2\pi} \oint_C [G(Z, T), G(T, Z)] \, \delta n_T \, ds_T ,$$

(89') $$\delta\nu(Z) = \frac{1}{2\pi} \oint_C [N(Z, T), N(T, Z)] \, \delta n_T \, ds_T \,.$$

One recognizes from (89) and (89') that $\gamma(Z)$ and $\nu(Z)$ change monotonically if $\delta n \geq 0$, i.e., if B diminishes. $\gamma(Z)$ decreases with B, while $\nu(Z)$ increases with shrinking domain. In view of (27), we have

(90) $$\frac{1}{2\pi} (\nu(Z) - \gamma(Z)) = K(Z, Z) > 0$$

which contains also a more detailed information on the increase of $K(Z, Z)$ with decreasing domain.

We may easily construct other functionals which vary monotonically with the domain B. A simple procedure, known already in the case of the Laplace equation (Schiffer [9]), is the following: consider the expression

(91) $$m(Z, W) = 2G(Z, W) - \gamma(Z) - \gamma(W).$$

It has obviously the variation formula

(91') $$\delta m(Z, W) = \frac{1}{2\pi} \oint_C [G(Z, T) - G(W, T), G(T, Z) - G(T, W)] \, \delta n_T \, ds_T \,,$$

which proves that $m(Z, W)$ increases with decreasing domain. Analogous expressions may be built by means of Neumann's function and the ν's. One understands easily that the variation formulas derived in this paragraph lead to numerous inequalities if one applies the above methods.

Another type of variational problems is connected with the variation of Green's and Neumann's functions if the coefficient $P(Z)$ in (1) changes. We give here as an example the variational formula for Green's function. Consider the differential equation

(1') $$\Delta\varphi - p\varphi = 0 \qquad\qquad (p(Z) > 0),$$

where $p(Z)$ is a positive continuous function of Z in a domain B. Let $g(Z, W)$ and $G(Z, W)$ be Green's functions for equations (1') and (1) respectively, corresponding to the domain B. With $T \equiv (\xi, \eta)$, one has by Green's identity

(92)
$$\iint_B [G(Z, T)\Delta_T g(T, W) - g(T, W)\Delta_T G(Z, T)] \, d\xi \, d\eta$$

$$= \iint_B (p(T) - P(T))G(Z, T)g(T, W) \, d\xi \, d\eta = 2\pi[G(Z, W) - g(Z, W)].$$

Denoting

(93) $$\delta P = p - P = \epsilon\kappa(Z),$$

where $\kappa(Z)$ is bounded in B, and

(93')
$$\delta G(Z, W) = g(Z, W) - G(Z, W) + o(\epsilon),$$

we get from (92) the following form for the first variation of G:

(92')
$$\delta G(Z, W) = - \frac{1}{2\pi} \iint_B \delta P(T) G(Z, T) G(T, W)\, d\xi\, d\eta.$$

It is easily seen that Green's function is positive in the whole domain B. Thus we derived the result:

Green's function decreases at every point of B if $P(Z)$ is increased at every point of B.

6. Several important auxiliary functions. We shall show in this section how one may express several other important functions in terms of the kernel function of an orthonormal system. We mention, for example, the function $U(Z)$ which is of class Λ and has on the boundary C of the domain B considered the constant value 1. This function plays an important role in estimates with respect to functions $\varphi(Z)$ of class Λ with given maximum M in $B + C$. In fact, consider the function

(94)
$$V(Z) = \varphi(Z) - M U(Z),$$

which is also of class Λ. It is on C non-positive, and the same holds therefore throughout B. For if $V(Z)$ were positive in B, it would possess there a positive maximum in the interior, say in Z_0. There one had obviously $\partial V/\partial x = \partial V/\partial y = 0$ and $\partial^2 V/\partial x^2 \le 0$, $\partial^2 V/\partial y^2 \le 0$. But in view of (1) we have $\partial^2 V/\partial x^2 + \partial^2 V/\partial y^2 = PV > 0$ which is contradictory to our assumption. Hence, $V(Z) \le 0$ throughout B. Thus we have the inequality

(95)
$$\varphi(Z) \le M U(Z).$$

Analogously, if the minimum m of $\varphi(Z)$ in $B + C$ is given, we find

(95')
$$m U(Z) \le \varphi(Z).$$

It is now easy to express this comparison function in terms of the kernel function. In fact, according to §3, we obtain the function $U(Z)$ by projecting the constant 1 into the space of solutions of (1) by means of the kernel function. Hence, we get for $W \equiv (u, v)$

(96)
$$U(Z) = D_W\{K(Z, W), 1\} = \iint_B P(W) K(Z, W)\, du\, dv.$$

One might improve the estimates (95) and (95') by introducing in the case of an N-fold connected domain B a set of N functions $\omega_r(Z)$, $\nu = 1, 2, \cdots, N$, of the following type (see [4]):

(a) $\omega_\nu(Z)$ is of class Λ.

(b) $\omega_\nu(Z)$ vanishes on every boundary continuum C_μ, except on C_ν, where it has the value 1. Obviously, the $\omega_\nu(Z)$ represent the generalization of the harmonic measures of the boundary continua C_ν with respect to the point Z. One has the relation

$$(97) \qquad \sum_{\nu=1}^{\Lambda} \omega_\nu(Z) = U(Z).$$

In view of (17) we may express $\omega_\nu(Z)$ in the following form:

$$(98) \qquad \omega_\nu(Z) = - \oint_{C_\nu} \frac{\partial}{\partial n_W} K(Z, W) \, ds_W.$$

We consider now the expressions $D\{\omega_\nu(Z), \omega_\mu(Z)\}$ and calculate them by means of Green's identity (16). In view of our assumptions concerning the boundary values of $\omega_\nu(Z)$, we get

$$(99) \qquad D\{\omega_\nu, \omega_\mu\} = - \oint_{C_\nu} \frac{\partial \omega_\mu}{\partial n} \, ds = - \oint_{C_\mu} \frac{\partial \omega_\nu}{\partial n} \, ds = p_{\mu\nu} = p_{\nu\mu}.$$

The matrix $(P_{\mu\nu})$ leads to a positive-definite quadratic form. For, in view of (99),

$$(100) \qquad \sum_{\mu,\nu=1}^{N} p_{\mu\nu} x_\mu x_\nu = D_Z\left\{ \sum_{\nu=1}^{N} x_\nu \omega_\nu(Z), \sum_{\nu=1}^{N} x_\nu \omega_\nu(Z) \right\} > 0$$

if $\sum_{\nu=1}^{N} x_\nu^2 > 0$. In particular we see that the matrix $(p_{\mu\nu})$ has a non-vanishing determinant and, therefore, an inverse matrix.

Every function of class Λ which has on C continuous derivatives may be split up in the form

$$(101) \qquad \varphi(Z) = \sum_{\nu=1}^{N} c_\nu \omega_\nu(Z) + \chi(Z),$$

where $\chi(Z)$ is of class Λ and satisfies the following N conditions:

$$(102) \qquad \oint_{C_\nu} \frac{\partial \chi}{\partial n} \, ds = 0 \qquad\qquad (\nu = 1, 2, \cdots, N).$$

Every function $\omega_\nu(Z)$ is orthonormal to every function $\chi(Z)$; for

$$(103) \qquad D\{\omega_\nu(Z), \chi(Z)\} = - \oint_{C_\nu} \frac{\partial \chi}{\partial n} \, ds = 0.$$

Thus we split the linear space of all Λ-functions into two orthogonal spaces. If we want to give to the condition (102) a sense even if χ has no continuous derivatives on C, we have just to define for an arbitrary $\varphi \in \Lambda$

$$(103') \qquad \oint_{C_\nu} \frac{\partial \varphi}{\partial n} \, ds = - D\{\omega_\nu, \varphi\}.$$

This role of the $\omega_\nu(Z)$ in defining line integrals over particular boundary curves C_ν for the whole class Λ shows their interest for the general theory.

In particular, we get for the kernel function the representation

$$(104) \qquad K(Z, W) = \sum_{\mu, \nu=1}^{N} c_{\mu\nu}\omega_\mu(Z)\omega_\nu(W) + \sum_{\lambda=1}^{\infty} \chi_\lambda(Z)\chi_\lambda(W),$$

where the $\chi_\lambda(Z)$ are a complete orthonormal system in the subspace of all functions of class Λ which satisfy the N conditions (102).

In order to determine the coefficients $c_{\mu\nu}$, we introduce (104) into (98). We get

$$(105) \qquad \sum_{\mu, \nu=1}^{N} c_{\mu\nu}\omega_\mu(Z) \oint_{C_\rho} \frac{\partial\omega_\nu(W)}{\partial n_W} ds_W = -\omega_\rho(Z),$$

and, in view of (99),

$$(105') \qquad \sum_{\mu, \nu=1}^{N} c_{\mu\nu}p_{\nu\rho}\omega_\mu(Z) = \omega_\rho(Z).$$

Since the $\omega_\nu(Z)$ are obviously linearly independent, we conclude from (105') that the matrix $(c_{\mu\nu})$ is the inverse of the matrix $(p_{\mu\nu})$.

Another important function is the solution $A(Z, W)$ of (1) which appears in the definition (19) of a fundamental solution. We want to indicate a few of its properties here. Our theory of the Green's and Neumann's function is, until now incomplete insofar as we did not succeed in expressing this important solution in terms of the kernel function of an orthonormal system. This has one reason in the fact that $A(Z, W)$ does not depend on the domain B considered while $K(Z, W)$, $G(Z, W)$ and $N(Z, W)$ do so. Thus we had to take a construction of the fundamental solutions from sources which do not belong to our theory of orthonormal functions.

Of course, we know that $A(Z, W)$ may be developed into a bilinear series

$$(106) \qquad A(Z, W) = \sum_{\mu, \nu=1}^{\infty} \alpha_{\mu\nu}\varphi_\mu(Z)\varphi_\nu(W) \qquad\qquad (\alpha_{\mu\nu} = \alpha_{\nu\mu}),$$

where the $\{\varphi_\nu(Z)\}$ are a complete orthonormal system for the class Λ. But the question of determining the $\alpha_{\mu\nu}$ for a given system $\{\varphi_\nu(Z)\}$ has not yet been solved. We can, however, point out some properties of this series. In view of the requirement (90') there holds identically in $Z \ \varepsilon \ B$

$$(107) \qquad \sum_{\mu, \nu=1}^{\infty} \alpha_{\mu\nu}\varphi_\mu(Z)\varphi_\nu(Z) = 1,$$

an interesting quadratic relation between the linearly independent functions of a complete orthonormal system. Applying the Laplace operator to this equation and interchanging summation and differentiation (permitted because of the uniform convergence of the series (107)), we get in view of (6)

(108) $$\sum_{\mu,\nu=1}^{\infty} \alpha_{\mu\nu}(\mathrm{grad}\ \varphi_\mu(Z),\ \mathrm{grad}\ \varphi_\nu(Z)) = -\ P(Z).$$

This is a representation of the characteristic coefficient $P(Z)$ of differential equation (1) in terms of the functions of a complete orthonormal system.

7. Generalizations of the theory. It is clear that the above theory permits several generalizations: We shall indicate here now three possible extensions of our results, namely: (a) dropping the very particular type of differential equation considered until now; (b) permitting an arbitrary number of independent variables; and (c) considering partial differential equations of higher order than the second.

(a) For this purpose let $\mathbf{u} \equiv (u_1,\ u_2,\ u_3)$ and $\mathbf{v} \equiv (v_1,\ v_2,\ v_3)$ be two sets of variables and

(109) $$Q(\mathbf{u},\ \mathbf{v}) = \sum_{i,k=1}^{3} a_{ik}u_iv_k$$

a symmetric positive-definite quadratic form. The coefficients a_{ik} may be continuously differentiable functions of the point Z in a given domain B of the (x, y)-plane.

We introduce now into the linear space Ω of all functions $f(Z)$ which have continuous derivatives in B, the metric based on the scalar product

(110) $$\theta\{f,\ g\} = \iint_B Q(\mathbf{f},\ \mathbf{g})\ dx\ dy$$

where $\mathbf{f}$ and $\mathbf{g}$ are vectors defined by

(111) $$\mathbf{f} \equiv \left(\frac{\partial f}{\partial x},\ \frac{\partial f}{\partial y},\ 1\right), \qquad \mathbf{g} \equiv \left(\frac{\partial g}{\partial x},\ \frac{\partial g}{\partial y},\ 1\right).$$

The integral $\theta\{f,\ f\}$ leads to a variational problem, the Euler-Lagrange equation of which is the self-adjoint differential equation

(112)
$$M(\varphi) = \frac{\partial}{\partial x}\left(a_{11}\frac{\partial\varphi}{\partial x} + a_{12}\frac{\partial\varphi}{\partial y} + a_{13}\varphi\right)$$
$$+ \frac{\partial}{\partial y}\left(a_{21}\frac{\partial\varphi}{\partial x} + a_{22}\frac{\partial\varphi}{\partial y} + a_{23}\varphi\right) - a_{33}\varphi = 0.$$

Obviously the solution of (112) will be closely related with the metric (110) and we will be able to carry over nearly all our former considerations. We might transform (112) by well-known methods into a simpler canonical form. However, it is important that we are able to develop our theory without this transformation since only for this reason will it be possible to generalize our considerations to differential equations of the same type with more than two variables.

We choose a complete system $\{\varphi_\nu(Z)\}$ of solutions of (112) and orthonormalize them by the requirement

$$(113) \qquad \theta\{\varphi_\nu, \varphi_\mu\} = \delta_{\nu\mu}.$$

Then we introduce a kernel function

$$(114) \qquad K(Z, W) = \sum_{\nu=1}^{\infty} \varphi_\nu(Z)\varphi_\nu(W),$$

which is again independent of the particular choice of the orthonormal system and may be characterized by the integro-differential relation

$$(115) \qquad \varphi(Z) = \theta_W\{K(Z, W), \varphi(W)\},$$

valid for every solution of (112) which is twice differentiable in B and for which $\theta\{\varphi, \varphi\} < \infty$. Its existence follows from the same arguments as were used in the special case (1). (See [4].)

Green's identity applied on the integral $\theta\{f, \varphi\}$, where φ is a solution of (112), with continuous derivatives on C, and f of class Ω, still continuous on C, yields

$$(116) \qquad \theta\{f, \varphi\} = - \oint_C Q(\varphi, \mathbf{n}) \cdot f \, ds$$

where $\mathbf{n}$ is a vector defined on the boundary C by

$$(117) \qquad \mathbf{n} = (\cos(n, x), \cos(n, y), 1).$$

A fundamental solution $S(Z, W)$ of (112) is a function which satisfies in Z the differential equation everywhere in B except at the point $Z = W$. Here $S(Z, W)$ has a logarithmic infinity such that

$$(118) \qquad \lim_{\rho \to 0} \oint_{|z-w|=\rho} Q(S(Z, W), \mathbf{n}_Z) \, ds_Z = 2\pi.$$

We define Green's function as that fundamental solution of (112) which vanishes on the boundary C of B, while Neumann's function is a fundamental solution of (112) satisfying on C (which is again supposed to consist of smooth curves) the requirement

$$(119) \qquad Q(\mathbf{N}(Z, W), \mathbf{n}_Z) \equiv 0.$$

As in §2, one proves at once the identity

$$(120) \qquad K(Z, W) = \frac{1}{2\pi}(N(Z, W) - G(Z, W))$$

and so the main result has been generalized for this case. By means of (120) Neumann's function with respect to the general differential equation (112) may be defined even for domains B which have a very irregular boundary C.

(b) It is obvious that our assumption of two independent variables played nowhere an important role and was only made for the sake of simplicity. We may develop exactly the same theory for any number of dimensions and get analogous results.

However, one must be careful with certain numerical factors. For example, one has to write in the three-dimensional case our fundamental identity (27) in the form

$$(121) \qquad K(Z, W) = \frac{1}{4\pi} (N(Z, W) - G(Z, W))$$

and analogous changes are necessary in the variation formulas of §5.

(c) Finally, one might inquire how these considerations could be generalized to differential equations of higher order. We do not intend to deal with this question here in its full extent, but we shall show here in an important particular case that relations between the kernel function and Green's function exist in fact for higher order differential equations.

We consider the fourth order partial differential equation

$$(122) \qquad \Delta\Delta\varphi = 0$$

which plays an important role in the theory of elasticity. It is well known that all its solutions may be written in the form

$$(123) \qquad \varphi(Z) = \text{Re} \{\bar{z}f(z) + g(z)\} \qquad (z = x + iy),$$

where $f(z)$ and $g(z)$ are arbitrary analytic functions of the complex variable z.

Consider a complete system of functions

$$(124) \qquad F_\nu(z, \bar{z}) = \bar{z}f_\nu(z) + g_\nu(z)$$

such that every function of this type regular in the domain B considered and L^2-integrable may be represented by a series of these $F_\nu(z, \bar{z})$ which converges uniformly in every interior partial domain of B. We orthonormalize this system by the condition

$$(125) \qquad \iint\limits_B \overline{F_\nu}F_\mu \, dx \, dy = \delta_{\nu\mu}$$

and obtain again a kernel function

$$(126) \qquad K(Z, W) = \sum_{\nu=1}^{\infty} F_\nu(z, \bar{z})\overline{F_\nu(w, \bar{w})}.$$

The kernel function is a solution of (122) and satisfies the two conditions

$$(127) \qquad K(Z, W) = \overline{K(W, Z)}$$

and

$$(128) \qquad F(z, \bar{z}) = \iint\limits_{B} K(Z, W) F(w, \bar{w}) \, du \, dv \qquad\qquad (w = u + iv; Z \, \varepsilon \, B)$$

for every function

$$(124') \qquad\qquad F(z, \bar{z}) = \bar{z} f(z) + g(z)$$

which is L^2-integrable in B. One shows easily that these conditions conversely determine the kernel function in a unique way.

We derive now from the Green's function corresponding to the differential equation (122) a new solution of (122) which satisfies conditions (127) and (128) and is therefore identical with the kernel function. For this purpose, let us recall the fact that $G(z, \bar{z}; w, \bar{w})$ is a solution of (122) throughout B, except that at $z = w$ it has a singularity such that

$$(129) \qquad G(z, \bar{z}; w, \bar{w}) = G(Z, W) = |z - w|^2 \log \frac{1}{|z - w|} + H(z, \bar{z}; w, \bar{w})$$

where $H(z, \bar{z}; w, \bar{w})$ is a regular solution at this point. It is symmetric in z and w and satisfies the boundary conditions

$$(130) \qquad\qquad G(Z, W) = 0, \qquad \frac{\partial}{\partial n_z} G(Z, W) = 0$$

for $Z \, \varepsilon \, C$ and W fixed in B.

Consider now the expression

$$(131) \qquad\qquad M(Z, W) = \frac{\partial^4 G(Z, W)}{\partial z^2 \, \partial \bar{w}^2}$$

where the differential operators are defined in (67). This function obviously satisfies the symmetry relation

$$(132) \qquad\qquad M(Z, W) = \overline{M(W, Z)}$$

It is regular throughout B since the singular term in the representation (129) of Green's function just cancels out by differentiation, and is a solution of (122).

Let us evaluate now the integral

$$(133) \qquad\qquad I(z, \bar{z}) = \iint\limits_{B} \frac{\partial^4 G}{\partial z^2 \, \partial \bar{w}^2} F(w, \bar{w}) \, du \, dv$$

by partial integration. Since the function $\partial^3 G / \partial z^2 \partial \bar{w}$ has at the point $z = w$ a development

$$(134) \qquad\qquad \frac{\partial^3 G}{\partial z^2 \, \partial \bar{w}} = \frac{1}{2} \left(\frac{1}{z - w} \right) + \frac{\partial^3 H}{\partial z^2 \, \partial \bar{w}},$$

we have to use Green's formula with respect to the domain B from which a circle of radius ρ around $z = w$ has been eliminated. We call this domain B_ρ and obtain, in view of (124')

$$
\begin{aligned}
(135) \qquad \iint\limits_{B_\rho} \frac{\partial^4 G}{\partial z^2 \, \partial \overline{w}^2} \, F(w, \overline{w}) \, du \, dv &= \frac{1}{2i} \oint\limits_C \frac{\partial^3 G}{\partial z^2 \, \partial \overline{w}} \, F(w, \overline{w}) \, dw \\
&- \frac{1}{2i} \oint\limits_{|w-z|=\rho} \frac{\partial^3 G}{\partial z^2 \, \partial \overline{w}} \, F(w, \overline{w}) \, dw - \iint\limits_{B_\rho} \frac{\partial^3 G}{\partial z^2 \, \partial \overline{w}} \, f(w) \, du \, dv.
\end{aligned}
$$

For $w \; \varepsilon \; C$ we have $\partial G/\partial u = \partial G/\partial v = 0$ because of the boundary conditions (130) and therefore, differentiating with respect to z,

$$
(136) \qquad \frac{\partial^3 G}{\partial z^2 \, \partial \overline{w}} \equiv 0
$$

for $w \; \varepsilon \; C$, z fixed in B. Thus the first right-hand integral in (135) vanishes. As $\rho \to 0$ we get finally from (133) and (135)

$$
(137) \qquad I(z, \overline{z}) = -\frac{\pi}{2} F(z, \overline{z}) - \iint\limits_B \frac{\partial^3 G}{\partial z^2 \, \partial \overline{w}} \, f(w) \, du \, dv.
$$

The last integral may be transformed again by partial integration. We obtain

$$
(138) \qquad \iint\limits_B \frac{\partial^3 G}{\partial z^2 \, \partial \overline{w}} \, f(w) \, du \, dv = \frac{1}{2i} \oint\limits_C \frac{\partial^2 G}{\partial z^2} \, f(w) \, dw - \iint\limits_B \frac{\partial^2 G}{\partial z^2} \frac{\partial}{\partial \overline{w}} \, f(w) \, du \, dv.
$$

Here we no longer have to consider the singularity of $G(z, \overline{z}; w, \overline{w})$ since it does not lead to any infinity. In view of the boundary conditions the first right-hand integral vanishes, while the second drops out because of the Cauchy-Riemann equations. Thus, we get finally

$$
(139) \qquad F(z, \overline{z}) = \iint\limits_B \left(-\frac{2}{\pi} \frac{\partial^4 G}{\partial z^2 \, \partial \overline{w}^2} \right) F(w, \overline{w}) \, du \, dv.
$$

Hence, in view of the conditions which determine the kernel function in a unique way, we get finally

$$
(140) \qquad K(Z, W) = -\frac{2}{\pi} \frac{\partial^4 G}{\partial z^2 \, \partial \overline{w}^2}.
$$

This example, which allows numerous applications in the theory of the equation (122) shows that even for partial differential equations of higher order than the second, there exist relations between the kernel function and Green's function.

BIBLIOGRAPHY

1. S. BERGMAN, *Sur les fonctions orthogonales de plusieurs variables complexes*, New York, 1941.
2. S. BERGMAN, *Partial differential equations*, Brown University Publications, Providence, 1941.
3. S. BERGMAN, *A remark on the mapping of multiply-connected domains*, American Journal of Mathematics, vol. 68(1946), pp. 20–28.
4. S. BERGMAN, *Functions satisfying certain partial differential equations of elliptic type and their representation*, this Journal, vol. 14(1947), pp. 349–366.
5. COURANT-HILBERT, *Methoden der mathematischen Physik II*, Chapter 7, Berlin, 1937.
6. J. HADAMARD, *Mémoire sur le problème d'analyse relatif à l'équilibre des plaques élastiques encastrées*, Mémoires présentés par divers savants à l'Académie des Sciences de l'Institut de France, vol. 33(1908), pp. 128 S.
7. P. LÉVY, *Leçons d'Analyse Fonctionelle*, Paris, 1922.
8. W. T. MARTIN, *On a minimum problem in the theory of analytic functions of several variables*, Transactions of the American Mathematical Society, vol. 48(1940), pp. 351–357.
9. M. SCHIFFER, *Hadamard's formula and variation of domain functions*, American Journal of Mathematics, vol. 68(1946), pp. 417–448.
10. M. SCHIFFER, *The kernel function of an orthonormal system*, this Journal, vol. 13(1946), pp. 529–540.
11. V. VOLTERRA, *Theory of Functionals*, London, 1931.
12. W. WIRTINGER, *Über eine Minimalaufgabe im Gebiet der analytischen Funktionen*, Monatshefte für Mathematik und Physik, vol. 39(1932), pp. 377–384.

HARVARD UNIVERSITY.

Commentary on

[20] S. Bergman and M. Schiffer, *A representation of Green s and Neumann s functions in the theory of partial differential equations of second order*, Duke Math. J. **14** (1947), 609 638.

This carefully written paper continues the advances [19, Br] made by the authors separately in a broad program of expressing various important functions associated with a plane domain (e.g., Green's function, Neumann's function, harmonic measure, etc.) in terms of an orthonormal system of functions. The authors consider the elliptic equation

$$\Delta \varphi - P\varphi := \frac{\partial^2 \varphi}{\partial x^2} + \frac{\partial^2 \varphi}{\partial y^2} - P\varphi = 0$$

in a plane domain B, where $P(x,y) > 0$ is a given function continuous in B (the positivity of P is needed to allow use of the maximum principle), and study the Hilbert space of solutions with respect to the scalar product

$$\mathcal{D}[u,v] := \int_B (\nabla u \cdot \nabla v + Puv)\, dxdy.$$

This space is shown to have the reproducing kernel K that solves the extremal problem of minimizing the $\mathcal{D}$-norm among all solutions with a prescribed value at a given point. (This fact has now become a standard point of departure in studying reproducing kernels in spaces of solutions of linear elliptic PDEs.)

The paper discusses the problem of uniqueness of the reproducing kernel and establishes an elegant relationship between the reproducing kernel K and Green's and Neumann's functions, G and N, associated with the operator $\Delta - P$:

$$2\pi K = N - G.$$

The authors also study, following Hadamard, variations of the kernel functions K, N, and G in terms of perturbations of the domain. Possible extensions to more general elliptic operators, to the case of n variables, and to higher order equations, viz., the biharmonic equation, are outlined at the end of the paper.

Some of the topics studied here were developed further in [26, 27, G, GS]. For more recent developments, see the publications [B1, B2, B3, B4] of Steven Bell and the references therein, as well as subsequent papers by the same author.

References

[B1] Steven R. Bell, *The Cauchy Transform, Potential Theory, and Conformal Mapping*, CRC Press, 1992.

[B2] Steven R. Bell, *Complexity of the classical kernel functions of potential theory*, Indiana Univ. Math. J. **44** (1995), 1337 1369.

[B3] Steven R. Bell, *Simplicity of the Bergman, Szego and Poisson kernel functions*, Math. Res. Lett. **2** (1995), 267 277.

[B4] Steven R. Bell, *Recipes for classical kernel functions associated to a multiply connected domain in the plane*, Complex Variables Theory Appl. **29** (1996), 367 378.

[Br] S. Bergman, *A remark on the mapping of multiply-connected domains*, Amer. J. Math. **68** (1946), 20 28.

[G] P. R. Garabedian, *Schwarz s lemma and the Szego kernel function*, Trans. Amer. Math. Soc. **67** (1949), 1 35.

[GS] P. R. Garabedian and D. C. Spencer, *Complex boundary value problems*, Trans. Amer. Math. Soc. **73** (1952), 223 242.

DMITRY KHAVINSON

[23] (with S. Bergman) Kernel functions in the theory of partial differential equations of elliptic type

[23] Menahem Max Schiffer, S. Bergman. "Kernel functions in the theory of partial differential equations of elliptic type," in *Duke Mathematical Journal*, Volume **15** (1948), 535–566.

KERNEL FUNCTIONS IN THE THEORY OF PARTIAL DIFFERENTIAL EQUATIONS OF ELLIPTIC TYPE

By S. Bergman and M. Schiffer

Introduction. The theory of certain classes of partial differential equations of elliptic type may be developed by means of the concept of orthonormal systems of functions. Two important facts must be given in the definition of such systems:

    (a) The metric with respect to which the orthonormalization is to be carried out.

    (b) The class of functions from which the elements of the system are to be chosen.

For a large class of partial differential equations of elliptic type which are investigated in a given domain B, one may define in a most natural way a definite quadratic Dirichlet or energy integral, extended over the domain, which defines the metric to be used. It is an integral whose Euler-Lagrange equation (of the extremum problem related with it) coincides with the differential equation considered. This metric has been extensively used in the study of boundary value and eigenvalue problems connected with the equation [5; Chapter 7].

However, the class of functions on which this metric was applied was mostly the general class Ω of all differentiable functions for which the integral converges or a subclass Ω^0 of it defined by certain conditions on the boundary of the domain. Curiously enough, the important subclass Λ of Ω which consists of the solutions of the differential equation considered has been widely neglected. There is, however, an important paper of Zaremba [9] in which this subclass and an orthonormal system in it are studied in the special case of Laplace's equation, and where interesting results are obtained in this way. These investigations seem to have been the first along this line.

One important advantage of the study of orthonormal systems in the subclass Λ of the solutions was indicated in [2]. Namely, while in the wider systems as Ω and Ω^0 the bilinear kernel of the orthonormal system does not converge, it converges uniformly in every closed subdomain of B in the case of the class Λ. This fact makes the theory of the class Λ and its orthonormal systems much easier than that of the wider classes which had been preferred before. A close relation between the kernel function of the class Λ and important fundamental solutions of the differential equation was first established in the case of Laplace's equation [7]. These relations were generalized by us in two previous papers and the connection between kernel function, Green's and Neumann's function with respect to the differential equation were established [3], [4]. The practical and theoretical advantages of using this new concept were shown.

Received October 23, 1947. Paper done under contract with the Office of Naval Research.

535

In the present paper this theory is continued. After a short review of the main concepts in §1, we prove in §2 several new properties for the kernel function. The relations between this function and certain fundamental solutions of the differential equation permit us to supply simultaneously the concepts of the theory of orthonormal systems and of the fundamental solutions in studying both, thus leading to great flexibility of the method. In §3 the variation of the fundamental solutions and of the kernel function is studied in the case where the coefficient of the differential equation varies. The results are applied in §4 and §5 to the study of a non-linear partial differential equation of elliptic type. We did not try to make this investigation in a general and complete form, but wanted only to show the possibilities evolving from a complete theoretical and numerical understanding of the linear differential equations treated before. In §6 the possibility of different metrics applied to the same classes Λ, Ω^0 and Ω is studied and representations are obtained for new fundamental solutions. Finally in §7 certain of our initial assumptions are weakened and the effect on the theory is investigated.

We want to point out the close relation between our work on kernel functions and the paper of Aronszajn [1] on reproducing kernels. Our theory gives numerous examples for his general theory and might even show a possible extension of his abstract approach. It will be noted that Neumann's and Green's functions are reproducing kernels for the classes Ω and Ω^0; they do not belong themselves to these classes because of their singularities but are nevertheless closely related to them. These examples show in which way the theory of reproducing kernels might be generalized.

1. **Generalities.** Let B be a finite domain in the plane bounded by r smooth curves C_ν, $\nu = 1, 2, \cdots, r$, which form the boundary $C = \sum_{\nu=1}^r C_\nu$ of B. A point of B will be denoted by a capital letter such as Z, W, V. If Z has the Cartesian coordinates x, y, we shall write $Z \equiv (x, y)$ and $z = x + iy$ will denote the corresponding complex number. Let $P(Z)$ be a continuous function in the closed domain $B + C$ and let

$$(1) \qquad\qquad\qquad P(Z) > 0 \qquad\qquad\qquad (Z \,\varepsilon\, B + C).$$

If $\Delta\varphi(Z) = \partial^2\varphi/\partial x^2 + \partial^2\varphi/\partial y^2$ denotes Laplace's operator applied to a function φ, we consider the following partial differential equation of elliptic type

$$(2) \qquad\qquad\qquad \Delta\varphi(Z) = P(Z)\varphi(Z).$$

We seek all solutions of this differential equation which are continuous and continuously differentiable in B. In particular, we are interested in boundary value problems, such as determining a solution $\varphi(Z)$ of (2) which is continuous in $B + C$ and attains on C prescribed values (boundary value problem of the first kind) or which is continuously differentiable in $B + C$ and has on C prescribed derivatives in the direction of the interior normal (boundary value problem of the second kind).

The investigation of these problems is carried out by considering besides the differential equation (2) certain integrals extended over B which are quadratic in $\varphi(Z)$ and its partial derivatives, positive-definite, and for the variation problem of which the equation (2) is the Euler-Lagrange differential equation. In this way, we relate the problem of differential equations to the calculus of variations and to the concept of metrics in functional spaces.

The simplest form of an integral connected with the differential equation (2) is the Dirichlet integral

$$D\{\varphi, \varphi\} = \iint\limits_{B} \left[\left(\frac{\partial \varphi}{\partial x} \right)^2 + \left(\frac{\partial \varphi}{\partial y} \right)^2 + P\varphi^2 \right] dx \, dy$$

(3)

$$= \iint\limits_{B} \left[(\operatorname{grad} \varphi)^2 + P\varphi^2 \right] dx \, dy.$$

Denote by Ω the class of all functions $f(Z)$ which are continuous and continuously differentiable in B and for which the Dirichlet integral (3) exists in the Lebesgue sense. These functions form a vector space and we define the scalar product of each two elements thereof by

(4)
$$D\{f, g\} = \iint\limits_{B} \left[(\operatorname{grad} f, \operatorname{grad} g) + Pfg \right] dx \, dy.$$

The only element of Ω of length zero is the function $f \equiv 0$.

This metric has been considered by us in two previous papers [3], [4] and in order to facilitate the study of this paper, we shall indicate here shortly the main results obtained. We define as usual the concept of a fundamental solution of (2). We call $S(Z, W)$ a fundamental solution of (2) with respect to B if it is continuous and continuously differentiable with respect to Z everywhere in B except for the point W and if it satisfies the differential equation (2) everywhere except at W. At $Z = W$, however, $S(Z, W)$ shall become logarithmically infinite such that

(5)
$$S(Z, W) + (2\pi)^{-1} \log |z - w| = H(Z, W)$$

is continuous at W. In particular, there are the following two important fundamental solutions:

(a) Green's function $G(Z, W)$ is that fundamental solution which vanishes if $Z \in C$.

(b) Neumann's function $N(Z, W)$ is the fundamental solution with vanishing normal derivatives on C.

Both functions can be shown to be symmetric in Z and W, and, considered as functions of W, they are also fundamental solutions of (2) with respect to B. They are continuous and continuously differentiable in the closed domain $B + C$ [5; Chapter 7]. They satisfy the following conditions: If $f(Z)$ is an element of the class Ω which is continuous in $B + C$, we have

(6) $$D_W\{G(Z, W), f(W)\} = f(Z) - \oint_C f(W)(\partial G(Z, W)/\partial n_W)\, ds_W ,$$

(7) $$D_W\{N(Z, W), f(W)\} = f(Z).$$

Here, D_W denotes the Dirichlet integral (3) where integration and differentiation is to be performed with respect to the variable point W. $\partial/\partial n_W$ denotes the differentiation in the direction of the interior normal at the point $W \, \varepsilon \, C$. We may write (6) also in the form

(6') $$D_W\{G(Z, W), f(W)\} = f(Z) - F(Z),$$

where

(6'') $$F(Z) = \oint_C f(W) \frac{\partial}{\partial n_W} G(Z, W)\, ds_W$$

denotes by virtue of Green's formula a solution of (2) which assumes on the boundary C of B the same values as $f(Z)$.

We define further the kernel function

(8) $$K(Z, W) = N(Z, W) - G(Z, W)$$

which is symmetric in Z and W, a solution of (2) in B and finite even for $Z = W$. From (7) and (6') we derive

(9) $$D_W\{K(Z, W), f(W)\} = F(Z).$$

This means that the scalar multiplication of $f(W)$ with the kernel function results in a solution of (2) which has the same boundary values as $f(W)$.

The space Ω may be decomposed into two subspaces Ω^0 and Λ which have only the element $f \equiv 0$ in common and are orthogonal to each other:

The subspace Ω^0 of Ω consists of all functions $f^0(Z)$ in Ω which vanish on the boundary C. The subspace Λ of Ω consists of all functions $\varphi(Z) \, \varepsilon \, \Omega$ which are solutions of (2). It is well known that

(10) $$\Omega = \Omega^0 + \Lambda$$

and that

(10') $$D\{\varphi, f^0\} = 0 \qquad\qquad \text{for all } \varphi \, \varepsilon \, \Lambda, \, f^0 \, \varepsilon \, \Omega^0.$$

We deduce from (6), (7) and (9) that

(11) $$D_W\{N(Z, W), f(W)\} = f(Z) \qquad\qquad (f \, \varepsilon \, \Omega),$$

(11') $$D_W\{G(Z, W), f^0(W)\} = f^0(Z) \qquad\qquad (f^0 \, \varepsilon \, \Omega^0),$$

(11'') $$D_W\{K(Z, W), \varphi(W)\} = \varphi(Z) \qquad\qquad (\varphi \, \varepsilon \, \Lambda).$$

Each one of the three functions N, G and K operates, therefore, as a unit multiplier in a certain function space or (using the notation of Aronszajn) is a reproducing kernel. This property allows an important application.

Consider at first the class Λ of solutions of (2). It is always possible to determine a denumerable set $\{\varphi_\nu(Z)\}$ of elements which satisfy the conditions of orthonormality

$$(12) \qquad\qquad D\{\varphi_\nu, \varphi_\mu\} = \delta_{\nu\mu},$$

where $\delta_{\nu\mu}$ equals 1 for $\nu = \mu$ and equals 0 for $\nu \neq \mu$, and by means of which every element $\varphi(Z) \, \varepsilon \, \Lambda$ may be developed into a series

$$(13) \qquad\qquad \varphi(Z) = \sum_{\nu=1}^{\infty} a_\nu \varphi_\nu(Z) \qquad\qquad (a_\nu = D\{\varphi, \varphi_\nu\}),$$

which converges uniformly in each closed subdomain of B. We call the system $\{\varphi_\nu(Z)\}$ a complete orthonormal system for the class Λ with respect to the metric (4). The set of numbers a_ν defined in (13) are called the Fourier coefficients of $\varphi(Z)$ with respect to the system $\{\varphi_\nu(Z)\}$. The kernel function $K(Z, W)$ is, for a fixed Z, an element of Λ and has by (11') the Fourier coefficient $\varphi_\nu(Z)$. Hence, by (13) we have

$$(14) \qquad\qquad K(Z, W) = \sum_{\nu=1}^{\infty} \varphi_\nu(Z)\varphi_\nu(W),$$

this series converging uniformly in each closed subdomain of B. It is, therefore, possible to represent the difference between the two important fundamental solutions $N(Z, W)$ and $G(Z, W)$ as a series which is numerically accessible.

The analogous formulas (11) and (11'') suggest a similar approach to Neumann's and Green's function themselves. One can readily introduce complete orthonormal systems $\{f_\nu(Z)\}$ and $\{f_\nu^0(Z)\}$ in Ω and Ω^0, respectively. The Fourier coefficients of $N(Z, W)$ and $G(Z, W)$ are, in view of (11) and (11'), just $f_\nu(W)$ and $f_\nu^0(W)$. However, $N(Z, W)$ and $G(Z, W)$ do not belong to the corresponding classes Ω and Ω^0 because of their singularities and we cannot apply to them the development theorems as we did for the kernel function. It is possible to overcome these difficulties and to represent $N(Z, W)$ and $G(Z, W)$ by means of complete orthonormal systems with respect to the metric (4) in the function spaces Ω and Ω^0 respectively [4].

2. **The kernel function.** As was already pointed out in §1, the representation of the kernel function by a development (14) in terms of a complete orthonormal system in Λ leads to important applications for the numerical treatment of the theory of the differential equation (2). It should be pointed out, however, that this same formula leads immediately to theoretical consequences which are by no means obvious directly.

Consider for example two points $Z \, \varepsilon \, B$ and $W \, \varepsilon \, B$ and construct the expression

$$(15) \qquad L(Z, W) = K(Z, Z) + K(W, W) - 2K(Z, W).$$

Applying (14), we find immediately

$$(15') \qquad L(Z, W) = \sum_{\nu=1}^{\infty} [\varphi_\nu(Z) - \varphi_\nu(W)]^2 \geq 0.$$

This yields the inequality

$$(15'') \qquad K(Z, W) \leq \tfrac{1}{2}\{K(Z, Z) + K(W, W)\}.$$

This result follows in a quite trivial way from the identity (14), but is of great use and interest if considered as an inequality for Green's and Neumann's functions.

We might have applied (11'') to the particular function $\varphi(Z) = K(Z, T)$ with $T \varepsilon D$. We obtain immediately

$$(11''') \qquad D_W\{K(Z, W), K(W, T)\} = K(Z, T),$$

whence by the Schwarz inequality,

$$(15''') \qquad K(Z, T)^2 \leq K(Z, Z)K(T, T).$$

This is, of course, an improvement of (15''). However, the ultimate reason for inequality (15'') seems to lie in the nature of $K(Z, T)$ as a kernel function.

On the other hand, we may increase our knowledge of the kernel function by using all results from the well-developed theories of the fundamental solutions of (2). We give, as an example, the following application, which will be used decisively later on. We formulate the following theorem.

THEOREM I. *The kernel function is non-negative in B.*

In view of (8), this statement is equivalent to

THEOREM II.

$$(16) \qquad N(Z, W) \geq G(Z, W) \qquad\qquad (Z, W \varepsilon B).$$

In order to prove this inequality it will be sufficient to prove that for $W \varepsilon B$ fixed and $Z \varepsilon C$ we have always $N(Z, W) \geq 0$. For, the difference $K(Z, W) = N(Z, W) - G(Z, W)$ is a solution of (2) and as such it is non-negative in B if it is non-negative on the boundary of the domain considered. For, otherwise, there would be a negative minimum in an interior point of B and there $\partial^2 K/\partial x^2 \geq 0$, $\partial^2 K/\partial y^2 \geq 0$ which stands in contradiction to the differential equation (2) and our assumption.

Since we will later on make frequent use of the same consideration, we formulate it in the form of the following principle:

Every solution of (2) *which is non-negative on C is non-negative in B.*

We prove the non-negative behavior of $N(Z, W)$ on C indirectly. Suppose that for a fixed $W \varepsilon B$, $N(Z, W)$ were negative somewhere in B. Divide B into that part where $N > 0$ and the other part where $N < 0$. Neither part is empty since, at W, N becomes positive infinite, and by assumption there exist points where N is negative. The boundary line of each part consists either of arcs of C or of the smooth level lines $N(Z, W) = 0$. Consider now the part B_- where N is negative and consider the integral

$$(17) \qquad J = \iint\limits_{B_-} [(\partial N(Z, W)/\partial x)^2 + (\partial N(Z, W)/\partial y)^2 + P(Z)N(Z, W)^2] \, dx \, dy.$$

This integral exists since the pole W of $N(Z, W)$ does not lie in B_- and is obviously positive. We transform this integral by means of Green's integral identity. We obtain

$$(18)$$
$$J = -\oint_{c_-} N(Z, W)(\partial N(Z, W)/\partial n_z) \, ds_z$$

$$- \iint\limits_{B_-} N(Z, W)[\Delta_z N(Z, W) - PN(Z, W)] \, dx \, dy,$$

where c_- denotes the boundary of B_-. Now, we remark that $N\partial N/\partial n = 0$ everywhere on c_- since c_- consists of the level lines $N = 0$ or of arcs of C where $\partial N/\partial n = 0$. The area integral over B vanishes since $N(Z, W)$ is a solution of (2). Hence, we find $J = 0$ which is impossible in view of (17). Thus, our assumption was wrong; there cannot exist points of B where $N(Z, W)$ is negative. Hence, $N(Z, W)$ is non-negative in $B + C$ which proves all our assertions.

From Theorem I and the differential equation (2) we conclude further that we have throughout B always $\Delta_z K(Z, W) \geq 0$, $i.e.$, that the kernel function is subharmonic in B.

Our last result may easily be generalized if we consider a larger class of fundamental solutions. Let us define a continuous non-negative function $\lambda(Z)$ on C. Consider the fundamental solution $R(Z, W)$ of (2) in B which satisfies on C the boundary conditions:

$$(19) \qquad \partial R(Z, W)/\partial n_z = \lambda(Z)R(Z, W) \qquad (Z \, \varepsilon \, C; \, W \, \varepsilon \, B).$$

$R(Z, W)$ is sometimes called Robin's function and plays an analogous role in the boundary value problem of the third kind as does Green's function in that of the first kind and Neumann's functions in that of the second kind. Green's and Neumann's functions may be considered as limit cases of Robin's function, namely for the limits $\lambda \to \infty$ and $\lambda \to 0$ respectively.

We want to show now, that each Robin function is non-negative in B. For this purpose, we assume as before that there were points in B where R is negative and divide B into two parts B_+ and B_- where R is positive and negative, respectively. Each of these domains is bounded by smooth arcs in B on which $R = 0$ and by arcs of C where $-\lambda R + \partial R/\partial n = 0$. Now, construct as before the integral

$$(17') \qquad J^* = \iint\limits_{B-} [(\partial R(Z, W)/\partial x)^2 + (\partial R(Z, W)/\partial y)^2 + P(Z)R(Z, W)^2] \, dx \, dy$$

which is obviously positive. Using Green's integral identity, we find on the other hand

$$J^* = -\oint\limits_{c-} R(Z, W)(\partial R(Z, W)/\partial n_Z) \, ds_Z$$

$(18')$

$$-\iint\limits_{B-} R(Z, W)[\Delta_Z R(Z, W) - PR(Z, W)] \, dx \, dy.$$

The area integral vanishes since $R(Z, W)$ is a solution of (2); on the boundary c_- of B_- we have clearly $R\partial R/\partial n \geq 0$. Hence, by $(18')$ we find for J^* a nonpositive value while $(17')$ obviously requires $J^* > 0$. Thus, we obtain a contradiction and have proved

$$(20) \qquad\qquad\qquad R(Z, W) \geq 0 \qquad\qquad\qquad (Z, W \varepsilon B).$$

Since the function $R(Z, W) - G(Z, W)$ is regular everywhere in B and nonnegative on the boundary C, (20) implies the further inequality

$$(20') \qquad\qquad Q(Z, W) = R(Z, W) - G(Z, W) \geq 0$$

everywhere in B. $Q(Z, W)$ is constructed in a very similar way to $K(Z, W)$; we will show later that $Q(Z, W)$ is, in fact, a kernel function of a complete orthonormal system in B with respect to another metric connected with the differential equation (2). Again, we may assert that $Q(Z, W)$ is subharmonic in B.

Finally, we make the following extension of the previous results: Let $R_1(Z, W)$ and $R_2(Z, W)$ be two Robin functions satisfying the boundary conditions

$$(21) \qquad \partial R_i(Z, W)/\partial n_Z = \lambda_i(Z)R_i(Z, W) \qquad\qquad (Z \varepsilon C; W \varepsilon B; i = 1, 2)$$

Suppose that on C we have

$$(22) \qquad\qquad\qquad \lambda_1(Z) \leq \lambda_2(Z).$$

Then we want to prove that we have everywhere in B

$$(23) \qquad\qquad\qquad R_1(Z, W) \geq R_2(Z, W).$$

Obviously this theorem will contain all previous results as particular cases.

We proceed again indirectly. Suppose there are points $Z \varepsilon B$ where $R_1 < R_2$. Call that part of B where $R_1(Z, W) < R_2(Z, W)$ again B_- and its boundary c_-. c_- is composed of the smooth arcs $R_1 = R_2$ and of arcs of C. The integral

$$\iint\limits_{B-} [(\partial(R_1 - R_2)/\partial x)^2 + (\partial(R_1 - R_2)/\partial y)^2 + P(Z)(R_1 - R_2)^2] \, dx \, dy$$

$$(24) \qquad = -\oint\limits_{c-} (R_1 - R_2)[\partial(R_1 - R_2)/\partial n_Z] \, ds_Z$$

$$- \iint\limits_{B-} (R_1 - R_2)[\Delta_Z(R_1 - R_2) - P(R_1 - R_2)] \, dx \, dy$$

is by its left-hand representation obviously positive. On the right side the area integral vanishes because of the differential equation (2) satisfied by $R_1 - R_2$, while the line integral is only to be extended over the arcs of c_- belonging to C. But here we have, by virtue of (20), (21) and (22) and in view of the definition of c_- ,

$$(25) \qquad R_1 - R_2 < 0, \qquad \partial(R_1 - R_2)/\partial n = \lambda_1 R_1 - \lambda_2 R_2 \le \lambda_2(R_1 - R_2) < 0\cdot$$

Thus, the value of the line integral is non-positive which leads to a contradiction with our above statement. Our assumption that somewhere $R_1 < R_2$ is impossible; hence, we have proved that, everywhere in B, (23) is valid.

3. **The variation of the kernel function.** The kernel function $K(Z, W)$ for a given domain B is a functional of the coefficient $P(Z)$ of the differential equation. In this section, we shall investigate its dependence on $P(Z)$. For this purpose, we will establish formulas for the variation of Green's and Neumann's functions for the case where $P(Z)$ is varied infinitesimally and shall obtain an analogous variation formula for $K(Z, W)$ by means of (8).

We define a continuous function $p(Z)$ in $B + C$ and introduce a real number ϵ which will serve as an infinitesimal parameter. Let

$$(26) \qquad \delta P(Z) = \epsilon p(Z)$$

and, in addition to the differential equation (2), consider the equation

$$(2') \qquad \Delta\varphi = [P(Z) + \delta P(Z)]\varphi.$$

We choose ϵ so small that $P + \delta P > 0$ in $B + C$; then our entire previous theory holds for equation $(2')$ also. Let $G(Z, W) + \delta G(Z, W)$ be Green's function for B with respect to the differential equation $(2')$. It is known that $\delta G(Z, W) \to 0$ as $\epsilon \to 0$. In order to determine $\delta G(Z, W)$ we consider the integral

$$(27) \qquad \iint\limits_{B} [G(Z, W)\Delta_Z(G(Z, V) + \delta G(Z, V))$$

$$- (G(Z, V) + \delta G(Z, V))\Delta_Z G(Z, W)] \, dx \, dy$$

$$= G(V, W) - (G(V, W) + \delta G(V, W)) = -\delta G(V, W).$$

Introducing the values for $\Delta_z G(Z, W)$ into the left-hand integral and for $\Delta_z(G(Z, V) + \delta G(Z, V))$ as obtained from the differential equations (2) and (2'), respectively, we obtain

$$\delta G(V, W)$$

(28)
$$= -\iint_B [\delta P(Z)G(Z, V)G(Z, W) + \delta P(Z)\delta G(Z, V)G(Z, W)]\, dx\, dy.$$

This formula shows that $\delta G(V, W)$ is of the order of magnitude of ϵ and that we may write

(28')
$$\delta G(V, W) = -\iint_B \delta P(Z)G(Z, V)G(Z, W)\, dx\, dy + o(\epsilon),$$

where $\lim_{\epsilon \to 0} \epsilon^{-1} o(\epsilon) = 0$.

In the same way and with analogous notations we derive immediately

(29)
$$\delta N(V, W) = -\iint_B \delta P(Z)N(Z, V)N(Z, W)\, dx\, dy + o(\epsilon).$$

Applying now the definition (8) of the kernel function, we find

$$\delta K(V, W)$$

(30)
$$= -\iint_B \delta P(Z)[N(Z, V)N(Z, W) - G(Z, V)G(Z, W)]\, dx\, dy + o(\epsilon).$$

Suppose now that we have two differential equations of type (2),

(2'')
$$\Delta \varphi = P_1(Z)\varphi, \qquad \Delta \varphi = P_2(Z)\varphi,$$

and everywhere in $B + C$

(31)
$$0 < P_1(Z) < P_2(Z).$$

Consider the auxiliary differential equations

(32)
$$\Delta \varphi = [P_1(Z) + t(P_2(Z) - P_1(Z))]\varphi \qquad (0 \leq t \leq 1)$$

which are also of type (2). Denote by $G(V, W; t)$, $N(V, W; t)$ and $K(Z, W; t)$ the Green, Neumann and kernel function of B with respect to (32). Then we have in view of (28'), (29) and (30):

(33)
$$\partial G(V, W; t)/\partial t = -\iint_B (P_2(Z) - P_1(Z))G(Z, V; t)G(Z, W; t)\, dx\, dy,$$

(33')
$$\partial N(V, W; t)/\partial t = -\iint_B (P_2(Z) - P_1(Z))N(Z, V; t)N(Z, W; t)\, dx\, dy,$$

$$\partial K(V, W; t)/\partial t = - \iint_B (P_2(Z) - P_1(Z))[N(Z, V; t)N(Z, W; t)$$

(33'')

$$- G(Z, V; t)G(Z, W; t)] \, dx \, dy.$$

Now, we know that everywhere in B

(34) $\qquad G(Z, W; t) \geq 0, \qquad N(Z, W; t) \geq 0, \qquad N(Z, W; t) \geq G(Z, W; t).$

This shows that the derivatives of all three functions are negative which leads to

THEOREM III. *If two differential equations (2'') are given and their coefficients satisfy the inequality (31), then the Green, the Neumann and the kernel functions with respect to the first differential equation are each larger than the corresponding functions with respect to the second equation.*

This theorem is of great use in the theory of the differential equation (2) and will be applied in the next section to prove an important result. A more immediate application of it leads to the following estimates. Suppose that in B

(35) $$0 < m \leq P(Z) \leq M$$

and consider the two auxiliary differential equations

(36) $$\Delta \varphi = m\varphi, \qquad \Delta \varphi = M\varphi$$

which are of type (2) but have constant coefficients. Let G_m, N_m, K_m and G_M, N_M, K_M be the three fundamental functions for both equations. Then we have for the corresponding functions with respect to equation (2) the inequalities

(37) $\qquad G_M \leq G \leq G_m, \qquad N_M \leq N \leq N_m, \qquad K_M \leq K \leq K_m.$

For the sake of completeness we want to remark that Theorem III may be extended to all Robin functions. In fact, we have

(38) $$\delta R(V, W) = - \iint_B \delta P(Z)R(Z, V)R(Z, W) \, dx \, dy + o(\epsilon)$$

as can be proved in just the same way as before and making use of the fact that $R(V, W)$ and $R(V, W) + \delta R(V, W)$ satisfy on C the same boundary conditions. From this one easily derives the monotonic behavior of $R(V, W)$ for increasing coefficient $P(Z)$, as is stated in

THEOREM IV. *If in the differential equation (2) the coefficient $P(Z)$ increases everywhere in B, all Robin's functions $R(Z, W)$ decrease.*

If we further apply the inequality (23) of the last section we see from (38) that $R(Z, W)$ is the more susceptible to a change of $P(Z)$ the smaller its corresponding function $\lambda(Z)$ is. This shows further that the generalized kernel functions $Q(Z, W)$, defined in (20'), also decrease if the function $P(Z)$ is increased everywhere in B. For, as already pointed out, we may consider Green's function as the limit case of a Robin function for $\lambda(Z) \to \infty$. This shows that Green's function varies less than any Robin function under an increase of the coefficient $P(Z)$.

Finally, we consider another important auxiliary function connected with the differential equation (2). We denote by $U(Z)$ that solution of (2) which has the value 1 at the boundary C of B. By means of (9) we obtain a simple representation of $U(Z)$ in terms of the kernel function:

$$(39) \qquad U(Z) = D_W\{K(Z, W), 1\} = \iint\limits_B P(W)K(Z, W)\, du\, dv, \qquad W \equiv (u, v),$$

since $D_W\{K(Z, W), 1\}$ is that solution of (2) which has on C the same values as 1. $U(Z)$ plays an important role in estimating the solutions of (2) by means of their boundary values. In fact, let $\varphi(Z)$ be an arbitrary solution of (2) which assumes values $\leq \mu$ on C. Then $\mu U(Z) - \varphi(Z)$ is non-negative on C and is a solution of (2). According to the principle of §2 that a solution of (2) which is non-negative on C is non-negative in B, we conclude that everywhere in B

$$(40) \qquad\qquad\qquad \varphi(Z) \leq \mu U(Z).$$

$U(Z)$ is, of course, non-negative in B and because of the differential equation (2) it is subharmonic in B. It attains, therefore, its maximum at the boundary and we have throughout B

$$(40') \qquad\qquad\qquad 0 \leq U(Z) \leq 1.$$

Let now $U_1(Z)$ and $U_2(Z)$ be the corresponding solutions with respect to the two differential equations (2″) where the coefficients $P_1(Z)$ and $P_2(Z)$ satisfy the inequality (31). We want to prove

THEOREM V. *If the coefficients of the differential equations* (2″) *satisfy the inequality* $0 < P_1(Z) \leq P_2(Z)$, *the corresponding U-functions satisfy* $U_1(Z) \geq U_2(Z)$.

In fact, we know that in view of the positive character of $U(Z)$ on C each $U(Z)$ is non-negative in B. The difference $U_1(Z) - U_2(Z)$ is zero on C; if it were negative somewhere in B there would necessarily exist a point $Z_0 \varepsilon B$ where it attained its minimum. At this point one has the necessary minimum conditions

$$\frac{\partial}{\partial x}(U_1 - U_2) = \frac{\partial}{\partial y}(U_1 - U_2) = 0;$$

$$(41)$$

$$\frac{\partial^2(U_1 - U_2)}{\partial x^2} \geq 0, \qquad \frac{\partial^2(U_1 - U_2)}{\partial y^2} \geq 0.$$

Hence adding together the last two inequalities we have

(41') $$\Delta(U_1 - U_2) = P_1U_1 - P_2U_2 \geq 0$$

at Z_0. On the other hand, we have in view of (31)

(41'') $$0 \leq P_1U_1 - P_2U_2 \leq P_2(U_1 - U_2)$$

at Z_0. This shows that at the minimum point the difference is non-negative in contradiction to our assumption. This shows that always $U_1(Z) \geq U_2(Z)$ as asserted.

If we send in particular $P(Z) \to 0$ uniformly in $B + C$ the function $U(Z)$ tends everywhere to the constant value 1, as is easily seen by considering the limit case of harmonic functions.

Theorem V might have been obtained by the following investigation which is of considerable interest in itself. Consider a variation δP of the coefficient $P(Z)$ and denote the corresponding variation of $U(Z)$ by δU. We have, by virtue of (39),

(42)
$$\delta U(Z) = \iint_B \delta P(W)K(Z, W)\, du\, dv + \iint_B P(W)\delta K(Z, W)\, du\, dv$$
$$+ \iint_B \delta P(W) \cdot \delta K(Z, W)\, du\, dv.$$

Now using (30) in order to express $\delta K(Z, W)$ we find (with $T \equiv (r, s)$)

(42')
$$\delta U(Z) = \iint_B \delta P(W)[K(Z, W) - N(Z, W) \iint_B P(T)N(W, T)\, dr\, ds$$
$$+ G(Z, W) \iint_B P(T)G(W, T)\, dr\, ds]\, du\, dv + o(\epsilon).$$

There arises now the question as to the meaning of the functions

(42'') $$\alpha(Z) = \iint_B P(T)N(Z, T)\, dr\, ds, \qquad \beta(Z) = \iint_B P(T)G(Z, T)\, dr\, ds$$

which appear in our variation formula. Considering the definition of Green's and Neumann's functions for B with respect to the differential equation (2), we obtain

(43) $$\Delta\alpha - P\alpha = -P, \qquad \partial\alpha/\partial n = 0 \qquad \text{(on } C),$$

(43') $$\Delta\beta - P\beta = -P, \qquad \beta = 0 \qquad \text{(on } C).$$

Hence, $1 - \alpha(Z)$ and $1 - \beta(Z)$ are solutions of (2), the first of which has the normal derivative zero on C while the second is constantly one on C. From these facts and the relation $\alpha(Z) - \beta(Z) = U(Z)$ which results from (39), we conclude

$$\alpha(Z) = \iint_B P(T)N(Z, T)\, dr\, ds \equiv 1,$$

(43'')

$$\beta(Z) = \iint_B P(T)G(Z, T)\, dr\, ds \equiv 1 - U(Z).$$

These results are, by the way, simple consequences of formulas (7) and (6'), since we have obviously

(43''') $\alpha(Z) = D_T\{N(Z, T), 1\}, \qquad \beta(Z) = D_T\{G(Z, T), 1\}.$

Thus after a slight transformation and the use of (8), (42') becomes

(44) $$\delta U(Z) = -\iint_B \delta P(W)U(W)G(Z, W)\, du\, dv + o(\epsilon)$$

which gives a concise variation formula for $U(Z)$ and a new proof for Theorem V.

4. On normal families of solutions. Consider the family **F** of all functions $\varphi(Z)$ in the class Λ with respect to a given differential equation (2) and which satisfy the inequality

(45) $$D\{\varphi, \varphi\} \leq 1.$$

We want to prove that the family **F** of all such $\varphi(Z)$ is normal, *i.e.*, that one may select from each infinite set of elements of **F** a sequence $\varphi_n(Z)$ which converges uniformly in each closed subdomain of B.

In fact, applying identity (11'') of §1 we have

$$\varphi(Z) = D_W\{K(Z, W), \varphi(W)\},$$

(46) $$\partial\varphi(Z)/\partial x = D_W\{\partial K(Z, W)/\partial x, \varphi(W)\},$$

$$\partial\varphi(Z)/\partial y = D_W\{\partial K(Z, W)/\partial y, \varphi(W)\};$$

and, therefore, by means of Schwarz's inequality and of (45)

$$\varphi(Z)^2 \leq K(Z, Z),$$

(47) $$\left(\frac{\partial\varphi}{\partial x}\right)^2 \leq D_W\left\{\frac{\partial K(Z, W)}{\partial x}, \frac{\partial K(Z, W)}{\partial x}\right\},$$

$$\left(\frac{\partial\varphi}{\partial y}\right)^2 = D_W\left\{\frac{\partial K(Z, W)}{\partial y}, \frac{\partial K(Z, W)}{\partial y}\right\}.$$

The right sides of the last two inequalities are finite because of the definition (8) of the kernel function. This shows that all functions of the family $\mathbf{F}$ are uniformly bounded and equicontinuous. But these two properties are sufficient in order to ensure the normality of $\mathbf{F}$.

It is easy to show that the function $\psi(Z) = \lim_{n\to\infty} \varphi_n(Z)$ also satisfies the differential equation (2) at each interior point of B. In fact, let Z_0 be an interior point of B and let Γ_0 be a circle around Z_0 which is still entirely in B. If $\gamma_0(Z, W)$ denotes Green's function of Γ_0 with respect to Laplace's differential equation, we have in view of (2)

$$(48) \qquad \varphi_n(Z) = \iint\limits_{\Gamma_0} \gamma_0(Z, W)P(W)\varphi_n(W) \, du \, dv + h_n(Z) \qquad (W \equiv (u, v)),$$

where $h_n(Z)$ is a harmonic function in Γ_0 which has the same values as $\varphi_n(Z)$ on the circumference. Because of the uniform convergence of the $\varphi_n(Z)$ in the closed circular area, we have

$$(48') \qquad \psi(Z) = \iint\limits_{\Gamma_0} \gamma_0(Z, W)P(W)\psi(W) \, du \, dv + h(Z)$$

where $h(Z)$ is again harmonic in Γ_0. Hence, we proved

$$(49) \qquad \Delta\psi(Z) = P(Z)\psi(Z)$$

for each interior point $Z \,\varepsilon\, B$.

Let us now extend this result. Consider all differential equations of type (2), the coefficients of which are uniformly bounded from both sides in B, say

$$(50) \qquad 0 < m \le P(Z) \le M \qquad\qquad (Z \,\varepsilon\, B).$$

Consider all solutions $\varphi(Z)$ of all possible equations (2) satisfying (50) for which (45) holds. It should be remarked that $D\{\varphi, \varphi\}$ depends on $P(Z)$ and that, for each particular $\varphi(Z)$, that $P(Z)$ should be taken which corresponds to it. Let $\mathbf{F}^+$ denote the family of all functions $\varphi(Z)$ thus obtained for all continuous functions $P(Z)$ which satisfy (50). We want to prove

THEOREM VI. *The family $\mathbf{F}^+$ is normal.*

In fact, we find for every function $\varphi(Z)$ of $\mathbf{F}^+$ by means of the identity (11''), by Schwarz's inequality and because of (45), that

$$(51) \qquad \varphi(Z)^2 \le K(Z, Z),$$

where $K(Z, W)$ is the kernel function with respect to the particular differential equation satisfied by $\varphi(Z)$. Introducing the kernel function $K_m(Z, W)$ with respect to the differential equation

$$(52) \qquad \Delta\varphi = m\varphi,$$

we find in view of (37) the inequality

(53) $$\varphi(Z)^2 \le K_m(Z, Z)$$

which gives a uniform upper bound for all functions of $\mathbf{F}^+$.

In order to show the equicontinuity of $\mathbf{F}^+$, we consider again an interior point $Z_0 \ \varepsilon \ B$, and a circle Γ_0 around it which is entirely in B with its Green's function $\gamma_0(Z, W)$ with respect to the Laplace equation. Let A denote the maximum modulus of all values which the functions of $\mathbf{F}^+$ assume in this circle. In view of the differential equation (2) satisfied by $\varphi(Z)$, we may write

(54) $$\varphi(Z) = \iint_{\Gamma_0} \gamma_0(Z, W)P(W)\varphi(W) \, du \, dv + h(Z) \qquad (Z \ \varepsilon \ \Gamma_0),$$

where $P(Z)$ is the coefficient of the particular differential equation satisfied by $\varphi(Z)$ and $h(Z)$ is that harmonic function in Γ_0 which has the same values as $\varphi(Z)$ on the circumference. Since $| \varphi(Z) | \le A$ in Γ_0 , the same inequality holds for $h(Z)$ by virtue of the maximum principle. Differentiating (54) with respect to x, we obtain

(55) $$\partial\varphi/\partial x = \iint_{\Gamma_0} (\partial\gamma_0(Z, W)/\partial x)P(W)\varphi(W) \, du \, dv + \partial h/\partial x.$$

Now it is well known that $\partial h/\partial x$ is uniformly bounded for a fixed circle $\Gamma_1 \subset \Gamma_0$. On the other hand, we have

(56) $$\left| \iint_{\Gamma_0} (\partial\gamma_0(Z, W)/\partial x)P(W)\varphi(W) \, du \, dv \right| \le MA \iint_{\Gamma_0} | \partial\gamma(Z, W)/\partial x | \, du \, dv$$

which establishes a uniform bound for $| \partial\varphi/\partial x |$ in $\Gamma_1 \subset \Gamma_0$. The same procedure leads to a uniform estimate for $| \partial\varphi/\partial y |$. Thus, the equicontinuity of the family $\mathbf{F}^+$ is established, which finishes the proof of its normality.

We might define other classes of normal families which will be of use in later applications. We prescribe, on the boundary C of B, an arbitrary continuous function $\rho(Z)$. We consider all solutions of differential equations (2) with the restrictions (50) which assume on the boundary C the given values $\rho(Z)$. All these solutions form a normal family $\mathbf{F}^*$.

In order to prove this assertion, we introduce the functions $U_m(Z)$ and $U_M(Z)$ which satisfy respectively the differential equations

(57) $$\Delta U_m = mU_m , \qquad \Delta U_M = MU_M$$

and have on C the boundary value 1. By Theorem V, U_m is nowhere smaller and U_M nowhere larger than the U-functions of any permissible differential equation (2). Let

(58) $$\alpha \le \rho(Z) \le \beta \qquad (Z \ \varepsilon \ C).$$

For each function $\varphi(Z)$ of the family $\mathbf{F}^*$ we have by virtue of the principle of §2

(59) $$\alpha U(Z) \leq \varphi(Z) \leq \beta U(Z) \qquad (Z \; \varepsilon \; B),$$

where $U(Z)$ satisfies the same differential equation as $\varphi(Z)$. Using now the comparison functions $U_m(Z)$ and $U_M(Z)$, we obtain uniform bounds for all functions of $\mathbf{F}^*$ in the closed domain $B + C$.

Let now $\gamma(Z, W)$ be Green's function of B with respect to Laplace's equation. Then we may represent each $\varphi(Z) \; \varepsilon \; \mathbf{F}^*$ in the form

(60) $$\varphi(Z) = \iint\limits_{B} \gamma(Z, W)P(W)\varphi(W) \, du \, dv + h(Z)$$

where $h(Z)$ is always the same harmonic function with boundary values $\rho(Z)$ on C. Differentiating (60) with respect to x, we obtain

(60') $$\frac{\partial \varphi}{\partial x} = \iint\limits_{B} \frac{\partial \gamma(Z, W)}{\partial x} P(W)\varphi(W) \, du \, dv + \frac{\partial h}{\partial x}$$

and an analogous formula holds for $\partial \varphi / \partial y$. Since all $\varphi(W) \; \varepsilon \; \mathbf{F}^*$ are uniformly bounded in $B + C$, the same is true for $\partial \varphi / \partial x$ and $\partial \varphi / \partial y$. This shows that the functions of $\mathbf{F}^*$ form a normal family in $B + C$; i.e., from each infinite set of functions of the family one can always select a subsequence which converges uniformly in $B + C$. The limit function will, therefore, have the same boundary values $\rho(Z)$ as all elements of the family.

Let us consider finally a family $\mathbf{P}$ of functions $P(Z)$ which are equicontinuous in $B + C$ and satisfy the inequalities (50). Let $\mathbf{F}^{**}$ be the family of all solutions $\varphi(Z)$ of (2) with coefficients $P(Z) \; \varepsilon \; \mathbf{P}$ which assume on the boundary C of B prescribed boundary values $\rho(Z)$. We may again select from each infinite set of functions $\varphi(Z) \; \varepsilon \; \mathbf{F}^{**}$ a uniformly convergent subsequence $\varphi_n(Z)$. We have for each element of this sequence

(61) $$\varphi_n(Z) = \iint\limits_{B} \gamma(Z, W)P_n(W)\varphi_n(W) \, du \, dv + h(Z),$$

where $P_n(Z)$ is the coefficient of (2) corresponding to $\varphi_n(Z)$ and $h(Z)$ is always the same harmonic function. Since the $P_n(W)$ are bounded and equicontinuous they form themselves a normal family in $B + C$. Hence, we may select a subsequence $P_{n_\nu}(Z)$ among the $P_n(Z)$ which converges uniformly in $B + C$ to a continuous function $P(Z)$. Let us denote $\lim_{\nu \to \infty} \varphi_{n_\nu}(Z) = \psi(Z)$. Then, we have by virtue of (61)

(62) $$\psi(Z) = \iint\limits_{B} \gamma(Z, W)P(W)\psi(W) \, du \, dv + h(Z)$$

which shows that $\psi(Z)$ itself is a solution of an equation (2) with $P(Z)$ satisfying the inequalities (50).

5. **Non-linear differential equations.** The considerations of the last section are of great importance in the investigation of solutions of certain non-linear differential equations of elliptic type.   Let us consider at first a function $P(x, y; t) = P(Z; t)$ which is defined for $Z \equiv (x, y) \; \varepsilon \; B + C$ and all real values of t.   Let $P(Z; t)$ be continuous for $Z \; \varepsilon \; B + C$ and satisfy a Lipschitz condition

$$(63) \qquad\qquad | P(Z; t) - P(Z; t_1) | \leq \kappa \cdot | t - t_1 |$$

and suppose further that the following inequality is satisfied by $P(Z; t)$:

$$(63') \qquad\qquad 0 < m \leq P(Z; t) \leq M \qquad (Z \; \varepsilon \; B + C; t \text{ real}).$$

We consider now the differential equation

$$(64) \qquad\qquad \Delta\varphi = P(Z; \varphi) \cdot \varphi$$

and seek a solution $\varphi(Z)$ of (64) which assumes prescribed continuous boundary values $\rho(Z)$ on C.

We may try to construct the required solution as follows.   We start with an arbitrary function $\varphi_0(Z)$ which is twice continuously differentiable in B and has the required boundary values on C.   Then we begin a recursive procedure based on the formula

$$(65) \qquad\qquad \Delta\varphi_n(Z) = P(Z; \varphi_{n-1}) \cdot \varphi_n(Z),$$

where $\varphi_n(Z) = \rho(Z)$ on C. Since the solutions $\varphi_n(Z)$ belong to the family $\mathbf{F^*}$ defined in the last section, we know that they are uniformly bounded in $B + C$. Again, because of (65), we have the following representation for φ_n :

$$(65') \qquad \varphi_n(Z) = \iint\limits_{B} \gamma(Z, W) P(W; \varphi_{n-1}(W)) \varphi_n(W) \, du \, dv + h(Z).$$

Now let us denote

$$(66) \qquad\qquad \underset{Z \varepsilon B+C}{\text{Max}} \; | \varphi_n(Z) - \varphi_{n-1}(Z) | = \mu_n .$$

We find easily from (65') the following recursive inequality for μ_n :

$$(67) \qquad\qquad \mu_n \leq MJ\mu_n + \kappa A J \mu_{n-1} ,$$

where κ and M are defined in (63) and (63'), A is the upper bound for the moduli of all $\varphi_n(Z)$ in $B + C$, and

$$(67') \qquad\qquad J = \underset{Z \varepsilon B+C}{\text{Max}} \left\{ \iint\limits_{B} \gamma(Z, W) \, du \, dv \right\}.$$

J is a functional of the domain B and becomes small for small domains B. Let us now assume J to be so small that

(68) $$0 < \frac{\kappa A J}{1 - M J} < 1.$$

In this case the recursive procedure converges geometrically in $B + C$ and we obtain in the limit a function $\varphi(Z)$ which satisfies the condition

(65'') $$\varphi(Z) = \iint\limits_{B} \gamma(Z, W) P(W; \varphi(W)) \varphi(W) \, du \, dv + h(Z)$$

which shows that $\varphi(Z)$ is a solution of the differential equation (64) which, on C, assumes just the required boundary values $\rho(Z)$.

It should be remarked that from a practical point of view the recursive formula

(69) $$\Delta \varphi_n = P(Z; \varphi_{n-1}) \varphi_{n-1} ,$$

where $\varphi_n(Z) = \rho(Z)$ on C, would have been much easier to apply. For in each step of the recursion we would have had to integrate an equation of the Poisson-Laplace type instead of an equation of the type (65) which is of the form of a wave-amplitude equation. However, under our assumptions, it is not possible to prove the convergence of this procedure. Moreover, the theory of orthogonal functions makes it relatively easy to carry out the numerical work connected with the recursive formula (65).

Let us now study a more general type of non-linear differential equation. Consider a function $P(x, y; t, t', t'') = P(Z; t, t', t'')$ which is defined for arbitrary $Z \, \varepsilon \, B + C$ and real t, t', t''. We assume P to be continuous in Z and satisfy the Lipschitz condition

(70) $$| P(Z; t, t', t'') - P(Z; t_1, t_1', t_1'') |$$
$$< \kappa \, \text{Max}\{| t - t_1 |, | t' - t_1' |, | t'' - t_1'' |\}$$

as well as the inequality

(70') $$0 < m \leq P(Z; t, t', t'') \leq M$$

for $Z \, \varepsilon \, B; t, t', t''$ real. We seek solutions of the corresponding differential equation

(71) $$\Delta \varphi = P(Z; \varphi, \partial \varphi / \partial x, \partial \varphi / \partial y) \cdot \varphi$$

which are continuous in $B + C$ and have arbitrarily prescribed continuous boundary values $\rho(Z)$ on C.

Again we define a recursive process based upon the formula

(72) $$\Delta \varphi_n = P(Z; \varphi_{n-1}, \partial \varphi_{n-1} / \partial x, \partial \varphi_{n-1} / \partial y) \varphi_n = P_{n-1}(Z) \varphi_n$$

where $\varphi_n(Z) = \rho(Z)$ on C, and $\varphi_0(Z)$ is an arbitrary function which is twice differentiable in B and has the required boundary values on C. We have once more

$$(72') \qquad \varphi_n(Z) = \iint\limits_B \gamma(Z, W) P_{n-1}(W) \varphi_n(W) \, du \, dv + h(Z).$$

Let

$$(73) \qquad \underset{Z \in B+C}{\text{Max}} \left\{ \left| \varphi_n(Z) - \varphi_{n-1}(Z) \right|, \left| \frac{\partial \varphi_n}{\partial x} - \frac{\partial \varphi_{n-1}}{\partial x} \right|, \left| \frac{\partial \varphi_n}{\partial y} - \frac{\partial \varphi_{n-1}}{\partial y} \right| \right\} = \mu_n ,$$

A be the upper bound for the moduli of all elements in $\mathbf{F}^*$, and

$$(74) \quad J = \underset{Z \in B+C}{\text{Max}} \left\{ \iint\limits_B \gamma(Z, W) \, du \, dv, \iint\limits_B \frac{\partial \gamma(Z, W)}{\partial x} \, du \, dv, \iint\limits_B \left| \frac{\partial \gamma(Z, W)}{\partial y} \right| \, du \, dv \right\};$$

we derive from (72'), (70), (70') and (71) the inequality

$$(75) \qquad \mu_n \le \mu_n M J + \kappa J A \mu_{n-1} .$$

Hence, on condition that

$$(76) \qquad 0 < \frac{JA\kappa}{1 - MJ} < 1,$$

we again have a uniform convergence of the $\varphi_n(Z)$ and their first partial derivatives in $B + C$, and the limit function

$$(72'') \qquad \varphi(Z) = \iint\limits_B \gamma(Z, W) P\left(W; \varphi, \frac{\partial \varphi}{\partial u}, \frac{\partial \varphi}{\partial v} \right) \varphi(W) \, du \, dv + h(Z),$$

which proves that $\varphi(Z)$ is a solution of (71) and has just the boundary values required.

In a similar way existence theorems for the boundary value problem of the second kind might be established. In this case, a continuous function $\nu(Z)$ would be defined on C and the solutions required must satisfy the condition

$$(77) \qquad \partial \varphi(Z)/\partial n = \nu(Z) \qquad\qquad (Z \in C).$$

In order to carry out the above considerations, a new auxiliary function $V(Z)$ has to be introduced which plays a role similar to that of $U(Z)$ above. $V(Z)$ is a solution of the differential equation (2) which has the normal derivative 1 everywhere on C. Since, by virtue of (11''), Green's integral identity and (8),

$$(78) \qquad \begin{aligned} V(Z) = D_W\{K(Z, W), V(W)\} &= -\oint\limits_C K(Z, W) \frac{\partial V}{\partial n_W} \, ds_W \\ &= -\oint\limits_C N(Z, W) \, ds_W \end{aligned}$$

we recognize by virtue of Theorem II of §2 that $V(Z)$ is negative throughout

B and from Theorem III of §3 we infer that $V(Z)$ increases if the coefficient $P(Z)$ decreases.

We may apply formula (29) in order to get an explicit variation formula for $V(Z)$. In fact, we have

$$\delta V(Z) = \oint_C \left[\iint_B \delta P(W) N(Z, W) N(T, W) \, du \, dv \right] ds_T + o(\epsilon)$$

(78′)

$$= \iint_B \delta P(W) N(Z, W) \left[\oint_C N(T, W) \, ds_T \right] du \, dv + o(\epsilon).$$

Applying (78) again, we obtain finally

(78″) $$\delta V(Z) = - \iint_B \delta P(W) V(W) N(Z, W) \, du \, dv + o(\epsilon).$$

This variation formula is very similar to the variation formula (44) of $U(Z)$ and describes the monotonic dependence of $V(Z)$ on $P(Z)$ in explicit form. Since $-V(Z)$ is non-negative in B and is a solution of (2), it is a subharmonic function.

Now let $\varphi(Z)$ be a solution of (2) for which the prescribed values $\nu(Z)$ of the normal derivative satisfy an inequality

(79) $$\alpha \leq \nu(Z) \leq \beta.$$

The functions

(80) $$A_1(Z) = \varphi(Z) - \alpha V(Z), \qquad A_2(Z) = \beta V(Z) - \varphi(Z)$$

have both non-negative normal derivatives on C and, therefore, we have

(81) $$A_i(Z) = - \oint_C N(Z, W) \frac{\partial A_i(W)}{\partial n_W} \, ds_W \leq 0 \qquad (Z \, \epsilon \, B; i = 1, 2).$$

Hence, we may estimate the required solution $\varphi(Z)$ by the inequalities

(82) $$\beta V(Z) \leq \varphi(Z) \leq \alpha V(Z).$$

This estimate and the monotonic behavior of $V(Z)$ in dependence on $P(Z)$ permits us to develop a theory for the boundary value problem of the second kind exactly in the same way as we did for that of the first kind in the case of non-linear differential equations of the above type.

6. **The generalized kernel functions.** In the previous sections we considered systems of functions which are orthonormalized with respect to the metric (4). This metric is closely related to the differential equation (2) since the Euler-Lagrange formula with respect to the integral (3) coincides with (2). There exists, however, an infinity of integrals which are positive-definite, homogeneous

and quadratic in a function $\varphi(Z)$ and its first partial derivatives and possess equation (2) as Euler-Lagrange equation. In fact, every integrand which differs from the integrand of (3) by a divergence term leads to the same variational equation (2).

Thus, let us introduce two arbitrary functions $a(Z)$ and $b(Z)$ which are continuous and continuously differentiable in $B + C$; suppose further that on C

$$(83) \qquad a(Z) \cos (n, x) + b(Z) \cos (n, y) = -\lambda(Z) < 0.$$

Then, the integral

$$(84) \qquad \theta\{\varphi, \psi\} = \iint_B \left[\frac{\partial}{\partial x} (a\varphi\psi) + \frac{\partial}{\partial y} (b\varphi\psi) \right] dx \, dy,$$

taken for a pair of functions $\varphi(Z)$ and $\psi(Z)$ of the class Ω, depends in reality only on the values of $\varphi(Z)$ and $\psi(Z)$ at the boundary C of the domain considered. If φ and ψ are still continuous on C, we may write in view of (83)

$$(84') \qquad \theta\{\varphi, \psi\} = \oint_C \lambda(Z)\varphi(Z)\psi(Z) \, ds_Z$$

and, in particular, we find

$$(84'') \qquad \theta\{\varphi, \varphi\} \geq 0$$

for every function $\varphi(Z) \, \epsilon \, \Omega$ which is still continuous on C. Every function of class Ω may be approximated uniformly in each closed subdomain of B by functions which are still continuous on C and such that even the first derivatives of the approximating functions converge uniformly to the corresponding derivatives of the approximated function in this subdomain. Hence, we may easily extend the inequality (84'') to all functions $\varphi(Z)$ of the class Ω.

Consider now the metric

$$(85) \qquad \vartheta\{\varphi, \psi\} = D\{\varphi, \psi\} + \theta\{\varphi, \psi\}$$

defined for each pair of functions of the class Ω. $\vartheta\{\varphi, \psi\}$ is homogeneous and linear in φ, φ_x, φ_y and in ψ, ψ_x, ψ_y; the corresponding quadratic expression $\vartheta\{\varphi, \varphi\}$ is positive-definite and only for $\varphi \equiv 0$ one has $\vartheta\{\varphi, \varphi\} = 0$. The Euler-Lagrange equation connected with the integral $\vartheta\{\varphi, \varphi\}$ is again (2). For functions $\varphi(Z)$ and $\psi(Z)$ which are continuous in $B + C$ one may write (85) in the form

$$(85') \qquad \vartheta\{\varphi, \psi\} = \iint_B \left[(\text{grad } \varphi, \text{grad } \psi) + P\varphi\psi \right] dx \, dy + \oint_C \lambda(Z)\varphi(Z)\psi(Z) \, ds_Z .$$

Consider now Robin's function $R(Z, W)$ defined in §2 which corresponds to the $\lambda(Z)$ of our metric. Using Green's identity we find

$$\vartheta_W\{R(Z, W), \varphi(W)\} = \varphi(Z) - \oint_C \varphi(W) \frac{\partial R(Z, W)}{\partial n_W} ds_W$$

(86)

$$+ \oint_C \lambda(W)\varphi(W)R(Z, W) ds_W ,$$

if $\varphi(W)$ is an arbitrary function of class Ω which is continuous in $B + C$. Applying the fundamental property (19) of $R(Z, W)$, we obtain

(87) $$\vartheta_W\{R(Z, W), \varphi(W)\} = \varphi(Z).$$

This identity, valid at first only for functions $\varphi(Z)$ ε Ω which are continuous in $B + C$, may, by the usual considerations, be extended to the whole class Ω. This shows that Robin's function has for the metric (85) the same reproducing property in the class Ω as Neumann's function has with respect to the metric (4).

Green's function $G(Z, W)$ yields for each function $\varphi(Z)$ ε Ω, continuous in $B + C$

(88) $$\vartheta_W\{G(Z, W), \varphi(W)\} = \varphi(Z) - \oint_C \varphi(W) \frac{\partial G(Z, W)}{\partial n_W} ds_W$$

and reproduces therefore every function of the class Ω^0. If, on the other hand, $\varphi(Z)$ is of the class Λ we have

(88') $$\vartheta_W\{G(Z, W), \varphi(W)\} \equiv 0 \qquad (\varphi(Z) \ ε \ \Lambda).$$

This shows that in the metric (85) the classes Ω^0 and Λ are still orthogonal to each other.

Subtracting the equations (87) and (88') and introducing the difference function (20'), we obtain

(89) $$\vartheta_W\{Q(Z, W), \varphi(W)\} = \varphi(Z) \qquad (\varphi(Z) \ ε \ \Lambda).$$

This shows that the difference between the Robin and the Green function is a reproducing kernel for the class Λ with respect to the metric (85). We remark that $Q(Z, W)$ is regular in B since the logarithmic infinities of $R(Z, W)$ and $G(Z, W)$ just cancel each other. $Q(Z, W)$ will be called the generalized kernel function of B with respect to the differential equation (2) and the metric (85).

Subtracting the equations (87) and (88), we find

(89') $$\vartheta_W\{Q(Z, W), \varphi(W)\} = \oint_C \varphi(W) \frac{\partial G(Z, W)}{\partial n_W} ds_W .$$

This formula, valid for each $\varphi(Z)$ ε Ω which is continuous in $B + C$, is the complete analogue of (9) and shows that the function $\varphi(Z)$ is transformed by the kernel $Q(Z, W)$ into the solution of (2) which has the same boundary values on C.

It is obvious that $Q(Z, W)$ may be represented as the kernel function of a complete system of solutions of (2) which are orthonormalized with respect to the metric (85). This procedure leads to a very efficient method for computing Robin's functions if Green's function is already known. In this fashion, boundary value problems of the third kind are numerically accessible. On the other hand, from our new representation of $Q(Z, W)$ as the kernel of an orthonormal system, we obtain important theoretical results. We mention as an example the inequality

$$(90) \qquad Q(Z, W) \leq \tfrac{1}{2}[Q(Z, Z) + Q(W, W)]$$

which is obtained in the same way as (15'').

For the given metric (85) there exists an infinity of complete orthonormal systems $\{\varphi_r(Z)\}$ in Λ. One of the most interesting systems is that which is at the same time orthogonal with respect to the metric (4). We determine this system in the following way; we ask for a system $\{\varphi_r(Z)\}$ which is complete in Λ and for which we also have

$$(91) \qquad D\{\varphi_r, \varphi_\mu\} = \kappa_r \delta_{r\mu}, \qquad \theta\{\varphi_r, \varphi_\mu\} = \delta_{r\mu}.$$

Since every solution $\varphi(Z) \in \Lambda$ may be composed linearly of the elements of $\{\varphi_r(Z)\}$, we may express our requirement also in the following form:

$$(91') \qquad D\{\varphi_r, \varphi\} = \kappa_r \theta\{\varphi_r, \varphi\}$$

for every $\varphi(Z) \in \Lambda$. Let us assume that all $\varphi_r(Z)$ are continuous in $B + C$ and consider only functions $\varphi(Z)$ with the same property. Then, we transform (91') by integrating by parts into

$$(91'') \qquad \oint_C \varphi \left[\frac{\partial \varphi_r(Z)}{\partial n_z} + \kappa_r \lambda(Z) \varphi_r(Z) \right] ds_z = 0$$

for every $\varphi(Z) \in \Lambda$ continuous in $B + C$. Since the values of $\varphi(Z)$ may be prescribed continuously on C but otherwise arbitrarily, this requirement leads to

$$(92) \qquad \partial \varphi_r(Z)/\partial n_z = -\kappa_r \lambda(Z) \varphi_r(Z) \qquad \qquad \text{(on } C).$$

We may, therefore, construct a system of the required type if we can find a complete set of solutions of (2) which satisfy the boundary conditions (92). The existence of such a set has been proved in the case of the Laplace equation by Stekloff [8]. It can be established by the method of integral equations for the more general case of the differential equation (2). (See [6].) It is easy to see that the set of solutions $\{\varphi_r(Z)\}$, which may be called Stekloff's functions, is simultaneously orthogonal with respect to both metrics (4) and (85).

Let us illustrate our result by specializing to the case $\lambda = \text{const.}$ In this case, Stekloff's functions are defined by the boundary conditions

$$(93) \qquad \partial \varphi_r/\partial n = -\kappa_r \varphi_r \qquad \qquad \text{(on } C);$$

there exists a denumerable set of eigenvalues κ_ν and the corresponding functions $\{\varphi_\nu(Z)\}$ form a complete orthogonal set in Λ with respect to (4) and (85). We have now

$$(93') \qquad D\{\varphi_\nu , \varphi_\nu\} = -\oint_C \varphi_\nu \frac{\partial \varphi_\nu}{\partial n} \, ds = \kappa_\nu \oint_C \varphi_\nu^2 \, ds = \kappa_\nu \lambda^{-1} \theta\{\varphi_\nu , \varphi_\nu\}$$

and, therefore

$$(93'') \qquad \vartheta\{\varphi_\nu , \varphi_\nu\} = (1 + \lambda\kappa_\nu^{-1}) D\{\varphi_\nu , \varphi_\nu\}.$$

We normalize the $\{\varphi_\nu(Z)\}$ by the requirement

$$(94) \qquad D\{\varphi_\nu , \varphi_\nu\} = 1.$$

Then, the functions

$$(94') \qquad \psi_\nu(Z) = (1 + \lambda\kappa_\nu^{-1})^{-\frac{1}{2}} \varphi_\nu(Z)$$

will form an orthonormal system with respect to the metric (85). Hence, we have

$$(95) \qquad Q(Z, W) = R(Z, W) - G(Z, W) = \sum_{\nu=1}^{\infty} \frac{\kappa_\nu}{\kappa_\nu + \lambda} \varphi_\nu(Z)\varphi_\nu(W).$$

Thus, we may represent all Robin's functions $R(Z, W)$ with constant factor λ by the same system of Stekloff's functions.

We mention finally an interesting relation between the different kernel functions. In fact, we have by virtue of (89),

$$(96) \qquad \vartheta_W\{Q(Z, W), K(W, T)\} = K(Z, T)$$

since $K(Z, T)$ is of class Λ. On the other hand, we have by the definition (85) of $\vartheta\{\varphi, \psi\}$ and because of (11'')

$$(97) \qquad K(Z, T) = Q(Z, T) + \oint_C \lambda(W)Q(Z, W)K(W, T) \, ds_W .$$

Since $\lambda(Z)$, $Q(Z, W)$ and $K(W, T)$ are all non-negative, this shows that

$$(97') \qquad K(Z, T) \geq Q(Z, T)$$

and leads to the estimate

$$\oint_C \lambda(W)Q(Z, W)Q(W, T) \, ds_W \leq K(Z, T) - Q(Z, T)$$

$$(98)$$

$$\leq \oint_C \lambda(W)K(Z, W)K(W, T) \, ds_W .$$

In the same way, we obtain the following relation between functions $Q_1(Z, W)$ and $Q_2(Z, W)$ belonging to different functions $\lambda_1(Z)$ and $\lambda_2(Z)$ respectively:

$$(99) \qquad Q_1(Z, T) - Q_2(Z, T) = \oint_C [\lambda_2(W) - \lambda_1(W)]Q_1(Z, W)Q_2(W, T)\, ds_W .$$

This formula shows again that $Q(Z, T)$ decreases with increasing λ. Using the definition of $Q(Z, T)$ and the fact that Green's function vanishes at the boundary, we obtain from (99) the following formula for Robin's functions:

$$(100) \qquad R_1(\dot{Z}, T) - R_2(Z, T) = \oint_C [\lambda_2(W) - \lambda_1(W)]R_1(Z, W)R_2(W, T)\, ds_W$$

which might have been obtained directly by the usual application of Green's formula.

7. Differential equations with non-definite coefficient.

In all previous considerations we used the fact that the coefficient $P(Z)$ of the differential equation (2) satisfies the inequality (1). It is of importance to investigate how much of this assumption might be deleted without affecting our main results.

Let us study, therefore, the differential equation

$$(101) \qquad \Delta\varphi(Z) = [P(Z) - k]\varphi(Z),$$

where $P(Z)$ has the same properties as before and k is a positive constant. It is obvious that every differential equation of type (2) with non-definite coefficient may be written in this form if this coefficient is continuous in $B + C$.

The metric which corresponds to the differential equation (101) is given by

$$(102) \qquad D_k\{\varphi, \psi\} = \iint_B [(\text{grad } \varphi, \text{grad } \psi) + (P - k)\varphi\psi]\, dx\, dy.$$

We recognize here the first difficulty in extending our previous methods by the fact that the integral $D_k\{\varphi, \varphi\}$ is not positive-definite. In order to overcome this difficulty, we introduce the integral

$$(103) \qquad H\{\varphi, \psi\} = \iint_B \varphi\psi\, dx\, dy$$

which is defined for every pair of functions φ and ψ of the class Ω. Then, we might express the integral (102) in the form

$$(102') \qquad D_k\{\varphi, \psi\} = D\{\varphi, \psi\} - kH\{\varphi, \psi\}$$

where $D\{\varphi, \psi\}$ is defined by (4) with the function $P(Z)$ appearing in (101).

The relation between the integrals $D\{\varphi, \psi\}$ and $H\{\varphi, \psi\}$ is very well known from the theory of eigenvalues connected with the differential equation (101) and we shall be able to investigate the metric (102) by means of its representation in the form (102′).

It is well known that for all functions $v(Z)$ of the class Ω for which

(104) $$H\{v, v\} = 1$$

holds, one has the inequality

(104′) $$D\{v, v\} \geq \kappa_1 > 0.$$

The minimum value κ_1 is assumed for the function $v_1(Z)$ which satisfies

(105) $$\Delta v_1 = (P - \kappa_1)v_1, \qquad \partial v_1/\partial n = 0 \qquad \text{(on } C\text{)}.$$

$v_1(Z)$ is the first eigenfunction of the differential equation

(105′) $$\Delta v = [P(Z) - \kappa]v$$

with the boundary condition

(105″) $$\partial v/\partial n = 0 \qquad \text{(on } C\text{)},$$

and κ_1 is the first eigenvalue. The extremum property (104′) may well serve as another definition of the first eigenfunction. If the N first eigenfunctions $v_\nu(Z)$ are known, we may define the $(N + 1)$-st by the requirement that

(106) $$D\{v, v\} = \text{Minimum}$$

among all functions of class Ω which satisfy

(106′) $$H\{v, v\} = 1, \qquad H\{v, v_\nu\} = 0 \qquad (\nu = 1, 2, \cdots, N).$$

The extremum function is $v_{N+1}(Z)$ and the corresponding minimum value of (106) is κ_{N+1}. By definition, one has $0 < \kappa_1 \leq \kappa_2 \leq \kappa_3 \leq \cdots$.

Let us now assume that the constant k in (101) is different from all the κ_ν. Suppose, further, that

(107) $$\kappa_1 \leq \kappa_2 \leq \cdots \leq \kappa_N < k < \kappa_{N+1} \leq \kappa_{N+2} \leq \cdots.$$

Consider a system of m functions $f_\nu(Z)$ ε Ω for which we have

(108) $$D_k\{f_\nu, f_\mu\} = -\delta_{\nu\mu} \qquad (\nu, \mu = 1, 2, \cdots, m).$$

We want to show, that in this case it is necessary that

(108′) $$m \leq N.$$

In fact, suppose there were $N + 1$ functions $f_\nu(Z)$ ε Ω for which (108) were valid. One might find a linear combination of the $f_\nu(Z)$

(109) $$f(Z) = \sum_{\nu=1}^{N+1} \alpha_\nu f_\nu(Z)$$

such that

(109') $$H\{f, f\} = 1, \qquad H\{f, v_\nu\} = 0 \qquad (\nu = 1, 2, \cdots, N).$$

Then one would have in view of the minimum property of $v_{N+1}(Z)$

(110) $$D\{f, f\} \geq \kappa_{N+1}$$

and, therefore, by virtue of (102') and (107)

(110') $$D_k\{f, f\} = D\{f, f\} - kH\{f, f\} > 0.$$

But from (108) and (109) follows

(111) $$D_k\{f, f\} = - \sum_{\nu=1}^{N+1} \alpha_\nu^2 < 0$$

which contradicts (110'). Hence, as was asserted, inequality (108') must hold.

Let us introduce next the class of all functions $\varphi(Z) \; \varepsilon \; \Omega$ which satisfy the differential equation (101) and denote it by Λ_k. There may exist functions in Λ_k which have a negative or vanishing norm $D_k\{\varphi, \varphi\}$. It might even happen that there exists a function $\psi(Z) \; \varepsilon \; \Lambda_k$ which is orthogonal to every element of Λ_k, itself included. In fact, we have if φ and ψ are continuous and continuously differentiable in $B + C$

(112) $$D_k\{\varphi, \psi\} = - \oint_C \psi \frac{\partial \varphi}{\partial n} ds = - \oint_C \varphi \frac{\partial \psi}{\partial n} ds.$$

If, therefore, ψ is a solution of (101) which vanishes on C or has there vanishing normal derivatives then in fact we have

(112') $$D_k\{\varphi, \psi\} = 0$$

for every φ continuously differentiable in $B + C$, and this identity may be easily extended to the whole class Λ_k. On the other hand, the existence of $\psi \; \varepsilon \; \Lambda_k$ which has vanishing normal derivatives on C means that k is an eigenvalue for the differential equation (105') with the boundary condition (105'') while the existence of $\psi \; \varepsilon \; \Lambda_k$ which vanishes on C means that k is an eigenvalue for the problem

(113) $$\Delta \varphi = [P(Z) - \rho]\varphi,$$

where $\varphi = 0$ on C. Now we want to prove the following theorem.

THEOREM VII. *If k is different from the eigenvalues of the differential equations* (105'), (105'') *and* (113), *the only function $\psi(Z) \; \varepsilon \; \Lambda_k$ for which $D_k\{\varphi, \psi\} = 0$ holds for every $\varphi \; \varepsilon \; \Lambda_k$ is $\psi(Z) \equiv 0$.*

In fact, since k is no eigenvalue for either boundary value problem there exist Neumann's and Green's function for B with respect to the differential equation (101). We denote again their difference by $K(Z, W)$ and call it the kernel function of B with respect to (101). $K(Z, W)$ is obviously an element of Λ_k for a fixed W; further one again has the identity

(114) $$D_k\{K(Z, W), \varphi(W)\} = \varphi(Z)$$

for every $\varphi \; \varepsilon \; \Lambda_k$. Now, if there existed a $\psi(Z) \; \varepsilon \; \Lambda_k$ which is orthogonal to every $\varphi(Z) \; \varepsilon \; \Lambda_k$, then by this property and in view of (114)

(115) $$\psi(Z) = D_k\{K(Z, W), \psi(W)\} \equiv 0$$

which proves our theorem.

An immediate consequence of our theorem is that there exists at least one function $\varphi_1(Z)$ in Λ_k with non-vanishing norm; we normalize it according to the fact that its norm is positive or negative by the requirement

(116) $$D_k\{\varphi_1, \varphi_1\} = \pm 1.$$

Next, consider the subspace of Λ_k which is orthogonal to $\varphi_1(Z)$. According to Theorem VII, there exists at least one function $\varphi_2(Z)$ with non-vanishing norm which will be assumed normalized, too. We choose a further function $\varphi_3(Z)$ with non-vanishing norm which is orthogonal to $\varphi_1(Z)$ and $\varphi_2(Z)$ and continue this procedure. Since it is easily shown that there exist complete bases in Λ_k, we are sure that we obtain finally a complete orthogonal system $\{\varphi_\nu(Z)\}$ for this class, satisfying the requirements

(116′) $$D_k\{\varphi_\nu, \varphi_\mu\} = \pm \delta_{\nu\mu} ;$$

we are sure by inequality (108′) that the number of functions $\varphi_\nu(Z)$ with negative norm is at most N.

Every function $\varphi(Z)$ of class Λ_k may be developed into a series

(117) $$\varphi(Z) = \sum_{\nu=1}^{\infty} \alpha_\nu \varphi_\nu(Z) \qquad (\alpha_\nu = \pm D_k\{\varphi, \varphi_\nu\}),$$

where the sign in the Fourier coefficient has the same value as in the formula (116′). In particular, we conclude from (114) and (117) that

(118) $$K(Z, W) = \sum_{\nu=1}^{\infty} \pm \varphi_\nu(Z)\varphi_\nu(W).$$

We are thus able to construct the difference between Neumann's and Green's functions in terms of a complete orthonormal system even in the case that the coefficient of our differential equation is not definite. The number of functions of the system which do not have a positive norm is bounded and may be estimated by considering the eigenvalues of the differential equation. One may reduce the new type of metric to the old one by the following device; suppose the orthogonal system $\{\varphi_\nu(Z)\}$ contains m functions $\varphi_1(Z), \cdots, \varphi_m(Z)$ with negative norm while the rest of the system has positive norms. The system $\varphi_{m+1}(Z), \varphi_{m+2}(Z), \cdots$ is then a complete orthogonal basis for this subspace of Λ_k which is orthogonal to the m-dimensional space of $\varphi_1(Z), \cdots, \varphi_m(Z)$; in this space $D_k\{\varphi, \psi\}$ is a definite metric and all considerations valid for such metrics may be applied to it. This shows that the increase in difficulty encountered by reducing our assumption with respect to $P(Z)$ is not too considerable and that our theory permits an easy extension to the case of non-definite coefficient $P(Z) - k$.

There exists, as before, an infinity of different orthogonal systems $\{\varphi_\nu(Z)\}$ with respect to Λ_k and the metric (102). We obtain a particularly interesting system of this type by discussing the following problem:

Consider the class of all functions $\chi(Z)$ ε Λ_k which satisfy the condition

$$(119) \qquad\qquad H\{\chi, \chi\} = 1$$

and determine those functions of this class which make

$$(120) \qquad\qquad D_k\{\chi, \chi\} = \text{Minimum}.$$

There exist functions $\chi(Z)$ of this class for which the minimum is really obtained as can easily be shown by the methods of §4. Let $\chi_1(Z)$ be one of those functions and let

$$(120') \qquad\qquad D_k\{\chi_1, \chi_1\} = \sigma_1 .$$

Now consider a sequence of functions $\chi_\nu(Z)$ ε Λ_k which are defined recursively as follows: $\chi_{N+1}(Z)$ is that function of class Λ_k which satisfies

$$(121) \qquad H\{\chi, \chi\} = 1, \qquad H\{\chi, \chi_\nu\} = 0 \qquad (\nu = 1, 2, \cdots, N)$$

and makes

$$(122) \qquad\qquad D_k\{\chi, \chi\} = \text{Minimum} = \sigma_{N+1} .$$

The existence of the $\chi_\nu(Z)$ is again proved by considerations of the type used in §4 and we find, in the usual way the following relations between the $\chi_\nu(Z)$:

$$(123) \qquad\qquad H\{\chi_\nu, \chi_\mu\} = \delta_{\nu\mu}, \qquad D_k\{\chi_\nu, \chi_\mu\} = \sigma_\nu\delta_{\nu\mu} .$$

Thus, we obtained in the $\{\chi_\nu(Z)\}$ a system which is simultaneously orthogonal in the H-metric and in the D_k-metric. The constants σ_ν are functionals of the domain B and of $P(Z) - k$. The kernel function appears now in the form

$$(124) \qquad\qquad K(Z, W) = \sum_{\nu=1}^{\infty} \sigma_\nu^{-1}\chi_\nu(Z)\chi_\nu(W).$$

From (124) and the first equation of (123), we obtain the following integral equation for the determination of all $\chi_\nu(Z)$:

$$(125) \qquad \chi_\nu(Z) = \sigma_\nu H_W\{K(Z, W), \chi_\nu(W)\} = \sigma_\nu \iint_B K(Z, W)\chi_\nu(W) \, du \, dv.$$

All these relationships show the great analogy between the eigenfunctions $v_\nu(Z)$ of the problem (105'), (105''), the eigenfunctions $u_\nu(Z)$ of the problem (113) and the functions $\chi_\nu(Z)$ of our new orthogonal system. In fact, we may say that $v_\nu(Z)$, $u_\nu(Z)$ and $\chi_\nu(Z)$ solve the same extremum problem when the functions considered are those from the classes Ω, Ω^0 and Λ_k, respectively. Each system of functions satisfies the corresponding integral equation

$$v_\nu(Z) = \kappa_\nu \iint\limits_B N(Z, W)v_\nu(W) \, du \, dv,$$

(126)
$$u_\nu(Z) = \rho_\nu \iint\limits_B G(Z, W)u_\nu(W) \, du \, dv,$$

$$\chi_\nu(Z) = \sigma_\nu \iint\limits_B K(Z, W)\chi_\nu(W) \, du \, dv,$$

where the kernel of the integral equation is always the reproducing kernel of the class considered. This shows the great symmetry between the classes Ω, Ω^0 and Λ_k and their corresponding reproducing kernels.

8. **Generalizations.** The greatest part of our previous considerations may be extended without any change to a more general class of self-adjoint partial differential equations of elliptic type. These possible generalizations were already discussed in [3]. In particular, one recognizes easily that the restriction to the case of two variables and two dimensional domains B is not essential for the success of our method.

The method of orthogonal functions and their kernel functions is applicable to all types of equations which are the variational equations for certain positive definite and quadratic integrals. It is not even necessary that these equations be differential equations. Consider, for example, a function of four variables $E(x, y; u, v) = E(Z; W)$ which is symmetric in Z and W, continuous in $B + C$ and such that

(127)
$$\iiiint\limits_{Z, W \varepsilon B} E(Z, W)f(Z)f(W) \, dx \, dy \, du \, dv > 0$$

for every $f(Z) \not\equiv 0$ of class Ω. Let $P(Z)$ be positive and continuous in $B + C$. Then, the metric

$$T\{f, g\} = \iint\limits_B [(\mathrm{grad}\ f, \mathrm{grad}\ g) + Pfg] \, dx \, dy$$

(128)
$$+ \iiiint\limits_{Z, W \varepsilon B} E(Z, W)f(Z)g(W) \, dx \, dy \, du \, dv$$

is defined for each pair of functions $f(Z)$ and $g(Z)$ of class Ω and

(128')
$$T\{f, f\} \geq 0.$$

The variational equation belonging to the extremum problem with respect to $T\{f, f\}$ has the form

$$(129) \qquad \Delta\varphi(Z) = P(Z)\varphi(Z) + \iint\limits_{B} E(Z, W)\varphi(W) \, du \, dv$$

and is an integro-differential equation.

It is not difficult to construct more complex metrics which lead to interesting integro-differential equations. Every case for which the existence of a Green's and Neumann's function has been ensured permits just the same treatment as the partial differential equations considered till now. We do not enter here into a more detailed discussion of this subject and mentioned this type of problem only in order to show the wide field of applications for the theory of orthogonal functions.

BIBLIOGRAPHY

1. N. ARONSZAJN, *La théorie des noyaux reproduisants et ses applications, Première partie,* Proceedings of the Cambridge Philosophical Society, vol. 39(1943), pp. 133–153.
2. STEFAN BERGMAN, *Functions satisfying certain partial differential equations of elliptic type and their representation,* this Journal, vol. 14(1947), pp. 349–366.
3. S. BERGMAN AND M. SCHIFFER, *A representation of Green's and Neumann's functions in the theory of partial differential equations of second order,* this Journal, vol. 14(1947), pp. 609–638.
4. S. BERGMAN AND M. SCHIFFER, *On Green's and Neumann's functions in the theory of partial differential equations,* Bulletin of the American Mathematical Society, vol. 53(1947), pp. 1141–1151.
5. R. COURANT AND D. HILBERT, *Methoden der mathematischen Physik II,* Berlin, 1937.
6. D. HILBERT, *Grundzüge einer allgemeinen Theorie der linearen Integralgleichungen,* Leipzig, 1912.
7. MENAHEM SCHIFFER, *The kernel function of an orthonormal system,* this Journal, vol. 13 (1946), pp. 529–540.
8. W. STEKLOFF, *Sur la théorie des fonctions fondamentales,* Comptes Rendus de l'Académie des Sciences, Paris, vol. 128(1899), pp. 984–987.
9. S. ZAREMBA, *Sur le calcul numérique des fonctions demandées dans le problème de Dirichlet et le problème hydrodynamique,* Bulletin international de l'Académie des Sciences de Cracovie, (1909), 1, pp. 125–195.

HARVARD UNIVERSITY.

Commentary on

[23] S. Bergman and M. Schiffer, *Kernel functions in the theory of partial differential equations of elliptic type*, Duke Math. J. **15** (1948), 535 566.

The authors continue their study of the kernels initiated in their earlier paper [20]. Let B be a bounded, finitely connected domain in the plane with smooth boundary, and let P be a positive continuous function defined on the closed region $\overline{B}$. It is shown that the reproducing kernel K associated with the Hilbert space of solutions of the elliptic equation $\Delta\varphi - P\varphi = 0$ in B under the scalar product

$$\mathcal{D}[u,v] := \int_B [\nabla u \cdot \nabla v + Puv]\,dxdy$$

is nonnegative and subharmonic in B.

The Robin function $R(z,w)$ is introduced as the solution of the mixed boundary value problem

$$(\Delta - P)R(z,w) = \delta_w, \quad \frac{\partial R}{\partial n_z} = \lambda(z)R \text{ on } \partial B,$$

where δ_w denotes the point mass at $w \in B$ and λ is a given continuous nonnegative function defined on ∂B; it plays a role analogous to that of Green's and Neumann's functions in their respective boundary value problems.

The authors also study the variation of the fundamental solutions when the coefficients of the equation vary. This allows them to apply some of their results to certain nonlinear elliptic PDEs. At the end of the paper, they relax some of the initial assumptions, notably the assumption that $P > 0$ (which allows application of the maximum principle), and study the equation

$$\Delta\varphi - (P-k)\varphi = 0,$$

where k is a constant different from the Neumann or Dirichlet eigenvalues for the operator $\Delta - P$. Further results along these alleys of investigation can be traced in [A, B, D].

References

[A] Maynard G. Arsove, *Normal families of subharmonic functions*, Proc. Amer. Math. Soc. **7** (1956), 115 126.

[B] Stefan Bergman, *On singularities of solutions of certain differential equations in three variables*, Trans. Amer. Math. Soc. **85** (1957), 462 488.

[D] Philip Davis, *An application of doubly orthogonal functions to a problem of approximation in two regions*, Trans. Amer. Math. Soc. **72** (1952), 104 137.

DMITRY KHAVINSON

[25] Faber polynomials in the theory of univalent functions

[25] Faber polynomials in the theory of univalent functions. *Bull. Amer. Math. Soc.* **54** (1948), 503–517.

FABER POLYNOMIALS IN THE THEORY OF UNIVALENT FUNCTIONS

MENAHEM SCHIFFER

Introduction. The Faber polynomials play an important role in the theory of univalent functions. Grunsky [1][1] succeeded in establishing a set of conditions for a given function which are necessary and in their totality sufficient for the univalency of this function, and in these conditions the coefficients of the Faber polynomials play an important role. Schiffer [2] gave a differential equation for univalent functions solving certain extremum problems with respect to the coefficients of such functions; in this differential equation appears again a polynomial which is just the derivative of a Faber polynomial (cf. Schiffer [3]; see also Schaeffer-Spencer [4]).

It seems, therefore, of interest to study these Faber polynomials more closely, in particular their dependence on the given function with respect to which they are defined, their variation with the latter and certain characteristic inequalities for them and their coefficients. This investigation is carried out in the present paper. In §1 we establish a generating function for all Faber polynomials with respect to a given function. In §2 we establish variation formulas for the Faber polynomials and their coefficients. In §3 we solve certain extremum problems with respect to the coefficients of these polynomials and find again all the inequalities which have been established by Grunsky. In §4 we use our method in order to generalize our results and to find inequalities for the Faber polynomials themselves.

1. Identities for Faber polynomials. Consider a function $f(z)$ which has in the neighborhood of $z = \infty$ a development of the form:

$$(1) \qquad f(z) = z + c_0 + c_1 z^{-1} + c_2 z^{-2} + \cdots .$$

$F_m(t)$ is called the mth Faber polynomial with respect to $f(z)$ if it is a polynomial of degree m in t and if we have at $z = \infty$ a development:

$$(2) \qquad F_m[f(z)] = z^m + \sum_{n=1}^{\infty} c_{mn} z^{-n}.$$

The existence and uniqueness of all Faber polynomials with respect to a given $f(z)$ are easily shown by recursion.

Presented to the Society, September 5, 1947; received by the editors July 14, 1947.
[1] Numbers in brackets refer to the bibliography at the end of the paper.

503

We shall now construct a generating function for all $F_m(t)$. For this purpose, we consider the function

$$(3) \qquad\qquad U(z, w) = \log \frac{f(z) - f(w)}{z - w}$$

which is regular for z and w in the neighborhood of infinity. We choose the principal branch of the logarithm, such that if either z or w converge to infinity, U converges to zero, as follows immediately from (1). Hence, we have the following series development for the function $U(z, w)$ of two complex variables:

$$(4) \qquad U(z, w) = \log [f(z) - f(w)] - \log (z - w) = \sum_{m,n=1}^{\infty} d_{mn} z^{-m} w^{-n}.$$

If we construct next

$$(5) \qquad\qquad \log \frac{f(z) - t}{z} = - \sum_{m=1}^{\infty} \frac{1}{m} F_m(t) z^{-m},$$

where the development is valid again in a neighborhood of $z = \infty$ and the $F_m(t)$ are polynomials of degree m, we find from (5) and (4):

$$(6) \qquad \begin{aligned} - \sum_{m=1}^{\infty} \frac{1}{m} F_m[f(w)] z^{-m} &= \log \frac{f(z) - f(w)}{z - w} + \log \left(1 - \frac{w}{z}\right) \\ &= - \sum_{m=1}^{\infty} \frac{1}{m} \left(w^m - \sum_{n=1}^{\infty} m d_{mn} w^{-n}\right) z^{-m}. \end{aligned}$$

A comparison of equal powers of z on both sides gives finally

$$(7) \qquad\qquad F_m[f(w)] = w^m - \sum_{n=1}^{\infty} m d_{mn} w^{-n}.$$

This shows that the coefficients $F_m(t)$ are in fact just the Faber polynomials defined above and that we possess in the left-hand side of (5) a generating function for them.

Comparing (2) and (7), we conclude

$$(8) \qquad\qquad c_{mn} = - m d_{mn}.$$

Now we have, because of the symmetry of $U(z, w)$ in z and w,

$$(9) \qquad\qquad d_{mn} = d_{nm},$$

whence

$$(10) \qquad\qquad n c_{mn} = m c_{nm},$$

an identity previously proved by Grunsky [1] and Schur [5].

Differentiating (5) with respect to t, we obtain:

$$(11) \qquad \frac{1}{f(z) - t} = \sum_{m=1}^{\infty} \frac{1}{m} F_m'(t) z^{-m},$$

a generating function for the derivatives of Faber's polynomials which will be of use later on.

Numerous further relations might be obtained, as for example the following: Differentiating (11) with respect to t we find:

$$(12) \qquad \frac{1}{(f(z) - t)^2} = \sum_{m=1}^{\infty} \frac{1}{m} F_m''(t) z^{-m} = \sum_{m,n=1}^{\infty} \frac{1}{mn} F_m'(t) F_n'(t) z^{-(m+n)},$$

whence

$$(13) \qquad \frac{1}{k} F_k''(t) = \sum_{m+n=k} \frac{1}{mn} F_m'(t) F_n'(t).$$

For our further developments, however, formulas (5), (10) and (11) will be sufficient.

2. A variation formula for the Faber polynomials. Let us suppose now that the function $f(z)$ is regular and univalent in a domain D of the z-plane which contains the point at infinity and is bounded by a finite number of proper continua. It will, therefore, map D conformally upon a domain Δ of the ζ-plane. If ζ_0 is an arbitrary point in the ζ-plane which does not belong to Δ and ρ any positive constant, there exist infinitely many functions which are univalent in Δ and have in the domain $|\zeta - \zeta_0| > 4\rho$ a development of the form:

$$(14) \qquad v(\zeta) = \zeta + \frac{a\rho^2}{\zeta - \zeta_0} + \frac{b\rho^3}{(\zeta - \zeta_0)^2} + \frac{c\rho^4}{(\zeta - \zeta_0)^3} + \cdots,$$

where $|a| \leqq 4^2$, $|b| \leqq 4^3$, $|c| \leqq 4^4$, $\cdots$ (see Schiffer [2]).

The function

$$(15) \qquad f^*(z) = v[f(z)] = f(z) + \frac{a\rho^2}{f(z) - \zeta_0} + \cdots$$

is again regular and univalent in the initial domain D and has still a development of the form (1) at infinity. It maps the domain D upon a new domain Δ^* in the ζ-plane. If ρ is small enough, Δ^* will be very near to Δ and we may conceive the mapping (14) as a small variation of this domain. Variations of the type (15) which transform a univalent function $f(z)$ into a function $f^*(z)$ of the same type are of great value in solving extremum problems within the class of normalized univalent functions.

Suppose now that to a given univalent function $f(z)$ the nth Faber polynomial $F_n(t)$ has been determined. If we subject $f(z)$ to a variation (15), how does the corresponding Faber polynomial change? In order to settle this question, we make use of the generating function (5). We have in view of (15)

$$
\begin{aligned}
(16) \quad -\sum_{m=1}^{\infty} \frac{1}{m} F_m^*(t) z^{-m} &= \log \frac{f^*(z) - t}{z} \\
&= \log \frac{f(z) - t}{z} + \frac{a\rho^2}{[f(z) - t][f(z) - \zeta_0]} + o(\rho^2),
\end{aligned}
$$

where $o(\rho^2)$ shall denote henceforth a term containing at least ρ^3 as a factor. We transform (16) by making again use of identity (5) and applying (11):

$$
\begin{aligned}
(17) \quad -\sum_{m=1}^{\infty} \frac{1}{m} F_m^*(t) z^{-m} &= -\sum_{m=1}^{\infty} \frac{1}{m} F_m(t) z^{-m} \\
&\quad + a\rho^2 \sum_{m=1}^{\infty} \frac{1}{m} \frac{F_m'(t) - F_m'(\zeta_0)}{t - \zeta_0} z^{-m} + o(\rho^2).
\end{aligned}
$$

Comparison of equal powers of z on both sides gives, therefore, the following variation formula for $F_m(t)$:

$$
(18) \quad F_m^*(t) = F_m(t) - a\rho^2 \frac{F_m'(t) - F_m'(\zeta_0)}{t - \zeta_0} + o(\rho^2).
$$

Let us now consider the value $F_m[f(w)]$ as a functional of $f(z)$ for a fixed point $w \in D$. Its variation is given by:

$$
(19) \quad F_m^*[f^*(w)] = F_m^*[f(w)] + a\rho^2 \frac{F_m'[f(w)]}{f(w) - \zeta_0} + o(\rho^2),
$$

whence, in view of (18),

$$
(20) \quad F_m^*[f^*(w)] = F_m[f(w)] + a\rho^2 \frac{F_m'(\zeta_0)}{f(w) - \zeta_0} + o(\rho^2).
$$

Using the series development (2) for $F_m[f(w)]$, an analogous representation for $F_m^*[f^*(w)]$, and the identity (11) for the second right-hand term, we obtain by comparison of the coefficients of z^{-n} on both sides:

$$(21) \qquad c_{mn}^* = c_{mn} + a\rho^2 \cdot \frac{1}{n} F_m'(\zeta_0)F_n'(\zeta_0) + o(\rho^2).$$

This variation formula for the c_{mn} exhibits, of course, again the symmetry of the matric nc_{mn}.

Finally, let us differentiate (15) with respect to z; we obtain:

$$(22) \qquad f'^*(z) = f'(z)\left[1 - \frac{a\rho^2}{(f(z) - \zeta_0)^2}\right] + o(\rho^2),$$

which may be written as a variation formula for $\log f'(z)$ as follows:

$$(23) \qquad \log f'^*(z) = \log f'(z) - \frac{a\rho^2}{(f(z) - \zeta_0)^2} + o(\rho^2),$$

and which will be used in this form later on. By differentiating (20) with respect to w and applying (22) we get after elementary transformations:

$$(24) \quad F_m'^*[f^*(w)] = F_m'[f(w)] + a\rho^2 \frac{F_m'[f(w)] - F_m'(\zeta_0)}{(f(w) - \zeta_0)^2} + o(\rho^2).$$

This is a variation formula for the functional $F_m'[f(w)]$ and is interesting in that it contains only the functional itself in addition to $f(w)$ and ζ_0.

3. Application of the variation formulas in extremum problems. We use now the formula (21) in order to solve certain extremum problems with respect to the family Φ of all functions (1) which are univalent in D. Let $x \equiv (x_1, x_2, \cdots, x_N)$ denote a vector of N complex numbers, not all zero, and consider the quadratic form

$$(25) \qquad Q(x, x) = \sum_{m,n=1}^{N} nc_{mn}x_m x_n.$$

For each given function $f(z)$ of the family Φ this form has a well defined complex value. We ask now the following question:

What is the maximum modulus of $Q(x, x)$, if $f(z)$ is a function of the family Φ, and for which functions $f(z) \in \Phi$ is this maximum attained?

It follows easily from the theory of normal families that this question is significant and that there exists at least one $f(z) \in \Phi$ for which the maximum value is obtained. We may, therefore, restrict ourselves to the task of characterizing these extremal functions and computing the corresponding value of $|Q(x, x)|$.

Suppose that we know already an extremal function $f(z)$ and the corresponding value of $Q(x, x) \cdot f(z)$ will map D conformally upon a domain Δ in the ζ-plane. Let ζ_0 be a point in this plane which does not belong to Δ and subject Δ to a variation (14) as described in §2. Under this variation Δ is transformed into a domain Δ^* which is obtained from D by means of a mapping with a function $f^*(z)$, given by (15). Since $f^*(z)$ also belongs to the family Φ, its corresponding value $|Q^*(x, x)|$ cannot be larger than the maximal $|Q(x, x)|$. On the other hand, we may easily compute $|Q^*(x, x)|$ by means of formula (21). We have, in view of this formula,

$$(26) \qquad Q^*(x, x) = Q(x, x) + a\rho^2 \left(\sum_{n=1}^{N} x_n F_n'(\zeta_0) \right)^2 + o(\rho^2).$$

The extremal property of $|Q(x, x)|$ implies

$$(27) \qquad |Q^*(x, x)| = \left| Q(x, x) + a\rho^2 \left(\sum_{n=1}^{N} x_n F_n'(\zeta_0) \right)^2 + o(\rho^2) \right|$$

$$\leq |Q(x, x)|,$$

which is equivalent to

$$(28) \qquad \mathrm{Re}\left\{ a\rho^2 Q^{-1}(x, x) \left(\sum_{n=1}^{N} x_n F_n'(\zeta_0) \right)^2 + o(\rho^2) \right\} \leq 0.$$

This inequality has to be fulfilled for every function $v(\zeta)$ as defined in (14). We now make use of the following lemma (Schiffer [2]):

Let Δ be a domain in the ζ-plane whose complement consists of continua Γ_ν ($\nu = 1, \cdots, c$) and $\sigma(\zeta) \neq 0$ a function analytic in each of the Γ_ν. Let, further, the inequality

$$(29) \qquad\qquad \mathrm{Re}\left\{ a\rho^2 \sigma(\zeta_0) \right\} + o(\rho^2) \leq 0$$

hold for every function (14), univalent in Δ. Then each Γ_ν is an analytic curve with the parametric representation $\zeta(s)$ and satisfying for properly chosen (real) parameter s the differential equation

$$(30) \qquad\qquad \zeta'(s)^2 \sigma\{\zeta(s)\} = 1.$$

Application of this lemma to the particular inequality (28) leads to the result: The extremal function $f(z)$ maps the domain D upon a domain Δ in the ζ-plane which is bounded by analytic slits $\zeta(s)$ each of which satisfies the differential equation

$$(31) \qquad \zeta'(s)^2 Q(x, x)^{-1} \left(\sum_{m=1}^{N} x_m F_m'[\zeta(s)] \right)^2 = 1.$$

We see now that our particular extremum problem leads to a differential equation (30) with a term $\sigma(\zeta)$ which is a complete square. Thus, this differential equation becomes immediately integrable in closed form. Among all extremum problems arising in the theory of univalent functions, the particular class of those problems with the above property plays an important role, since these problems permit a particularly easy solution.

We conclude from (31) that the boundary curves Γ, of Δ satisfy the requirement

$$(32) \qquad \mathrm{Im}\left\{Q(x,\,x)^{-1/2}\sum_{m=1}^{N}x_mF_m(\zeta)\right\} = k_\nu,$$

$$\zeta \in \Gamma_\nu;\ k_\nu = \text{constant};\ \nu = 1, 2, \cdots, c.$$

The function

$$(32') \qquad S(z) = Q(x,\,x)^{-1/2}\sum_{m=1}^{N}x_mF_m[f(z)]$$

is regular in D, except for an Nth order pole at infinity and possesses constant imaginary parts on each boundary continuum of D.

In order to exploit these conditions, we introduce the following class of functions with respect to D. We define associate pairs of functions $A_m(z)$ and $B_m(z)$, having at $z = \infty$ the respective developments

$$(33) \qquad A_m(z) = z^m + \sum_{n=1}^{\infty}a_{mn}z^{-n},$$

$$(33') \qquad B_m(z) = \sum_{n=1}^{\infty}b_{mn}z^{-n}$$

which are regular in D except for the pole of $A_m(z)$ at infinity. On each boundary continuum of D we require:

$$(34) \qquad A_m(z) = \overline{B_m(z)} + \text{const.}$$

where the complex constant depends on m and the particular boundary continuum. The existence and uniqueness of these function-pairs is easily proved (cf. Grunsky [1]).

We construct now the function

$$(35) \quad T(z) = Q(x,\,x)^{-1/2}\sum_{m=1}^{N}x_mA_m(z) + \overline{Q(x,\,x)^{-1/2}}\sum_{m=1}^{N}\bar{x}_mB_m(z).$$

This function has at $z = \infty$ the same principal part as $S(z)$ and, in view of the characteristic property (34) of $A_m(z)$ and $B_m(z)$, it has on each boundary continuum of D a constant imaginary part. The function $S(z) - T(z)$ is, therefore, regular throughout D and has on each boundary continuum a constant imaginary part. Hence, it follows by the usual considerations that this function is a constant, that is,

$$
\begin{aligned}
Q(x,\,x)^{-1/2} \sum_{m=1}^{N} x_m F_m[f(z)] &= Q(x,\,x)^{-1/2} \sum_{m=1}^{N} x_m A_m(z) \\
&\quad + \overline{Q(x,\,x)^{-1/2}} \sum_{m=1}^{N} \bar{x}_m B_m(z) + C.
\end{aligned}
$$

(36)

Comparing in this equation the terms in z^0, we find from (2), (33), and (33′) at once that $C = 0$. Let us introduce the real parameter δ by the definition

(37) $$e^{i\delta} = \operatorname{sgn} Q(x;\,x);$$

then (36) may be written in the form

(36′) $$\sum_{m=1}^{N} x_m F_m[f(z)] = \sum_{m=1}^{N} x_m A_m(z) + e^{i\delta} \sum_{m=1}^{N} \bar{x}_m B_m(z).$$

Let us now apply the formulas (2), (33), and (33′); comparing the coefficients of z^{-n} on both sides yields:

(38) $$\sum_{m=1}^{N} x_m c_{mn} = \sum_{m=1}^{N} (x_m a_{mn} + e^{i\delta} \bar{x}_m b_{mn}).$$

Multiply the nth equation with $n x_n$ and add all resulting equations for $n = 1$ till $n = N$. We obtain:

(39) $$\sum_{m,n=1}^{N} n c_{mn} x_m x_n = \sum_{m,n=1}^{N} n a_{mn} x_m x_n + e^{i\delta} \sum_{m,n=1}^{N} n b_{mn} \bar{x}_m x_n,$$

whence:

(39′) $$\left| \sum_{m,n=1}^{N} n c_{mn} x_m x_n \right| \leq \left| \sum_{m,n=1}^{N} n a_{mn} x_m x_n \right| + \sum_{m,n=1}^{N} n b_{mn} \bar{x}_m x_n.$$

The last right-hand side term is non-negative, since it can be shown that $(n b_{mn})$ is a positive-definite Hermitian matric; we mention here that $(n a_{mn})$ is a symmetric matric (cf. Grunsky [1]).

The significant feature of (39′) is that the right-hand terms depend only upon D and not upon the mapping function $f(z)$. The inequality (39'), proved for the extremal function, holds, therefore, a fortiori for every other admissible function also.

The form of (39) leads us to the following extremum problem:

What is the maximum modulus of $P(x,x) = Q(x,x) - \sum_{m,n=1}^{N} n a_{mn} x_m x_n$, if $f(z)$ is a function of the family Φ?

One sees easily that the above considerations have to be repeated and that the same sequence of formulas appears, except that instead of $Q(x, x)$ we have always to introduce $P(x, x)$. We obtain finally, by putting

$$(40) \qquad\qquad e^{i\gamma} = \text{sgn } P(x, x),$$

the equation

$$(41) \qquad \sum_{m,n=1}^{N} n(c_{mn} - a_{mn}) x_m x_n = e^{i\gamma} \sum_{m,n=1}^{N} n b_{mn} \bar{x}_m x_n$$

for the coefficients c_{mn} belonging to the extremal function $f(z)$. Hence, we have in general:

$$(42) \qquad \left| \sum_{m,n=1}^{N} n(c_{mn} - a_{mn}) x_m x_n \right| \leqq \sum_{m,n=1}^{N} n b_{mn} \bar{x}_m x_n.$$

These are just Grunsky's inequalities, derived here by variational methods. We have obtained, moreover, the following equation for extremal functions:

$$(42') \qquad \sum_{m=1}^{N} x_m F_m[f(z)] = \sum_{m=1}^{N} x_m A_m(z) + e^{i\gamma} \sum_{m=1}^{N} \bar{x}_m B_m(z),$$

which shows that $f(z)$ is an algebraic function of the A_m and B_m $(m = 1, \cdots, N)$.

4. Inequalities for the Faber polynomials. The extremum problems of the last paragraph were so easily solved since they led to a variational differential equation with a complete square expression. It will, therefore, be useful to discuss other extremum problems of the same class.

Consider, for example, the functional

$$(43) \quad R(w, x) = \sum_{m,n=1}^{N} n c_{mn} x_m x_n + 2 \sum_{m=1}^{N} x_m F_m[f(w)] - \log f'(w),$$

which depends upon the function $f(z)$, assumed to be of the class Φ, and upon a fixed point $w \in D$. In view of (20), (21) and (23), we have for $R(w, x)$ the following variation formula:

$$(44) \quad R^*(w, x) = R(w, x) + a\rho^2 \left(\sum_{m=1}^{N} x_m F'_m(\zeta_0) + \frac{1}{f(w) - \zeta_0} \right)^2 + o(\rho^2).$$

We see that $R(w, x)$ has a variation formula containing again a perfect square term and we expect, therefore, that the connected extremum problem will permit an easy solution. Therefore, we propose the following problem:

If $f(z)$ is an arbitrary function of the class Φ, what is the maximum modulus of $R(w, x)$ and for which function $f(z)$ is this extremum attained?

Assuming that $f(z)$ is the desired extremum function and that it maps the original domain D upon a domain Δ, we find, by reiterating literally the conclusions of the last paragraph, that Δ is bounded by analytic slits $\zeta(s)$, each of which satisfies the differential equation

$$(45) \quad R(w, x)^{-1} \cdot \zeta'(s)^2 \left[\sum_{m=1}^{N} x_m F'_m[\zeta(s)] + \frac{1}{f(w) - \zeta(s)} \right]^2 = 1.$$

This equation may be integrated and yields

$$(46) \quad \operatorname{Im} \left\{ R(w, x)^{-1/2} \left[\sum_{m=1}^{N} x_m F_m(\zeta) - \log(\zeta - f(w)) \right] \right\} = \text{const.}$$

along each boundary continuum of Δ. The function

$$(47) \quad U(z) = R(w, x)^{-1/2} \left[\sum_{m=1}^{N} x_m F_m[f(z)] - \log(f(z) - f(w)) \right]$$

is regular throughout D, except for an Nth order pole at infinity and a logarithmic pole at $z = w$. It has a constant imaginary part along each boundary continuum of D.

In order to make use of this fact as in the last paragraph, we have to define another pair of functions. Let $A(z, w)$ and $B(z, w)$ be univalent functions of z in D which map it upon the whole ζ-plane, slit along circular arcs with center at the origin, and along straight segments pointing to the origin respectively, and such that $w \in D$ corresponds to the origin in the ζ-plane. We suppose at infinity the developments:

$$(48) \quad \log A(z, w) = \log(z - w) + \sum_{n=1}^{\infty} \alpha_n(w) z^{-n},$$

$$(48') \quad \log B(z, w) = \log(z - w) + \sum_{n=1}^{\infty} \beta_n(w) z^{-n}.$$

On each boundary continuum of D, $\log A$ will have constant real parts while $\log B$ will possess constant imaginary parts. The functions

$$(49) \quad \phi(z, w) = \frac{1}{2} \log [A(z, w) \cdot B(z, w)] = \log (z - w) + \sum_{n=1}^{\infty} a_n(w) z^{-n}$$

and

$$(49') \quad \psi(z, w) = \frac{1}{2} \log \frac{B(z, w)}{A(z, w)} = \sum_{n=1}^{\infty} b_n(w) z^{-n}$$

are, therefore, regular in D with the exception of the logarithmic poles at w and ∞ for $\phi(z, w)$. On each boundary continuum of D, we have obviously:

$$(50) \quad \phi(z, w) = \overline{\psi(z, w)} + \text{const.}$$

where the constant depends on the particular boundary continuum and on w.

There exist, of course, close relations between the functions ϕ, ψ on the one hand, and the functions $A_m(z)$, $B_m(z)$ on the other. In fact, let us consider the integral around the whole boundary C of D:

$$(51) \quad \frac{1}{2\pi i} \oint_C A_m(z) \phi'(z, w) dz = A_m(w) + m a_m(w) - w^m; \quad \phi' = \frac{d}{dz} \phi.$$

The above equation is an elementary consequence of Cauchy's residue theorem. On the other hand, we have in view of (34), (50) and the single-valuedness in D of all functions concerned:

$$
(52) \quad
\begin{aligned}
\frac{1}{2\pi i} \oint_C A_m(z) \phi'(z, w) dz &= \frac{1}{2\pi i} \oint_C \overline{B_m(z)} \; \overline{d\psi(z, w)} \\
&= -\overline{\frac{1}{2\pi i} \oint_C B_m d\psi}.
\end{aligned}
$$

Again, we apply the residue theorem which shows that the above integral is zero, B_m and ψ being regular throughout D. Hence, we have proved:

$$(52') \quad A_m(w) = - m a_m(w) + w^m.$$

Using (33) and (49), we find, for fixed w and z near infinity, the development

$$(53) \quad \phi(z, w) = \log (z - w) - \sum_{m,n=1}^{\infty} m^{-1} a_{mn} w^{-n} z^{-m}.$$

Since (na_{mn}) is a symmetric matric, we see that $\phi(z, w) - \log (z - w)$ is a symmetric analytic function of z and w.

Let us consider next the following consequence of the residue theorem:

$$(54) \qquad \frac{1}{2\pi i} \oint_C A_m(z)\psi'(z, w)dz = mb_m(w).$$

In view of (34) and (50), we may write this integral also in the form:

$$(55) \qquad \frac{1}{2\pi i} \oint_C A_m(z)d\psi = \frac{1}{2\pi i} \oint_C \overline{B_m(z) \, d\phi}$$

$$= -\overline{\frac{1}{2\pi i} \oint_C B_m(z)\phi'(z, w)dz},$$

which gives, again because of the residue theorem,

$$(56) \qquad mb_m(w) = - \overline{B_m(w)}.$$

Therefore, in view of (33') and (49'):

$$(57) \quad \psi(z, w) = - \sum_{m,n=1}^{\infty} \frac{1}{m} b_{mn}\bar{w}^{-n}z^{-m} = - \sum_{m,n=1}^{\infty} \frac{1}{mn} nb_{mn}\bar{w}^{-n}z^{-m}.$$

Using the Hermitian character of mb_{nm}, we find finally:

$$(57') \qquad \psi(z, w) = - \sum_{m,n=1}^{\infty} \frac{1}{n} b_{nm}\bar{w}^{-n}z^{-m};$$

(57) and (57') show that ψ is analytic in z and $\bar{w}$.

Let us return now to the function $U(z)$, defined in (47). We prove easily, as in the last paragraph, the identity:

$$R(w, x)^{-1/2}\left[\sum_{m=1}^{N} x_m F_m[f(z)] - \log (f(z) - f(w)) \right]$$

$$(58) \qquad = R(w, x)^{-1/2}\left[\sum_{m=1}^{N} x_m A_m(z) - \phi(z, w) \right]$$

$$+ \overline{R(w, x)^{-1/2}}\left[\sum_{m=1}^{N} x_m B_m(z) - \psi(z, w) \right].$$

Putting again

$$(58') \qquad e^{i\tau} = \operatorname{sgn} R(w, x),$$

we may write (58) in the form:

$$\sum_{m=1}^{N} x_m F_m[f(z)] - \log(f(z) - f(w))$$

(59)
$$= \sum_{m=1}^{N} x_m A_m(z) - \phi(z, w) + e^{i\tau} \left[\sum_{m=1}^{N} \bar{x}_m B_m(z) - \psi(z, w) \right].$$

Comparing the coefficients of z^{-n} on both sides, we find in view of (2), (5), (33), (33'), (49), (49'), (52') and (56):

$$\sum_{m=1}^{N} x_m c_{\tilde{m}n} + \frac{1}{n} F_n[f(w)]$$

(60)
$$= \sum_{m=1}^{N} x_m a_{mn} + \frac{1}{n} A_n(w) + e^{i\tau} \left[\sum_{m=1}^{N} \bar{x}_m b_{mn} + \frac{1}{n} \overline{B_n(w)} \right].$$

Multiplying with nx_n and summing up for $1 \leqq n \leqq N$, we get:

$$\sum_{m,n=1}^{N} n c_{mn} x_m x_n + \sum_{m=1}^{N} x_m F_m[f(w)] = \sum_{m,n=1}^{N} n a_{mn} x_m x_n$$

(61)
$$+ \sum_{m=1}^{N} A_m(w) x_m + e^{i\tau} \left[\sum_{m,n=1}^{N} n b_{mn} \bar{x}_m x_n + \sum_{m=1}^{N} \overline{B_m(w)} \, x_m \right].$$

On the other hand, we obtain from (59) in the limit $z \to w$:

$$\sum_{m=1}^{N} x_m F_m[f(w)] - \log f'(w) = \sum_{m=1}^{N} x_m A_m(w) - \chi(w)$$

(62)
$$+ e^{i\tau} \left[\sum_{m=1}^{N} \bar{x}_m B_m(w) - \psi(w, w) \right],$$

where $\chi(w) = \lim_{z \to w} [\phi(z, w) - \log(z - w)]$. Adding together (61) and (62) yields:

$$\sum_{m,n=1}^{N} n c_{mn} x_m x_n + 2 \sum_{m=1}^{N} x_m F_m[f(w)] - \log f'(w)$$

(63)
$$= \sum_{m,n=1}^{N} n a_{mn} x_m x_n + 2 \sum_{m=1}^{N} x_m A_m(w) - \chi(w)$$

$$+ e^{i\tau} \left[\sum_{m,n=1}^{N} n b_{mn} \bar{x}_m x_n + 2 \operatorname{Re} \left\{ \sum_{m=1}^{N} \bar{x}_m B_m(w) \right\} - \psi(w, w) \right].$$

From (63) we derive:

$$
(64)\quad | R(w, x) | \leq \left| \sum_{m,n=1}^{N} na_{mn}x_m x_n + 2\sum_{m=1}^{N} x_m A_m(w) - \chi(w) \right|
$$

$$
+ \left| \sum_{m,n=1}^{N} nb_{mn}\bar{x}_m x_n \right.
$$

$$
\left. + 2\,\mathrm{Re}\left\{ \sum_{m=1}^{N} \bar{x}_m B_m(w) \right\} - \psi(w, w) \right|.
$$

Since the right-hand side depends only on D and not on the particular function $f(z)$, this inequality, proved for the maximum of $|R(w, x)|$ only, holds a fortiori for the rest of the family Φ.

As before, we may improve our result and prove:

$$
(65)\quad \left| R(w, x) - \sum_{m,n=1}^{N} na_{mn}x_m x_n - 2\sum_{m=1}^{N} x_m A_m(w) + \chi(w) \right|
$$

$$
\leq \left| \sum_{m,n=1}^{N} nb_{mn}\bar{x}_m x_n + 2\,\mathrm{Re}\left\{ \sum_{m=1}^{N} \bar{x}_m B_m(w) \right\} - \psi(w, w) \right|.
$$

This result contains of course Grunsky's inequalities as particular cases. If, on the other hand, we put $x_1 = x_2 = \cdots = x_N = 0$, we get from (65)

$$
(66)\qquad | \log f'(w) + \chi(w) | \leq | \psi(w, w) |
$$

or, in view of (53) and (57'), the following inequality which is valid as long as all series concerned are convergent:

$$
(66')\quad \left| \log f'(w) + \sum_{m,n=1}^{\infty} \frac{1}{m} a_{mn}w^{-(m+n)} \right| \leq \sum_{m,n=1}^{\infty} \frac{1}{n} b_{nm}\bar{w}^{-n}w^{-m}.
$$

In the particular case that D is the domain $|z| > 1$, we have $a_{mn} = 0$, $b_{mn} = \delta_{mn}$, whence:

$$
(66'')\qquad | \log f'(w) | \leq \log \frac{|w|^2}{|w|^2 - 1},
$$

a well known inequality.

It would be of interest to derive estimates for $F_n[f(w)]$ without combining this functional with other terms. But since the variational differential equation in this problem is much more complicated, no easy solution is to be expected. To indicate the difficulty of these questions, let us remark that the determination of exact bounds for $|F_n'[f(w)]|$, even only in the case that D is the exterior of the unit circle, would solve simultaneously the Bieberbach problem for functions which are univalent inside the unit circle. On the other hand, the determination of exact bounds for the coefficients c_{mn} in the Faber development would lead to such bounds for the c_{1n}, that is, the coefficients of univalent functions of the type (1). Since this problem is yet unsolved, we see that little progress is at the moment to be expected in the general question of the c_{mn}. It appears, therefore, particularly interesting that for the functionals discussed above such an easy and complete answer is possible. Numerous other functionals of such variational behavior might be constructed and new inequalities be established.

BIBLIOGRAPHY

1. H. Grunsky, *Koeffizientenbedingungen für schlicht abbildende meromorphe Funktionen*, Math. Zeit. vol. 45 (1939) pp. 29–61.

2. M. Schiffer, *A method of variation within the family of simple functions*, Proc. London Math. Soc. (2) vol. 44 (1938) pp. 432–449.

3. ———, *Variation of the Green function and theory of the p-valued functions*, Amer. J. Math. vol. 65 (1943) pp. 341–360.

4. A. C. Schaeffer and D. C. Spencer, *The coefficients of schlicht functions*, Duke Math. J. vol. 10 (1943) pp. 611–635.

5. I. Schur, *On Faber polynomials*, Amer. J. Math. vol. 67 (1945) pp. 33–41.

HEBREW UNIVERSITY AND
    HARVARD UNIVERSITY

Commentary on

[25] *Faber polynomials in the theory of univalent functions*, Bull. Amer. Math. Soc. **54** (1948), 503 517.

Closely related to the class S is the class Σ of functions

$$g(z) = z + b_0 + b_1 z^{-1} + b_2 z^{-2} + \cdots$$

analytic and univalent in $\Delta = \{z : |z| > 1\}$, except for a simple pole at infinity with residue 1. Such a function g maps Δ conformally onto the complement of a compact simply connected set E. Corresponding to each function $g \in \Sigma$, the *Faber polynomials* F_n are defined by the generating relation

$$\frac{zg'(z)}{g(z) - w} = \sum_{n=0}^{\infty} F_n(w) z^{-n}, \qquad z \in \Delta, \; w \in E.$$

(1)

It can be verified that $F_n(w)$ is a monic polynomial of degree n and that its composition with g has the form

$$F_n(g(z)) = z^n + \sum_{k=1}^{\infty} c_{nk} z^{-k}.$$

(2)

When Issai Schur, Schiffer's most inspirational teacher in Berlin, died in Tel Aviv in 1941, a group of mathematicians at the Hebrew University undertook the task of editing his posthumous papers for publication. Schiffer edited Schur's paper [S], which gave some detailed algebraic properties of Faber polynomials, including a new proof of the symmetry relation $kc_{nk} = nc_{kn}$ for the coefficients in (2), a result due to Grunsky [Gr]. In the same paper, Grunsky had generalized the area principle to derive the sharp inequalities

$$\left| \sum_{n=1}^{N} \sum_{k=1}^{N} kc_{nk}\lambda_n\lambda_k \right| \leq \sum_{n=1}^{N} n|\lambda_n|^2, \qquad \lambda_n \in \mathbb{C},$$

(3)

now known as the *Grunsky inequalities*.

Evidently inspired by Schur's work, Schiffer then turned his attention to Faber polynomials

and produced his important paper [25]. There he introduced a new generating relation

$$\log \frac{g(z) - g(\zeta)}{z - \zeta} = -\sum_{n=1}^{\infty} \sum_{k=1}^{\infty} \frac{1}{n} c_{nk} z^{-n} \zeta^{-k},$$

(4)

from which Grunsky's symmetry relation follows as an immediate corollary. The relations (2) and (4) allowed Schiffer to apply the method of boundary variation, as developed in [5], to obtain variational formulas for the Faber polynomials and for the coefficients c_{nk}. He was then able to give a new proof of the Grunsky inequalities (3) by a variational method. Schiffer concludes his paper [25] by applying the variational formulas to derive sharp inequalities for functions $g \in \Sigma$ and their Faber polynomials. In particular, he deduces that

$$|\log g'(z)| \leq \log \frac{|z|^2}{|z|^2 - 1}, \qquad |z| > 1,$$

for all functions $g \in \Sigma$, a special case of the so-called Goluzin inequalities.

Schiffer's paper [25] was among the first applications of Faber polynomials to geometric function theory. Subsequently, many authors have discovered similar connections. For instance, Pommerenke [P1] applied bounds on Faber polynomials to estimate the rate at which the n-diameter of a continuum approaches the transfinite diameter as $n \to \infty$. The books [Go, P2, D] give further information.

The Faber polynomials were introduced by Georg Faber [F] in an effort to generalize the Taylor series expansion to noncircular domains. For a bounded Jordan domain D with analytic boundary curve, Faber developed an apparatus by which any function f analytic in D can be expanded into a series $f(w) = \sum_{n=0}^{\infty} a_n F_n(w)$, where F_n are the Faber polynomials of D. The article by Curtiss [C] gives an elementary overview of basic principles.

Since the time of Faber, an extensive theory of Faber polynomials and Faber series has emerged.

For a comprehensive account, see the expository article [Su1] and the monograph [Su2] by Suetin.

References

[C] J. H. Curtiss, *Faber polynomials and the Faber series*, Amer. Math. Monthly **78** (1971), 577 596.

[D] Peter L. Duren, *Univalent Functions*, Springer-Verlag, 1983.

[F] Georg Faber, *Über polynomische Entwickelungen*, Math. Ann. **57** (1903), 389 408.

[Go] G. M. Goluzin, *Geometric Theory of Functions of a Complex Variable*, second edition, Izdat. "Nauka", Moscow 1966; English transl., American Mathematical Society, 1969.

[Gr] Helmut Grunsky, *Koef"zientenbedingungen für schlicht abbildende meromorphe Funktionen*, Math. Z. **45** (1939), 29 61.

[P1] Ch. Pommerenke, *Über die Faberschen Polynome schlichter Funktionen*, Math. Z. **85** (1964), 197 208.

[P2] Ch. Pommerenke, *Univalent Functions*, Vandenhoeck & Ruprecht, 1975.

[S] Issai Schur, *On Faber polynomials*, Amer. J. Math. **67** (1945), 33 41.

[Su1] P. K. Suetin, *Basic properties of Faber polynomials*, Uspehi Mat. Nauk **19** (1964), no. 4, 125 154; English transl., Russian Math. Surveys **19** (1964), no. 4, 121 149.

[Su2] P. K. Suetin, *Series of Faber Polynomials*, "Nauka", Moscow, 1984; English transl., Gordon and Breach, 1998.

PETER DUREN

[26] (with P. R. Garabedian) Identities in the theory of conformal mapping

[26] (with P. R. Garabedian) Identities in the theory of conformal mapping. *Trans. Amer. Math. Soc.* **65** (1949), 187–238.

IDENTITIES IN THE THEORY OF CONFORMAL MAPPING

BY

P. R. GARABEDIAN AND M. SCHIFFER[1]

Introduction. In the theory of conformal mapping numerous canonical domains are considered upon which a given domain may be mapped. The functions performing this map are functions of the domain considered and might be called domain functions. Numerous relations between domain functions of different types are known; very many of these functions may be constructed from a few fundamental ones, such as Green's and Neumann's functions of the domain and the harmonic measures of the boundary continua. But these fundamental functions themselves are also closely interrelated and permit numerous identities. It is of interest to organize the system of relations between the domain functions into a simple form. This is convenient for the theory of variation of domain functions with their domain; in fact, we obtain often in extremum problems relative to domain functions several different characterizations of the extremum domain, depending on the type of variation applied in the investigation. It is, therefore, essential to be able to reduce one type of equation to another by means of the various identities for domain functions.

An understanding of all identities between domain functions may be obtained by sustained application of Schottky's theory of multiply-connected domains [15][2]. Schottky proved that there is a close relation between the mapping theory of these domains and the theory of closed Riemann surfaces; the identities among domain functions have their complete analogue in the theory of Abelian integrals and might be proved by means of the latter.

It seems, however, that a theory will be of interest which operates only with concepts of conformal mapping and the geometric properties of the functions considered. The functions which prove in such a theory to be the more basic domain functions may be expected to have importance in the general study of conformal mapping, too. In fact, it will be seen that one of the most fundamental functions in the theory will be a kernel function. This type of function has been studied from various points of view recently [1, 2, 13]. The development in this paper gives further a new understanding of variation formulas which have been applied frequently in conformal mapping; it allows us also to carry out variational methods in extremum problems with additional conditions, such as the invariance of conformal type.

Presented to the Society, April 17, 1948; received by the editors January 8, 1948.

[1] Paper done under contract with the Office of Naval Research.

[2] Numbers in brackets refer to the bibliography at the end of the paper.

187

It is difficult to say how much of the methods for obtaining identities applied in this paper is new. Several authors considered similar domain functions and investigated them by analogous methods; in particular, Grunsky used in his papers combinations of domain functions which are closely related to those applied here [6, 7]. However, we felt for ourselves the need to carry out a systematic study of these formal identities and to establish a more or less unified method. The identities derived in this paper are far from complete, but we hope that the methods for obtaining them have been worked out clearly enough to facilitate the establishment of further identities which might be required.

1. **The fundamental domain functions.** Let D be a domain in the complex z-plane bounded by n proper continua C_ν ($\nu = 1, 2, \cdots, n$) which form together the boundary $C = \sum_{\nu=1}^{n} C_\nu$ of D. For the sake of simplicity and without any loss of generality, we shall assume that D is bounded and that each C_ν is a smooth closed curve. We define two univalent functions $\Phi(z; u, v)$ and $\Psi(z; u, v)$ by the following requirements: (a) $\Phi(z; u, v)$ maps D upon the entire complex plane slit along concentric circular arcs around the origin so that the point $u \in D$ corresponds to the origin and the point $v \in D$ to infinity. The residue of the simple pole at v is 1. (b) $\Psi(z; u, v)$ maps D upon the entire complex plane slit along rectilinear segments directed towards the origin so that the point $u \in D$ corresponds to the origin and the point $v \in D$ to infinity. The residue of the simple pole at v is 1.

Existence and uniqueness of these two domain functions is well known. Let us consider the logarithms of both functions of $z \in D$. We have

$$(1) \qquad \log \Phi(z; u, v) = \log \frac{z - u}{z - v} + F(z; u, v),$$

where $F(z; u, v)$ is regular and single-valued in D, and because of the condition on the residue at $z = v$, we have

$$(1') \qquad \log (v - u) + F(v; u, v) = 0.$$

If the point z lies in the boundary continuum C_ν, we have by definition of $\Phi(z; u, v)$

$$(2) \qquad \log \Phi(z; u, v) = \kappa_\nu(u, v) + i r_\nu(z; u, v), \qquad u, v \in D, z \in C_\nu,$$

where κ_ν and r_ν are both real-valued nonanalytic functions of their arguments.

In the same way, we have for $\Psi(z; u, v)$ the equations

$$(3) \qquad \log \Psi(z; u, v) = \log \frac{z - u}{z - v} + G(z; u, v),$$

with regular and single-valued $G(z; u, v)$ for $z \in D$, and

$$(3') \qquad \log (v - u) + G(v; u, v) = 0.$$

For $z \in C_\nu$, we have in this case

(4) $$\log \Psi(z; u, v) = s_\nu(z; u, v) + i\lambda_\nu(u, v), \qquad u, v \in D, z \in C_\nu.$$

Here $s_\nu(z; u, v)$ and $\lambda_\nu(u, v)$ are again real-valued functions of their arguments.

Consider the class Ω of all functions $f(z)$ which are analytic, regular and single-valued in D and for which the integral

(5) $$(f, f^+) = \int\int_D |f'(z)|^2 dx dy, \qquad z = x + iy, \qquad f'(z) = \frac{d}{dz} f(z),$$

taken in the Lebesgue sense, exists and is finite[3]. We may introduce a metric into the linear space of all functions of class Ω by defining the scalar product

(6) $$(f, g^+) = \int\int_D f'(z)(g'(z))^+ dx dy$$

of two arbitrary elements $f(z)$ and $g(z)$ in Ω. We may easily transform the surface integral (6) into a line integral by means of Green's theorem. We obtain

(7) $$(f, g^+) = \frac{1}{2i} \oint_C f'(z)(g(z))^+ dz = -\frac{1}{2i} \oint_C f(z)(g'(z))^+ dz^+.$$

Of course, not for every pair of functions of class Ω may this integration by parts really be carried out; it is, however, permissible if $f(z)$ and $g(z)$ are both continuously differentiable in $D+C$. On the other hand, our second formula for the scalar product permits the extension of the metric to functions which are meromorphic in D and such that the representation (7) is defined.

In particular, we consider now the expressions $(f, (\log \Phi)^+)$ and $(f, (\log \Psi)^+)$. These are well defined by (7), since $\log \Phi$ and $\log \Psi$ are differentiable and single-valued on C. We have

(8) $$(f(z), (\log \Phi(z; u, v))^+) = \frac{1}{2i} \oint_C f'(z)(\log \Phi(z; u, v))^+ dz,$$

which leads, because of (2) and the single-valuedness of $f(z)$ in D, to

(8′) $$(f(z), (\log \Phi(z; u, v))^+) = -\frac{1}{2i} \oint_C f'(z) \log \Phi(z; u, v) dz$$
$$= \frac{1}{2i} \oint_C f(z) d \log \Phi(z; u, v).$$

By virtue of the residue theorem this leads finally to

_____

[3] In this paper the conjugate of a complex term A will be denoted by A^+.

(9) $(f(z), (\log \Phi(z; u, v))^+) = \pi[f(u) - f(v)]$.

Analogously, we have

(10) $(f(z), (\log \Psi(z; u, v))^+) = \dfrac{1}{2i} \oint_C f'(z)(\log \Psi(z; u, v))^+ dz$,

and, in view of (4),

$$(f(z), (\log \Psi(z; u, v))^+) = \frac{1}{2i} \oint_C f'(z) \log \Psi(z; u, v) dz$$

(10′)

$$= -\frac{1}{2i} \oint_C f(z) d \log \Psi(z; u, v)$$

and finally

(11) $(f(z), (\log \Psi(z; u, v))^+) = - \pi[f(u) - f(v)]$.

From formulas (9) and (11) two interesting combinations of $\log \Phi$ and $\log \Psi$ present themselves because of their simple properties:

(12) $P(z; u, v) = 2^{-1}[\log \Phi(z; u, v) - \log \Psi(z; u, v)]$,

(13) $Q(z; u, v) = 2^{-1}[\log \Phi(z; u, v) + \log \Psi(z; u, v)]$.

The function $P(z; u, v)$ is regular and single-valued for $z \in D$, since the logarithmic poles at u and v cancel by subtraction. Because of (1′) and (3′) one has further

(12′) $P(v; u, v) = 2^{-1}[F(v; u, v) - G(v; u, v)] = 0$,

and from (9) and (11) we conclude

(14) $(f(z), (P(z; u, v))^+) = \pi[f(u) - f(v)]$.

The identity (14) has been derived for functions $f(z) \in \Omega$ which are continously differentiable in $D+C$; but by the usual considerations of approximation it may be easily extended to the whole class Ω. Since $P(z; u, v)$ belongs itself to the class Ω we may use in (14) either definition (6) or (7) for the scalar product.

The function $Q(z; u, v)$ has logarithmic poles at $z=u$ and $z=v$ and is not even single-valued in D. Each determination of it is, however, single-valued on C; it changes only by an integral multiple of $2\pi i$ if the point z makes a circuit around u or v. From (9) and (11) we conclude

(15) $(f(z), (Q(z; u, v))^+) = 0$.

In order to extend this identity to the whole class Ω, we have to replace the line integral occurring in the definition (7) of the scalar product by an integral over the interior of the domain D. For this purpose consider the improper integral

$$(15') \quad J(f, Q^+) = \int\int_D f'(z)(Q'(z; u, v))^+ dxdy, \quad Q'(z; u, v) = \frac{d}{dz}Q(z; u, v).$$

If $f(z)$ is continuously differentiable over $D+C$ we may evaluate this integral by integration by parts. We obtain easily

$$J(f, Q^+) = -\frac{1}{2i}\oint_C f(z)(Q'(z; u, v))^+ dz^+ + \pi[f(v) - f(u)]$$
$$(15'') \qquad = (f, Q^+) + \pi[f(v) - f(u)].$$

This formula permits us to define the scalar product (f, Q^+) by means of a surface integral which has a meaning for every function $f(z) \in \Omega$. It is now possible to extend the identity (15) to the whole class Ω.

The most important relation between the two functions $P(z; u, v)$ and $Q(z; u, v)$ results from the equations (2) and (4). We obtain from them easily the equation

$$(16) \qquad P(z; u, v) = -(Q(z; u, v))^+ + (k_\nu(u, v))^+,$$
$$k_\nu(u, v) = \kappa_\nu(u, v) + i\lambda_\nu(u, v),$$

if z lies on the boundary continuum C_ν and u, v are arbitrary points in D. We shall make use of this relation in a systematic way and we shall derive from it numerous identities for the functions P, Q, and other closely related functions. It is this formula which makes the pair of domain functions P and Q the most convenient basis for an investigation of the fundamental domain functions connected with conformal mapping.

We make a first application of (16) in order to determine the norm of $Q(z; u, v)$ in our metric (7). We find

$$(17) \qquad (Q(z; u, v), (Q(z; u, v))^+) = \frac{1}{2i}\oint_C (Q(z; u, v))^+ dQ(z; u, v),$$

and using (16) and remarking that

$$(17') \qquad \oint_{C_\nu} dQ(z; u, v) = 0,$$

we obtain

$$(18) \qquad (Q(z; u, v), (Q(z; u, v))^+) = \frac{1}{2i}\oint_C P(z; u, v)d(P(z; u, v))^+$$
$$= -(P(z; u, v), (P(z; u, v))^+).$$

Finally, in view of (14) and (12'), we arrive at

(18') $(Q(z; u, v), (Q(z; u, v))^+) = - (P(z; u, v), (P(z; u, v))^+) = - \pi P(u; u, v).$

Since the norm of every function $f(z) \in \Omega$ is non-negative by virtue of definition (5), and since P is of this class, we conclude

(18'') $\qquad P(u; u, v) \geq 0, \qquad (Q(z; u, v), (Q(z; u, v))^+) \leq 0.$

We see that the function $Q(z; u, v)$ with logarithmic poles at u and v has a nonpositive norm.

Returning to the class Ω we remark that the only functions $f(z) \in \Omega$ with vanishing norm are the constants. In order to build up a theory of orthonormal systems it is important to reduce the class Ω to a subclass in which the vanishing of (f, f^+) implies $f \equiv 0$. Such a subclass can be defined in different ways. The most obvious one is to distinguish the point $v \in D$ and to define as subclass $\Omega_v \subset \Omega$ the class of those functions $f(z) \in \Omega$ which vanish at the point v. Because of (12') the function $P(z; u, v)$ belongs to this class. Now it is obvious that if $f(z) \in \Omega_v$ has the norm zero it is a constant, and since it vanishes at v it must be identically zero. From (14) we deduce further

(14') $\qquad (f(z), (P(z; u, v))^+) = \pi f(u), \qquad f(z) \in \Omega_v.$

It is now always possible to determine a complete orthonormal system for the class Ω_v; that is, there exists a system $\{f_\nu(z)\}$ of functions in Ω_v which satisfies the conditions

(19) $\qquad (f_\nu(z), (f_\mu(z))^+) = \delta_{\nu\mu}, \qquad \delta_{\nu\mu} = \begin{cases} 1, & \nu = \mu, \\ 0, & \nu \neq \mu, \end{cases}$

and such that every function $f(z) \in \Omega_v$ may be developed in a Fourier series

(20) $\qquad f(z) = \sum_{\nu=1}^{\infty} a_\nu f_\nu(z), \qquad a_\nu = (f, f_\nu^+),$

which converges uniformly in every closed subdomain of D.

Since $P(z; u, v)$ is of the class Ω_v, we may apply to it the development (20) and, in view of (14'), we find

(21) $\qquad P(z; u, v) = \pi \sum_{\nu=1}^{\infty} f_\nu(z)(f_\nu(u))^+.$

This shows that the function $P(z; u, v)$ is the kernel function of every complete orthonormal system with respect to Ω_v. It is remarkable that the kernel function is independent of the choice of the particular system and that it coincides with a function defined in terms of canonical mapping functions.

From (21) it follows further that $P(z; u, v)$ is anti-analytic in u. The dependence of $P(z; u, v)$ on v is much more complicated, since v enters already into the definition of the basic system $\{f_\nu(z)\}$. We shall see later that the class Ω may be reduced in a way which does not distinguish the point v, and that the kernel function of the class obtained is anti-analytic in u and v. The importance of (21) for an effective computational construction of $P(z; u, v)$ needs not to be stressed. The formula is still more interesting because it is possible to compute the second fundamental domain function $Q(z; u, v)$ by simple integration from the function $P(z; u, v)$.

In fact, let us represent $Q(z; u, v)$ in the form

$$(22) \qquad Q(z; u, v) = \log \frac{z - u}{z - v} + Q_1(z; u, v),$$

where $Q_1(z; u, v)$ is of class Ω. Applying now (15) to the particular function $P(z; a, b)$ of class Ω and using the symmetry law $(f, g^+) = (g, f^+)^+$, we find

$$(23) \qquad \left(\log \frac{z - u}{z - v}, \ (P(z; a, b))^+ \right) + (Q_1(z; u, v), (P(z; a, b))^+) = 0.$$

On the other hand, application of (14) to the particular function $Q_1(z; u, v)$ leads to

$$(23') \qquad (Q_1(z; u, v), (P(z; a, b))^+) = \pi [Q_1(a; u, v) - Q_1(b; u, v)].$$

From (23) and (23') we derive

$$(24) \qquad Q_1(a; u, v) - Q_1(b; u, v) = -\frac{1}{\pi} \left(\log \frac{z - u}{z - v}, \ (P(z; a, b))^+ \right)$$

and by definition (22) finally

$$
\begin{aligned}
(25) \qquad & Q(a; u, v) - Q(b; u, v) \\
& = \log \frac{a - u}{a - v} - \log \frac{b - u}{b - v} - \frac{1}{\pi} \left(\log \frac{z - u}{z - v}, \ (P(z; a, b))^+ \right).
\end{aligned}
$$

This formula determines $Q(z; u, v)$ up to an additive constant. We shall see later that the combination $Q(a; u, v) - Q(b; u, v)$ is of great interest for the general theory because of its symmetry properties. We note further that the scalar product in (25) is to be understood in the sense of a contour integral, since $\log ((z-u)/(z-v))$ has singularities in D. One recognizes that the function $Q(a; u, v) - Q(b; u, v)$ is analytic in all four variables; another proof for this fact will be given later as application of a more general theory.

2. **Some extremum problems in conformal mapping.** Consider the family $\mathfrak{F}$ of all functions $F(z)$ which are regular in D except for logarithmic poles at the points $z = u$ and $z = v$ and which have the form

$$(26) \qquad F(z) = \log \frac{z - u}{z - v} + F_1(z),$$

where $F_1(z)$ is of class Ω. We assume further that

$$(26') \qquad\qquad \log\,(v - u) + F_1(v) = 0,$$

that is, that the function $\exp\,\{F(z)\}$ has a simple pole at $z=v$ with residue 1. In view of (22), we may write every function $F(z) \in \mathfrak{F}$ also in the form

$$(26'') \qquad\qquad F(z) = Q(z;\, u,\, v) + F_2(z),$$

where $F_2(z)$ is again of class Ω, and since $\exp\,\{Q(z;\, u,\, v)\}$ has at $z=v$ the residue 1, we have

$$(26''') \qquad\qquad F_2(v) = 0.$$

We may now consider the norms

$$(27) \quad \begin{aligned} (F, F^+) &= (Q(z;\, u,\, v),\, (Q(z;\, u,\, v))^+) + (Q(z;\, u,\, v),\, (F_2(z))^+) \\ &\quad + (F_2(z),\, (Q(z;\, u,\, v))^+) + (F_2,\, F_2^+), \end{aligned}$$

which are well defined for the family $\mathfrak{F}$ because of $(15'')$ and $(18')$. We have in view of (15) and $(18')$

$$(28) \qquad\qquad (F, F^+) = -\,\pi P(u;\, u,\, v) + (F_2,\, F_2^+).$$

Since $F_2(z)$ is of class Ω_v and has, therefore, a non-negative norm, we obtain the result:

THEOREM I. *For every function* $F(z) \in \mathfrak{F}$ *we have* $(F, F^+) \geqq -\pi P(u;\, u,\, v)$. *Equality holds only for* $F(z) = Q(z;\, u,\, v)$.

If $F(z)$ is continuously differentiable in $D+C$, we may write

$$(29) \qquad\qquad -\,(F, F^+) = -\,\frac{1}{2i}\,\oint_C (F(z))^+ dF(z),$$

and this expression has an easy geometric interpretation. The function $\zeta = F(z)$ maps D conformally (but not necessarily univalently) upon a domain Δ, and the curves C_ν are transformed into new sets of boundary curves Γ_i $(i = 1, 2, \cdots)$; each set is obtained from a basic set $\Gamma_1, \Gamma_2, \cdots, \Gamma_n$ by simple translation by an integral multiple of $2\pi i$. The set Γ_ν $(\nu = 1, \cdots, n)$ encloses a certain part of the ζ-plane and its area is $-(F, F^+)$. The function $\exp\,\{F(z)\}$ maps D upon a domain bounded by n curves and we may call $-(F, F^+)$ the logarithmic area of the continua enclosed by them. Thus, we have the result:

THEOREM Ia. *If* $F(z) \in \mathfrak{F}$, *the function* $\exp\,\{F(z)\}$ *maps* D *upon a domain whose complement has a logarithmic area not greater than* $\pi P(u;\, u,\, v)$. *Only the mapping* $\exp\,\{Q(z;\, u,\, v)\}$ *yields the maximal value for the logarithmic area of the complement.*

We shall prove that the function $\exp\,\{Q(z;\, u,\, v)\}$ is univalent in D. Since the above theorem is a fortiori true for univalent mappings, we obtain the interesting corollary:

THEOREM Ib. *If D is mapped univalently so that $z=u$ corresponds to the origin, $z=v$ to infinity, and so that the residue of the pole at $z=v$ is 1, then the logarithmic area of the complement of the image domain is not greater than $\pi P(u; u, v)$. This upper limit is attained by the univalent function* $\exp\{Q(z; u, v)\}$.

In order to prove the univalency of $\exp\{Q(z; u, v)\}$, we return to the definition (13) of $Q(z; u, v)$. If z lies on a fixed boundary continuum C_ν the abscissa of the point $Q(z; u, v)$ is determined by $\log \Psi(z; u, v)$ and the ordinate by $\log \Phi(z; u, v)$ up to a fixed additive constant. By its geometric definition $\log \Psi$ attains every value on C_ν twice and so does $\log \Phi$. This shows easily that $Q(z; u, v)$ maps C_ν upon a curve without self-intersection which is cut by each parallel to the real or imaginary axis exactly twice. Since the imaginary part of $\log \Phi$ varies on C_ν by less than 2π, it follows that $\exp\{Q(z; u, v)\}$ maps each C_ν in a one-to-one manner upon a simple closed curve. It is further clear that the mapping covers the point at infinity and the origin exactly once. If we can prove that $\exp\{Q(z; u, v)\}$ has nowhere in D a vanishing derivative, it follows from elementary topological considerations that this function is univalent in D.

For the last step in our reasoning we remark that the function

$$(30) \quad \log S(z; u, v) = [\cos \alpha \log \Phi(z; u, v) + i \sin \alpha \log \Psi(z; u, v)]e^{-i\alpha}, \quad \alpha \text{ real,}$$

is the logarithm of a univalent function in D. In fact, the right-hand side of (30) is a multivalued function with periods $m \cdot 2\pi i$. The boundary continuum C_ν is mapped by $\log S$ upon a rectilinear segment, which is transformed into a logarithmic spiral slit if we consider the map by $S(z; u, v)$. This mapping function has been thoroughly investigated [4, 6]. We use here only the fact that the derivative of S is not zero for any $z \in D$. Hence, we find

$$(31) \qquad J(z; u, v) = \frac{\Phi'(z; u, v)}{\Phi(z; u, v)} \cdot \frac{\Psi(z; u, v)}{\Psi'(z; u, v)} \neq -i \tan \alpha, \qquad z \in D, \alpha \text{ real.}$$

This shows that $J(z; u, v)$ is never imaginary for $z \in D$. For $z=u$ and $z=v$ one has $J=1$. This shows that

$$(32) \qquad\qquad\qquad \operatorname{Re}\{J(z; u, v)\} > 0, \qquad\qquad\qquad z \in D.$$

For if there were a point $z \in D$ with negative $\operatorname{Re}\{J\}$ we could connect it by a continuous curve with $z=u$ and there would exist a point on this curve where $\operatorname{Re}\{J\}=0$. But this is a contradiction to (31), which proves finally (32).

In particular, we conclude from (32) that

$$(33) \qquad \frac{d}{dz}[a \log \Phi(z; u, v) + b \log \Psi(z; u, v)] \neq 0, \qquad z \in D; a, b > 0.$$

From our preceding considerations follows therefore easily:

THEOREM II. *Every function* $\exp \{ a \log \Phi(z; u, v) + b \log \Psi(z; u, v) \}$ *is univalent in* D, *if* $a > 0$, $b > 0$, *and* $a + b = 1$. *In particular*, $\exp \{ Q(z; u, v) \}$ *is univalent in* D.

It is now of interest to study the univalent functions of the above type in more detail and to characterize them by extremum properties. We may write in view of (12) and (13)

$$(34) \qquad \begin{aligned} \log T(z) &= a \log \Phi(z; u, v) + b \log \Psi(z; u, v) \\ &= (a - b)P(z; u, v) + (a + b)Q(z; u, v). \end{aligned}$$

For $a + b = 1$, we obtain in $T(z)$ a univalent function with $(\log T) \in \mathfrak{F}$, since $\exp \{ Q(z; u, v) \}$ has at $z = v$ the residue 1 and $P(v; u, v) = 0$. In the neighborhood of $z = u$, we have in view of (22)

$$(34') \qquad \begin{aligned} \log T(z) &= \log (z - u) - \log (u - v) + Q_1(u; u, v) \\ &\quad + (a - b)P(u; u, v) + O(|z - u|). \end{aligned}$$

Since $P(u; u, v) > 0$, we may by appropriate choice of a and b obtain for $\log T(z)$ a development

$$(34'') \qquad \log T(z) = \log (z - u) + \gamma + O(|z - u|)$$

with an arbitrarily prescribed possible value for $\mathrm{Re}\{\gamma\}$. For it is well known [12] that the largest possible value of $\mathrm{Re}\{\gamma\}$ is attained by $\log \Phi(z; u, v)$ and the smallest by $\log \Psi(z; u, v)$, corresponding to a choice of $a = 1$ or $b = 1$, respectively.

Consider now the subclass $\mathfrak{F}_\gamma$ of all functions $F(z) \in \mathfrak{F}$ which have at $z = u$ a given value $\mathrm{Re}\{\gamma\}$. They may be written in the form

$$(35) \qquad F(z) = \log T(z) + F_2(z),$$

where $T(z)$ is the uniquely defined combination of P and Q in the family $\mathfrak{F}_\gamma$ and $F_2(z)$ is of the class Ω. Obviously, we have by definition

$$(35') \qquad F_2(v) = 0, \qquad F_2(u) = \text{imaginary}.$$

Let us compute now the norm of $F(z)$. We have, in view of (34) and (35),

$$(36) \qquad \begin{aligned} (F, F^+) &= (Q + (a - b)P + F_2, (Q + (a - b)P + F_2)^+) \\ &= (Q, Q^+) + (Q, ((a - b)P + F_2)^+) + ((a - b)P + F_2, Q^+) \\ &\quad + ((a - b)P + F_2, ((a - b)P + F_2)^+). \end{aligned}$$

Applying now (18′), (15) and (14), we obtain

$$(37) \qquad \begin{aligned} (F, F^+) &= -\pi P(u; u, v) + (a - b)^2 \pi P(u; u, v) \\ &\quad + (a - b)\pi(F_2(u) + (F_2(u))^+) + (F_2, F_2^+). \end{aligned}$$

Taking account of (35′) and of the equation $a + b = 1$, we arrive finally at

$$(37')   \qquad (F, F^+) = - 4\pi ab P(u; u, v) + (F_2, F_2^+).$$

Since $(F_2, F_2^+) \geqq 0$ and equality holds only for $F_2 \equiv 0$, we have

$$(38)   \qquad - (F, F^+) \leqq 4\pi ab P(u; u, v),$$

and equality holds only for $F(z) = \log T(z)$. This leads to the theorem:

THEOREM IIa. *Among all univalent functions $f(z)$ in D vanishing at u and having a simple pole with residue 1 at v and a prescribed value $|f'(u)|$, the function* $\exp \{Q(z; u, v) + \alpha P(z; u, v)\}$, *for a suitable choice of α, maps D upon the domain whose complement has the maximum logarithmic area.*

These results are now closely related to an extremum problem of quite different type. Consider for this purpose the class $\mathfrak{S}$ of all functions which are regular and single-valued in D, vanish at v and have the value 1 at u. A typical function of this class is

$$(39)   \qquad f_0(z) = P(z; u, v) \cdot P(u; u, v)^{-1}.$$

Every other function of the class may be written in the form

$$(40)   \qquad f(z) = f_0(z) + \phi(z)$$

and, by definition, we have

$$(40')   \qquad \phi(u) = \phi(v) = 0.$$

Computing now the norm of $f(z)$, we find because of (39), (14) and (40')

$$(41)   \qquad (f, f^+) = \pi P(u; u, v)^{-1} + (\phi, \phi^+).$$

Since ϕ is regular and single-valued in D, its norm is non-negative and we arrive at the inequality

$$(41')   \qquad (f, f^+) \geqq \pi P(u; u, v)^{-1},$$

valid for every function $f(z)$ of our class. Equality holds only for the function $f_0(z)$. This leads to the result:

THEOREM III. *All functions $f(z)$ which are regular and single-valued in D and satisfy $f(u) = 1$ and $f(v) = 0$ map this domain upon an area which is at least $\pi P(u; u, v)^{-1}$. The extremum map is obtained by means of the function (39).*

Comparing theorems Ia and III we obtain the interesting result:

THEOREM IIIa. *For every domain D the product between the maximal logarithmic area of the complement and the minimal area under mappings of the family $\mathfrak{S}$ is exactly π^2.*

There is even a simple geometric relationship between the extremum domains for the two different mapping problems. In fact, we find in view of (16) that the boundary curves of our second extremum problem are similar to the reflected logarithmic images of the boundary curves occurring in our initial extremum problem. They are, therefore, simple nonintersecting closed curves, which shows that the extremum problem with respect to the class $\mathfrak{S}$ leads to a domain which is at most n times covered and that $P(z; u, v)$ is at most n-valent. We state, therefore, the following theorem.

THEOREM IV. *The function $Q(z; u, v)$ is the logarithm of a univalent function and has logarithmic poles at u and v; the function $P(z; u, v)$ is regular and at most n-valued, if n is the connectivity of the domain.*

It will be of interest to us to have also the following result. Let $f_{ij}(z)$ be a map of D on an annullus cut along concentric circular slits such that C_i and C_j, $i \neq j$, go respectively into the inner and outer boundaries of the annullus, and let $g_{ij}(z)$ be a map of D on an annullus cut along radial slits such that C_i and C_j, $i \neq j$, go respectively into the inner and outer boundaries. Let $\mathfrak{F}_{ij}$ be the general class of schlicht maps $h_{ij}(z)$ of D upon domains in annulli such that C_i and C_j are carried into the concentric circles bounding the annullus, and let A_h denote the logarithmic area of the continua inside the images of the C_k, $i \neq k \neq j$. Then for prescribed modulus ρ of the annullus, where ρ lies between the moduli corresponding to f_{ij} and g_{ij}, A_h is maximized by the schlicht map $h_{ij} = (f_{ij}g_{ij})^{1/2}(g_{ij}/f_{ij})^{\lambda/2}$, for suitable λ, $-1 \leqq \lambda \leqq 1$. If we pose the same problem without restriction on ρ the extremal function is $(f_{ij}g_{ij})^{1/2}$ and $\lambda = 0$. This extremal problem plays for the harmonic measures the role which those we have considered in detail play for Green's and Neumann's functions, for, as we shall see later, $\log f_{ij}$ is a linear combination of harmonic measures. We skip the proof of the result, since it is so much like those already given.

Finally, we remark that by means of $P(z; u, v)$ and $Q(z; u, v)$ it is possible to obtain interesting mapping functions in the following way: We have, in view of (16), on every boundary continuum C,

$$(42) \qquad P'(z; u, v)z' = - (Q'(z; u, v)z')^{+}, \qquad z' = dz/ds.$$

This shows that the quotient

$$(43) \qquad E(z; u, v) = P'(z; u, v)/Q'(z; u, v)$$

has on C always the modulus 1. Since $Q'(z; u, v) \neq 0$ for $z \in D$, $E(z; u, v)$ maps D upon the unit circle, of course covered several times. It is easy to determine the number of coverings. In fact, from (42) we derive that

$$(44) \qquad P'(z; u, v)Q'(z; u, v)z'^{2} < 0, \qquad z \in C.$$

By the argument principle we conclude, therefore, that the difference between the number of zeros and the number of poles of the left-hand product is $2n - 4$. Now, $Q' \neq 0$ in D, but has two poles at $z = u$ and $z = v$; hence $P'(z; u, v)$ has $2n - 2$ zeros and the function $E(z; u, v)$ maps D upon the unit circle covered

$2n$ times. The close relations between functions which have the modulus 1 on C and the domain functions $P(z; u, v)$ and $Q(z; u, v)$ will be discussed later.

3. **The method of contour integration.** In this section systematic use will be made of the equation (16), which connects the functions $P(z; u, v)$ and $Q(z; u, v)$ at the boundary C of D. It will be seen that this relation leads by means of the residue theorem to a number of important identities. The idea of procedure will become clear from our first application, which will, therefore, be carried out in more detail than the following ones.

We have by Cauchy's integral theorem

$$(45) \qquad \frac{1}{2\pi i} \oint_C P(z; u, v)P'(z; a, b)dz = 0, \qquad u, v, a, b \in D,$$

since $P(z; u, v)$ and $P(z; a, b)$ are regular in D. By means of (16) we may now replace in this contour integral P by Q; taking into consideration that $Q(z; a, b)$ does not change if z describes a continuum C_ν, that is,

$$(46) \qquad \oint_{C_\nu} Q'(z; a, b)dz = 0,$$

we obtain from (45), (16) and (46)

$$(47) \qquad \frac{1}{2\pi i} \oint_C Q(z; u, v)Q'(z; a, b)dz = 0.$$

The function $Q(z; u, v)$ is not single-valued in D, but is so in the domain D_γ obtained from D by performing a cut along a smooth curve γ from u to v. Applying now the residue theorem to D_γ and using equation (47), we find

$$(48) \qquad Q(a; u, v) - Q(b; u, v) + \frac{1}{2\pi i} \int_u^v 2\pi i Q'(z; a, b)dz = 0,$$

which leads to the symmetry rule

$$(49) \qquad Q(a; u, v) - Q(b; u, v) = Q(u; a, b) - Q(v; a, b).$$

This rule may be compared to the rule of interchange of parameter and argument in the theory of integrals on Riemann surfaces. It shows that $Q(u; a, b) - Q(v; a, b)$ is analytic in all four variables, a result which we obtained in a different way at the end of §1.

For the next application we start with the identity

$$(50) \qquad \frac{1}{2\pi i} \oint_C P(z; u, v)Q'(z; a, b)dz = P(a; u, v) - P(b; u, v),$$

which is an immediate consequence of the residue theorem and the analytic character of P and Q in D. By means of (16) we obtain

$$P(a; u, v) - P(b; u, v) = \frac{1}{2\pi i} \oint_C (Q(z; u, v)P'(z; a, b)dz)^+$$

(51)

$$= \left(\frac{1}{2\pi i} \oint_C P(z; a, b)Q'(z; u, v)dz\right)^+;$$

a new application of the residue theorem leads to the symmetry law

(52) $$P(a; u, v) - P(b; u, v) = (P(u; a, b))^+ - (P(v; a, b))^+.$$

This result shows that $P(u; a, b) - P(v; a, b)$ is analytic in u and v, but anti-analytic in a and b.

It is now clear how important pairs of functions are which have on C conjugate boundary values. It is possible to define another pair of functions with this property and to obtain relations between them and the functions P and Q. We start with two functions $A(z; u)$ and $B(z; u)$ which are both univalent in D and are there regular except for a simple pole at $z = u$ with residue 1. $A(z; u)$ maps D upon the complex plane slit along rectilinear segments parallel to the real axis, while $B(z; u)$ maps D upon the complex plane slit along rectilinear segments parallel to the imaginary axis. In both cases, the point u corresponds to infinity. Our requirements determine the functions $A(z; u)$ and $B(z; u)$ up to an additive constant [5].

One has by definition

(53) $$A(z; u) = \gamma_\nu(z) + i\alpha_\nu, \qquad z \in C_\nu,$$

where γ_ν and α_ν are real and depend on u, but α_ν does not depend on z. In the same way

(54) $$B(z; u) = \beta_\nu + i\delta_\nu(z), \qquad z \in C_\nu,$$

with real β_ν and $\delta_\nu(z)$ and β_ν independent of z.

In view of preceding considerations it is suggestive to define now the pair of functions

(55) $$M(z; u) = [A(z; u) - B(z; u)]/2$$

and

(56) $$N(z; u) = [A(z; u) + B(z; u)]/2.$$

Obviously, $M(z; u)$ is regular everywhere in D while $N(z; u)$ has at $z = u$ a simple pole with residue 1. Because of (53) and (54) one has for $z \in C_\nu$

(57) $$M(z; u) = (N(z; u))^+ + (l_\nu(u))^+,$$

where $l_\nu(u) = -i\alpha_\nu - \beta_\nu$ is a constant depending on ν and (not analytically) on u.

The functions $M(z; u)$ and $N(z; u)$ play an important role in the theory of conformal mapping. One shows by the methods of §2 that $N(z; u)$ is univalent in D and that even every combination

(58)
$$F(z; u) = N(z; u) + \lambda M(z; u), \qquad 0 \leqq \lambda \leqq 1,$$

is univalent in D. By means of these functions the following extremum problems can be solved:

I. Consider the family of all functions which are regular and single-valued in D and have at $u \in D$ the derivative 1. Determine that function of the family which maps D upon a domain of smallest area.

The solution of this problem is given by the function $M(z; u) M'(u; u)^{-1}$ (where $M'(z; u) = dM(z; u)/dz$). The minimum area is $\pi M'(u; u)^{-1}$. $M(z; u)$ is at most n-valent in D, as can be shown by reasoning similar to that in §2.

II. Consider the family of all functions which are univalent and single-valued in D and have at the point $u \in D$ a simple pole with residue 1. Each such function maps D upon a domain Δ; we seek a function of this class which yields a maximum area for the complement $C(\Delta)$ of Δ in the complex plane.

The solution of this problem is given by the function $N(z; u)$ [11, 14]. The maximum area is $\pi M'(u; u)$. We have again the remarkable fact that the minimum area of problem I and the maximum area of problem II have for every domain D the product π^2 [13].

III. Consider the family of all functions which are univalent and single-valued in D and have at the point $u \in D$ a development

(59)
$$f(z) = (z - u)^{-1} + k_0 + k_1(z - u) + \cdots,$$

with fixed Re $\{k_1\}$. Each such function maps D upon a domain Δ; we seek a function of this class which yields a maximum area for the complement $C(\Delta)$ of Δ (cf. [3]).

The solution is given by that function (58) which has the right value of Re $\{k_1\}$. The great analogy between the pairs M, N and P, Q is evident. We shall now show that they are in fact closely related, and this can again be done by our integration method.

Starting with the identity

(60)
$$\frac{1}{2\pi i} \oint_C M(z; a) P'(z; u, v) dz = 0, \qquad a, u, v \in D,$$

and applying the relations (16) and (57), we obtain easily

(60′)
$$\frac{1}{2\pi i} \oint_C N(z; a) Q'(z; u, v) dz = 0.$$

By virtue of the residue theorem we arrive finally at

(61)
$$Q'(a; u, v) = N(v; a) - N(u; a),$$

connecting the function $N(z; u)$ with the derivative of the function $Q(z; u, v)$. We recognize, in particular, the fact that $Q'(z; u, v)$ is analytic in all three arguments. This might also have been expected from (49) for the limit case $a \rightarrow b$.

Let us consider next the following instance of the residue theorem:

$$(62) \qquad \frac{1}{2\pi i} \oint_C N(z; a) P'(z; u, v) dz = P'(a; u, v), \qquad a, u, v \in D.$$

From (16) and (57) we conclude

$$(62') \qquad \frac{1}{2\pi i} \oint_C M(z; a) Q'(z; u, v) dz = (P'(a; u, v))^+,$$

and applying again the residue theorem, we obtain

$$(63) \qquad (P'(a; u, v))^+ = M(u; a) - M(v; a).$$

This shows that $P'(z; u, v)$ is analytic in z and anti-analytic in u and v.

Let us combine further M and N in the following equation:

$$(64) \qquad \frac{1}{2\pi i} \oint_C M(z; u) N'(z; v) dz = - M'(v; u), \qquad \begin{cases} M'(z; u) = \dfrac{d}{dz} M(z; u), \\[2mm] N'(z; u) = \dfrac{d}{dz} N(z; u). \end{cases}$$

This leads by virtue of (57) to

$$(64') \qquad \frac{1}{2\pi i} \oint_C N(z; u) M'(z; v) dz = (M'(v; u))^+,$$

and, by means of the residue theorem, to the symmetry law

$$(65) \qquad M'(u; v) = (M'(v; u))^+.$$

From the equation

$$(66) \qquad \frac{1}{2\pi i} \oint_C M(z; u) M'(z; v) dz = 0, \qquad u, v \in D,$$

we derive in the same way

$$(66') \qquad \frac{1}{2\pi i} \oint_C N(z; u) N'(z; v) dz = 0,$$

which leads again by the residue theorem to the analogous symmetry formula for N,

$$(67) \qquad N'(v; u) = N'(u, v).$$

From (65) and (67) we infer that $M'(u; v)$ is analytic in u and anti-analytic in v, while $N'(u; v)$ is symmetric and analytic in both arguments.

One recognizes again the great similarity between the above relations and the corresponding formulas in the theory of Abelian integrals. The functions P and Q correspond to the fundamental integrals of the third kind and M and N to the integrals of the second kind. The functions P and M, however, have no singularities at all in D. This is closely related to the result of Schottky [15] that a domain D of connectivity n corresponds to a *half* of a symmetric Riemann surface of genus $n-1$. P and M correspond, therefore, to fundamental integrals having their singularity in the missing half of the Riemann surface. Our method of integration along the boundary C of D is closely related to a procedure of Riemann for obtaining relations between Abelian integrals; it has been called the method of contour integration. The central role of the formulas (16) and (57) in this type of reasoning is obvious.

Pursuing the analogy between domain functions of a domain D and Abelian integrals on a Riemann surface, we introduce now functions which correspond to Abelian integrals of the first kind. For this purpose we define the harmonic measures of each boundary continuum C_ν at the point z with respect to the domain D. These functions $\omega_\nu(z)$, which play an important role in the general theory of functions, are defined as follows:

$\omega_\nu(z)$ is harmonic for $z \in D$ and has on C_μ the boundary values $\delta_{\mu\nu}$. One has obviously the relation

$$(68) \qquad \sum_{\nu=1}^{n} \omega_\nu(z) = 1,$$

since the left-hand side represents a function harmonic in D with boundary values 1 on C.

We may complete $\omega_\nu(z)$ to an analytic function $w_\nu(z)$ such that

$$(69) \qquad \omega_\nu(z) = \mathrm{Re}\ \{w_\nu(z)\}.$$

$w_\nu(z)$ is defined up to an additive imaginary constant; it is, in general, not single-valued in D, though it has a single-valued real part. If z describes a circuit around the contour C_μ, $w_\nu(z)$ increases by the period

$$(70) \qquad \oint_{C_\mu} w_\nu'(z)dz = -i \oint_{C_\mu} \frac{\partial \omega_\nu}{\partial n} ds = -2\pi i P_{\mu\nu},$$

where $\partial/\partial n$ denotes differentiation in the direction of the interior normal at $z \in C$.

Since the real part of $w_\nu(z)$ is constant on each boundary continuum C_μ of C, we have

$$(71) \qquad w_\nu'(z)dz = \text{imaginary} \qquad \text{for } z \text{ varying on } C.$$

This property will now be used in contour integration in the same way as (16) and (57).

Consider the integral

$$(72) \qquad \frac{1}{2\pi i} \oint_C P(z; u, v) w_\nu'(z) dz = 0, \qquad\qquad u, v \in D.$$

We may, in fact, apply Cauchy's theorem, since $w_\nu'(z)$ is single-valued and regular in D. Applying now (16) and (71), we obtain

$$(73) \qquad \frac{1}{2\pi i} \sum_{\mu=1}^n \oint_{C_\mu} [Q(z; u, v) - k_\mu(u, v)] w_\nu'(z) dz = 0.$$

The term $k_\mu(u, v)$ in (16) cannot be neglected now, since $w_\nu'(z)$ is not the derivative of a single-valued function. Using (70), we may put (73) into the form

$$(74) \qquad \frac{1}{2\pi i} \oint_C Q(z; u, v) w_\nu'(z) dz = - \sum_{\mu=1}^n k_\mu(u, v) P_{\mu\nu}.$$

The left-hand integral may be evaluated as before by applying Cauchy's theorem to a domain D_γ obtained from D by making a cut from u to v along a smooth curve γ. $Q(z; u, v)$ is single-valued in D_γ and has at the two borders of the cut a saltus of amount $2\pi i$. Hence, we find

$$(75) \qquad w_\nu(u) - w_\nu(v) = \sum_{\mu=1}^n k_\mu(u, v) P_{\mu\nu}, \qquad \nu = 1, 2, \cdots, n,$$

which shows an important relation between the constants k_μ appearing in (16) and the harmonic measures.

In the same fashion, we derive from

$$(76) \qquad \frac{1}{2\pi i} \oint_C M(z; u) w_\nu'(z) dz = 0, \qquad\qquad u \in D,$$

and from (57), (71)

$$(77) \qquad \frac{1}{2\pi i} \sum_{\mu=1}^n \oint_{C_\mu} [N(z; u) + l_\mu(u)] w_\nu'(z) dz = 0.$$

This leads because of (70) to

$$(78) \qquad w_\nu'(u) = \sum_{\mu=1}^n l_\mu(u) P_{\mu\nu},$$

a relation connecting the constants l_μ occurring in (57) with the derivatives of the harmonic measures.

Let us point out finally one interesting property of the harmonic measures with respect to the scalar product (f, g^+) defined in §1. Consider, for this purpose, the scalar product

(79) $$(f, w_\nu^+) = \iint_D f'(z)(w_\nu'(z))^+ dxdy$$

between $w_\nu(z)$ and an arbitrary function $f(z)$ of the class Ω. Integrating by parts, we obtain

(80) $$(f, w_\nu^+) = -\frac{1}{2i}\oint_C f(z)(w_\nu'(z)dz)^+ = \frac{1}{2i}\oint_C f(z)w_\nu'(z)dz$$

in view of (71). But the last integral vanishes because of Cauchy's theorem. Thus, we see that

(81) $$(f, w_\nu^+) = 0 \qquad\qquad \text{for every } f(z) \in \Omega,$$

that is, the functions $w_\nu(z)$ are orthogonal to all functions $f(z) \in \Omega$.

In order to compute the expressions

(82) $$(w_\nu, w_\mu^+) = \iint_D w_\nu'(z)(w_\mu'(z))^+ dxdy$$

we have to perform a slightly different method of partial integration. We find

(83) $$(w_\nu, w_\mu^+) = -\frac{1}{i}\oint_C \omega_\nu(z)(w_\mu'(z)dz)^+ = \frac{1}{i}\oint_{C_\nu} w_\mu'(z)dz = -2\pi P_{\mu\nu}.$$

This shows that the terms (w_ν, w_μ^+) are real and form, therefore, a symmetric matrix. The $P_{\mu\nu}$ form a symmetric matrix belonging to a semi-definite quadratic form. In fact, consider the analytic function

(84) $$w(z) = \sum_{\nu=1}^n \xi_\nu w_\nu(z)$$

with real, but otherwise arbitrary, coefficients ξ_ν. Its real part is single-valued in D and has on C_ν the boundary values ξ_ν. Since, in view of (83),

(85) $$(w, w^+) = \iint_D |w'(z)|^2 dxdy = -2\pi \sum_{\mu,\nu=1}^n \xi_\mu\xi_\nu P_{\mu\nu},$$

and this is a non-negative expression, our assertion is proved. The quadratic form may vanish only if $|w'(z)| \equiv 0$, that is, if $w(z)$ is constant. But this is possible only if all ξ_ν are equal. In this case, $w(z)$ is really a constant because of (68).

Another way of understanding the fact that

(86) $$\sum_{\mu,\nu=1}^n P_{\mu\nu} = 0$$

is the consideration of the periods of

$$(87) \qquad\qquad \sum_{\nu=1}^{n} w_{\nu}(z) = \text{const.}$$

for circuits around the boundary continuum C_μ. One finds in view of (70)

$$(87') \qquad\qquad \sum_{\nu=1}^{n} P_{\mu\nu} = 0, \qquad\qquad \mu = 1, 2, \cdots, n,$$

which leads immediately to (86).

It should be remarked that a system of equations of the form (75) or (78) determines the unknowns k_μ or l_μ only up to a common additive constant. In fact, if k_μ ($\mu=1, 2, \cdots, n$) is a solution of equations (75), then $k_\mu + k$ ($\mu=1, \cdots, n$) is, for arbitrary k, another solution in view of (87'). On the other hand, it follows from the semi-definite character of the matrix $(P_{\mu\nu})$ that this is the most general solution of the system of equations (75).

Let us investigate briefly the geometric meaning of the relation (75). It is evident from the definition of the $P_{\mu\nu}$ that when u makes a circuit about the μth boundary contour C_μ of D, $k_\nu(u, v)$ has a period

$$(70') \qquad\qquad - 2\pi i \delta_{\mu\nu} + \frac{2\pi i}{n} = \oint_{C_\mu} \frac{d k_\nu(u, v)}{du} \, du.$$

Hence the function

$$(75') \qquad\qquad K_{\mu\nu}(u, v) = k_\mu(u, v) - k_\nu(u, v)$$

has about C_i, $\mu \neq i \neq \nu$, no period, and about C_ν and C_μ the periods $2\pi i$ and $-2\pi i$ respectively. Hence we see in the usual fashion that $\exp\{k_\mu(u, v) - k_\nu(u, v)\}$ is a map $f_{\mu\nu}(u)$ of D upon an annullus cut along concentric circular slits, such as we introduced in §2. Since the mapping function $f_{\mu\nu}(u)$ is determined up to a constant factor, we conclude further that $K_{\mu\nu}(u, v)$ has the form

$$(75'') \qquad\qquad K_{\mu\nu}(u, v) = K_{\mu\nu}^{(1)}(u) - K_{\mu\nu}^{(1)}(v) + \text{const.},$$

a result which is also obvious from (75).

4. On Green's and Neumann's functions. We introduce now the two domain functions which play a central role in the theory of logarithmic potential of plane domains.

(a) Green's function $g(z; \zeta)$ is harmonic for $z \in D$ except for the point $\zeta \in D$, where $g(z; \zeta) + \log|z - \zeta|$ is harmonic. If z converges to the boundary C of D, Green's function tends to zero.

Green's function plays a decisive role in the first boundary value problem for harmonic functions. It is well known that under our assumption on D Green's function is still continuously differentiable on the boundary C of D. It satisfies the symmetry law

(88)
$$g(z; \zeta) = g(\zeta; z)$$

and is, therefore, also harmonic in ζ except for $\zeta = z$.

(b) Neumann's function $\gamma(z; \zeta)$ is harmonic for $z \in D$ except for the point $\zeta \in D$, where $\gamma(z; \zeta) + \log |z - \zeta|$ is harmonic. $\gamma(z; \zeta)$ is continuously differentiable in $D + C$ and on C one has

(89)
$$\frac{\partial \gamma(z; \zeta)}{\partial n} = \frac{2\pi}{L},$$

where L is the total length of the boundary curves C. Finally one normalizes Neumann's function by the requirement

(90)
$$\oint_C \gamma(z; \zeta) ds = 0.$$

Neumann's function serves for solving the second boundary value problem for harmonic functions. Again one has a symmetry law

(91)
$$\gamma(z; \zeta) = \gamma(\zeta; z).$$

It is useful to complete the harmonic functions $g(z; \zeta)$ and $\gamma(z; \zeta)$ to analytic functions of z. For this purpose, we introduce two functions $p(z; \zeta)$ and $\pi(z; \zeta)$ analytic in z and such that

(92)
$$g(z; \zeta) = \text{Re} \{p(z; \zeta)\}, \qquad \gamma(z; \zeta) = \text{Re} \{\pi(z; \zeta)\}.$$

These requirements fix p and π only up to an additive imaginary constant which may still depend on ζ. Both functions have obviously a logarithmic pole at $z = \zeta$; they are not single-valued in D. Besides the period $2\pi i$ caused by the logarithmic pole they have periods with respect to circuits of z around the boundary continua C_ν. In fact, in view of the Cauchy-Riemann equations we have

(93)
$$\oint_{C_\nu} dp(z; \zeta) = i \oint_{C_\nu} \frac{\partial}{\partial s} \text{Im} \{p(z; \zeta)\} ds = - i \oint_{C_\nu} \frac{\partial g(z; \zeta)}{\partial n} ds.$$

Now, obviously

(94)
$$\omega_\nu(\zeta) = \frac{1}{2\pi} \oint_{C_\nu} \frac{\partial g(z; \zeta)}{\partial n} ds$$

is that harmonic function of ζ which has on the boundary continuum C_μ the value $\delta_{\mu\nu}$ and is the harmonic measure of C_ν with respect to D at the point ζ, as defined in the last section. Hence, we may put (93) into the form

(95)
$$\oint_{C_\nu} dp(z; \zeta) = - 2\pi i \omega_\nu(\zeta).$$

The periods of $\pi(z; \zeta)$ follow from the expression

(96)
$$\oint_{C_\nu} d\pi(z; \zeta) = i \oint_{C_\nu} \frac{\partial}{\partial s} \operatorname{Im} \{\pi(z; \zeta)\} ds = - i \oint_{C_\nu} \frac{\partial \gamma(z; \zeta)}{\partial n} ds.$$

In view of (89) this yields

(97)
$$\oint_{C_\nu} d\pi(z; \zeta) = - 2\pi i \frac{L_\nu}{L},$$

where L_ν is the length of the boundary continuum C_ν. It is important to note that the periods of $\pi(z; \zeta)$ do not depend on ζ and that the function

(98)
$$q(z; v, \zeta) = \pi(z; v) - \pi(z; \zeta)$$

is, therefore, free from periods around the C_ν.

It is easily seen from (89) that the real part of $q(z; v, \zeta)$ has on C the normal derivative zero and its imaginary part is constant along each C_ν. Hence, one concludes by the usual reasoning that $\exp \{q(z; v, \zeta)\}$ is a single-valued function which maps D univalently upon the entire plane slit along rectilinear segments pointing towards the origin. The points ζ and v correspond to the origin and infinity respectively. Hence, this function is related to $\Psi(z; \zeta, v)$, defined in §1, by an equation

(99)
$$\log \Psi(z; \zeta, v) = \pi(z; v) - \pi(z; \zeta) + \text{const.}$$

The additive constant in (99) may be easily determined from the normalization of $\Psi(z; \zeta, v)$ to have the residue 1 at its simple pole v.

For our further developments it is important to notice that $dp(z; \zeta)/dz = p'(z; \zeta)$ and $d\pi(z; \zeta)/dz = \pi'(z; \zeta)$ are single-valued in D and have a simple pole with residue -1 at ζ. On the boundary C we have evidently

(100)
$$p'(z; \zeta)dz = \text{imaginary}, \qquad z \text{ varying on } C,$$

since the real part of $p(z; \zeta)$ is zero there. On the other hand, we obtain from the Cauchy-Riemann equations

(101)
$$\pi'(z; \zeta)dz = \frac{\partial \pi(z; \zeta)}{\partial s} ds = \left[\frac{\partial \gamma(z; \zeta)}{\partial s} - i \frac{\partial \gamma(z; \zeta)}{\partial n} \right] ds$$
$$= \left[\frac{\partial \gamma(z; \zeta)}{\partial s} - \frac{2\pi i}{L} \right] ds.$$

This leads to the remarkable relation

(101')
$$(\pi'(z; \zeta)dz)^+ = \pi'(z; \zeta)dz + \frac{4\pi i}{L} ds.$$

After these preparations we apply the method of contour integration to

$p(z; \zeta)$ and $\pi(z; \zeta)$. We start with the identity

(102) $$\frac{1}{2\pi i} \oint_C P(z; u, v) p'(z; \zeta) dz = - P(\zeta; u, v), \qquad u, v, \zeta \in D.$$

We transform the left-hand integral by means of (16) and (100); we obtain

(102') $$\frac{1}{2\pi i} \sum_{\mu=1}^{n} \oint_{C_\mu} [Q(z; u, v) - k_\mu(u, v)] p'(z; \zeta) dz = (P(\zeta; u, v))^+.$$

Because of the period-formula (95), we find

(102'') $$\frac{1}{2\pi i} \oint_C Q(z; u, v) p'(z; \zeta) dz + \sum_{\mu=1}^{n} k_\mu(u, v) \omega_\mu(\zeta) = (P(\zeta; u, v))^+.$$

Now, we evaluate the left-hand integral by applying Cauchy's theorem to the domain D_γ which is obtained by slitting D along a smooth curve γ from u to v. We find easily

(103) $$p(v; \zeta) - p(u; \zeta) = Q(\zeta; u, v) + (P(\zeta; u, v))^+ - \sum_{\mu=1}^{n} k_\mu(u, v) \omega_\mu(\zeta).$$

This formula expresses $p(v; \zeta)$ in terms of the functions P and Q; it is clear that by our definition only a formula for $p(v; \zeta) - p(u; \zeta)$ and not for $p(v; \zeta)$ alone was to be expected. For our definition left an arbitrary imaginary constant free which might depend on ζ. Hence, only a difference between $p(v; \zeta)$ and its value at a fixed point $u \in D$ is uniquely defined; this shows that (103) yields the maximum information which is to expected for $p(v; \zeta)$.

We apply now the same considerations to the function $\pi(z; \zeta)$. From

(104) $$\frac{1}{2\pi i} \oint_C P(z; u, v) \pi'(z; \zeta) dz = - P(\zeta; u, v), \qquad u, v, \zeta \in D,$$

we obtain by means of (16) and (101')

(104') $$\frac{1}{2\pi i} \sum_{\mu=1}^{n} \oint_{C_\mu} [Q(z; u, v) - k_\mu(u, v)] \left[\pi'(z; \zeta) z' + \frac{4\pi i}{L} \right] ds$$
$$= - (P(\zeta; u, v))^+, \qquad z' = \frac{dz}{ds}.$$

Hence, applying the period formula (97) we find after a simple transformation

(104'') $$\frac{1}{2\pi i} \oint_C Q(z; u, v) \pi'(z; \zeta) dz - \sum_{\mu=1}^{n} k_\mu(u, v) \frac{L_\mu}{L} + \frac{2}{L} \oint_C Q(z; u, v) ds$$
$$= - (P(\zeta; u, v))^+.$$

Using again Cauchy's formula with respect to D_γ, we arrive finally at the equation

(105)
$$\pi(v;\zeta) - \pi(u;\zeta) = Q(\zeta;u,v) - (P(\zeta;u,v))^+$$
$$+ \sum_{\mu=1}^{n} k_\mu(u,v)\frac{L_\mu}{L} - \frac{2}{L}\oint_C Q(z;u,v)ds,$$

which expresses $\pi(v;\zeta)$ in terms of the functions Q and P.

At this stage it becomes evident that we may obtain much simpler formulas by an appropriate change of normalization of the functions $P(z;u,v)$ and $Q(z;u,v)$. These functions were defined as the difference and the sum of the logarithms of two canonical mapping functions. The geometrical properties of these two mappings, namely mapping upon a circular or radial slit domain, determine the two functions only up to a constant factor. We determined this factor by the requirement that the residue at the pole $z=v$ be one. This normalization was quite natural in connection with the extremum problems of §2, which made a distinction between u and v. In our formal transformations of §§3 and 4, however, we made use only of the equations (16), and not of the requirements at the point $z=v$. From the normalization (90), which fixed an arbitrary additive constant without distinction of special points, and, in particular, from (105), we recognize that a new type of normalization is preferable and will lead to much simpler expressions for Neumann's function. In fact, we shall add to $P(z;u,v)$ and $Q(z;u,v)$ constant terms with respect to z which depend, however, on u and v such that

(106)
$$\oint_C P(z;u,v)ds_z = \oint_C Q(z;u,v)ds_z \equiv 0 \qquad \text{for every } u, v \in D.$$

We denote the renormalized functions again by $P(z;u,v)$ and $Q(z;u,v)$ in order to avoid an excessive number of letters; from now on P and Q will denote only the fundamental domain functions with the new normalization (106).

It is clear that the new functions P and Q satisfy again equations (16), but with new constants $k_\nu(u,v)$. Because of (106) we derive from (16) for the $k_\nu(u,v)$ the relation

(107)
$$\sum_{\nu=1}^{n} L_\nu k_\nu(u,v) \equiv 0, \qquad\qquad u, v \in D.$$

Returning to (105), which was proved from (16) and remains valid for the new functions P and Q, we find by virtue of (106) and (107)

(108)
$$\pi(v;\zeta) - \pi(u;\zeta) = Q(\zeta;u,v) - (P(\zeta;u,v))^+.$$

By means of (107) we are further able to invert the system of equations (75) for the $k_\mu(u,v)$ and to express these terms by means of $w_\nu(u) - w_\nu(v)$. In fact, taking the first $(n-1)$ equations (75) and eliminating $k_n(u,v)$ from them by means of (107) we find all other $k_\mu(u, v)$ as linear combinations of $w_\nu(u) - w_\nu(v)$. This shows that every $k_\mu(u, v)$ is a multiple-valued analytic function of u and v.

Finally, let us solve the equations (103) and (108) with respect to $Q(\zeta; u, v)$ and $(P(\zeta; u, v))^+$; we obtain

$$Q(\zeta; u, v) = \frac{1}{2} \left[\{ p(v; \zeta) - p(u; \zeta) \} + \{ \pi(v; \zeta) - \pi(u; \zeta) \} \right]$$

(109)

$$+ \frac{1}{2} \sum_{\mu=1}^{n} k_\mu(u, v) \omega_\mu(\zeta),$$

$$(P(\zeta; u, v))^+ = \frac{1}{2} \left[\{ p(v; \zeta) - p(u; \zeta) \} - \{ \pi(v; \zeta) - \pi(u; \zeta) \} \right]$$

(110)

$$+ \frac{1}{2} \sum_{\mu=1}^{n} k_\mu(u, v) \omega_\mu(\zeta).$$

These equations show that $Q(\zeta; u, v)$ is analytic in ζ, u and v and satisfies the symmetry law

(111)
$$Q(\zeta; u, v) = - Q(\zeta; v, u).$$

$P(\zeta; u, v)$ is analytic in ζ and anti-analytic in u, v and has the symmetry rule

(112)
$$P(\zeta; u, v) = - P(\zeta; v, u).$$

The analytic behavior of $P(z; u, v)$ in dependance on its arguments might also be understood from the following considerations: We had in §1 to restrict the class Ω in order to exclude from this class all nonvanishing constants. We did this by distinguishing a point $v \in D$ and requiring the vanishing of all functions of our subclass Ω_v at v. Our new normalization for P and Q leads naturally to a new subclass of Ω. Let Ω_1 be the class of all functions $f(z) \in \Omega$ which satisfy the condition

(113)
$$\oint_C f(z) ds = 0.$$

We have, however, to give a meaning to the integral (113) for functions which are not defined on C. This is possible by the remark that for every function $f(z)$ which is continuously differentiable in $D+C$ we have in view of (101')

$$(f, \pi^+) = - \frac{1}{2i} \oint_C f(z) (d\pi(z; \zeta))^+$$

(114)

$$= - \frac{1}{2i} \oint_C f(z) \pi'(z; \zeta) dz - \frac{2\pi}{L} \oint_C f(z) ds,$$

whence by virtue of the residue theorem

$$(114') \qquad \oint_C f(z)ds = -\frac{L}{2\pi}(f(z),(\pi(z;\zeta))^+) + \frac{L}{2}f(\zeta).$$

The right-hand side of (114') is defined for every function $f(z) \in \Omega$, since it may be expressed by an improper integral extended over D. This permits an interpretation of the condition (113) for every function $f(z) \in \Omega$ and a unique definition for the class Ω_1. It is obvious that the only constant in this class is $f(z) \equiv 0$. We introduce now the metric (6) or (7) into this class and remark that $(f, f^+) = 0$ implies, in Ω_1, $f(z) \equiv 0$.

The function $P(z; u, v)$ belongs to Ω_1 because of its normalization (106). It satisfies still the relation (14) for every $f(z) \in \Omega_1$. We introduce now a complete orthonormal system $\{f_\nu(z)\}$ for the class Ω_1, which is possible because of our exclusion of constants; developing $P(z; u, v)$ into a Fourier series with respect to this system, we obtain because of (14)

$$(115) \qquad P(z; u, v) = \pi \sum_{\nu=1}^{\infty} f_\nu(z)[(f_\nu(u))^+ - (f_\nu(v))^+].$$

This series converges uniformly in each closed subdomain of D; it shows that P is analytic in z and anti-analytic in u and v. The symmetry law (112) is also evident from (115).

Finally, we derive from (103) and (108) two remarkable identities. Differentiating (103) and (108) with respect to v and ζ and comparing the results, we find

$$(116) \qquad \frac{\partial^2 p}{\partial v \partial \zeta} = -\frac{\partial^2 \pi}{\partial v \partial \zeta} - \frac{1}{2}\sum_{\mu=1}^{n} \frac{\partial}{\partial v} k_\mu(u, v) w_\mu'(\zeta),$$

while the same reasoning after differentiation with respect to v and ζ^+ yields

$$(117) \qquad \frac{\partial^2 p}{\partial v \partial \zeta^+} = \frac{\partial^2 \pi}{\partial v \partial \zeta^+} - \frac{1}{2}\sum_{\mu=1}^{n} \frac{\partial}{\partial v} k_\mu(u, v)(w_\mu'(\zeta))^+.$$

These two relations between the second derivatives of Green's and Neumann's functions play a role in the theory of the kernel function of an orthonormal system [2].

It is perhaps worth remarking at this point that many of our identities obtained by contour integration and the boundary relation (16) can be derived directly by inspection of the geometric character of the mappings involved. The relations (116) and (117) offer a good opportunity of demonstrating this fact. Indeed, let $\zeta = \sigma + i\tau$. The functions

$$(99') \qquad \alpha(v; \zeta) = p(v; \zeta) + \sum_{\mu=1}^{n} k_\mu(u, v)\omega_\mu(\zeta),$$

$$(99'') \qquad \beta(v; \zeta) = \pi(v; \zeta).$$

have fixed periods about each C_ν which are independent of v, ζ. Hence, the derivatives with respect to σ and τ will have no periods. Furthermore, since $\text{Re} \{\alpha(v, \zeta)\}$ and $\text{Im} \{\beta(v, \zeta)\}$ have at the worst variations along the C_ν which are independent of ζ, the derivatives will have for α constant real part and for β constant imaginary part on each C_ν. Thus the partial derivatives $\partial\alpha/\partial\sigma$ and $-i\partial\beta/\partial\tau$ map D on domains bounded by vertical slits, and $-i\partial\alpha/\partial\tau$ and $\partial\beta/\partial\sigma$ map D on domains bounded by horizontal slits. We see, then, by the cancelation of the poles, that

$$(116') \qquad \frac{\partial p}{\partial\sigma} + \sum_{\mu=1}^{n} k_\mu(u, v) \frac{\partial\omega_\mu}{\partial\sigma} + i \frac{\partial\pi}{\partial\tau} = \text{const.},$$

$$(117') \qquad \frac{\partial p}{\partial\tau} + \sum_{\mu=1}^{n} k_\mu(u, v) \frac{\partial\omega_\mu}{\partial\tau} - i \frac{\partial\pi}{\partial\sigma} = \text{const.}$$

But these relations imply (116), (117), upon differentiation with respect to v. This method often gives insight and shorter proofs, but because of its unsystematic nature we do not investigate it further, with one exception at the end of §6.

Now, we have to consider an appropriate normalization for the functions $M(z; u)$ and $N(z; u)$, which were defined in §3 up to an additive constant. We require again the normalization

$$(106') \qquad \oint_C M(z; u)ds_z = \oint_C N(z; u)ds_z \equiv 0, \qquad \text{for every } u \in D,$$

which implies in view of (57)

$$(107') \qquad \sum_{\nu=1}^{n} L_\nu l_\nu(u) \equiv 0, \qquad\qquad u \in D.$$

The first advantage of this normalization is the fact that the $l_\nu(u)$'s, which were until now defined by (78) only up to an additive constant, are defined in a unique way and become linear combinations of the $w'_\nu(u)$, that is, analytic functions of u. Next, we may apply to $M(z; u)$, $N(z; u)$, $p(z; \zeta)$ and $\pi(z; \zeta)$ the method of contour integration and perform the same transformations as we did before with $P(z; u, v)$ and $Q(z; u, v)$. We obtain finally

$$(103') \qquad p'(z; \zeta) = N(\zeta; z) - (M(\zeta; z))^+ + \sum_{\nu=1}^{n} l_\nu(z)\omega_\nu(\zeta)$$

$$(108') \qquad \pi'(z; \zeta) = N(\zeta; z) + (M(\zeta; z))^+. \quad .$$

Solving these equations with respect to $N(\zeta; z)$ and $(M(\zeta; z))^+$, we obtain

$$(109') \qquad N(\zeta; z) = \frac{1}{2} [\pi'(z; \zeta) + p'(z; \zeta)] - \frac{1}{2} \sum_{\nu=1}^{n} l_\nu(z)\omega_\nu(\zeta),$$

$$(110') \qquad (M(\zeta; z))^+ = \frac{1}{2} \left[\pi'(z; \zeta) - p'(z; \zeta) \right] + \frac{1}{2} \sum_{\nu=1}^{n} l_\nu(z) \omega_\nu(\zeta),$$

which shows that because of our normalization $N(\zeta; z)$ is analytic in both arguments and $M(\zeta; z)$ is analytic in ζ and anti-analytic in z.

In summary, we have

$$(61) \qquad Q'(a; u, v) = N(v; a) - N(u; a),$$

$$(63) \qquad (P'(a; u, v))^+ = M(u; a) - M(v; a),$$

$$(65) \qquad M'(u; v) = (M'(v; u))^+,$$

$$(67) \qquad N'(u; v) = N'(v; u),$$

$$(75) \qquad w_\nu(u) - w_\nu(v) = \sum_{\mu=1}^{n} k_\mu(u, v) P_{\mu\nu}, \qquad\qquad \nu = 1, 2, \cdots, n,$$

$$(78) \qquad w_\nu'(u) = \sum_{\mu=1}^{n} l_\mu(u) P_{\mu\nu}, \qquad\qquad \nu = 1, 2, \cdots, n,$$

$$(103) \quad p(v; \zeta) - p(u; \zeta) = Q(\zeta; u, v) + (P(\zeta; u, v))^+ - \sum_{\mu=1}^{n} k_\mu(u, v) \omega_\mu(\zeta),$$

$$(108) \quad \pi(v; \zeta) - \pi(u; \zeta) = Q(\zeta; u, v) - (P(\zeta; u, v))^+,$$

$$(109) \quad \begin{aligned} Q(\zeta; u, v) &= \frac{1}{2} \left[\{ p(v; \zeta) - p(u; \zeta) \} + \{ \pi(v; \zeta) - \pi(u; \zeta) \} \right] \\ &\quad + \frac{1}{2} \sum_{\mu=1}^{n} k_\mu(u, v) \omega_\mu(\zeta), \end{aligned}$$

$$(110) \quad \begin{aligned} (P(\zeta; u, v))^+ &= \frac{1}{2} \left[\{ p(v; \zeta) - p(u; \zeta) \} - \{ \pi(v; \zeta) - \pi(u; \zeta) \} \right] \\ &\quad + \frac{1}{2} \sum_{\mu=1}^{n} k_\mu(u, v) \omega_\mu(\zeta), \end{aligned}$$

$$(109') \quad N(\zeta; z) = \frac{1}{2} \left[\pi'(z; \zeta) + p'(z; \zeta) \right] - \frac{1}{2} \sum_{\mu=1}^{n} l_\mu(z) \omega_\mu(\zeta),$$

$$(110') \quad (M(\zeta; z))^+ = \frac{1}{2} \left[\pi'(z; \zeta) - p'(z; \zeta) \right] + \frac{1}{2} \sum_{\mu=1}^{n} l_\mu(z) \omega_\mu(\zeta),$$

$$(115) \quad P(z; u, v) = \pi \sum_{\nu=1}^{\infty} f_\nu(z) \left[(f_\nu(u))^+ - (f_\nu(v))^+ \right].$$

5. **Schottky functions and related classes.** Schottky [15] was the first to consider the family $\Re$ of all functions $f(z)$ which are single-valued and meromorphic in D and have real boundary values on C. He developed an interest-

ing theory of conformal mapping of multiply-connected domains from the properties of this family and established by means of it the relation of this theory with the theory of closed Riemann surfaces. It is evident that functions $f(z) \in \mathfrak{R}$ are very useful in the method of contour integration.

In fact, let $f(z)$ be an arbitrary function of this family. We assume, for the sake of simplicity, that $f(z)$ has only simple poles. Let $z_\nu \ (\nu = 1, \cdots, N)$ be the coordinates of the poles and $r_\nu \ (\nu = 1, 2, \cdots, N)$ the corresponding residues. Consider now the integral

$$(118) \qquad \frac{1}{2\pi i} \oint_C P(z; u, v) f'(z) dz = - \sum_{\nu=1}^{N} r_\nu P'(z_\nu; u, v).$$

Using (16) and the fact that $f'(z)dz$ is real on C, we obtain

$$\sum_{\nu=1}^{N} r_\nu^+ (P'(z_\nu; u, v))^+ = - \frac{1}{2\pi i} \oint_C Q(z; u, v) f'(z) dz$$

$$(119)$$

$$= \frac{1}{2\pi i} \oint_C f(z) Q'(z; u, v) dz.$$

By means of the residue theorem we deduce from (119)

$$(120) \qquad \sum_{\nu=1}^{N} r_\nu^+ (P'(z_\nu; u, v))^+ = \sum_{\nu=1}^{N} r_\nu Q'(z_\nu; u, v) + f(u) - f(v).$$

Applying finally the formulas (61) and (63) we arrive at

$$f(v) - f(u) = \sum_{\nu=1}^{N} [r_\nu N(v; z_\nu) + r_\nu^+ M(v; z_\nu)]$$

$$(121)$$

$$- \sum_{\nu=1}^{N} [r_\nu N(u; z_\nu) + r_\nu^+ M(u; z_\nu)].$$

This proves the following theorem:

THEOREM V. *Every function $f(z) \in R$ may be developed in the form*

$$(122) \qquad f(z) = A + \sum_{\nu=1}^{N} [r_\nu N(z; z_\nu) + r_\nu^+ M(z; z_\nu)].$$

We have now to establish the additional conditions in order that an expression (122) be in fact a Schottky function. Obviously, the right-hand side represents a single-valued meromorphic function of $z \in D$ with simple poles at the z_ν. If z lies on the boundary continuum C_ρ we have by virtue of (57)

$$(123) \qquad f(z) = A + \sum_{\nu=1}^{N} [r_\nu N(z; z_\nu) + r_\nu^+ (N(z; z_\nu))^+] + \sum_{\nu=1}^{N} r_\nu^+ (l_\rho(z_\nu))^+.$$

If we require now that $f(z)$ be real on each C_ρ, we obtain the conditions

$$(124) \qquad \text{Im}\ \left\{ A + \sum_{\nu=1}^{N} r_\nu^+ (l_\rho(z_\nu))^+ \right\} = 0, \qquad \rho = 1, 2, \cdots, n.$$

Multiplying the ρth equation (124) with $P_{\rho\mu}$ and summing over ρ from 1 to n, we obtain in view of (78) and (87')

$$(125) \qquad \text{Im}\ \left\{ \sum_{\nu=1}^{N} r_\nu w_\mu'(z_\nu) \right\} = 0, \qquad \mu = 1, 2, \cdots, n.$$

One recognizes easily that every function (122) for which (125) is fulfilled is of the family $\Re$ under proper choice of the constant A. The similarity of the conditions (125) and an analogous result of Abel for single-valued functions on closed Riemann surfaces is evident.

Another family of functions which is closely connected with $\Re$ may be defined as follows. $\mathfrak{E}$ is the family of all functions $E(z)$ which are regular in D and which satisfy on the boundary C of D the condition

$$(126) \qquad |E(z)|^2 = E(z)(E(z))^+ = 1.$$

It is obvious that for every $E(z) \in \mathfrak{E}$ the expression $E(z) + E(z)^{-1}$ is an element of $\Re$.

We obtain now for every function $E(z) \in \mathfrak{E}$ a simple representation in terms of the functions $M(z; u)$ and $N(z; u)$ defined in §3. We assume, for the sake of simplicity, that $E(z)$ has only simple zero-points in D and that at each such point n_ν ($\nu = 1, 2, \cdots, N$) one has

$$(127) \qquad \lim_{z \to n_\nu} (z - n_\nu)^{-1} E(z) = r_\nu^{-1}, \qquad \nu = 1, 2, \cdots, N.$$

Because of (122), we have in view of (127)

$$(128) \qquad E(z) + E(z)^{-1} = A + \sum_{\nu=1}^{N} [r_\nu N(z; n_\nu) + r_\nu^+ M(z; n_\nu)]$$

with the additional condition, derived from (125),

$$(128') \qquad \text{Im}\ \left\{ \sum_{\nu=1}^{N} r_\nu w_\mu'(n_\nu) \right\} = 0, \qquad \mu = 1, 2, \cdots, n.$$

Since further the function $i[E(z) - E(z)^{-1}]$ is again of class $\Re$, we derive from (122), (127) and (125) the formulas

$$(129) \qquad E(z) - E(z)^{-1} = B - \sum_{\nu=1}^{N} [r_\nu N(z; n_\nu) - r_\nu^+ M(z; n_\nu)],$$

$$(130) \qquad \text{Re}\ \left\{ \sum_{\nu=1}^{N} r_\nu w_\mu'(n_\nu) \right\} = 0, \qquad \mu = 1, 2, \cdots, n.$$

Combining (128) and (129), we find the following result:

THEOREM VI. *Let $E(z)$ be a function of the class $\mathfrak{E}$ with simple zeros at the points n_ν and let r_ν be the residues of $E(z)^{-1}$ at the corresponding poles. Then we have the representations*:

$$(131) \qquad E(z) = A_1 + \sum_{\nu=1}^{N} r_\nu^+ M(z; n_\nu),$$

$$(132) \qquad E(z)^{-1} = A_2 + \sum_{\nu=1}^{N} r_\nu N(z; n_\nu),$$

and the zeros and residues are connected by the equations

$$(133) \qquad \sum_{\nu=1}^{N} r_\nu w_\mu'(n_\nu) = 0, \qquad\qquad \mu = 1, 2, \cdots, n.$$

These results might also have been proved directly without using the class $\mathfrak{R}$ by applying the method of contour integration and using the fact that because of (126) one has, on C, $(E(z))^+ = E(z)^{-1}$.

Since the function $E(z)$ vanishes at all points n_μ, we derive from (131) the $(N-1)$ equations

$$(134) \qquad \sum_{\nu=1}^{N} r_\nu^+ M(n_\mu; n_\nu) = \sum_{\nu=1}^{N} r_\nu^+ M(n_1; n_\nu), \qquad \mu = 2, 3, \cdots, N.$$

We want now to verify that if N points n_ν and N numbers r_ν can be found such that the $N+n-2$ equations (133) and (134) are satisfied then it is possible to determine constants A_1 and A_2 such that the function (131) is proportional to a function of class $\mathfrak{E}$ and that (132) is proportional to its inverse. In fact, construct formally two functions $E(z)$ and $E_1(z)$ which equal the right-hand sides of (131) and (132). In view of (134), we may choose A_1 in such a way that $E(z)$ vanishes at all points n_ν. Hence, the product $E(z) \cdot E_1(z)$ is regular in D, since the simple poles of $E_1(z)$ cancel against the zeros of $E(z)$. For $z \in C_\rho$ we have by virtue of (57)

$$(135) \qquad \begin{aligned} E(z) \cdot E_1(z) = &\left[A_1 + \sum_{\nu=1}^{N} r_\nu^+ (l_\rho(n_\nu))^+ + \sum_{\nu=1}^{N} r_\nu^+ (N(z; n_\nu))^+ \right] \\ &\cdot \left[A_2 + \sum_{\nu=1}^{N} r_\nu N(z; n_\nu) \right]. \end{aligned}$$

Now, we deduce from the character of the matrix $(P_{\mu\nu})$ and (133) and (78)

$$(135') \qquad \sum_{\nu=1}^{N} r_\nu^+ (l_\rho(n_\nu))^+ = \sum_{\nu=1}^{N} r_\nu^+ (l_1(n_\nu))^+, \qquad \rho = 2, 3, \cdots, n.$$

If we choose now

$$(135'')\qquad\qquad A_2 = A_1^+ + \sum_{\nu=1}^{N} r_\nu l_1(n_\nu)$$

we have non-negative boundary values for $E(z)E_1(z)$ on C. But since the only regular functions $f(z)\in\Re$ are real constants, we find $E(z)\cdot E_1(z)=k$, where k is a positive constant. This proves our assertion; we may always multiply the r_ν's by a common factor so that $E(z)$ becomes of class $\mathfrak{E}$.

If we introduce into the identity $E(z)\cdot E(z)^{-1}=1$ the expressions (131) and (132) and if we compute the product at a point n_μ, we find easily

$$(136)\qquad\qquad r_\mu \sum_{\nu=1}^{N} r_\nu^+ M'(n_\mu; n_\nu) = 1, \qquad\qquad \mu = 1, 2, \cdots, N.$$

This system of equations for the r_ν's must necessarily be a consequence of (133) and (134). Numerous other identities of similar form may be easily obtained. We shall restrict ourselves here to one example.

We may eliminate the constant A_1 in (131) by using the fact that $E(n_1)=0$, and this gives to (131) the form

$$(131')\qquad\qquad E(z) = \sum_{\nu=1}^{N} r_\nu^+ [M(z; n_\nu) - M(n_1; n_\nu)],$$

which is in view of (63) equivalent to

$$(131'')\qquad\qquad E(z) = \sum_{\nu=1}^{N} r_\nu^+ (P'(n_\nu; z, n_1))^+.$$

This is another standard representation for the class $\mathfrak{E}$ in which the constant term has been determined. Let now

$$(137)\qquad\qquad F(z) = \sum_{\mu=1}^{M} s_\mu^+ (P'(m_\mu; z, m_1))^+$$

be another function of class $\mathfrak{E}$. For the sake of simplicity, we assume that no zero of $E(z)$ coincides with one of $F(z)$. The function $E(z)\cdot F(z)$ is again of the same class $\mathfrak{E}$. Hence, multiplying on the one hand (131'') with (137) and using, on the other hand, the representation (131'') for the product, we obtain

$$(138)\qquad \begin{aligned} &\sum_{\mu=1}^{M}\sum_{\nu=1}^{N} r_\nu^+ s_\mu^+ (P'(n_\nu; z, n_1))^+ (P'(m_\mu; z, m_1))^+ \\ &= \sum_{\nu=1}^{N} r_\nu^+ (F(n_\nu)^{-1})^+ (P'(n_\nu; z, n_1))^+ + \sum_{\mu=1}^{M} s_\mu^+ (E(m_\mu)^{-1})^+ (P'(m_\mu; z, n_1))^+, \end{aligned}$$

an interesting relation between the left-hand second order expressions of P' and the linear terms on the right-hand side. Using the above procedure with respect to the product $E(z)^{-1}\cdot F(z)^{-1}$ gives an analogous identity in Q'.

We consider now the particular function (43) of class $\mathfrak{E}$ and apply to it the representation formula (131''). We obtain

$$(139) \qquad P'(z; u, v)Q'(z; u, v)^{-1} = \sum_{\nu=1}^{2n} r_\nu^+(P'(n_\nu; z, u))^+,$$

where the summation is extended over all $2n$ zeros n_ν of the left-hand expression and it is to be noted that $z=u$ and $z=v$ are zeros of this function. This formula shows that $Q'(z; u, v)$ may be expressed by means of $P'(z; u, v)$ and $P'(n_\nu; z, u)$. This result is of interest because $P(z; u, v)$ is a kernel function and, therefore, numerically easy to handle. The determination of the n_ν and r_ν will, however, be in general so difficult that little immediate use of (139) is to be expected.

Let us consider, further, the matrix $(M'(z_\nu; z_\mu))$ which occurred in the equation (136) for the residues r_ν. Because of (65) this matrix is Hermitian. In order to study it in more detail, we remark that for every function $f(z)$ which is single-valued and continuously differentiable in $D+C$ we have because of (57)

$$(140) \qquad \begin{aligned} (f(z), (M(z; u))^+) &= \frac{1}{2i}\oint_C f'(z)(M(z; u))^+dz \\ &= \frac{1}{2i}\oint_C f'(z)N(z; u)dz = \pi f'(u). \end{aligned}$$

By the usual reasoning we conclude for every function $f(z) \in \Omega$

$$(140') \qquad (f(z), (M(z; u))^+) = \pi f'(u).$$

In particular, we derive from (140') for every function of the type

$$(141) \qquad f(z) = A + \sum_{\nu=1}^{N} \lambda_\nu M(z; n_\nu),$$

which is obviously of the class Ω, the relation

$$(142) \qquad (f, f^+) = \int\int_D |f'(z)|^2 dxdy = \pi \sum_{\mu,\nu=1}^{N} \lambda_\nu \lambda_\mu^+ M'(n_\nu; n_\mu).$$

This shows that the Hermitian form with matrix $(M'(n_\nu; n_\mu))$ is positive-definite. If we choose in particular as $f(z)$ a function $E(z) \in \mathfrak{E}$, we find

$$(142') \qquad (E, E^+) = \int\int_D |E'(z)|^2 dxdy = \sum_{\mu,\nu=1}^{N} r_\nu r_\mu^+ M'(n_\nu; n_\mu) \cdot \pi.$$

We find from (136) that the value of this sum is $N \cdot \pi$. This result is easily understood; each function $E(z)$ maps D upon the unit circle covered N times and the area of this image domain is exactly $N \cdot \pi$.

There is another interesting representation for all functions $f(z) \in \mathfrak{R}$. Let

$$(143) \qquad w'(z) = \sum_{\nu=1}^{n} \xi_\nu w_\nu'(z), \qquad\qquad \xi_\nu \text{ real},$$

be an arbitrary linear combination of derivatives of harmonic measures with real coefficients ξ_ν. If $f(z) \in \mathfrak{R}$ with N simple poles z_ν and corresponding residues r_ν is given, consider the function $f(z)w'(z)$. We have, evidently, on C

$$(144) \qquad (f(z)w'(z)dz)^+ = - f(z)w'(z)dz, \qquad z \text{ varies on } C.$$

Consider now the equation

$$(145) \qquad \frac{1}{2\pi i} \oint_C M(z; u) f(z) w'(z) dz = \sum_{\nu=1}^{N} r_\nu w'(z_\nu) M(z_\nu; u).$$

Because of (57) and (144), we obtain

$$(146) \qquad \sum_{\nu=1}^{N} (r_\nu w'(z_\nu) M(z_\nu; u))^+ = \frac{1}{2\pi i} \sum_{\mu=1}^{n} \oint_{C_\mu} [N(z; u) + l_\mu(u)] f(z) w'(z) dz,$$

whence by means of the residue theorem

$$(146') \qquad \begin{aligned} f(u)w'(u) = &- \sum_{\nu=1}^{N} r_\nu w'(z_\nu) N(z_\nu; u) + \sum_{\nu=1}^{N} r_\nu^+ (w'(z_\nu) M(z_\nu; u))^+ \\ &- \sum_{\mu=1}^{n} l_\mu(u) a_\mu, \end{aligned}$$

where

$$(146'') \qquad a_\mu = \frac{1}{2\pi i} \oint_{C_\mu} f(z) w'(z) dz, \qquad \mu = 1, 2, \cdots, n,$$

are, because of (144), real constants. Considering, finally, (78) we find:

THEOREM VII. *Every function* $f(z) \in \mathfrak{R}$ *may be represented in the form*

$$(147) \quad f(z)w'(z) = \sum_{\nu=1}^{N} [r_\nu^+ (w'(z_\nu))^+ (M(z_\nu; z))^+ - r_\nu w'(z_\nu) N(z_\nu; z)] + \sum_{\mu=1}^{n} \lambda_\mu w_\mu'(z),$$

with real coefficients λ_μ.

We make the following application of this result:

Take an arbitrary function $w'(z)$ of the type (143); then we have in

$$(148) \qquad f(z) = w_i'(z)/w'(z)$$

a function of the class $\mathfrak{R}$. Suppose that all zeros z_ν of $w'(z)$ are simple; it follows easily from the argument principle that there are exactly $n-2$ of them in

D. We assume in the following considerations $n > 2$ so that the existence of zeros z_ν is assured. The corresponding residues of $f(z)$ are

$$(148') \qquad\qquad r_\nu = w_i'(z_\nu)/w''(z_\nu).$$

We apply now formula (147) with the function $w'(z)$ replaced there by $w_k'(z)$; we have

$$(149) \quad \frac{w_i'(z)\,w_k'(z)}{w'(z)} - \sum_{\mu=1}^{n} \lambda_{ik,\mu} w_\mu'(z) = \sum_{\nu=1}^{n-2} \left(\left[\frac{w_i'(z_\nu)\,w_k'(z_\nu)}{w''(z_\nu)}\right]\right)^+ (M(z_\nu;z))^+ \\ - \sum_{\nu=1}^{n-2} \frac{w_i'(z_\nu)\,w_k'(z_\nu)}{w''(z_\nu)} N(z_\nu;z).$$

Since there are $n-1$ linearly independent $w_i'(z)$, we may write down $n(n-1)/2$ equations (149) for a given denominator $w'(z)$. We may consider the $2n-4$ terms $M(z_\nu;z)$ and $N(z_\nu;z)$ as unknowns and the equations (149) as a system for their determination. The number of linearly independent equations (149) is of great importance from this point of view.

First we conclude from (149) that there are at most $3n-6$ linearly independent terms $w_i'(z)w_k'(z)$. In fact, choose $w'(z)$ in (149) as $w_3'(z)$. Take all equations (149) with $w_1'(z)w_k'(z)/w_3'(z)$ and $w_2'(z)w_k'(z)/w_3'(z)$ ($k=1,2,4,$ $\cdots, n-1$). These are exactly $2n-5$ equations. If now another equation (149) of the form $w_i'(z)w_k'(z)/w_3'(z)$ is considered with $i, k \neq 1, 2, 3$, we have $2n-4$ equations with $2n-4$ unknowns. Further, applying condition (125) to our functions $f(z)$ we have in each case

$$(150) \qquad \sum_{\nu=1}^{n-2} \left(\left[\frac{w_i'(z_\nu)\,w_k'(z_\nu)}{w''(z_\nu)}\right]\right)^+ - \sum_{\nu=1}^{n-2} \frac{w_i'(z_\nu)\,w_k'(z_\nu)}{w''(z_\nu)} = 0.$$

This shows that the rank of the matrix of our $2n-4$ equations is at most $2n-5$. There exist, therefore, constants such that

$$(151) \quad \alpha w_i'(z)w_k'(z) + \sum_{\nu=1}^{n-1} \beta_\nu w_1'(z)w_\nu'(z) + \sum_{\nu=1}^{n-1} \gamma_\nu w_2'(z)w_\nu'(z) \\ + \sum_{\nu=1}^{n-1} \delta_\nu w_3'(z)w_\nu'(z) = 0.$$

In the same fashion we show easily that each set of $3n-5$ terms $w_i'(z)w_k'(z)$ is linearly dependent and that all terms may be obtained as linear combinations of at most $3n-6$ basic independent products of this form.

THEOREM VIII. *Among the expressions $w_i'(z)w_k'(z)$ there are at most $3n-6$ linearly independent ones.*

Let us assume now that we can find $2n-5$ combinations of indices i and k for which the equations (149) are linearly independent. This assumption is in general fulfilled, as follows from a theorem of Noether (cf. [8, pp. 502–527]) in the theory of algebraic integrals. If we want to solve the system of equations, we have first to take notice of the relations (150) between the coefficients of this system. Since by virtue of the residue theorem

$$(152) \qquad \sum_{\nu=1}^{n-2} \frac{w_i'(z_\nu) w_k'(z_\nu)}{w''(z_\nu)} = \frac{1}{2\pi i} \oint_C \frac{w_i'(z) w_k'(z)}{w'(z)}\, dz,$$

we may put the equations (149) in the form

$$
\begin{aligned}
(149') \qquad & \frac{w_i'(z) w_k'(z)}{w'(z)} - \sum_{\mu=1}^{n} \lambda_{ik,\mu} w_\mu'(z) \\
&= \sum_{\nu=2}^{n-2} \left(\left[\frac{w_i'(z_\nu) w_k'(z_\nu)}{w''(z_\nu)} \right] \right)^+ \left((M(z_\nu; z))^+ - (M(z_1; z))^+ \right) \\
&\quad - \sum_{\nu=2}^{n-2} \frac{w_i'(z_\nu) w_k'(z_\nu)}{w''(z_\nu)} \left(N(z_\nu; z) - N(z_1; z) \right) \\
&\quad - (N(z_1; z) - (M(z_1; z))^+) \cdot \frac{1}{2\pi i} \oint_C \frac{w_i' w_k'}{w'}\, dz.
\end{aligned}
$$

Applying finally (103') we find

$$
\begin{aligned}
(153) \qquad & \frac{w_i'(z) w_k'(z)}{w'(z)} - \sum_{\mu=1}^{n} \lambda_{ik,\mu} w_\mu'(z) - \sum_{\mu=1}^{n} l_\mu(z)\omega_\mu(z_1) \cdot \frac{1}{2\pi i} \oint_C \frac{w_i w_k'}{w'}\, dz \\
&= \sum_{\nu=2}^{n-2} \left(\left[\frac{w_i'(z_\nu) w_k'(z_\nu)}{w''(z_\nu)} \right] \right)^+ \left((M(z_\nu; z))^+ - (M(z_1; z))^+ \right) \\
&\quad - \sum_{\nu=2}^{n-2} \frac{w_i'(z_\nu) w_k'(z_\nu)}{w''(z_\nu)} \left(N(z_\nu; z) - N(z_1; z) \right) \\
&\quad - \frac{p'(z; z_1)}{2\pi i} \oint_C \frac{w_i' w_k'}{w'}\, dz.
\end{aligned}
$$

From this system of equations in $2n-5$ unknowns with nonvanishing determinant we may compute the unknowns. We find:

THEOREM IX. *The functions* $p'(z; z_1)$, $\{N(z_\nu; z) - N(z_1; z)\}$ *and* $\{(M(z_\nu; z))^+ - (M(z_1; z))^+\}$ *are rational functions of the functions* $w_i'(z)$ $(i=1, \cdots, n-1)$ *with coefficients depending on* z_1.

For example, we have

$$(153') \qquad p'(z; z_1) = \frac{\sum a_{ik}(z_1) w_i'(z) w_k'(z)}{\sum_{\nu=1}^{n} \xi_\nu w_\nu'(z)},$$

where the summation in the numerator is to be extended over any basic system of the $w_i'(z)w_k'(z)$. The coefficients ξ_ν in the denominator depend also on z_1, since the denominator has to vanish at this point. We might choose as such a denominator the expression

$$(154) \qquad w'(z) = \frac{1}{i} \begin{vmatrix} w_1'(z) & w_2'(z) & w_3'(z) \\ w_1'(z_1) & w_2'(z_1) & w_3'(z_1) \\ (w_1'(z_1))^+ & (w_2'(z_1))^+ & (w_3'(z_1))^+ \end{vmatrix},$$

if it is not identically zero, which occurs only at exceptional points z_1.

It is possible to obtain additional information about the representation (153') in the following way. Introduce the differential operators

$$(155) \qquad \frac{\partial}{\partial z} = \frac{1}{2}\left(\frac{\partial}{\partial x} - i\frac{\partial}{\partial y}\right), \qquad \frac{\partial}{\partial z_1} = \frac{1}{2}\left(\frac{\partial}{\partial x_1} - i\frac{\partial}{\partial y_1}\right),$$

$$z = x + iy, \qquad\qquad\qquad z_1 = x_1 + iy_1.$$

Differentiating (153') with respect to z_1 and using the definition of $p'(z; z_1)$, we find

$$(156) \qquad \frac{\partial^2 g(z; z_1)}{\partial z \partial z_1} = \frac{\sum A_{\kappa\lambda\mu}(z_1)w_\kappa'(z)w_\lambda'(z)w_\mu'(z)}{\left[\displaystyle\sum_{\nu=1}^{n} \xi_\nu(z_1)w_\nu'(z)\right]^2}.$$

But this expression is symmetric with respect to z and z_1; hence we have here a rational expression in $w_\nu'(z)$ and $w_\nu'(z_1)$ for a second derivative of Green's function. We do not pursue these formal considerations; they show the importance of the method of contour integration for studying identities between domain functions. It appears that for the general domain D of connectivity $n > 2$ the derivatives of the harmonic measures are among the most important domain functions, since many others may be constructed rationally from them. The last results of this section are closely related to Noether's theorem [8] on algebraic integrals. In view of Schottky's theorem on the mapping of multiply-connected domains on Riemann surfaces they might even be derived directly from Noether's result. Our method leads to them directly by simple applications of Cauchy's theorem and seems to be more appropriate for the theory of conformal mapping.

6. **Theory of variation for the fundamental domain functions.** In this chapter the method of contour integration will be applied in order to study the variation of domain functions under an infinitesimal change of the domain D. We shall vary the domain D as follows; we choose a fixed point $z_0 \in D$ and consider the conformal mapping

$$(157) \qquad\qquad z^* = z + \frac{e^{i\phi}\rho^2}{z - z_0}, \qquad\qquad 0 \leq \phi < 2\pi, 0 < \rho.$$

This map is univalent in the domain $|z-z_0| > \rho$. If we choose ρ sufficiently small, the whole boundary C of D will lie in this domain of univalency and will, therefore, be mapped by (157) upon a set of n smooth curves C^*. These curves bound a new domain D^* which will be less different from D the smaller ρ is chosen. It is this domain D^* for which we want to compute the fundamental domain functions in terms of those belonging to D. We shall denote the domain functions of D^* by the same letters as those of D, but shall indicate their new domain by an asterisk. Thus $P^*(z; u, v)$ has with respect to D^* the same meaning as $P(z; u, v)$ has with respect to D.

In applying the method of contour integration, we shall have to eliminate from D a small circle of radius ϵ around z_0. We denote the circumference by γ_ϵ and we denote the remaining part of D by D_ϵ. By virtue of Cauchy's theorem, we have

$$
(158) \quad \frac{1}{2\pi i} \oint_C P^*(z^*(z); u^*, v^*) P'(z; a, b) dz
$$
$$
= \frac{1}{2\pi i} \oint_{\gamma_\epsilon} P^*(z^*(z); u^*, v^*) P'(z; a, b) dz,
$$

since $z^*(z)$ is a regular analytic function of z in D_ϵ. On γ_ϵ we may develop $P^*(z^*; u^*, v^*)$ into a series in powers of ρ^2/ϵ which will converge if $\epsilon \gg \rho^2$.

We find, in fact,

$$
P^*(z^*; u^*, v^*) = P^*(z; u^*, v^*) + \frac{e^{i\phi}\rho^2}{z - z_0} P^{*\prime}(z; u^*, v^*) + \cdots ;
$$
$$
(159)
$$
$$
P^{*\prime}(z; u^*, v^*) = \frac{d}{dz} P^*(z; u^*, v^*).
$$

Introducing this development into the right-hand integral of (158), remarking that $P^*(z; u^*, v^*)$ is regular in the interior of γ_ϵ and applying the residue theorem, we obtain

$$
(160) \quad \frac{1}{2\pi i} \oint_{\gamma_\epsilon} P^*(z^*(z); u^*, v^*) P'(z; a, b) dz
$$
$$
= e^{i\phi}\rho^2 P^{*\prime}(z_0; u^*, v^*) P'(z_0; a, b) + o(\rho^2),
$$

where $o(\rho^2)$ shall henceforth always denote a corrective term of higher order in ρ such that $\lim_{\rho \to 0} \rho^{-2} o(\rho^2) = 0$. It is obvious that (159) permits the computation of the higher terms in ρ, too; the remainder term may be estimated in terms of the values of $P^*(z; u^*, v^*)$ on γ_ϵ.

We apply now formula (16) for $P(z; a, b)$ and the corresponding relation for $P^*(z^*; u^*, v^*)$; in fact, if z lies on C, the varied point z^* will lie on C^*, so that both instances of (16) hold simultaneously. We find:

$$(161) \quad \frac{1}{2\pi i} \sum_{\mu=1}^{n} \oint_{C_\mu} [Q^*(z^*(z); u^*, v^*) - k_\mu^*(u^*, v^*)]Q'(z; a, b)dz$$

$$= - e^{-i\phi}\rho^2(P^{*\prime}(z_0; u^*, v^*)) + (P'(z_0; a, b))^+ + o(\rho^2).$$

On the other hand, we obtain from the residue theorem, applied to D_ϵ,

$$(162) \quad \frac{1}{2\pi i} \oint_c Q^*(z^*; u^*, v^*)Q'(z; a, b)dz$$
$$= Q^*(a^*; u^*, v^*) - Q^*(b^*; u^*, v^*) - Q(u; a, b) + Q(v; a, b)$$
$$+ \frac{1}{2\pi i} \oint_{\gamma_\epsilon} Q^*(z^*; u^*, v^*)Q'(z; a, b)dz.$$

From a development of $Q^*(z^*; u^*, v^*)$ on γ_ϵ such as we used in (159) for $P^*(z^*; u^*, v^*)$ we get

$$(163) \quad \frac{1}{2\pi i} \oint_{\gamma_\epsilon} Q^*(z^*; u^*, v^*)Q'(z; a, b)dz$$

$$= e^{i\phi}\rho^2 Q^{*\prime}(z_0; u^*, v^*)Q'(z_0; a, b) + o(\rho^2).$$

In view of the uniform convergence of $Q^*(z; u^*, v^*)$ and $P^*(z; u^*, v^*)$ to the functions $Q(z; u, v)$, $P(z; u, v)$ in each closed subdomain of D with $\rho \to 0$ and because of the symmetry rule (49), we obtain from (161), (162) and (163) the variation formula

$$(164) \quad \begin{aligned} &Q^*(u^*; a^*, b^*) - Q^*(v^*; a^*, b^*) = Q(u; a, b) - Q(v; a, b) \\ &- e^{i\phi}\rho^2 Q'(z_0; u, v)Q'(z_0; a, b) - e^{-i\phi}\rho^2(P'(z_0; u, v))^+(P'(z_0; a, b))^+ + o(\rho^2). \end{aligned}$$

This formula expresses the variation of the symmetric expression $Q(u; a, b)$ $-Q(v; a, b)$ in terms of the unvaried domain functions P and Q. One sees how the method of contour integration leads immediately to this important result.

Just in the same way, we obtain from the identities

$$(165) \quad \frac{1}{2\pi i} \oint_c Q^*(z^*; u^*, v^*)P'(z; a, b)dz$$

$$= P(v; a, b) - P(u; a, b) + \frac{1}{2\pi i} \oint_{\gamma_\epsilon} Q^*(z^*; u^*, v^*)P'(z; a, b)dz$$

and

$$(165') \quad \frac{1}{2\pi i} \oint_c P^*(z^*; u^*, v^*)Q'(z; a, b)dz$$

$$= P^*(a^*; u^*, v^*) - P^*(b^*; u^*, v^*) + \frac{1}{2\pi i} \oint_{\gamma_\epsilon} P^*(z^*; u^*, v^*)Q'(z; a, b)dz,$$

by use of the relations (16) on C the final result

(166)
$$P^*(u^*; a^*, b^*) - P^*(v^*; a^*, b^*) = P(u; a, b) - P(v; a, b)$$
$$- e^{i\phi}\rho^2 Q'(z_0; u, v)P'(z_0; a, b) - e^{-i\phi}\rho^2(P'(z_0; u, v)Q'(z_0; u, v))^+ + o(\rho^2).$$

Next we transform, by similar methods, the two identities

(167)
$$\frac{1}{2\pi i} \oint_C P^*(z^*; u^*, v^*)w'_\nu(z)dz = \frac{1}{2\pi i} \oint_{\gamma_\epsilon} P^*(z^*; u^*, v^*)w'_\nu(z)dz,$$

(167')
$$\frac{1}{2\pi i} \sum_{\mu=1}^n \oint_{C_\mu} [Q^*(z^*; u^*, v^*) - k_\mu^*(u^*, v^*)]w'_\nu(z)dz$$
$$= w_\nu(v) - w_\nu(u) + \sum_{\mu=1}^n k_\mu^*(u^*, v^*)P_{\mu\nu} + \frac{1}{2\pi i} \oint_{\gamma_\epsilon} Q^*(z^*; u^*, v^*)w'_\nu(z)dz.$$

Using series developments on γ_ϵ and the formulas (16) and (71), we obtain

(167'')
$$\sum_{\mu=1}^n k_\mu^*(u^*, v^*)P_{\mu\nu} = w_\nu(u) - w_\nu(v) - e^{i\phi}\rho^2 Q'(z_0; u, v)w'_\nu(z_0)$$
$$- e^{-i\phi}\rho^2(P'(z_0; u, v))^+(w'_\nu(z_0))^+ + o(\rho^2).$$

Because of (75) this yields

(168)
$$\sum_{\mu=1}^n [k_\mu^*(u^*, v^*) - k_\mu(u, v)]P_{\mu\nu} = - e^{i\phi}\rho^2 Q'(z_0; u, v)w'_\nu(z_0)$$
$$- e^{-i\phi}\rho^2(P'(z_0; u, v))^+(w'_\nu(z_0))^+ + o(\rho^2).$$

In view of the equations (78) and the character of the matrix $(P_{\mu\nu})$ this leads to

(168')
$$k_\mu^*(u^*, v^*) = k_\mu(u, v) - e^{i\phi}\rho^2 Q'(z_0; u, v)l_\mu(z_0)$$
$$- e^{-i\phi}\rho^2(P'(z_0; u, v))^+(l_\mu(z_0))^+ + A(z_0; u, v) + o(\rho^2),$$

where the function $A(z_0; u, v)$ does not depend on the index μ.

The $k_\mu(u, v)$ are by virtue of (75) closely related to the harmonic measures $\omega_\nu(z)$. The variation formulas for $\omega_\nu(z)$ and its periods $P_{\mu\nu}$ under variations (157) are well known. We have [12]

(169)
$$\omega_\nu^*(z^*) = \omega_\nu(z) + 2^{-1}e^{i\phi}\rho^2 p'(z_0; z)w'_\nu(z_0)$$
$$+ 2^{-1}e^{-i\phi}\rho^2(p'(z_0; z))^+(w'_\nu(z_0))^+ + o(\rho^2)$$

and

(170)
$$P_{\mu\nu}^* = P_{\mu\nu} + 2^{-1}e^{i\phi}\rho^2 w'_\nu(z_0)w'_\mu(z_0)$$
$$+ 2^{-1}e^{-i\phi}\rho^2(w'_\nu(z_0))^+(w'_\mu(z_0))^+ + o(\rho^2).$$

These formulas could also have been derived by our method of contour integration, but we omit this proof for the sake of brevity. They are very useful in the following considerations. We start with the two obvious equations

$$
(171) \quad \frac{1}{2\pi i} \oint_C P^*(z^*; u^*, v^*) dp(z; a) = - P^*(a^*; u^*, v^*)
$$

$$
+ e^{i\phi}\rho^2 P'(z_0; u, v) p'(z_0; a) + o(\rho^2),
$$

$$
(171') \quad \frac{1}{2\pi i} \sum_{\mu=1}^{n} \oint_{C_\mu} [Q^*(z^*; u^*, v^*) - k_\mu^*(u^*, v^*)] dp(z; a)
$$

$$
= - Q^*(a^*; u^*, v^*) + p(v; a) - p(u; a)
$$

$$
+ \sum_{\mu=1}^{n} k_\mu^*(u^*, v^*) \omega_\mu(a) + e^{i\phi}\rho^2 Q'(z_0; u, v) p'(z_0; a) + o(\rho^2).
$$

By virtue of the formulas (16) we obtain from these equations

$$
Q^*(a^*; u^*, v^*) + (P^*(a^*; u^*, v^*))^+
$$

$$
(172) \quad = p(v; a) - p(u; a) + \sum_{\mu=1}^{n} k_\mu^*(u^*, v^*) \omega_\mu(a)
$$

$$
+ e^{i\phi}\rho^2 Q'(z_0; u, v) p'(z_0; a) + e^{-i\phi}\rho^2 (P'(z_0; u, v))^+ (p'(z_0; a))^+ + o(\rho^2).
$$

Applying now the formula (103), we find

$$
p^*(v^*; a^*) - p^*(u^*; a^*)
$$

$$
(172') \quad = p(v; a) - p(u; a) + \sum_{\mu=1}^{n} k_\mu(u, v) [\omega_\mu(a) - \omega_\mu^*(a^*)]
$$

$$
+ e^{i\phi}\rho^2 Q'(z_0; u, v) p'(z_0; a) + e^{-i\phi}\rho^2 (P'(z_0; u, v))^+ (p'(z_0; a))^+ + o(\rho^2).
$$

By virtue of (169) we have, therefore, for the domain function

$$
(172'') \qquad\qquad D(v, u; a) = p(v; a) - p(u; a),
$$

the variational equation

$$
D^*(v^*, u^*; a^*)
$$

$$
(172''') \quad = D(v, u; a) + e^{i\phi}\rho^2 \left[Q'(z_0; u, v) - \frac{1}{2} \sum_{\mu=1}^{n} k_\mu(u, v) w_\mu'(z_0) \right] p'(z_0; a)
$$

$$
+ e^{-i\phi}\rho^2 \left[(P'(z_0; u, v))^+ - \frac{1}{2} \sum_{\mu=1}^{n} k_\mu(u, v) (w_\mu'(z_0))^+ \right] (p'(z_0; a))^+
$$

$$
+ o(\rho^2).
$$

Using again the notation (172'') and (103) we bring this result into the final form

$$
(173) \quad
\begin{aligned}
D^*(v^*, u^*; a^*) &= D(v, u; a) + e^{i\phi}\rho^2 p'(z_0; a) \frac{\partial}{\partial z_0} D(v, u; z_0) \\
&\quad + e^{-i\phi}\rho^2 (p'(z_0; a))^+ \frac{\partial}{\partial z_0{}^+} D(v, u; z_0) + o(\rho^2).
\end{aligned}
$$

If we are interested only in the variation of the real part of $D(v, u; a)$ we may derive from (172''') the simpler formula

$$
(174) \quad
\begin{aligned}
&\mathrm{Re}\,\{D^*(v^*, u^*; a^*) - D(v, u; a)\} \\
&\quad = \mathrm{Re}\,\bigg\{ e^{i\phi}\rho^2 p'(z_0; a) \bigg[Q'(z_0; u, v) + P'(z_0; u, v) \\
&\qquad\qquad - \frac{1}{2} \sum_{\mu=1}^{n} w_\mu'(z_0)(k_\mu(u, v) + (k_\mu(u, v))^+) \bigg] \bigg\} + o(\rho^2).
\end{aligned}
$$

Now one finds easily from (75) and (78) that

$$
(174') \quad \frac{1}{2} \sum_{\mu=1}^{n} w_\mu'(z_0)[k_\mu(u, v) + (k_\mu(u, v))^+] = \sum_{\nu=1}^{n} [\omega_\nu(u) - \omega_\nu(v)]l_\nu(z_0),
$$

whence by use of (61), (63) and (103') after easy transformations

$$
(175) \quad
\begin{aligned}
g^*(v^*; a^*) - g^*(u^*; a^*) &= g(v; a) - g(u; a) \\
&\quad + \mathrm{Re}\,\{e^{i\phi}\rho^2 p'(z_0; a)[p'(z_0; v) - p'(z_0; u)]\} + o(\rho^2).
\end{aligned}
$$

Sending finally u to the boundary C of D one obtains

$$
(175') \quad g^*(v^*; a^*) = g(v; a) + \mathrm{Re}\,\{e^{i\phi}\rho^2 p'(z_0; a)p'(z_0; v)\} + o(\rho^2).
$$

This is a known variation formula for Green's function [10, 12].

Subtracting the conjugate of equation (166) from equation (164) and using (108), we obtain a variation formula for Neumann's function:

$$
(176) \quad
\begin{aligned}
&\pi^*(b^*; u^*) - \pi^*(a^*; u^*) - \pi^*(b^*; v^*) + \pi^*(a^*; v^*) \\
&\quad = \pi(b; u) - \pi(a; u) - \pi(b; v) + \pi(a; v) \\
&\qquad - e^{i\phi}\rho^2 Q'(z_0; a, b)[Q'(z_0; u, v) - P'(z_0; u, v)] \\
&\qquad + \sigma^{-i\phi}\rho^2 (P'(z_0; a, b))^+ [(Q'(z_0; u, v))^+ - (P'(z_0; u, v))^+] + o(\rho^2),
\end{aligned}
$$

which leads by application of (108'), (61) and (63) to

$$
(176') \quad
\begin{aligned}
&\pi^*(b^*; u^*) - \pi^*(a^*; u^*) - \pi^*(b^*; v^*) + \pi^*(a^*; v^*) \\
&\quad = \pi(b; u) - \pi(a; u) - \pi(b; v) + \pi(a; v) \\
&\qquad - e^{i\phi}\rho^2 Q'(z_0; a, b)[\pi'(z_0; v) - \pi'(z_0; u)] \\
&\qquad + e^{-i\phi}\rho^2 (P'(z_0; a, b))^+ [(\pi'(z_0; v))^+ - (\pi'(z_0; u))^+] + o(\rho^2).
\end{aligned}
$$

Let us introduce the domain function $\log \Psi(z; u, v)$, defined in (99). Considering only the real parts in (176'), we obtain

$$
\log \left| \Psi^*(b^*; u^*, v^*) \right| - \log \left| \Psi^*(a^*; u^*, v^*) \right|
$$

(177)
$$
= \log \left| \Psi(b; u, v) \right| - \log \left| \Psi(a; u, v) \right|
$$

$$
+ \operatorname{Re} \left\{ e^{i\phi} \rho^2 \frac{\partial}{\partial z_0} \log \Psi(z_0; u, v) \frac{\partial}{\partial z_0} \log \Psi(z_0; a, b) \right\} + o(\rho^2).
$$

Numerous similar identities may be obtained by the same method. We applied in §3 the method of contour integration by pairing different domain functions of D. If one of the two domain functions is replaced by the corresponding domain function with respect to D^*, we get a variation formula. There is a similar procedure for Riemann surfaces which makes it possible to obtain variation formulas for the normal integrals and their periods by contour integration [12].

The variation formula (175') might have been obtained also by the following application of our formal identities. Suppose that $g^*(z; a)$ is Green's function for the domain D^*, obtained by the variation (157) from the original domain D. If z and a are different from z_0 we have by Taylor's formula

$$
(178) \qquad g^*(z^*; a^*) = g^*(z; a) + \operatorname{Re} \left\{ e^{i\phi} \rho^2 \left(\frac{p'(z; a)}{z - z_0} + \frac{p'(a; z)}{a - z_0} \right) \right\} + o(\rho^2).
$$

The first two terms on the right-hand side represent a harmonic function of z in D, except for the point z_0, where it has a simple pole, and the point a, where the first term has a logarithmic pole but the second term remains finite. If z lies on C we have further because of (178) small values of magnitude $o(\rho^2)$ for this expression. We shall now add to this function another harmonic function of z which has the boundary values 0 on C and cancels just the singularity of the second term at z_0. The sum function is then harmonic everywhere in D, except for the logarithmic pole at a, and has on C the order of magnitude $o(\rho^2)$. It is, therefore, up to a corrective term of order $o(\rho^2)$ identical with Green's function. The new harmonic function is

$$
(178') \qquad\qquad H(z, a; z_0) = - \operatorname{Re} \left\{ e^{i\phi} \rho^2 p'(z_0; z) p'(z_0; a) \right\}.
$$

In fact, because of (103') and (57) one has

$$
(178'') \qquad\qquad p'(w; z) \equiv 0 \qquad\qquad \text{for } w \in D, z \in C.
$$

Hence, this harmonic function vanishes on C and one finds by inspection that it cancels the singularity of the second right-hand term in (178). We obtain finally

$$
(178''') \quad g^*(z; a) + \operatorname{Re} \left\{ e^{i\phi} \rho^2 \left[\frac{p'(z; a)}{z - z_0} + \frac{p'(a; z)}{a - z_0} - p'(z_0; z) p'(z_0; a) \right] \right\}
$$

$$
= g(z; a) + o(\rho^2).
$$

This consideration shows the close connection between a variational problem and the problem of constructing a harmonic function with prescribed singularities which vanishes at the boundary of the domain. This is just the same problem which arises in the theory of the Schottky functions and it becomes obvious why the same method applies to these different fields.

7. Restricted variations. In the theory of domain functions one has sometimes to investigate an extremum problem in a class of domains which is defined by certain normalization conditions. If one wants to attack such a problem by variational methods, the question arises how to carry out the variation within the restricted class of domains considered. This is analogous to the case of isoperimetric problems in the classical calculus of variations, but may be much more difficult. Our results enable us now to overcome these difficulties in the case of two important types of restriction.

A typical case of the first type is the following. We wish to vary the domain D bounded by the continua $C_1, \cdots, C_n$ without changing the conformal type by means of a variation of the form

$$(179) \qquad z^* = z + \sum_{\iota=1}^{N} \frac{a_\iota \rho^2}{z - z_\iota}, \qquad \rho > 0, \; |a_\iota| \leqq 1,$$

to be understood in the same sense as the variation (157).

We assume that $N \geqq 3n - 6$, which is the number of conformal moduli of the domain, and we try to find the conditions upon the parameters a_ι, after the z_ι have been chosen, so that the conformal type of the domain is unchanged. For this purpose we merely pick a set of moduli, determine their variations by the methods of §6, and set these equal to zero.

A convenient set of moduli is obtained by use of a fixed linear combination $w(z)$ of the harmonic measures with real coefficients. We take $n-1$ of the periods of $w(z)$, the values of $\omega(z) = \text{Re}\{w(z)\}$ at the $n-2$ critical points $d_1, \cdots, d_{n-2}$ of $w(z)$ and the $n-3$ differences of Im $\{w(z)\}$ at the pairs of critical points $(d_2, d_1), \cdots, (d_{n-2}, d_1)$. These invariants are moduli, since if they are the same for two domains D and D^*, then the implicit equation

$$(180) \qquad w^*(z^*) = w(z)$$

can be solved to yield the map $z^*(z)$ of D upon D^*. Here the asterisks refer to quantities associated with D^*.

To obtain the variations of our moduli so chosen, we note that they are linear functions in the coefficients a_ι and recall (169). By application of (103′)

$$(169') \qquad \omega^*(z^*) = \omega(z) + \text{Re} \left\{ \sum_{\iota=1}^{N} a_\iota \rho^2 w'(z_\iota) [N(z; z_\iota) - (M(z; z_\iota))^+ \right.$$
$$\left. + \sum_{\nu=1}^{n} l_\nu(z_\iota)\omega_\nu(z)] \right\} + o(\rho^2),$$

whence wè obtain for the analytic functions $w^*(z^*)$ and $w(z)$

$$w^*(z^*) = w(z) + \sum_{t=1}^{N} a_t \rho^2 w'(z_t) N(z; z_t) - \sum_{t=1}^{N} a_t^+ \rho^2 (w'(z_t))^+ M(z; z_t)$$

(169'')

$$+ \sum_{t=1}^{N} \sum_{\nu=1}^{n} \mathrm{Re} \left\{ a_t \rho^2 w'(z_t) l_\nu(z_t) \right\} w_\nu(z) + iK + o(\rho^2),$$

where K is some real constant. Since $N(z; z_t)$ and $M(z; z_t)$ have no periods, we see that the conditions for the invariance of the $n-1$ periods of $w(z)$ which we have chosen are

$$(181) \qquad \mathrm{Re} \left\{ \sum_{t=1}^{N} a_t w'(z_t) w_\mu'(z_t) \right\} = o(1), \qquad \mu = 1, 2, \cdots, n-1,$$

upon application of (78). This is, of course, also a direct consequence of (170). Letting d_μ^* be the critical points of $\omega^*(z^*)$, we verify that

$$(182) \qquad \omega^*(d_\mu^*) = \omega^*(\bar{d}_\mu) + o(\rho^2), \qquad \bar{d}_\mu = z^*(d_\mu).$$

Hence we can conclude from (169) that the $\omega(d_\mu)$ are invariant when

$$(183) \qquad \mathrm{Re} \left\{ \sum_{t=1}^{N} a_t w'(z_t) p'(z_t; d_\mu) \right\} = o(1), \qquad \mu = 1, 2, \cdots, n-2.$$

Finally we note that when (181) is satisfied, the third term on the right in (169'') must be constant because of (78) and the character of the $P_{\mu\nu}$-matrix. Hence, under these circumstances we find for the remaining $n-3$ moduli the conditions of invariance

$$(184) \qquad \mathrm{Im} \left\{ \sum_{t=1}^{N} a_t w'(z_t) \left[\pi'(z_t; d_\mu) - \pi'(z_t; d_1) \right] \right\} = o(1),$$

$$\mu = 2, 3, \cdots, n-2,$$

by use of (108') in transforming (169''). The constant K drops out, since we take a difference $\mathrm{Im}\{w(d_\mu) - w(d_1)\}$ in each case. The $3n-6$ conditions (181), (183) and (184) are, then, necessary and sufficient that the variation (179) leave the conformal type of D unchanged, for suitably chosen terms $o(1)$.

If the determinant of the coefficients of $a_1, \cdots, a_{3n-6}$ does not vanish identically for $N = 3n-6$ in the equations (181), (183) and (184), then by the topological analogue of the implicit function theorem we may find numbers a_t so that these equations are actually fulfilled. The determinant will not vanish identically provided that the functions $w_1'(z), \cdots, w_{n-1}'(z)$, $p'(z, d_1), \cdots, p'(z, d_{n-2}), \{\pi'(z, d_2) - \pi'(z, d_1)\}, \cdots, \{\pi'(z, d_{n-2}) - \pi'(z, d_1)\}$ are linearly independent. Since $\pi'(z, d_\mu)$ and $p'(z, d_\mu)$ are the only functions in this set with poles at d_μ, a linear dependence relation must have the form

$$
L'(z) = \sum_{\mu=1}^{n-1} \alpha_\mu w'_\mu(z)
$$

(185)

$$
+ \sum_{\mu=2}^{n-2} \beta_\mu [\pi'(z; d_\mu) - \pi'(z; d_1) - p'(z; d_\mu) + p'(z; d_1)] \equiv 0.
$$

Thus for any function $f(z)$ regular in D and smooth on C

$$
0 = \oint_C f(z)(L'(z))^+ dz^+ = - \sum_{\mu=1}^{n-1} \alpha_\mu^+ \oint_C f(z) dw_\mu(z)
$$

(186)

$$
+ \sum_{\mu=2}^{n-2} \beta_\mu^+ \oint_C f(z)[\pi'(z; d_\mu) - \pi'(z; d_1) + p'(z; d_\mu) - p'(z; d_1)]dz
$$

$$
= - 4\pi i \sum_{\mu=2}^{n-2} \beta_\mu^+ [f(d_\mu) - f(d_1)]
$$

by our usual device. Since $[f(d_\mu) - f(d_1)]$ may be chosen arbitrarily, we have $\beta_\mu = 0$, $\mu = 2, \cdots, n-2$, and it follows from the linear independence of the $w'_\mu(z)$ that $\alpha_\mu = 0$, $\mu = 1, \cdots, n-1$. Thus we have the required independence and our result is completely proved. The points in the proof where we use smoothness of C do not require this assumption in any essential way, so that we can take C to consist of any n proper continua.

If we wish to consider extremum problems for conformal mappings leaving, say, the point at infinity in D fixed, we may derive from the Green's function $p(z, \infty)$ and the corresponding system of moduli in this case the $3n-4$ conditions of invariance of restricted conformal type

(187)
$$
\text{Re} \left\{ \sum_{t=1}^{N} a_t p'(z_t; \infty) w'_\mu(z_t) \right\} = o(1), \qquad \mu = 1, 2, \cdots, n-1,
$$

(188)
$$
\text{Re} \left\{ \sum_{t=1}^{N} a_t p'(z_t; \infty) p'(z_t; d_\mu) \right\} = o(1), \qquad \mu = 1, 2, \cdots, n-1,
$$

(189)
$$
\text{Im} \left\{ \sum_{t=1}^{N} a_t p'(z_t; \infty) [\pi'(z_t; d_\mu) - \pi'(z_t; d_1)] \right\} = o(1),
$$

$$
\mu = 2, 3, \cdots, n-1,
$$

where the d_μ are the $n-1$ finite critical points of $p(z, \infty)$. Similar conditions may be derived for the original Hadamard variational formulas which give the variation of the domain functions under an arbitrary variation of the boundary [12].

We see from our work on the functions $w'_i w'_k$ at the end of §5 that in general it will be sufficient to set the variations of the P_{ik} equal to zero to obtain a type-invariant variation. For by the methods of that section we can easily show that every domain function $f(z)$ for which $f(z)z'^2$ is real on C and which is regular in D may be composed linearly from $3n - 6$ linearly independent products $w_i'(z)w_k'(z)$. Now the analytic functions appearing in (181), (183), (184) are just of this type and, on the other hand, the variations of the P_{ik} are expressed in terms of the $w_i'w_k'$.

We solve a particular well known problem by this variational method in order to show how such proofs go. Consider the domain $\tilde{D}$ bounded by C_1 and C_2, where C_1 is assumed to be the outer boundary of D. In general the conformal type of $\tilde{D}$ will change when D is mapped conformally, and we minimize the modulus of $\tilde{D}$ for all conformal images of D. The mappings of D may be restricted in any number of ways to give a normal family without changing the problem, for example by fixing the derivative at a point. We take the modulus to be the period $\tilde{P}_{11}$ of the harmonic measure of C_1 in any domain $\tilde{D}$ obtained by mapping D.

We make a variation (179) of the extremal domain D satisfying (181), (183) and (184) and obtain from the minimum property of $\tilde{P}_{11}$ the inequality

$$(190) \qquad \tilde{P}_{11}^* - \tilde{P}_{11} = \mathrm{Re}\left\{ \sum_{t=1}^{N} a_t \rho^2 \tilde{w}_1'(z_t)^2 \right\} + o(\rho^2) \geqq 0,$$

where $\tilde{w}_1(z)$ is the harmonic measure function of C_1 in $\tilde{D}$.

The reader can verify by the familiar procedure leading to Lagrange multipliers in the calculus of variations that this implies the existence of real numbers λ_μ such that

$$(191) \qquad \begin{aligned} \tilde{w}_1'(z)^2 &= \sum_{\mu=1}^{n-1} \lambda_\mu w'(z) w_\mu'(z) + \sum_{\mu=1}^{n-2} \lambda_{\mu+n-1} w'(z) p'(z; d_\mu) \\ &\quad + i \sum_{\mu=2}^{n-2} \lambda_{\mu+2n-4} w'(z)[\pi'(z; d_\mu) - \pi'(z; d_1)]. \end{aligned}$$

It follows from this relation that the boundary components C_ν, $\nu > 2$, are analytic curves $z_\nu(s)$ satisfying the differential equations

$$(192) \qquad \begin{aligned} \tilde{w}_1'(z)^2 \cdot z_\nu'(s)^2 &= \sum_{\mu=1}^{n-1} \lambda_\mu \frac{\partial w}{\partial s} \frac{\partial w_\mu}{\partial s} + \sum_{\mu=1}^{n-2} \lambda_{\mu+n-1} \frac{\partial w}{\partial s} \frac{\partial p(z; d_\mu)}{\partial s} \\ &\quad + i \sum_{\mu=2}^{n-2} \lambda_{\mu+2n-4} \frac{\partial w}{\partial s} \left[\frac{\partial \pi(z; d_\mu)}{\partial s} - \frac{\partial \pi(z; d_1)}{\partial s} \right] = \mathrm{real}. \end{aligned}$$

For each separate arc we can bring this into the form

$$(193) \qquad [\partial \tilde{w}_1(z_\nu(t))/\partial t]^2 = \pm 1$$

by suitable choice of the real parameter t. Now there is a map of each small arc C_ρ of mapping radius ρ of C_ν, $\nu > 2$, on the exterior of a circle of radius ρ, of the form

$$(194) \qquad z^* = z + a\rho^2/(z - z_0) + b\rho^3/(z - z_0)^2 + \cdots, \qquad z_0 \in C_\rho,$$

with uniformly bounded coefficients a, b, $\cdots$. In each such map we obtain from (170) and the minimum property of D

(190') $$\text{Re} \left\{ a\rho^2 \bar{w}_1'(z_0)^2 \right\} + o(\rho^2) \geqq 0.$$

If the curve C_ρ is rotated until it has a horizontal tangent at z_0, we can choose $a = -1 + o(1)$ and we find

(190'') $$\text{Re} \left\{ \bar{w}_1'(z_0)^2 \right\} \leqq 0.$$

Hence the lower sign must hold in our differential equations and we have

(193') $$\bar{w}_1(z_\nu(t)) = \pm it + \text{const.},$$

so that the C_ν lie along the level curves of $\tilde{\omega}_1(z)$. We remark that $\bar{P}_{11}$ is unchanged by conformal mappings of $\bar{D}$, which are also mappings of D, and that in consequence there is no loss of generality in assuming C_1 and C_2 to be concentric circles. In this case the C_ν, $\nu > 2$, are concentric circular slits and the extremal map is a function $f_{12}(z)$ of the type introduced in §2.

Our progress, then, has been to find that for the extremal domain the C_ν, $\nu > 2$, are analytic curves without use of the boundary variation (194) introduced earlier [9]. We then use the boundary variation to complete the result, but in a very simple form. Thus we have reduced the work involved in the application of variational methods to a considerably more elementary and less involved set of ideas.

We come now to the second type of restricted variational problems which we desire to handle. Given are n closed smooth curves C_ν which are the boundary of a domain D in the complex plane which contains the point at infinity. We seek a continuum containing all the curves C_ν of minimum capacity. We require of variations considered that they transform each continuum C_ν into itself. This is, in fact, possible to obtain by means of variations

(195) $$z^* = z + (e^{i\phi}\rho^2 Q'(z; u, v) + e^{-i\phi}\rho^2 P'(z; u, v))(w_i'(z)w_k'(z))^{-1} + o(\rho^2).$$

According to (16) we have, namely, in C

(196) $$z'Q'(z; u, v)e^{i\phi} + z'P'(z; u, v)e^{-i\phi} = \text{real}, \qquad z' = dz/ds.$$

Since further $w_i'(z)z'$ and $w_k'(z)z'$ are imaginary on C, one sees that $z^* - z$ has the direction of the tangent vector at z to C, up to a term $o(\rho^2)$. By a correction term $o(\rho^2)$ it is possible to attain a complete preservation of the contours C_ν. This can be seen by making a variation

(197) $$z^{**} = z^* + \sum_{t=1}^{N} \frac{a_t}{z^* - z_t} \cdot o(\rho^2)$$

to bring the conformal type of our domain back to that of the original region, and then it is clear that there is a conformal map

(198) $$z^{***} = z^{**} + o(\rho^2)$$

of the domain of the z^{**}-plane back upon the original domain. For fixed $w(z)$ we have a freedom of choice of u and v in D and of arbitrary real ϕ and ρ. Thus, this type of variation permits the usual applications in extremum problems. We proceed to the complete solution of the above-mentioned problem.

Let $\tilde{D}$ be the domain bounded by the minimizing continuum, and let $\tilde{p}(z, \infty)$ be the corresponding analytic function for Green's function with pole at infinity. The function $\tilde{g}(z, \infty)$ has at ∞ an expansion

(199) $$\tilde{g}(z; \infty) = \log |z| + \gamma + o(1),$$

and the capacity of our continuum is by definition $e^{-\gamma}$. Hence we obtain upon making our variation the inequality

(200)
$$
\begin{aligned}
0 \geq \gamma^* - \gamma = \mathrm{Re} \Bigg\{ & e^{i\phi}\rho^2 \left[\frac{\tilde{p}'(u; \infty)^2}{w_i'(u)w_k'(u)} - \frac{\tilde{p}'(v; \infty)^2}{w_i'(v)w_k'(v)} \right] \\
& + \sum_{\mu=1}^{n-2} \frac{e^{i\phi}\rho^2 Q'(d_{i\mu}; u, v) + e^{-i\phi}\rho^2 P'(d_{i\mu}; u, v)}{w_i''(d_{i\mu})w_k'(d_{i\mu})} \tilde{p}'(d_{i\mu}; \infty)^2 \\
& + \sum_{\mu=1}^{n-2} \frac{e^{i\phi}\rho^2 Q'(d_{k\mu}; u, v) + e^{-i\phi}\rho^2 P'(d_{k\mu}; u, v)}{w_i'(d_{k\mu})w_k''(d_{k\mu})} \tilde{p}'(d_{k\mu}; \infty)^2 \Bigg\} + o(\rho^2),
\end{aligned}
$$

where the $d_{i\mu}$ are the critical points of $w_i(z)$ and the $d_{k\mu}$ are the critical points of $w_k(z)$. By the arbitrariness of ϕ and ρ and from the relations (61) and (63),

(200′)
$$
\begin{aligned}
& \frac{\tilde{p}'(u; \infty)^2}{w_i'(u)w_k'(u)} - \sum_{\mu=1}^{n-2} \left[\frac{\tilde{p}'(d_{i\mu}; \infty)^2}{w_i''(d_{i\mu})w_k'(d_{i\mu})} N(u; d_{i\mu}) \right. \\
& \qquad \left. - \left(\frac{\tilde{p}'(d_{i\mu}; \infty)^2}{w_i''(d_{i\mu})w_k'(d_{i\mu})} \right)^+ M(u; d_{i\mu}) \right] \\
& \quad - \sum_{\mu=1}^{n-2} \left[\frac{\tilde{p}'(d_{k\mu}; \infty)^2}{w_i'(d_{k\mu})w_k''(d_{k\mu})} N(u; d_{k\mu}) - \left(\frac{\tilde{p}'(d_{k\mu}; \infty)^2}{w_i'(d_{k\mu})w_k''(d_{k\mu})} \right)^+ M(u; d_{k\mu}) \right] \\
& = \frac{\tilde{p}'(v; \infty)^2}{w_i'(v)w_k'(v)} - \sum_{\mu=1}^{n-2} \left[\frac{\tilde{p}'(d_{i\mu}; \infty)^2}{w_i''(d_{i\mu})w_k'(d_{i\mu})} N(v; d_{i\mu}) \right. \\
& \qquad \left. - \left(\frac{\tilde{p}'(d_{i\mu}; \infty)^2}{w_i''(d_{i\mu})w_k'(d_{i\mu})} \right)^+ M(v; d_{i\mu}) \right] \\
& \quad - \sum_{\mu=1}^{n-2} \left[\frac{\tilde{p}'(d_{k\mu}; \infty)^2}{w_i'(d_{k\mu})w_k''(d_{k\mu})} N(v; d_{k\mu}) - \left(\frac{\tilde{p}'(d_{k\mu}; \infty)^2}{w_i'(d_{k\mu})w_k''(d_{k\mu})} \right)^+ M(v; d_{k\mu}) \right] \\
& = \mathrm{const.},
\end{aligned}
$$

since each side of the equation is a function of a different variable. We conclude that $\tilde{p}'(u; \infty)^2$ is analytic throughout D; in fact we have shown that

$$(200'') \qquad \tilde{p}'(u;\ \infty)^2/w_i'(u)w_k'(u) \doteq S(u),$$

where $S(u)$ is a Schottky function in D.

From the fact that $\tilde{p}'(u;\ \infty)^2$ is analytic in D, we can conclude that each slit $\tilde{C}_\alpha$ of the boundary of $\tilde{D}$ in D is an analytic curve, and that in the mapping of D upon the unit circle taking the point at infinity into the origin each side of any segment of the slit $\tilde{C}_\alpha$ is taken into an arc of the same length as that of the image of the other side of the slit. This is because the distortion of the mapping on the boundary $\tilde{C}_\alpha$ is in both cases $|\tilde{p}'(z;\ \infty)|$. The function $\tilde{p}'(z;\ \infty)$ can be extended to be analytic on a two-sheeted Riemann surface covering D; it assumes values symmetric in the origin at points lying over one another. From these facts one can carry out a further discussion of the nature of the $\tilde{C}_\alpha$.

We remark that this problem could be solved equally well with the restricted variation leaving D conformally invariant for maps preserving the point at infinity with fixed distortion. But this method would have led us to a set of Lagrange multipliers which are determined explicitly by the tangential variation in terms of the various domain functions. Thus the importance of the relations we have derived becomes apparent.

As a final application of our identities we derive the conditions for unchanged conformal type under variations (179) from a second point of view in the case of a smoothly bounded domain D. We start from the assumption that the parameters a_t, z_t have been chosen in the right way and that D is mapped upon a domain D^* which is conformally equivalent to it. There exists, therefore, a univalent function in D which maps D upon D^*. Let

$$(201) \qquad \zeta = z + \rho^2 k_\rho(z)$$

be the mapping function, where the factor ρ^2 measures the order of the second term. If we invert (179) into

$$(179') \qquad z_1 = v^{-1}(z^*) = z^* - \sum_{t=1}^{N} \frac{a_t \rho^2}{z^* - z_t} + o(\rho^2)$$

this gives a map of D^* upon D except for a certain neighborhood of the $z_t \in D^*$. In particular, the boundary C^* of D^* will be transformed into the boundary of D by means of (179'). Therefore, the composite map

$$(202) \qquad v^{-1}(\zeta) = z + \rho^2 k_\rho(z) - \sum_{t=1}^{N} \frac{a_t \rho^2}{z - z_t} + o(\rho^2)$$

transforms each C_ν into itself. Thus, having chosen a_t and z_t in the right way, we have in the limit $\rho \to 0$ the following condition on C:

$$(202') \qquad \left(\lim_{\rho \to 0} k_\rho(z) - \sum_{t=1}^{N} \frac{a_t}{z - z_t} \right) z'^{-1} = \text{real}, \qquad z' = \frac{dz}{ds}.$$

On the other hand, if we can determine a function $k(z)$ which is regular in

$D+C$ and such that

(202'') $$k(z) - \sum_{t=1}^{N} \frac{a_t}{z - z_t} = \lambda(z)z', \qquad \lambda(z) = \text{real}, z \in C,$$

we can chose a correction term $o(\rho^2)$ so that

(179'') $$z^* = z + \sum_{t=1}^{N} \frac{a_t \rho^2}{z - z_t} + o(\rho^2)$$

maps D upon a conformally equivalent domain D^*.

Thus our problem is reduced to the determination of a function $k(z)$ of the required type. Choose for this purpose a function $w'(z)$ of type (143); evidently

(203) $$f(z) = \frac{1}{i} w'(z)\left[k(z) - \sum_{t=1}^{N} \frac{a_t}{z - z_t}\right]$$

is a Schottky function. It has simple poles at the N points z_t and the condition (125) connecting poles and residues of a function of class $\Re$ yields

(204) $$\text{Re}\left\{\sum_{t=1}^{N} a_t w'(z_t) w'_\mu(z_t)\right\} = 0, \qquad \mu = 1, 2, \cdots, n-1.$$

Now we may apply the representation (122) for $f(z)$ and find

(205) $$\frac{1}{i} w'(z)\left[k(z) - \sum_{t=1}^{N} \frac{a_t}{z - z_t}\right]$$
$$= A - \frac{1}{i}\sum_{t=1}^{N}(a_t w'(z_t)N(z; z_t) - a_t^+(w'(z_t))^+ M(z; z_t)).$$

We have further to satisfy the condition that the function (205) vanishes at all the $n-2$ zeros d_ν of $w'(z)$. We may transform these conditions by using the formulas (109') and (110') to eliminate $N(z; z_t)$ and $M(z; z_t)$; we obtain

(206) $$\text{Re}\left\{A - \frac{1}{i}\sum_{t=1}^{N} a_t w'(z_t)\pi'(z_t; d_\nu)\right\} = 0, \qquad \nu = 1, 2, \cdots, n-2,$$

(206') $$\text{Im}\left\{A - \frac{1}{i}\sum_{t=1}^{N} a_t w'(z_t)\left[p'(z_t; d_\nu) - \sum_{\mu=1}^{n} l_\mu(z_\nu)\omega_\mu(d_\nu)\right]\right\} = 0,$$
$$\nu = 1, 2, \cdots, n-2.$$

From (204), (78) and (107') we conclude that

(206'') $$\text{Re}\left\{\sum_{t=1}^{N} a_t w'(z_t)l_\mu(z_t)\right\} = 0, \qquad \mu = 1, 2, \cdots, n.$$

Sending z in (205) to the boundary C of D and using (57) and (206''), we find that A is real. This leads to the conditions:

$$(207) \quad \text{Im} \left\{ \sum_{i=1}^{N} a_i w'(z_i) \left[\pi'(z_i; d_\nu) - \pi'(z_i; d_1) \right] \right\} = 0, \quad \nu = 2, 3, \cdots, n-2,$$

$$(208) \quad \text{Re} \left\{ \sum_{i=1}^{N} a_i w'(z_i) p'(z_i; d_\nu) \right\} = 0, \quad \nu = 1, 2, \cdots, n-2.$$

We have in (204), (207) and (208) exactly our $3n-6$ conditions (181), (183) and (184) for the parameters a_i, z_i of our variation after passage to the limit $\rho \to 0$. It is interesting that the fullfillment of these conditions for a particular function $w'(z)$ guarantees the equations for every function of this type.

BIBLIOGRAPHY

1. S. Bergman, *Partial differential equations*, Brown University Publications, Providence, 1941.

2. S. Bergman and M. Schiffer, *A representation of Green's and Neumann's functions in the theory of partial differential equations of second order*, Duke Math. J. vol. 14 (1947) pp. 609–638.

3. B. Epstein, *Some inequalities relating to conformal mapping upon canonical slit-domains*, Bull. Amer. Math. Soc. vol. 53 (1947) pp. 813–819.

4. H. Groetzsch, *Ueber die Verzerrung bei schlichter konformer Abbildung mehrfach zusammenhaengender schlichter Bereiche* III, Berichte der Saechsische Akademie der Wissenschaften, Leipzig, vol. 83 (1931) pp. 283–297.

5. ———, *Ueber das Parallelschlitztheorem der konformen Abbildung schlichter Bereiche*, Berichte der Saechsischen Akademie der Wissenschaften, Leipzig, vol. 84 (1932) pp. 15–36.

6. H. Grunsky, *Neue Abschaetzungen zur konformen Abbildung ein- und mehrfach zusammenhaengender Bereiche*, Schriften des mathematischen Seminars und des Instituts für angewandte Mathematik an der Universität Berlin vol. 1 (1932) pp. 94–140.

7. ———, *Koeffizientenbedingungen fuer schlicht abbildende meromorphe Funktionen*, Math. Zeit. vol. 45 (1939) pp. 29–61.

8. K. Hensel and G. Landsberg, *Theorie der algebraischen Functionen einer Variabeln und ihre Anwendung auf algebraische Kurven und algebraische Integrale*, Leipzig, 1902.

9. M. Schiffer, *A method of variation within the family of simple functions*, Proc. London Math. Soc. (2) vol. 44 (1938) pp. 432–449.

10. ———, *Variation of the Green function and theory of the p-valued functions*, Amer. J. Math. vol. 65 (1943) pp. 341–360.

11. ———, *The span of multiply connected domains*, Duke Math. J. vol. 10 (1943) pp. 209–216.

12. ———, *Hadamard's formula and variation of domain functions*, Amer. J. Math. vol. 68 (1946) pp. 417–448.

13. ———, *The kernel function of an orthonormal system*, Duke Math. J. vol. 13 (1946) pp. 529–540.

14. ———, *An application of orthonormal functions in the theory of conformal mapping*, Amer. J. Math. vol. 70 (1948) pp. 147–156.

15. E. Schottky, *Ueber die konforme Abbildung mehrfach zusammenhaengender ebener Flaechen*, J. Reine Angew. Math. vol. 83 (1877) pp. 300–351.

HARVARD UNIVERSITY,
CAMBRIDGE, MASS.

Commentary on

[26] P.R. Garabedian and M. Schiffer, *Identities in the theory of conformal mapping*, Trans. Amer. Math. Soc. **65** (1949), 187–238.

Paul Garabedian and Max Schiffer both arrived at Harvard in 1946, Garabedian as a very young Ph.D. student under Lars Ahlfors and Schiffer as a Research Lecturer, his first appointment in the United States. For an account of how they first met, told in Garabedian's own words, see [G] elsewhere in these *Selecta*. The first of their 11 joint papers, [26] was an impressive debut to a long and fruitful collaboration that continued until 1967.

The conformal mappings referred to in the title are of multiply connected domains, and the identities, of which there are many, arise primarily via contour integration, often employing the residue theorem and applied to combinations of two of the so-called canonical mappings. Interpretations accompanying some of the identities lead the authors to formulate and solve several interesting geometric extremal problems. And, as might be expected, in Schiffer's hands no good formula goes unvaried, so the authors also study how the identities fare under an interior variation.

It is not obvious that one should expect such an outpouring of identities, though the authors make contact with old work of Schottky, which they say (generously) at least anticipates the possibility through identities for Abelian integrals on the double of a Riemann surface with boundary. Nor is it obvious which mappings are amenable to the systematic approach proposed by the authors for finding identities, since for multiply connected domains, one has a choice of conformal mappings onto a variety of canonical domains bounded by slits of the same geometric character (e.g., horizontal, vertical, radial, or concentric circular slits); see, for instance, [N].

Let D be a bounded, multiply connected domain with boundary C consisting of n smooth curves $C_1, \ldots, C_n$. Let $u, v \in D$ and let $\Phi(z; u, v)$ and $\Psi(z; u, v)$ denote the mappings of D onto a circular slit domain and a radial slit domain, respectively, each mapping having a simple zero at u and a simple pole at v with residue 1. Next, let

$$P(z; u, v) = \frac{1}{2}(\log \Phi(z; u, v) - \log \Psi(z; u, v)),$$

$$Q(z; u, v) = \frac{1}{2}(\log \Phi(z; u, v) + \log \Psi(z; u, v)).$$

These are the functions most intensively studied, since their boundary values are usefully related for the purposes of contour integration. It is shown that $P(z; u, v)$ is at most n-valent, while for a convex combination $F(z; u, v) = a \log \Phi(z; u, v) + b \log \Psi(z; u, v)$, $a, b > 0$, $a + b = 1$, the mapping $\exp F$ is univalent. In particular, $\exp Q$ is univalent.

Examples of identities are

$$P(a; u, v) - P(b; u, v) = \overline{P(u; a, b)} - \overline{P(v; a, b)},$$

$$Q(a; u, v) - Q(b; u, v) = Q(u; a, b) - Q(v; a, b).$$

For an example of a geometric extremal problem suggested by and solved through identities, consider the family of conformal mappings $f(z; u, v)$ of D with zero at u and pole at v with residue 1. The logarithmic area $-(1/2i) \int_C \bar{f} df$ of the complement of the image domain is not greater than $\pi P(u; u, v)$, and equality holds for $\exp Q$. The functions P and Q also appear in connection with Green's function for D and the harmonic measures of the boundary curves, in the form

$$p(v; \zeta) - p(u; \zeta) = Q(\zeta; u, v) + \overline{P(\zeta; u, v)}$$

$$- \sum_{\mu=1}^{n} k_\mu(u, v) \omega_\mu(\zeta).$$

Here the real part of p is Green's function, k_μ is the value of $Q + \bar{P}$ on the boundary curve C_μ, and $\mathrm{Re}\, \omega_\mu$ is the harmonic measure of C_μ. There is an analogous identity for Neumann's function. This is just a sample. The reader will find much more, including identities based on the canonical mappings onto complements of horizontal and vertical slits.

It is a pleasure to watch the authors calculate; but it is the framework for the calculations that the authors want the reader to appreciate, namely, the (then, fairly new) ideas in function theory deriving from the development of the Bergman kernel. In fact, the authors make explicit contact with the kernel function by showing that $P(z; u, v)$ is the Bergman kernel for the space of analytic functions in D that vanish at v and have square integrable derivatives. In general, the authors consider the scalar product

$$(f, g) = \iint_D f'(z)\overline{g'(z)}\, dx\, dy,$$

where by Green's theorem one can replace the area integral by

$$(f, g) = \frac{1}{2i} \int_C f'(z)\overline{g(z)}\, dz.$$

Since the contour integral may exist when the area integral does not, it can serve as an alternate definition of the scalar product, and this is what the authors use. From this point of view, which was gaining currency at the time, it becomes natural to ask: What is the scalar product of X with Y? (This is the source of the identities.) What is the norm of Z? (Being nonnegative, this is the source of the extremal problems for areas.) Still, it is one thing to ask such questions, particularly the right questions, and quite another to find the answers.

The Bergman kernel had been around since the 1920s, and its role in conformal mapping, even for multiply connected domains, had been developed to some extent; see, for example, the early lecture notes of Bergman [B1][1] and his monograph [B2]. Schiffer's own summary is in his appendix [34] to Courant's book [C]; and in [8], he provided a proof by variational methods of the existence of a conformal mapping of a multiply connected domain onto the disk with concentric circular slits. In 1952, Bergman and Schiffer joined Garabedian at Stanford, and the three remained colleagues there until 1959, when Garabedian moved to the Courant Institute.

References

[B1] Stefan Bergman, *Partial Differential Equations, Advanced Topics*, Brown University, 1941.

[B2] Stefan Bergman, *The Kernel Function and Conformal Mapping*, second revised edition, American Mathematical Society, 1970.

[C] R. Courant, *Dirichlet s Principle, Conformal Mapping, and Minimal Surfaces*, Interscience, 1950.

[G] Paul R. Garabedian, *Recollections of Menahem Max Schiffer*, these Selecta, Vol. 1.

[N] Zeev Nehari, *Conformal Mapping*, McGraw-Hill, 1952; Dover edition, 1975.

BRAD OSGOOD

---

[1] The title notwithstanding, these notes are not at all about partial differential equations, and they contain an interesting chapter by D.C. Spencer on distortion in conformal mapping.

[28] (with A. C. Schaeffer and D. C. Spencer) The coefficient regions of schlicht functions

[28] Menahem Max Schiffer, A.C. Schaeffer, D.C. Spencer. "The coefficient regions of schlicht functions," in *Duke Mathematical Journal*, Volume **16** (1949), 493–527.

THE COEFFICIENT REGIONS OF SCHLICHT FUNCTIONS

By A. C. Schaeffer, M. Schiffer and D. C. Spencer

1. We denote by $\mathfrak{F}$ the family of all functions $f(z) = z + a_2 z^2 + a_3 z^3 + \cdots$ which are regular and schlicht in the circle $|z| < 1$. Let $\mathfrak{V}_n$ be the $2n$-dimensional region composed of all points $(a_2, a_3, \cdots, a_{n+1})$ belonging to the different elements of $\mathfrak{F}$, and let $\mathfrak{S}_n$ be the boundary of $\mathfrak{V}_n$. Since the family $\mathfrak{F}$ is compact, $\mathfrak{V}_n$ is a closed domain. (A closed domain is a closed set which is the closure of a domain.)

The coefficient problem for schlicht functions is that of determining $\mathfrak{V}_n$ —that is to say, $\mathfrak{S}_n$ —for all values $n = 1, 2, \cdots$. This problem was first seriously considered by Peschl [7] in 1937. At that time the most penetrating characterization of schlicht functions was provided by the method of Löwner [6]; Peschl applied this method to the coefficient problem and obtained qualitative results and also the 2-dimensional region of (a_2, a_3) when both are real. In the following years variational methods were introduced into the theory of schlicht functions [2], [3], [4], [8], [9], [11], [14], [15], [16], [18] (the systematic development of the variational method began with the paper [14]) and their development has provided new tools for the investigation of schlicht functions. In particular, the coefficient problem has been studied using variational methods [10], [12], [13], [19].

Since there is some overlapping of results presented here with those in [13] and [17], the repetition should be justified. The results of this paper are subsequent to [13] (despite different dates of submission) and each is subsequent to [17]. One of our main purposes is to bring the various ideas together under a general scheme and for the sake of completeness and clarity a certain amount of repetition seems unavoidable.

A combination of Löwner's method with variational methods was proposed in [17] and Löwner's differential equation was used in a somewhat different way in the coefficient problem (see [13]). It turns out that there is quite a remarkable duality relationship between the two ways in which the Löwner method has been combined with the variational approach. We show that the method adopted in [17] leads to the theory of the characteristic curves of the partial differential equation defining the boundary $\mathfrak{S}_n$ of $\mathfrak{V}_n$. It is interesting from the methodological point of view that the method of variations leads immediately to the partial differential equation for $\mathfrak{S}_n$ and that the classical theory of characteristic curves leads then necessarily to Löwner's differential equation for the schlicht functions associated with them. The method of the characteristics enables us to connect every point on $\mathfrak{S}_n$ with the distinguished "Koebe" point

Received August 17, 1948. This paper represents work carried out under sponsorship of the Office of Naval Research.

in $\mathfrak{S}_n$ which belongs to the function $f(z) = z/(1 - z)^2 = z + 2z^2 + 3z^3 + \cdots$. Integrating a system of ordinary differential equations we are able to compute curves on $\mathfrak{S}_n$ all of which finally terminate in the Koebe point.

The second method works with a dual system of differential equations which is formally closely related to the above system; but the curves obtained lie entirely in $\mathfrak{B}_n$ and connect the points on $\mathfrak{S}_n$ with the origin of the coefficient space. Both methods seem to be equally applicable and interesting and may be developed simultaneously.

The above considerations may be extended to the coefficients $a_\nu^{(\mu)}$ which are defined by

$$f(z)^\mu = \sum_{k=0}^{\infty} a_{k+\mu}^{(\mu)} z^{k+\mu} \qquad (a_\mu^{(\mu)} = 1).$$

If we let $\mathfrak{B}_n^{(\mu)}$ denote the region of points $(a_{1+\mu}^{(\mu)}, a_{2+\mu}^{(\mu)}, \cdots, a_{n+\mu}^{(\mu)})$ and $\mathfrak{S}_n^{(\mu)}$ its boundary, it turns out that the characteristic equations have a form which is independent of μ. The different spaces obtained for different values of μ arise from the different choices of initial values; but the formal theory is in each case the same and the consideration of all values for μ leads to some new information in the case $\mu = 1$.

Since we want to emphasize here the general outlines of the theory we shall often omit details; but we have tried to show in this paper the mutual interdependence of the different methods applied in the theory of schlicht functions.

2. We begin by investigating the structure of $\mathfrak{S}_n$ from the point of view of the characteristic curves. We shall not assume any knowledge of Löwner's method but shall be led to it automatically.

If $w = f(z)$ maps the unit circle $|z| < 1$ onto a domain in the w-plane we may obtain neighboring elements of f in $\mathfrak{F}$ by considering schlicht functions $f^{\blacktriangle}(z)$ which map the unit circle into domains $\mathfrak{D}^{\blacktriangle}$ which are arbitrarily near $\mathfrak{D}$ in the Fréchet sense. One method of obtaining such elements is the following. We choose N interior points w_ν ($\nu = 1, 2, \cdots, N$) in $\mathfrak{D}$ and consider the function

$$(2.1) \qquad w^{\blacktriangle} = w\left[1 + \sum_{\nu=1}^{N} C_\nu \rho^2/(w - w_\nu)\right] \qquad (w_\nu = f(z_\nu); \nu = 1, 2, \cdots, N).$$

For given C_ν, w_ν and for ρ sufficiently small, this function maps the boundary $\mathfrak{B}$ of $\mathfrak{D}$ in a one-to-one fashion onto a new boundary continuum $\mathfrak{B}_1$ which determines a new domain $\mathfrak{D}_1$. The interior of the unit circle is then mapped onto $\mathfrak{D}_1$ by means of the following schlicht function [16]:

$$(2.2) \qquad f_1(z) = f(z) + \rho^2 \sum_{\nu=1}^{N} \left\{ C_\nu \left(-\frac{zf'(z)f(z_\nu)}{z_\nu f'(z_\nu)^2(z - z_\nu)} + \frac{f(z)}{f(z) - f(z_\nu)} \right) \right.$$

$$\left. + C_\nu^* \frac{z^2 f'(z)f^*(z_\nu)}{z^* f'^*(z_\nu)^2(1 - z_\nu^* z)} \right\} + o(\rho^2).$$

(Here and elsewhere the asterisk attached to a symbol indicates the complex conjugate.) We now normalize $f_1(z)$ in such a way that the derivative at $z = 0$ becomes equal to 1 and we obtain in this way an element $f^\blacktriangle$ of $\mathfrak{F}$ which is of the form

$$(2.3) \quad f^\blacktriangle(z) = f(z) + \rho^2 \sum_{\nu=1}^{N} \left\{ C_\nu \left(-\frac{zf'(z)f(z_\nu)}{z_\nu f'(z_\nu)^2(z - z_\nu)} - \frac{f(z)f(z_\nu)}{z_\nu^2 f'(z_\nu)^2} \right. \right.$$

$$\left. \left. + \frac{f(z)^2}{f(z_\nu)(f(z) - f(z_\nu))} \right) + C_\nu^* \frac{z^2 f'(z)f^*(z_\nu)}{z_\nu^* f'^*(z_\nu)^2(1 - z_\nu^* z)} \right\} + o(\rho^2).$$

Denoting the coefficients of $f^\blacktriangle$ by $a_k^\blacktriangle$ we find

$$(2.4) \quad a_{k+1}^\blacktriangle = a_{k+1} + \rho^2 \sum_{\nu=1}^{N} C_\nu \left[\frac{f(z_\nu)}{z_\nu^2 f'(z_\nu)^2} \left(ka_{k+1} + \sum_{m=1}^{k} \frac{ma_m}{z_\nu^{k+1-m}} \right) - \frac{1}{f(z_\nu)} S_{k+1}\left(\frac{1}{f(z_\nu)} \right) \right]$$

$$+ \rho^2 \sum_{\nu=1}^{N} C_\nu^* \frac{f^*(z_\nu)}{z_\nu^{*2} f'^*(z_\nu)^2} \sum_{m=1}^{k} ma_m z_\nu^{*k+1-m} + o(\rho^2).$$

Here $S_{k+1}(x)$ is a polynomial in x of degree k which is defined by the generating function

$$(2.5) \quad f(z)/(1 - xf(z)) = \sum_{k=0}^{\infty} (a_{k+1} + S_{k+1}(x))z^{k+1}.$$

Let

$$(2.6) \quad U_{k+1}(z) = \frac{f(z)}{z^2 f'(z)^2} \left(ka_{k+1} + \sum_{m=1}^{k} \frac{ma_m}{z^{k+1-m}} \right) - \frac{1}{f(z)} S_{k+1}\left(\frac{1}{f(z)} \right),$$

$$(2.6') \quad V_{k+1}(z) = \frac{f(z)}{z^2 f'(z)^2} \sum_{m=1}^{k} ma_m^* z^{k+1-m}.$$

To each point $(a_2, a_3, \cdots, a_{n+1})$ of $\mathfrak{B}_n$ we have a whole neighborhood of points $(a_2^\blacktriangle, a_3^\blacktriangle, \cdots, a_{n+1}^\blacktriangle)$ given by the formula

$$(2.7) \quad a_{k+1}^\blacktriangle = a_{k+1} + \rho^2 \left\{ \sum_{\nu=1}^{N} C_\nu U_{k+1}(z_\nu) + \sum_{\nu=1}^{N} C_\nu^* V_{k+1}^*(z_\nu) \right\} + o(\rho^2),$$

where $k = 1, 2, \cdots, N$. Here the C_ν are absolutely arbitrary and the z_ν are subject only to the condition that $|z_\nu| < 1$. In general, this neighborhood will contain a sufficiently small full $2n$-sphere around the point. However, if at a certain point $a = (a_2, a_3, \cdots, a_{n+1})$ in $\mathfrak{B}_n$ the matrix

$$(2.8) \quad \begin{pmatrix} U_p(z_q) & V_p^*(z_s) \\ V_r(z_q) & U_r^*(z_s) \end{pmatrix} = M \qquad (p, r = 2, \cdots, n+1; q, s = 1, \cdots, n)$$

has a rank less than $2n$ for every choice of the z_ν, then the variation of a is restricted to a lower dimensional manifold. Let δa_{k+1} denote the term of order

ρ^2 in (2.7) and let us assume at first that the rank of M is $2n - 1$ for a general choice of the z_ν. In this case, all vectors $\delta a = (\delta a_2, \delta a_3, \cdots, \delta a_{k+1})$ will lie in a fixed $(2n - 1)$-dimensional hyperplane for all possible variations of the type (2.1). From the theory of linear equations we conclude that there exist constants λ_k, λ_k' such that for every $|z| < 1$ we have

$$(2.9) \qquad \sum_{k=1}^{n} \lambda_{k+1} U_{k+1}(z) + \sum_{k=1}^{n} \lambda_{k+1}' V_{k+1}(z) = 0.$$

Since in the matrix M the U_k and V_k appear in the same relations as the V_k^* and the U_k^*, we have also

$$(2.9') \qquad \sum_{k=1}^{n} \lambda_{k+1} V_{k+1}^*(z) + \sum_{k=1}^{n} \lambda_{k+1}' U_{k+1}^*(z) = 0.$$

Thus the vectors $(\lambda_2, \cdots, \lambda_{n+1}, \lambda_2', \cdots, \lambda_{n+1}')$ and $(\lambda_2'^*, \cdots, \lambda_{n+1}'^*, \lambda_2^*, \cdots, \lambda_{n+1}^*)$ satisfy both the linear equations (2.9) and (2.9'); but since the matrix M is of rank $2n - 1$ they can only differ by a constant factor. One recognizes easily that we may assume without restriction of generality that this factor is one, that is,

$$(2.10) \qquad \lambda_{k+1} = \lambda_{k+1}'^*.$$

Substituting from (2.6) and (2.6') we obtain a differential equation for $w = f(z)$ which describes the behavior of schlicht functions connected with exceptional points $a \subset \mathfrak{B}_n + \mathfrak{S}_n$ for which the rank of M is $2n - 1$. We find

$$(2.11) \qquad \left(\frac{z}{w}\frac{dw}{dz}\right)^2 P_n(w) = Q_n(z) \qquad\qquad (w = f(z)),$$

where

$$(2.12) \qquad P_n(w) = \sum_{k=1}^{n} \lambda_{k+1} S_{k+1}(1/w),$$

$$(2.12') \qquad Q_n(z) = \sum_{k=1}^{n} \left\{ \lambda_{k+1}\left(ka_{k+1} + \sum_{m=1}^{k}\frac{ma_m}{z^{k+1-m}}\right) + \lambda_{k+1}^* \sum_{m=1}^{k} ma_m^* z^{k+1-m} \right\}.$$

We see that all these functions $f(z)$ are still analytic for $|z| = 1$ except for finitely many singular points and that they map the unit circle on a domain in the w-plane which is bounded by a finite number of analytic arcs. This fact will play an important role in our further considerations.

We have next to consider the case where the rank of M is less than $2n - 1$ for every choice of the z_ν. By the same consideration as above we conclude that there are two linearly independent vectors λ_k, λ_k^* and λ_k', $\lambda_k'^*$ for which simultaneously

$$(2.13) \quad \sum_{k=1}^{n} [\lambda_{k+1} U_{k+1}(z) + \lambda_{k+1}^* V_{k+1}(z)] = 0, \quad \sum_{k=1}^{n} [\lambda_{k+1}' U_{k+1}(z) + \lambda_{k+1}'^* V_{k+1}(z)] = 0.$$

In this case we have two different and independent differential equations for $f(z)$. Dividing one by the other, we see that $f(z)$ is an algebraic function. Thus, the points in $\mathfrak{B}_n + \mathfrak{S}_n$ for which the rank of M is smaller than $2n - 1$ belong to a very special class of schlicht functions; one can easily see that the dimension of the manifold of all such singular points is less than $2n - 1$.

The study of points $a \subset \mathfrak{B}_n + \mathfrak{S}_n$ for which the rank of M is less than $2n$ is of great interest since all points $a \subset \mathfrak{S}_n$ are of this type. In fact, no point in $\mathfrak{S}_n$ can have a $2n$-dimensional neighborhood. If $\mathfrak{S}_n$ has a tangential hyperplane at the point a, the rank of M is necessarily $2n - 1$; since we have in view of (2.7) and (2.9)

$$(2.14) \qquad \mathrm{Re}\left\{ \sum_{k=1}^{n} \lambda_{k+1}\delta a_{k+1} \right\} = 0,$$

we obtain in this case a geometrical interpretation for the coefficients λ_k. The vector λ_k, λ_k^* represents the normal of the tangential hyperplane in question.

Not every point a, however, for which the rank of M is less than $2n$ is a point of $\mathfrak{S}_n$. The point a might show a singular behavior under variations of the special type (2.1) but fill a $2n$-dimensional neighborhood under more general variations. Thus, we are led to consider the most general variations of the function $f(z)$ connected with the point a. Let Γ be the image curve of the unit circle in the map $w = f(z)$; because of the analytic behavior of $f(z)$ for $|z| = 1$, Γ is composed of a finite number of analytic arcs. Let $\zeta = e^{i\varphi}$; then $\omega = f(\zeta)$ will describe the boundary curve Γ if ζ runs over the unit circle. We make now a deformation of Γ by shifting every point $\omega \subset \Gamma$ along a vector $\delta\omega$ which points into the direction of the normal and varies continuously along Γ. According to a well-known formula of Julia [5], we have for the variation of the corresponding mapping function $f(z)$

$$(2.15) \qquad \delta f(z) = zf'(z) \oint_{\Gamma} \frac{\zeta + z}{\zeta - z}\left[\frac{1}{2\pi i}\frac{\delta\omega\, d\omega}{f'(\zeta)^2 \zeta^2} \right].$$

$\omega = f(\zeta)$ describes Γ in the positive sense and the term in brackets is negative if $\delta\omega$ points into the domain, positive if it points out of it.

We remark at first that the variation (2.15) is, in general, not a permissible variation within the family $\mathfrak{F}$. One sees easily that the normalization at the origin is not preserved; it may, however, be easily reestablished by multiplication with an appropriate factor. Thus, we arrive finally at a variation of $f(z)$ which keeps it in $\mathfrak{F}$ and leads to the following variation of the coefficients a_k of $f(z)$:

$$(2.16) \qquad \delta a_k = (k - 1)a_k J_0 + 2 \sum_{\nu=1}^{k-1} \nu a_\nu J_{k-\nu}$$

with

$$(2.17) \qquad J_k = \oint_{\Gamma} \frac{1}{\zeta^k}\left[\frac{1}{2\pi i}\frac{\delta\omega\, d\omega}{f'(\zeta)^2 \zeta^2} \right].$$

From (2.12′) and (2.16), (2.17) we derive the formula

$$(2.18) \qquad \text{Re} \left\{ \sum_{k=1}^{n} \lambda_{k+1} \delta a_{k+1} \right\} = \text{Re} \left\{ \oint_{\Gamma} Q_n(\zeta) \left[\frac{1}{2\pi i} \frac{\delta \omega \, d\omega}{f'(\zeta)^2 \zeta^2} \right] \right\}.$$

The last variation which we still need is of the form

$$(2.19) \qquad f^{\blacktriangle}(z) = e^{-i\epsilon} f(e^{i\epsilon} z),$$

(ϵ real, small) which leads to the variation of the coefficients a_k:

$$(2.20) \qquad \delta a_k = i(k - 1) a_k \epsilon.$$

Thus we find under this variation

$$(2.20') \qquad \text{Re} \left\{ \sum_{k=1}^{n} \lambda_{k+1} \delta a_{k+1} \right\} = \text{Re} \left\{ i\epsilon \sum_{k=1}^{n} k \lambda_{k+1} a_{k+1} \right\}.$$

From (2.14), (2.18) and (2.20′) we can now derive a set of necessary conditions for the point a to lie in $\mathfrak{S}_n$. In fact, suppose the expression $\sum_{k=1}^{n} k\lambda_{k+1}a_{k+1}$ were not real. Then we might obtain for the term (2.20′) positive and negative values by appropriate choice of a real ϵ. The special variations (2.1) led to vectors δa which covered the whole $(2n - 1)$-dimensional hyperplane $\text{Re} \left\{ \sum_{k=1}^{n} \lambda_{k+1} \delta a_{k+1} \right\} = 0$. The additional variation (2.19) would give vectors δa which lead out of this plane on both sides; since the variations may be added by superposition, we see that in this case a whole $2n$-dimensional neighborhood of a would be covered and a could not lie in $\mathfrak{S}_n$. Thus, we are led to the first additional condition:

$$(2.21) \qquad \sum_{k=1}^{n} k\lambda_{k+1}a_{k+1} = \text{real}.$$

Then we may write $Q_n(z)$ in the more symmetrical form

$$(2.12'') \qquad Q_n(z) = \sum_{k=1}^{n} \left\{ \lambda_{k+1} \left(\tfrac{1}{2} k a_{k+1} + \sum_{m=1}^{k} \frac{m a_m}{z^{k+1-m}} \right) \right. $$
$$\left. + \lambda_{k+1}^{*} \left(\tfrac{1}{2} k a_{k+1}^{*} + \sum_{m=1}^{k} m a_m^{*} z^{k+1-m} \right) \right\}.$$

This shows that $Q_n(z)$ is real for $|z| = 1$ and, hence, we derive from (2.18) for the case of a Julia variation

$$(2.18') \qquad \text{Re} \left\{ \sum_{k=1}^{n} \lambda_{k+1} \delta a_{k+1} \right\} = \oint_{\Gamma} Q_n(\zeta) \left[\frac{1}{2\pi i} \frac{\delta \omega \, d\omega}{f'(\zeta)^2 \zeta^2} \right].$$

In view of the differential equation (2.11) we may write this also in the form

$$(2.18'' \qquad \text{Re} \left\{ \sum_{k=1}^{n} \lambda_{k+1} \delta a_{k+1} \right\} = \frac{1}{2\pi i} \oint_{\Gamma} P_n(\omega) \frac{\delta \omega \, d\omega}{\omega^2}.$$

From (2.18′) we conclude at first that the image domain of the unit circle has no exterior points. For otherwise we might vary the boundary curve Γ at a certain point ω in both directions and would obtain positive and negative values for Re $\{\sum_{k=1}^{n} \lambda_{k+1} \delta a_{k+1}\}$. This would lead, as before, to the conclusion that a is an interior point of $\mathfrak{B}_n$. Hence:

The function $f(z)$ belonging to a point $a \subset \mathfrak{S}_n$ must map the unit circle upon a slit domain. The slits Γ consist of a finite number of analytic arcs.

In the same fashion we see that $Q_n(\zeta)$ must have all over Γ the same sign, for otherwise the expression (2.18′) could again obtain both signs. Thus, we have for the right choice of the sign of the λ_k:

$$(2.22) \qquad\qquad Q_n(\zeta) \geq 0 \qquad\qquad (|\zeta| = 1)$$

and therefore for every Julia variation

$$(2.23) \qquad\qquad \text{Re} \left\{ \sum_{k=1}^{n} \lambda_{k+1} \delta a_{k+1} \right\} \leq 0.$$

We found the following necessary conditions for a function $f(z)$ to be associated with a point $a \subset \mathfrak{S}_n$. It must satisfy the differential equation (2.11) with a non-negative $Q_n(\zeta)$ for $|\zeta| = 1$ and with a real term (2.21); it must map the unit circle upon a slit domain.

If, on the other hand, a function $f(z)$ satisfies all these conditions, the corresponding point a is a boundary point of the coefficient domain $\mathfrak{B}_n$ and if the rank of the matrix M is exactly $2n - 1$ there exists a tangential hyperplane at this point with the normal vector λ_k, λ_k^*. This is easily concluded from the fact that the most general variation of the image domain may be approximated arbitrarily by the variations used above.

Thus, the boundary $\mathfrak{S}_n$ of $\mathfrak{B}_n$ consists of points a belonging to functions which are schlicht in $|z| < 1$, satisfy a differential equation (2.11) and map the unit circle onto the w-plane slit along analytic arcs with the equation

$$(2.24) \qquad \left(\frac{1}{w} \frac{dw}{d\tau}\right)^2 \sum_{k=1}^{n} \lambda_{k+1} S_{k+1}\left(\frac{1}{w}\right) + 1 = 0,$$

where τ is a suitably chosen real parameter (see [14]).

3. We can now derive a partial differential equation for $\mathfrak{S}_n$. We observe that if $f(z)$ is a function corresponding to a boundary point $a \subset \mathfrak{S}_n$, it maps the unit circle upon slits in the w-plane. The tip of a slit corresponds to a value $z = 1/\kappa = \kappa^*$ on the unit circle $|z| = 1$. Here $f'(z)$ must vanish at least in the first order, and so $1/\kappa$ is at least a second order zero for $Q_n(z)$. Hence, we have

$$Q_n(1/\kappa) = Q_n^\sim(\lambda_\nu, a_\nu, 1/\kappa)$$

$$(3.1) \qquad = \sum_{k=1}^{n} \left\{ \lambda_{k+1}\left(\tfrac{1}{2}k a_{k+1} + \sum_{m=1}^{k} m a_m \kappa^{k+1-m}\right) \right.$$

$$\left. + \lambda_{k+1}^*\left(\tfrac{1}{2}k a_{k+1}^* + \sum_{m=1}^{k} m a_m^* \kappa^{*k+1-m}\right) \right\} = 0$$

and at the same time

$$\kappa \frac{\partial \tilde{Q}_n(\lambda_\nu, a_\nu, 1/\kappa)}{\partial \kappa} = \sum_{k=1}^{n} \left\{ \lambda_{k+1} \sum_{m=1}^{k} m(k + 1 - m) a_m \kappa^{k+1-m} \right.$$

(3.2)

$$\left. - \lambda_{k+1}^* \sum_{m=1}^{k} m(k + 1 - m) a_m^* \kappa^{*k+1-m} \right\} = 0.$$

If we eliminate κ from these two equations we obtain a certain equation between the coordinates $(a_2, a_3, \cdots, a_{n+1})$ of the boundary point and the components of the normal vector at the point a with respect to $\mathfrak{S}_n$. At points of regularity, the surface $\mathfrak{S}_n$ may be represented by an equation

(3.3) $$F(a_2, a_2^*, a_3, a_3^*, \cdots, a_{n+1}, a_{n+1}^*) = 0$$

where F is a real-valued differentiable function. Writing

(3.4) $$F_k = \partial F / \partial a_k$$

we see that F satisfies the following partial differential equation of the first order:

$$\sum_{k=1}^{n} \left\{ F_{k+1} \left(\tfrac{1}{2} k a_{k+1} + \sum_{m=1}^{k} m a_m \kappa^{k+1-m} \right) \right.$$

(3.5)

$$\left. + F_{k+1}^* \left(\tfrac{1}{2} k a_{k+1}^* + \sum_{m=1}^{k} m a_m^* \kappa^{*k+1-m} \right) \right\} = 0,$$

where κ is to be eliminated by the condition

(3.6) $$\kappa \partial \tilde{Q}_n(F_\nu, a_\nu, 1/\kappa) / \partial \kappa = 0.$$

The differential equation (3.5) which may also be written in the form

(3.5') $$\tilde{Q}_n(F_\nu, a_\nu, 1/\kappa) = 0$$

becomes a complicated non-linear partial differential equation of the first order if we introduce κ into it by means of elimination from (3.6). It is much easier to investigate the differential equation by determining its characteristics. The equations of these characteristics are in general given by the following system of ordinary differential equations:

$$\frac{da_{k+1}}{dt} = 2 \frac{\partial \tilde{Q}_n}{\partial F_{k+1}}, \qquad \frac{dF_{k+1}}{dt} = -2 \frac{\partial \tilde{Q}_n}{\partial a_{k+1}},$$

(3.7)

$$\frac{dF}{dt} = 2 \sum_{k=1}^{n} \left(\frac{\partial \tilde{Q}_n}{\partial F_{k+1}} F_{k+1} + \frac{\partial \tilde{Q}_n}{\partial F_{k+1}^*} F_{k+1}^* \right).$$

In the special case which we are considering, Q_n depends on F_ν and a_ν in two different ways, directly and by means of κ which is a complicated function of

F_ν and a_ν. However, since $\partial Q_n^{\sim}/\partial \kappa = 0$ we may neglect the latter dependence when establishing the system of characteristic equations.

We find

$$(3.8) \qquad da_{k+1}/dt = ka_{k+1} + 2 \sum_{\nu=1}^{k} (k + 1 - \nu)a_{k+1-\nu}\kappa^\nu,$$

$$(3.9) \qquad dF_{k+1}/dt = -kF_{k+1} - 2(k + 1) \sum_{\nu=1}^{n-k} F_{k+1+\nu}\kappa^\nu,$$

and, finally,

$$(3.10) \qquad dF/dt = 2Q_n^{\sim}(F_\nu, a_\nu, 1/\kappa).$$

Here, κ is determined from the algebraic equation

$$(3.2') \qquad \sum_{k=1}^{n} \left\{ F_{k+1} \sum_{m=1}^{k} m(k + 1 - m)a_m \kappa^{k+1-m} \right.$$
$$\left. - F_{k+1}^* \sum_{m=1}^{k} m(k + 1 - m)a_m^* \kappa^{*k+1-m} \right\} = 0.$$

We have to choose a root of $(3.2')$ which has the modulus $|\kappa| = 1$, and (3.8), (3.9), $(3.2')$ form a closed system of equations which permits us to compute a piece of a characteristic.

Writing $\tau_{k+1} = \frac{1}{2}ka_{k+1} + \sum_{\nu=1}^{k} (k + 1 - \nu)a_{k+1-\nu}\kappa^\nu$, we may interpret the equation (3.5) to mean that the vector τ_k, τ_k^* is tangential to the surface. If we let $\kappa = e^{i\theta}$ it is clear that the vector $d\tau_k/d\theta$, $d\tau_k^*/d\theta$ will also be tangential to the surface and this is the geometrical meaning of $(3.2')$. Equation (3.5) may also be interpreted as the strip condition for the system of characteristics (3.8), (3.9) and $(3.2')$, as is obvious from (3.10).

The system of ordinary differential equations (3.8), (3.9) with the condition $(3.2')$ is of order $2n + 1$ since it contains the $2n + 1$ variables a_2, a_3, $\cdots$, a_{n+1}, F_2, F_3, $\cdots$, F_{n+1} and κ. It is possible, however, to introduce new unknowns which are combinations of the previous ones and which are determined by an analogous system of the order $n + 1$ only. We may do this by writing

$$(3.11) \qquad Q_n(z) = \sum_{k=-n}^{n} B_k/z^k \qquad (B_{-k} = B_k^*),$$

where

$$(3.12) \qquad B_k = \sum_{\nu=0}^{n-k} (\nu + 1)a_{\nu+1}F_{k+\nu+1},$$

$$(3.12') \qquad B_0 = \sum_{\nu=0}^{n} \nu a_{\nu+1}F_{\nu+1}.$$

By (3.8) and (3.9)

$$(3.13) \qquad dB_k/dt = -kB_k - 2 \sum_{\nu=1}^{n-k} (k - \nu)B_{k+\nu}\kappa^{\nu},$$

$$(3.13') \qquad dB_0/dt = 2 \sum_{\nu=1}^{n} \nu B_{\nu}\kappa^{\nu}.$$

The system (3.13) consists of n ordinary differential equations with n dependent variables $B_k(t)$ and the function $\kappa(t)$ which is defined by (3.2'). This latter equation may also be written in the form

$$(3.2'') \qquad \sum_{k=-n}^{n} kB_k\kappa^k = 0.$$

We see that (3.13) and (3.2'') form a closed system of $n + 1$ equations which permits a continuous determination of the $B_k(t)$ and $\kappa(t)$. This system does not give the whole information concerning the boundary surface $\mathfrak{S}_n$ but yields intermediate results with greater ease since it has only half the order of the complete system.

We observe that

$$(3.14) \qquad \begin{aligned} dB_{-k}/dt &= -kB_{-k} - 2 \sum_{\nu=1}^{n-k} (k - \nu)B_{-(k+\nu)}\kappa^{-\nu} \\ &= -kB_{-k} - 2 \sum_{\nu=-n}^{-(k+1)} (2k + \nu)B_{\nu}\kappa^{k+\nu}. \end{aligned}$$

On the other hand,

$$(3.14') \qquad \sum_{\nu=-n}^{n} (2k + \nu)B_{\nu}\kappa^{\nu} = 0$$

by (3.2'') and the fact that $Q_n(1/\kappa) = \sum_{\nu=-n}^{n} B_{\nu}\kappa^{\nu} = 0$. Combining (3.14) with (3.14') we obtain

$$dB_{-k}/dt = -kB_{-k} + 2 \sum_{\nu=-k}^{n} (2k + \nu)B_{\nu}\kappa^{k+\nu}$$

$$(3.14'') \qquad \begin{aligned} &= kB_{-k} + 2 \sum_{\nu=-k+1}^{n} (2k + \nu)B_{\nu}\kappa^{k+\nu} \\ &= kB_{-k} + 2 \sum_{\nu=1}^{n+k} (k + \nu)B_{-k+\nu}\kappa^{\nu}. \end{aligned}$$

Thus, the formula (3.13) is valid for all B_k with positive, vanishing or negative subscript.

The whole system of differential equations (3.13) may be condensed into one formula. For we see easily that $Q = Q_n(z, t)$ satisfies the partial differential equation

$$(3.15) \qquad \frac{\partial Q}{\partial t} = z \frac{\partial Q}{\partial z} \frac{1 + \kappa z}{1 - \kappa z} + \frac{4\kappa z}{(1 - \kappa z)^2} Q$$

which is equivalent to the whole system of differential equations (3.13).

Instead of reducing the partial differential equation to a system of ordinary differential equations (3.13) by comparing the coefficient of z^k on both sides, we may perform a similar transformation by splitting up $Q(z, t)$ into a product of linear functions of z and studying the variation of the roots of Q with t. Let $z_i = z_i(t)$ be a point where $Q_n(z_i, t) = 0$ and let us investigate the variation of z_i with t. We have by (3.15)

$$(3.16) \qquad \frac{dQ(z_i(t), t)}{dt} = \frac{\partial Q}{\partial z}\Big|_{z=z_i} \frac{dz_i}{dt} + \frac{\partial Q}{\partial t} = \left[\frac{dz_i}{dt} + z_i \frac{1 + \kappa z_i}{1 - \kappa z_i}\right] \frac{\partial Q}{\partial z}\Big|_{z=z_i} = 0.$$

Hence, if $(\partial Q/\partial z)|_{z=z_i} \neq 0$

$$(3.16') \qquad d(\log z_i)/dt = -(1 + \kappa z_i)/(1 - \kappa z_i).$$

In a similar fashion we may investigate the higher roots of Q. If, for example, z_i is a double root of Q we have by (3.15)

$$(3.17) \qquad \frac{d}{dt}\left(\frac{\partial Q}{\partial z}\right)_{z=z_i} = \left[\frac{dz_i}{dt} + z_i \frac{1 + \kappa z_i}{1 - \kappa z_i}\right] \frac{\partial^2 Q}{\partial z^2}\Big|_{z=z_i},$$

and hence, differentiating the identity (3.16) again with respect to t,

$$(3.17') \qquad \frac{d^2 Q(z_i(t), t)}{dt^2} = \left[\frac{dz_i}{dt} + z_i \frac{1 + \kappa z_i}{1 - \kappa z_i}\right] \frac{\partial^2 Q}{\partial z^2}\Big|_{z=z_i} = 0.$$

Since $(\partial^2 Q/\partial z^2)|_{z=z_i} \neq 0$ this leads once more to (3.16'). In the same way roots of any order may be treated. It is evident from our consideration that the order of each root z_i remains unchanged for varying t as long as $\kappa z_i \neq 1$.

If $|z_i| < 1$, we conclude from (3.16')

$$(3.18) \qquad d(\log |z_i|)/dt = -\operatorname{Re}\{(1 + \kappa z_i)/(1 - \kappa z_i)\} < 0$$

and we see that the modulus of any root of Q which is interior to the unit circle is decreasing in t. If, however, $|z_i| = 1$ then

$$(3.18') \qquad d(\log |z_i|)/dt = 0.$$

Thus as we move along a characteristic in the direction of increasing t the roots of Q on $|z| = 1$ remain there and the other roots move further away from $|z| = 1$.

It is often useful in numerical considerations to operate with the term $p_i = \frac{1}{2}(|z_i| + |z_i|^{-1})$. One recognizes easily from (3.16') that it satisfies the differential inequality

$$(3.19) \qquad p_i - 1 \leq dp_i/dt \leq p_i + 1.$$

Since $p_i \geq 1$ we see that p_i is a monotonic function of t the variation of which is easily estimated by means of (3.19).

We study the variation of $Q(z, t)$ best by setting

$$(3.20) \qquad Q(z, t) = B_n^* z^{-n} \prod_{i=1}^{2n} (z - z_i(t))$$

and introducing this representation into (3.15). Using logarithmic derivatives, we find

$$(3.21) \qquad \frac{dB_n^*/dt}{B_n^*} - \sum_{i=1}^{2n} \frac{dz_i/dt}{z - z_i} = \frac{1 + \kappa z}{1 - \kappa z} \left[\sum_{i=1}^{2n} \frac{z}{z - z_i} - n \right] + \frac{4\kappa z}{(1 - \kappa z)^2}.$$

Letting $z \to \infty$ and comparing both sides we find

$$(3.22) \qquad d(\log B_n^*)/dt = -n.$$

Comparing the residues at all points z_i which are different from $1/\kappa$ we have

$$(3.16'') \qquad d(\log z_i)/dt = -(1 + \kappa z_i)/(1 - \kappa z_i).$$

Further let $1/\kappa$ be a root of multiplicity k. Then comparing all terms $(z - 1/\kappa)^{-2}$ in (3.21) we find

$$(3.22) \qquad 2k = 4, \quad k = 2$$

on the assumption that all derivatives with respect to t are finite. If, therefore, $1/\kappa$ is a root of higher order than the second, obviously some derivative in (3.21) cannot be finite and we see that this leads to a singularity for the characteristics.

Let us assume at first that we are at a regular point of the characteristic. Then, the point $1/\kappa$ is a double zero of $Q(z, t)$ and comparing the residue at this point on both sides of (3.21), we find

$$(3.23) \qquad \frac{2d\kappa/dt}{\kappa^2} = -\frac{2}{\kappa} \sum_{z_i \neq 1/\kappa} \frac{1}{1 - \kappa z_i} + \frac{2(n - 1)}{\kappa},$$

that is,

$$(3.23') \qquad d(\log \kappa)/dt = -\tfrac{1}{2} \sum_{z_i \neq 1/\kappa} (1 + \kappa z_i)/(1 - \kappa z_i).$$

We remark that this result may also be obtained by observing that in view of (3.20) and the symmetry of $Q(z, t)$ we have

$$(3.24) \qquad B_n^* \kappa^{-2} \prod_{z_i \neq 1/\kappa} z_i = B_n$$

and hence

$$(3.25) \qquad \kappa^{-2} \prod_{z_i \neq 1/\kappa} z_i = \text{constant}.$$

This yields (3.23') immediately if (3.16') is known.

We have now in (3.16') and (3.23') a closed non-linear system of differential equations of the first order for the unknowns $z_i(t)$ and $\kappa(t)$. It is equivalent to the system of differential equations for the B_k since the $z_i(t)$ are just the roots of an algebraic equation with the B_k as coefficients. Its integration does not lead, therefore, to a full knowledge of the surface $\mathfrak{S}_n$ which is described by the original characteristic equations. It is, however, useful to obtain partial information by employing a less complicated system.

The case that κ^{-1} is of higher than the second order with respect to $Q(z, t)$ leads, in view of (3.11), to the conditions

$$(3.26) \qquad \sum_{k=-n}^{n} B_k \kappa^k = \sum_{k=-n}^{n} k B_k \kappa^k = \sum_{k=-n}^{n} k^2 B_k \kappa^k = 0.$$

This shows that the equation (3.2″) cannot be solved in a unique way locally in t. We have different possibilities of continuing the characteristic and κ as a function of t has a singularity at this point.

If we set $\kappa = e^{i\theta}$ and $z_i = e^{i\varphi_i}$, we obtain from (3.16′) the differential equation

$$(3.27) \qquad d\varphi_i/dt = - \cot \tfrac{1}{2}(\theta + \varphi_i).$$

This shows that if $\varphi_i > -\theta$ the angle φ_i will decrease and that it will increase for $\varphi_i < -\theta$. Hence all roots z_i on the unit circle move toward $1/\kappa = e^{-i\theta}$ as t increases. If one of these points z_i moves into κ^{-1} we have a singularity of the above type. A further increase in t then causes z_i (which is a zero of even order) to bifurcate into two roots each of half-order. One moves into $|z| < 1$ and the other moves into $|z| > 1$ and is the image of the first, thus preserving the symmetry of Q with respect to $|z| = 1$. Two roots will in general approach $1/\kappa$ simultaneously from either side. A more detailed explanation of this behavior is given in §6.

4. Let us consider a point a of $\mathfrak{S}_n$ where the function Q has several distinct roots on the unit circumference. These roots are at least of order 2 and so they satisfy

$$(4.1) \qquad Q = 0,$$

$$(4.2) \qquad z\partial Q/\partial z = 0.$$

Now these roots are of essentially two types; those that correspond to tips of slits in the $(w = f(z))$-plane and those that correspond to side points of the slits. If a zero maps into a side point of a slit it corresponds generally to a critical point of the slit—a point where two analytic arcs come together at an angle different from π.

The selection of one of these roots corresponding either to a tip or to a side point for the role of $1/\kappa$ will in general define a characteristic direction at the point a. Hence, if Q has several roots on $|z| = 1$ there are several characteristics through the point. Let the root selected be designated by $1/\kappa$, and assume for simplicity that $1/\kappa$ is not a root of order higher than 2, for in that case we have seen that κ has a singularity at a. More generally, we may select N roots $1/\kappa_j$ and then we obtain an $(N-1)$-parameter family of characteristics through the point. In fact, we introduce N real parameters α_j and consider the system of equations

$$
(4.1')\quad \sum_{k=1}^{n} \left\{ F_{k+1} \sum_{j=1}^{N} \alpha_j \left(\tfrac{1}{2} k a_{k+1} + \sum_{m=1}^{k} m a_m \kappa_j^{k+1-m} \right) \right.
$$
$$
\left. + F_{k+1}^* \sum_{j=1}^{N} \alpha_j \left(\tfrac{1}{2} k a_{k+1} + \sum_{m=1}^{k} m a_m^* \kappa^{*k+1-m} \right) \right\} = 0,
$$

$$
(4.2')\quad Q_n^{\sim}(F_\nu, a_\nu, \kappa_j^{-1}) = z \frac{\partial Q_n^{\sim}(F_\nu, a_\nu, z)}{\partial z} \bigg|_{z=1/\kappa_j} = 0 \qquad (j = 1, \cdots, N).
$$

For each fixed parameter vector α we may consider $(4.1')$ and $(4.2')$ as an algebraic first order partial differential equation for the unknown function $F = F(a_2, a_2^*, \cdots, a_{n+1}, a_{n+1}^*)$. In fact, from $(4.2')$ we may determine each κ_j as an algebraic function of the a_ν and F_ν. Introducing these algebraic expressions into $(4.1')$ we obtain the partial differential equation for F. The corresponding characteristic equations are

$$
da_{k+1}/dt = k a_{k+1} \sum_{j=1}^{N} \alpha_j + 2 \sum_{\nu=1}^{k} (k + 1 - \nu) a_{k+1-\nu} \sum_{j=1}^{N} \alpha_j \kappa_j^\nu,
$$

$$
(4.3)\quad dF_{k+1}/dt = -k F_{k+1} \sum_{j=1}^{N} \alpha_j - 2(k + 1) \sum_{\nu=1}^{n-k} F_{k+1-\nu} \sum_{j=1}^{N} \alpha_j \kappa_j^\nu,
$$

$$
dF/dt = 2 \sum_{j=1}^{N} \alpha_j Q_n^{\sim}(F_\nu, a_\nu, 1/\kappa_j).
$$

If we consider again the coefficients $B_k(t)$ as defined by (3.12) we obtain for them the system of differential equations

$$
(4.4)\quad \frac{dB_k}{dt} = -k B_k \sum_{j=1}^{N} \alpha_j - 2 \sum_{\nu=1}^{n-k} (k - \nu) B_{k+\nu} \sum_{j=1}^{N} \alpha_j \kappa_j^\nu \qquad (k = -n, \cdots, n).
$$

with the additional set of conditions

$$
(4.4')\quad \sum_{k=-n}^{n} B_k \kappa_j^k = 0, \qquad \sum_{k=-n}^{n} k B_k \kappa_j^k = 0 \qquad (j = 1, \cdots, N).
$$

This set of differential equations is equivalent to the one partial differential equation for $Q(z, t)$, namely

$$
(4.5)\quad \frac{\partial Q}{\partial t} = z \frac{\partial Q}{\partial z} \sum_{j=1}^{N} \alpha_j \frac{1 + \kappa_j z}{1 - \kappa_j z} + 4zQ \sum_{j=1}^{N} \frac{\alpha_j \kappa_j}{(1 - \kappa_j z)^2}.
$$

The variation of the slit according to (4.5) means geometrically that we have a slit with N terminal tips which are growing or being annihilated simultaneously with relative speeds depending on the choice of the α_i . When $\alpha_i < 0$, the corresponding tip is growing; when $\alpha_i > 0$, it is being annihilated.

In order to show that the characteristic curves determined by the system (4.3) lie on $\mathfrak{S}_n$ we have to reason as follows. Suppose that the point a belongs to a function $w = f(z)$ which maps the circle $|z| < 1$ upon a domain $\mathfrak{D}$ bounded by a slit with N proper terminal tips. That is, all zeros of Q on $|z| = 1$ (where Q is defined by the differential equation (2.11) for $w = f(z)$) are double zeros and map into tips in the w-plane. All neighboring points $a^{\blacktriangle}$ in $\mathfrak{S}_n$ then belong to functions $f^{\blacktriangle}(z)$ which satisfy differential equations of the same type as that for $f(z)$, but with slightly varied coefficients. Hence they map $|z| < 1$ upon domains $\mathfrak{D}^{\blacktriangle}$ bounded by a topologically equivalent "tree" of slits which has the same number N of terminal tips. Thus the point a is embedded in a surface which is characterized by a differential equation (4.1'), (4.2') for every fixed set of real parameter values α_i . For each such given set there exists exactly one characteristic which passes through that point and lies on the surface; it is determined by the set of equations (4.3) and so we are sure that if we follow a characteristic curve according to a system (4.3) we remain in $\mathfrak{S}_n$.

This type of reasoning, however, breaks down if a belongs to a function $f(z)$ of $\mathfrak{S}_n$ of the following class. Its corresponding function Q has N separate roots on $|z| = 1$ but one or more of these map not into terminal tips but into side points as mentioned at the beginning of this section. In this case, it is easily seen that in a portion of the neighborhood of a there are points $a^{\blacktriangle}$ belonging to functions $f^{\blacktriangle}(z)$ which map on slits with N or more tips and no zeros going into side points while another part of the neighborhood contains points $a^{\blacktriangle}$ belonging to functions $f^{\blacktriangle}(z)$ with fewer tips and for which the corresponding $Q^{\blacktriangle}$ has fewer than N roots on $|z| = 1$, no one of which goes into a side point. At the points $a^{\blacktriangle}$ of the second kind the differential equation (4.1'), (4.2') no longer holds. We can, therefore, only state that we stay on $\mathfrak{S}_n$ if we follow a characteristic (4.3) through a which leads to neighboring points of the first kind. If, on the other hand, we follow a characteristic through a in the wrong direction we may well leave $\mathfrak{S}_n$.

Geometrically, this means that if we are at a side point of the slit corresponding to a we may prolong a slit out from the side point but we cannot remove more of the slit structure at the side point without destroying its connectivity. This reasoning shows that the points a of $\mathfrak{S}_n$ belonging to such exceptional functions $f(z)$ lie on the intersection manifold of two surfaces of $\mathfrak{S}_n$ of quite different type. The one surface consists of points a obtained by prolonging slits from the side zeros in as many different directions as possible, thus eliminating the side zeros by placing them at tips. This surface is accessible by traveling out from the original point along characteristics. The other surface is not accessible by characteristics beginning at the original point, but corresponds to trees which are topologically the same as the original one minus side zeros. The structure of these two surfaces is quite different and this is reflected in the fact that the corresponding differential equations (4.1'),

(4.2′) are different. In the special case $\mathfrak{S}_2$ there are exactly two such surfaces which together form all of $\mathfrak{S}_2$. The analytic representation of these two surfaces has been given in [10], and their different structure is there obvious.

5. Now we generalize the above considerations by introducing in place of $f(z)$ the function

$$(5.1) \qquad f(z)^\mu = \sum_{k=0}^{\infty} a_{k+\mu}^{(\mu)} z^{k+\mu} \qquad\qquad (a_\mu^{(\mu)} = 1),$$

where μ is any real number, not necessarily an integer. Let $\mathfrak{B}_n^{(\mu)}$ denote the region of variability of the point $(a_{1+\mu}^{(\mu)}, a_{2+\mu}^{(\mu)}, \cdots, a_{n+\mu}^{(\mu)})$ and let $\mathfrak{S}_n^{(\mu)}$ be its boundary. When $\mu = 1$ we obtain the coefficient region $\mathfrak{B}_n^{(1)} = \mathfrak{B}_n$ considered above. We want to show in this section that the structure of $\mathfrak{S}_n^{(\mu)}$ is quite analogous to that of $\mathfrak{S}_n$ and may be studied in exactly the same fashion.

Let us define the polynomial $S_{k+\mu}^{(\mu)}(x)$ of degree k in x by the generating function

$$(5.2) \qquad f(z)^\mu / (1 - xf(z)) = \sum_{k=0}^{\infty} (a_{k+\mu}^{(\mu)} + S_{k+\mu}^{(\mu)}(x)) z^{k+\mu},$$

that is,

$$(5.2') \qquad S_{k+\mu}^{(\mu)}(x) = \sum_{\nu=1}^{k} a_{k+\mu}^{(\nu+\mu)} x^\nu.$$

The variation $f^\blacktriangle(z)$ of $f(z)$ defined by (2.1) and (2.3) may be used to define a corresponding variation formula for $f(z)^\mu$. It is easily verified that the formula analogous to (2.7) is

$$(5.3) \qquad (a_{k+\mu}^{(\mu)})^\blacktriangle = a_{k+\mu}^{(\mu)} + \rho^2 \left\{ \sum_{\nu=1}^{N} C_\nu U_{k+\mu}^{(\mu)}(z_\nu) + \sum_{\nu=1}^{N} C_\nu^* V_{k+\mu}^{(\mu)*}(z_\nu) \right\} + o(\rho^2),$$

$k = 1, 2, \cdots, n$, where

$$(5.4) \qquad U_{k+\mu}^{(\mu)}(z) = \frac{f(z)}{z^2 f'(z)^2} \left(k a_{k+\mu}^{(\mu)} + \sum_{\nu=0}^{k-1} \frac{(\nu + \mu) a_{\nu+\mu}^{(\mu)}}{z^{k-\nu}} \right) - \frac{\mu}{f(z)} S_{k+\mu}^{(\mu)} \left(\frac{1}{f(z)} \right),$$

$$(5.4') \qquad V_{k+\mu}^{(\mu)}(z) = \frac{f(z)}{z^2 f'(z)^2} \sum_{\nu=0}^{k-1} (\nu + \mu) a_{\nu+\mu}^{(\mu)} z^{k-\nu}.$$

By a repetition of the reasoning given above we find a differential equation for the function $w = f(z)$ belonging to the point $a^{(\mu)}$ of $\mathfrak{S}_n^{(\mu)}$, namely

$$(5.5) \qquad (zw^{-1} \, dw/dz)^2 P_n^{(\mu)}(w) = Q_n^{(\mu)}(z).$$

Here

$$(5.6) \qquad P_n^{(\mu)}(w) = \mu \sum_{k=1}^{n} \lambda_{k+\mu}^{(\mu)} S_{k+\mu}^{(\mu)}(1/w),$$

$$Q_n^{(\mu)}(z) = \sum_{k=1}^{n} \left\{ \lambda_{k+\mu}^{(\mu)} \left(k a_{k+\mu}^{(\mu)} + \sum_{\nu=0}^{k-1} \frac{(\nu+\mu) a_{\nu+\mu}^{(\mu)}}{z^{k-\nu}} \right) \right.$$

$$(5.6')$$

$$\left. + \lambda_{k+\mu}^{(\mu)*} \sum_{\nu=0}^{k-1} (\nu+\mu) a_{\nu+\mu}^{(\mu)*} z^{k-\nu} \right\}.$$

Similarly, the Julia variation (2.15) gives

$$(5.7) \qquad \operatorname{Re}\left\{ \sum_{k=1}^{n} \lambda_{k+\mu}^{(\mu)} \delta a_{k+\mu}^{(\mu)} \right\} = \oint_{\Gamma} Q_n^{(\mu)}(\zeta) \left[\frac{1}{2\pi i} \frac{\delta\omega\, d\omega}{f'(\zeta)^2 \zeta^2} \right]$$

and as in the case $\mu = 1$, we see that $Q_n^{(\mu)}(\zeta) \geq 0$ on $|\zeta| = 1$ with at least one zero there.

The derivation of the characteristic equations is the same as in the case $\mu = 1$. Let $1/\kappa$ be a zero of $Q_n^{(\mu)}(z)$ which corresponds to a terminal tip of a slit and let $\mathfrak{S}_n^{(\mu)}$ be represented at a point of regularity by an equation $F^{(\mu)}(a_{1+\mu}^{(\mu)}, a_{1+\mu}^{(\mu)*}, \cdots, a_{n+\mu}^{(\mu)}, a_{n+\mu}^{(\mu)*}) = 0$. Writing $F_{k+\mu}^{(\mu)} = \partial F^{(\mu)}/\partial a_{k+\mu}^{(\mu)}$ we obtain the characteristic equations

$$(5.8) \qquad da_{k+\mu}^{(\mu)}/dt = k a_{k+\mu}^{(\mu)} + 2 \sum_{\nu=1}^{k} (k+\mu-\nu) a_{k+\mu-\nu}^{(\mu)} \kappa^\nu,$$

$$(5.9) \qquad dF_{k+\mu}^{(\mu)}/dt = -k F_{k+\mu}^{(\mu)} - 2(k+\mu) \sum_{\nu=1}^{n-k} F_{k+\mu-\nu}^{(\mu)} \kappa^\nu,$$

$$(5.10) \qquad dF^{(\mu)}/dt = 2 Q_n^{(\mu)\sim}(F_{\nu+\mu}^{(\mu)}, a_{\nu+\mu}^{(\mu)}, 1/\kappa),$$

where

$$Q_n^{(\mu)\sim}(F_{\nu+\mu}^{(\mu)}, a_{\nu+\mu}^{(\mu)}, 1/\kappa) = \sum_{k=1}^{n} \left\{ F_{k+\mu}^{(\mu)} \left(\tfrac{1}{2} k a_{k+\mu}^{(\mu)} + \sum_{\nu=0}^{k-1} (\nu+\mu) a_{\nu+\mu}^{(\mu)} \kappa^{k-\nu} \right) \right.$$

$$(5.11)$$

$$\left. + F_{k+\mu}^{(\mu)*} \left(\tfrac{1}{2} k a_{k+\mu}^{(\mu)*} + \sum_{\nu=0}^{k-1} (\nu+\mu) a_{\nu+\mu}^{(\mu)*} \kappa^{*k-\nu} \right) \right\}.$$

Equations (5.8), (5.9) and (5.10) generalize (3.8), (3.9) and (3.10).

We again define a new variable $B_k^{(\mu)}$ by the equation

$$(5.12) \qquad Q_n^{(\mu)} = \sum_{k=-n}^{n} B_k^{(\mu)}/z^k \qquad\qquad (B_{-k}^{(\mu)} = B_k^{(\mu)*}),$$

where

$$(5.13) \qquad B_k^{(\mu)} = \sum_{\nu=0}^{n-k} (\nu + \mu) a_{\nu+\mu}^{(\mu)} F_{k+\mu+\nu}^{(\mu)} \qquad (k = 1, \cdots, n),$$

$$(5.13') \qquad B_0^{(\mu)} = \sum_{\nu=0}^{n} \nu a_{\nu+\mu}^{(\mu)} F_{\mu+\nu}^{(\mu)}.$$

By (5.8) and (5.9)

$$(5.14) \qquad dB_k^{(\mu)}/dt = -kB_k^{(\mu)} - 2 \sum_{\nu=1}^{n-k} (k - \nu) B_{k+\nu}^{(\mu)} \kappa^\nu,$$

and it may be shown as before that this differential equation holds for $k = -n$, $-n + 1, \cdots, n$. κ^{-1} is again determined as a root of the equation

$$(5.14') \qquad z \, dQ_n^{(\mu)}(z)/dz = 0$$

and (5.14), (5.14') form a closed system of ordinary differential equations which permits computation of all κ and B values from given initial values. This system may be condensed into one partial differential equation for $Q_n^{(\mu)} = Q_n^{(\mu)}(z, t)$:

$$(5.15) \qquad \frac{\partial Q_n^{(\mu)}}{\partial t} = z \frac{\partial Q_n^{(\mu)}}{\partial z} \frac{1 + \kappa z}{1 - \kappa z} + \frac{4\kappa z}{(1 - \kappa z)^2} Q_n^{(\mu)}.$$

We see that all $Q_n^{(\mu)}$ satisfy the same partial differential equation (3.15) whatever the value of μ.

Let us write

$$(5.16) \qquad \alpha_{k+\mu}^{(\mu)} = e^{(\mu-1)t} a_{k+\mu}^{(\mu)}, \qquad \phi_{k+\mu}^{(\mu)} = e^{-(\mu-1)t} F_{k+\mu}^{(\mu)}.$$

Equations (5.8) and (5.9) become

$$(5.8') \qquad d\alpha_{k+\mu}^{(\mu)}/dt = (k + \mu - 1)\alpha_{k+\mu}^{(\mu)} + 2 \sum_{\nu=1}^{k} (k + \mu - \nu)\alpha_{k+\mu-\nu}^{(\mu)} \kappa^\nu,$$

$$(5.9') \qquad d\phi_{k+\mu}^{(\mu)}/dt = -(k + \mu - 1)\phi_{k+\mu}^{(\mu)} - 2(k + \mu) \sum_{\nu=1}^{n-k} \phi_{k+\mu+\nu}^{(\mu)} \kappa^\nu.$$

Let us now define

$$(5.17) \qquad \alpha_\nu^{(\mu)} = 0 \qquad\qquad (\nu < \mu),$$

$$(5.18) \qquad \phi_\nu^{(\mu)} = 0 \qquad\qquad (\nu > n + \mu).$$

Letting $m = k + \mu$, the equations (5.8') and (5.9') can be written

$$(5.8'') \qquad d\alpha_m^{(\mu)}/dt = (m - 1)\alpha_m^{(\mu)} + 2 \sum_{\nu=1}^{m} (m - \nu)\alpha_{m-\nu}^{(\mu)} \kappa^\nu,$$

$$(5.9'') \qquad d\phi_m^{(\mu)}/dt = -(m - 1)\phi_m^{(\mu)} - 2m \sum_{\nu=1}^{n} \phi_{m+\nu}^{(\mu)} \kappa^\nu,$$

and we observe that these equations remain the same for every value μ. The system (5.8″) and (5.9″) has the same form with respect to each coefficient region $\mathfrak{B}_n^{(\mu)}$, independent of μ.

In particular, since the matrix $\alpha_s^{(r)}$ ($r, s = 2, \cdots , p$) is triangular, $\alpha_s^{(r)} = 0$ for $s < r$, $\alpha_r^{(r)} = 1$, we see that the solutions $(\alpha_2^{(\mu)}, \alpha_3^{(\mu)}, \cdots , \alpha_p^{(\mu)})$ ($\mu = 2, 3, \cdots , p$) of the equations (5.8″) for $m = 2, 3, \cdots , p$ always constitute a fundamental system of solutions. Thus

$$(5.19) \qquad \alpha_m = \alpha_m^{(1)} = c_2\alpha_m^{(2)} + c_3\alpha_m^{(3)} + \cdots + c_p\alpha_m^{(p)} \qquad (m = 2, \cdots , p),$$

where the c_j ($j = 2, 3, \cdots , p$) are constants, independent of t. Taking $p = n + 1$, we have

$$(5.19') \quad \alpha_m = \alpha_m^{(1)} = c_2\alpha_m^{(2)} + c_3\alpha_m^{(3)} + \cdots + c_{n+1}\alpha_m^{(n+1)} \quad (m = 2, \cdots , n + 1),$$

a relation which connects $\mathfrak{B}_n = \mathfrak{B}_n^{(1)}$ with the succeeding higher coefficient bodies. More generally, raising superscripts by $\mu - 1$

$$(5.19'') \qquad \alpha_m^{(\mu)} = c_1\alpha_m^{(\mu+1)} + c_2\alpha_m^{(\mu+2)} + \cdots + c_n\alpha_m^{(\mu+n)}.$$

This shows how we may descend from the coefficient body $\mathfrak{B}_n^{(\mu)}$ to the coefficient body $\mathfrak{B}_n^{(1)}$; if we were able to solve the coefficient problem completely for large positive integers μ we might by the above recursion formulas obtain complete information about $\mathfrak{B}_n^{(1)}$.

From (5.8″) and (5.9″) we derive for any value of μ (not necessarily an integer) and an integer l

$$(5.20) \qquad \frac{d}{dt} \sum_{k=1}^{n} \alpha_{k+\mu}^{(\mu+l)}\phi_{k+\mu}^{(\mu)} = 2\mu\alpha_\mu^{(\mu+l)} \sum_{k=1}^{n} \phi_{k+\mu}^{(\mu)}\kappa^k.$$

Consider first the case $l = 0$. We define $\phi_\mu^{(\mu)}$ by the formula

$$(5.21) \qquad \alpha_\mu^{(\mu)}\phi_\mu^{(\mu)} = - \sum_{k=1}^{n} \alpha_{k+\mu}^{(\mu)}\phi_{k+\mu}^{(\mu)} , \qquad \phi_\mu^{(\mu)} = e^{-(\mu-1)t} \sum_{k=1}^{n} \alpha_{k+\mu}^{(\mu)}\phi_{k+\mu}^{(\mu)} .$$

Then one sees easily

$$(5.21) \qquad d\phi_\mu^{(\mu)}/dt = -(\mu - 1)\phi_\mu^{(\mu)} - 2\mu \sum_{\nu=1}^{n} \phi_{\nu+\mu}^{(\mu)}\kappa^\nu,$$

and so (5.9″) is still valid under this definition. Thus when $l = 0$, (5.20) may be written in the form

$$(5.22) \qquad \frac{d}{dt} \sum_{k=0}^{n} \alpha_{k+\mu}^{(\mu)}\phi_{k+\mu}^{(\mu)} = 0.$$

If $l \geq 1$, we have $\alpha_\mu^{(\mu+l)} = 0$ by definition (5.17) and so we conclude from (5.20)

$$(5.23) \qquad \frac{d}{dt} \sum_{k=0}^{n} \alpha_{k+\mu}^{(\mu+l)}\phi_{k+\mu}^{(\mu)} = 0 \qquad (l = 0, 1, 2, \cdots).$$

Thus, we find

$$(5.24) \qquad \sum_{k=0}^{n} \alpha_{k+\mu}^{(\mu+1)} \phi_{k+\mu}^{(\mu)} = \text{constant}$$

along each characteristic. This relation shows that the vectors $\alpha^{(\mu+1)}$ and $\phi^{(\mu)}$ vary in a contragredient manner if the point a moves along a characteristic. Introducing into (5.24) the definitions (5.16), we obtain

$$(5.24') \qquad \sum_{k=0}^{n} a_{k+\mu}^{(\mu+1)} F_{k+\mu}^{(\mu)} = c_l^{(\mu)} e^{-lt},$$

where the $c_l^{(\mu)}$ are constants.

By virtue of (5.2') and (5.6) we have

$$(5.25) \qquad P_n^{(\mu)}(w) = \sum_{l=1}^{n} A_l^{(\mu)} w^{-l},$$

where

$$(5.26) \qquad A_l^{(\mu)} = \mu \sum_{k=l}^{n} a_{k+\mu}^{(l+\mu)} F_{k+\mu}^{(\mu)}.$$

Comparing (5.24') and (5.26), we find

$$(5.27) \qquad A_l^{(\mu)} = \mu^{-1} e^{-lt} c_l^{(\mu)}.$$

This result has a very important interpretation. Setting $w = e^{-t}\omega$, we may write

$$(5.28) \qquad P_n^{(\mu)}(e^{-t}\omega) = \mu^{-1} \sum_{l=1}^{n} c_l^{(\mu)} \omega^{-l},$$

and this expression is clearly independent of t. Let $f(z, t)$ be the schlicht function of the family $\mathfrak{F}$ which belongs to the point $a(t)$ on $\mathfrak{S}_n^{(\mu)}$ and define

$$(5.29) \qquad g(z, t) = e^t f(z, t) = e^t(z + a_2(t)z^2 + \cdots).$$

Since $f(z, t)$ satisfies the differential equation (5.5), we have for $\omega = g(z, t)$ the equation

$$(5.30) \qquad \left(\frac{z}{\omega}\frac{d\omega}{dz}\right)^2 P_n^{(\mu)}(e^{-t}\omega) = Q_n^{(\mu)}(z).$$

The function $\omega(z)$ maps the unit circle upon a schlicht domain in the ω-plane which is bounded by an analytic slit $\omega(\tau)$ (τ = real parameter) which satisfies the differential equation

$$(5.31) \qquad \left(\frac{1}{\omega}\frac{d\omega}{d\tau}\right)^2 \sum_{l=1}^{n} c_l^{(\mu)} \omega^{-l} + 1 = 0.$$

Since the parameter t does not occur in this differential equation the slits in the w-plane corresponding to the different maps $g(z, t)$ for varying t are always part of the same slit structure Γ defined by (5.31). Γ may be defined here as the set of all loci (5.31) obtained by starting at $w = \infty$. (For the purpose of investigating the dependence of ω on the coefficients $c_i^{(\mu)}$, it is better to define Γ as the set of all the loci (5.31) which can be obtained starting from any zero of $\sum_1^n c_i^{(\mu)} w^{-l}$, including $w = \infty$. The Γ-structure and its relation to coefficient regions is discussed in detail in [13].) If $w = f(z)$ is a function of $\mathfrak{S}_n^{(\mu)}$ which maps $|z| < 1$ onto the w-plane minus a subcontinuum γ of Γ and if $Q^{(\mu)}$ is the corresponding function (5.6'), then any zero z_i of $Q^{(\mu)}$ which does not map into a tip corresponds necessarily to a zero of $\sum_1^n c_i^{(\mu)} w^{-l}$. For if not then $f'(z_i) \neq 0$—contradicting the differential equation (5.5). Hence we may classify the zeros of Q into two groups: (i) those that map into tips and (ii) those that map into zeros of $\sum_1^n c_i^{(\mu)} w^{-l}$. We call zeros of type (i) "tip zeros" and those of type (ii) "side zeros". The side zeros map into knots of Γ. From a zero of $\sum_1^n c_i^{(\mu)} w^{-l}$ of multiplicity q there issue exactly $q + 2$ branches of Γ. If a subcontinuum γ contains a knot of Γ but does not contain pieces of all possible branches issuing from the knot, we say that γ is incomplete at the knot.

This fact now explains the dependence of $f(z, t)$ upon t. All conformal maps $f(z, t)$ are obtained by mapping the unit circle upon the w-plane slit along part of a fixed Γ-structure and normalizing by multiplication with an appropriate factor e^{-t}. This meaning of $f(z, t)$ is obvious if Löwner's theory is known; here we obtain the interpretation as a theorem on characteristics and their integrals and arrive at a new approach to the Löwner theory. We may say that a characteristic curve on $\mathfrak{S}_n^{(\mu)}$ corresponds to one fixed Γ slit-structure in the w-plane.

The interpretation of a characteristic curve on $\mathfrak{S}_n^{(\mu)}$ as image of a slit-structure in the w-plane illustrates clearly the general result of the characteristic theory that any two surfaces which satisfy the same partial differential equation intersect along a characteristic. Let us consider, in fact, the boundaries between any two regular pieces of $\mathfrak{S}_n^{(\mu)}$. These can be of two different types: (1) an intersection manifold of the type discussed above which is "one-sidedly" accessible by characteristics; (2) sets which correspond to Γ-structures satisfying two independent differential equations (5.31). Since the Γ-structure is invariant along a characteristic, the property of satisfying two differential equations persists and a characteristic beginning at a point of (2) must remain on (2). As for (1), a characteristic beginning at one of its points a will proceed along it provided that the terminal slits present at a are alone varied, no new slits being generated out of knots. The characteristic may leave the given intersection manifold of the two surfaces at a singularity produced by a zero z_i of $Q^{(\mu)}$ on $|z| = 1$ striking $1/\kappa$ (the case discussed at the end of §3).

We remark that the singular manifolds (2) are not necessarily of type (1). For it is easy to construct such a manifold [13]. Let

$$(5.32) \qquad \psi(w) = \frac{c_1}{w} + \frac{c_2}{w^2} + \cdots + \frac{c_k}{w^k} \qquad (k \geq 1).$$

Starting at $w = \infty$ prolong the locus Re $\psi(w) = 0$ until it has mapping radius unity and let $w = f(z)$ map $|z| < 1$ onto its exterior. On this locus $\psi'(w)\, dw$ is pure imaginary and, if m is a non-negative integer, $\psi(w)^{2m}$ is real. Let

$$(5.33) \qquad \left[\psi(w)^{2m} \psi'(w) z\, \frac{dw}{dz} \right]^2 = \left(\frac{z}{w}\, \frac{dw}{dz} \right)^2 P(w),$$

where $P(w) = A_{2(2m+1)}/w^{2(2m+1)} + \cdots + A_n/w^n$, $n = 2(2m+1)k$. By the reflection principle

$$(5.34) \qquad \left(\frac{z}{w}\, \frac{dw}{dz} \right)^2 P(w) = Q(z),$$

where $Q(z) = \sum_{-n}^{n} B_k z^{-k}$ and $Q(z) \geq 0$ on $|z| = 1$ with at least one zero there. The locus Re $\psi(w) = 0$ defines a Γ-structure. It is clear, rechoosing the c_i in (5.32) if necessary, that we can find a subcontinuum γ of mapping radius unity which is complete at each of its knots. The resulting function $f(z)$ defines a point of (2) which does not belong to (1).

We mention finally that the system of equations

$$(5.24'') \qquad \sum_{k=1}^{n} a_{k+\mu}^{(\mu+l)} F_{k+\mu}^{(\mu)} = c_l^{(\mu)} e^{-lt} \qquad\qquad (l = 1, 2, \cdots, n)$$

may serve to eliminate the $F_{k+\mu}^{(\mu)}$ by means of the $a_{k+\mu}^{(\mu)}$ and the integration constants $c_l^{(\mu)}$. Thus, given the constants $c_l^{(\mu)}$, we may express the components of the normal vector at a with respect to $\mathfrak{S}_n^{(\mu)}$ in terms of the coordinates of the point a. The different characteristics are distinguished by the different sets of constants $c_l^{(\mu)}$.

6. For simplicity, let us return to the case $\mu = 1$ and let a be a given boundary point of $\mathfrak{B}_n$, that is, a given point of $\mathfrak{S}_n$. If $f_1(z)$ is a function belonging to a, then f_1 extremalizes a certain form

$$(6.1) \qquad \mathrm{Re} \sum_{k=1}^{n} \lambda_{k+1} a_{k+1}.$$

If there were another function f_2 belonging to a, f_2 would extremalize the same form. Because f_1 and f_2 belong to the same point, they have the same coefficients a_{k+1} ($k = 1, 2, \cdots, n$) and it follows from (5.13), (5.13') and (5.26) (with $\mu = 1$, $\lambda_{k+1} = F_{k+1}$) that f_1 and f_2 satisfy the same differential equation (5.5) (with $\mu = 1$). It can then be shown that $f_1 = f_2$ identically. (For details see [13].) Thus there is a one-to-one correspondence between points of $\mathfrak{S}_n$ and boundary functions. This result may also be deduced from a theorem of Teichmüller [20] which implies in addition that the two-dimensional cross-section of $\mathfrak{B}_n$ for which $a_2 = a_2^{(0)}$, $a_3 = a_3^{(0)}$, $\cdots$, $a_n = a_n^{(0)}$, $a_\nu^{(0)}$ fixed, is convex.

Next, given a point $a = a(0)$ of $\mathfrak{S}_n$ let us start at this point at $t = 0$ and prolong a characteristic from it. To the point a there corresponds a unique Γ-structure defined by (5.31) (with $\mu = 1$), and this structure remains the same

as we proceed along the characteristic. Select a double zero $1/\kappa$ of Q at $t = 0$
which corresponds to a tip, and prolong the characteristic in the direction of t
increasing. The slits corresponding to $g(z, t)$ form a subcontinuum γ_t of Γ, and
γ_t decreases continually as t increases. Speaking intuitively, we may say that
Γ is being continuously annihilated and the point being annihilated at the
instant t is the image of $1/\kappa(t)$ by $g(z, t)$. The portion γ_t of Γ maps onto $|z| =$
1 by the inverse function $g^{-1}(w, t)$ and the remainder of Γ maps into $|z| < 1$
by g^{-1}. Let us define

$$(6.2) \qquad h(z, t) = g^{-1}\{g(z, 0), t\} = e^{-t}(z + b_2(t)z^2 + \cdots).$$

This function maps $|z| < 1$ onto the interior of the unit circle minus a set of
slits which corresponds to the portion of Γ destroyed from $t = 0$ up to time t.
The point of Γ being destroyed at time t is recreated at $1/\kappa(t)$ and pushed into
the interior of the unit circle. Thus the image of the destroyed Γ-structure
appears in the interior of the unit circle, entering by way of $1/\kappa$.

Whenever a knot of the Γ-structure is destroyed, the behavior of κ is singular.
We have already remarked that in this case κ has an algebraic singularity, but
now the geometrical meaning becomes apparent. A knot of Γ can only be destroyed if all terminal branches issuing from it have been destroyed except one
(by a terminal branch we here mean a branch ending in a free tip). In addition
to the terminal branch there is of course another branch issuing from the knot
which connects it with infinity. Let us suppose then that of the $q + 2$ possible
branches issuing from the knot, q have been destroyed. If we imagine that we
are at the tip of the terminal branch looking toward the knot, we see to the
right of the knot an angular opening $2\pi(r + 1)/(q + 2)$ corresponding to r
missing branches and to the left we see an angular opening $2\pi(l + 1)/(q + 2)$
corresponding to l missing branches on the left of the knot, where $r + l = q$.
The two sides of the knot corresponding to these two angles map into two
zeros z_r, z_l of the function Q where z_r is of multiplicity $2r$, z_l of multiplicity $2l$.
Looking at $1/\kappa$ from inside the unit circle z_r is to the right of $1/\kappa$, z_l to the left.
To z_r there is attached the image of that portion of the Γ-tree which belongs to
the r missing branches and a similar structure is attached to z_l. As t increases
and as we annihilate more and more of the one remaining branch, z_r and z_l
move toward $1/\kappa$ and strike it at the moment when all of the remaining branch
is destroyed. A moment later the knot appears recreated just below $1/\kappa$ and
plainly corresponds to a zero of Q of order q. (When a zero of Q is created in
$|z| < 1$, a twin (the image with respect to $|z| = 1$) is of course also created
in $|z| > 1$.) Thus the zeros of the function

$$(6.3) \qquad \Pi_n(w) = \sum_{k=1}^{n} c_k w^{-k}$$

which defines Γ by (5.31) finally become zeros of Q in $|z| < 1$.

Let us suppose, for example, that the subcontinuum γ_0 of Γ at $t = 0$ is complete at each of its knots. Then all zeros of $Q(z, 0)$ on $|z| = 1$ are double zeros which map into tips. When the first terminal branch is destroyed, the corresponding zero becomes a side zero. When an adjacent terminal branch is destroyed, the corresponding tip zero merges with this side zero. Q now has one quadruple side zero. The zeros of Q on $|z| = 1$ thus coalesce into groups and migrate toward $1/\kappa$. At $1/\kappa$ the groups fuse and form one new internal zero of Q and a corresponding external one (the image with respect to $|z| = 1$).

In the course of destroying Γ, we meet finitely many knots and each of these produces a singularity in κ. At these instants t_0 we make an algebraic substitution $t - t_0 = \tau^2$ and then $d\kappa/d\tau$ will remain finite.

Instead of destroying the branches one at a time, we may also annihilate several simultaneously as described in §4. For simplicity, however, let us suppose that we annihilate only one branch at a time. Whenever the subcontinuum γ_t of Γ is incomplete at a knot, we are at a point of $\mathfrak{S}_n$ corresponding to a one-sidedly accessible edge-manifold. We remain on manifolds of this type unless all such knots become destroyed, at which time we may move once more into a regular surface belonging to $\mathfrak{S}_n$. When the destruction of a branch leads to a side point, we are forced to choose a new zero of Q as $1/\kappa$. This choice defines a new characteristic direction at the point and we proceed along it. In general the characteristic along which we approached the point will make an angle different from π with the characteristic by which we leave it.

As t tends to infinity γ_t shrinks and eventually in the limit all of Γ is destroyed. Now removal of the points at 0 and ∞ may divide Γ into several components $\Gamma_1, \Gamma_2, \cdots, \Gamma_N$ and we then have the choice of destroying the Γ_ν one at a time or simultaneously. In either case we see from (5.27) that A_k tends to zero like e^{-kt}. From the differential equation for B_k, namely

$$(6.4) \qquad d(e^{kt}B_k)/dt = -2e^{kt} \sum_{\nu=1}^{n-k} (k + \nu)B_k\kappa',$$

we see that B_k tends to zero at least as fast as e^{-kt}, but may go to zero faster because of cancelling effects. Suppose that $k_0 \geq 1$ is the least k for which $A_k(0) \neq 0$; divide both sides of the differential equation (2.11) by $e^{-k_0 t}$ and let t tend to infinity. It is easily seen that in the limit we obtain the differential equation

$$(6.5) \qquad \left(\frac{z}{w}\frac{dw}{dz}\right)^2 \frac{A_{k_0}(0)}{w^{k_0}} = \sum_{-k_0}^{k_0} \frac{B_k}{z^k}.$$

This differential equation is satisfied only by star-like schlicht functions belonging to $\mathfrak{S}_n$. By a star-like schlicht function is meant a schlicht function which maps $|z| < 1$ onto a domain having the property that any point of it can be joined to the origin by a straight-line segment lying entirely in the domain. Hence starting at any point $a = a(0)$ of $\mathfrak{S}_n$ the progressive destruction of Γ eventually takes us to the portion $\mathfrak{S}_n^*$ of $\mathfrak{S}_n$ which corresponds to star-like functions. We see also that if we destroy the components Γ_1, Γ_2, $\cdots$, Γ_N one at a time we can only arrive at the point of $\mathfrak{S}_n^\triangle$ belonging to

$$(6.6) \qquad f(z) = z/(1 - e^{i\theta}z)^2.$$

It is thus of some interest to characterize $\mathfrak{S}_n^\triangle$, and this is easily done. For it corresponds to all star-like functions

$$(6.7) \qquad f(z) = z/\prod_{\nu=1}^{n}(1 - e^{i\theta_\nu}z)^{2\mu_\nu},$$

where the θ_ν are arbitrary and the μ_ν ($\nu = 1, 2, \cdots, n$) satisfy

$$(6.7') \qquad \mu_\nu = m_\nu/k, \qquad \sum_{\nu=1}^{n}\mu_\nu = 1.$$

Here m_ν and k are integers, $m_\nu \geq 0$, $1 \leq k \leq n$.

If we proceed from $a = a(0)$ in the direction of t decreasing, the situation is similar.

We may also reverse the above process by beginning with functions (6.6) or (6.7) and prolonging characteristics from them. Let us consider for simplicity the "Koebe point" on $\mathfrak{S}_n$ which corresponds to the particular function

$$(6.6') \qquad w = z/(1 - z)^2 = \sum_{k=0}^{\infty}(k + 1)z^{k+1}.$$

This point is of high degeneracy with respect to $\mathfrak{S}_n$ as the rank of the matrix (2.8) is exactly n. In fact, the function (6.6') satisfies all differential equations

$$(6.8) \quad \left(\frac{z}{w}\frac{dw}{dz}\right)^2 \sum_{k=1}^{n}\lambda_{k+1}S_{k+1}\left(\frac{1}{w}\right) = \sum_{k=1}^{n}\lambda_{k+1}\left\{k(k+1) + \sum_{m=1}^{k}m^2\left(\frac{1}{z^{k+1-m}} + z^{k+1-m}\right)\right\}$$

with arbitrary real λ_k. The characteristic equations of §3 have at the initial moment the form

$$(6.9) \qquad \frac{da_{k+1}}{dt} = 0, \quad \text{Im}\left\{\frac{dF_{k+1}}{dt}\right\} = 0, \quad \frac{d\kappa/dt}{\kappa}\sum_{k=-n}^{n}k^2 B_k\kappa^k + \sum_{k=-n}^{n}k\frac{dB_k}{dt}\kappa^k = 0.$$

Since by (3.12) the B_k have real initial values and κ has the initial value -1, we conclude from (6.9) that if $\sum_{-n}^{n}k^2 B_k\kappa^k \neq 0$, then

$$(6.10) \qquad d\kappa/dt = 0.$$

It may be shown by continued differentiation that all higher derivatives of κ vanish as well so that κ is constantly -1. This leads easily to the consequence that the characteristic considered does not lead out of the Koebe point and that we are stuck at the function (6.6'). This point thus appears as a singular point with respect to the general theory of the characteristic differential equations.

The same holds for all rotated Koebe points on $\mathfrak{S}_n$ corresponding to the more general functions (6.6). We know that by successive slit annihilation we may connect each point on $\mathfrak{S}_n$ by a characteristic curve with a Koebe point and that we may even consider the whole surface $\mathfrak{S}_n$ as the integral conoid of all characteristics through the Koebe points. But the Koebe points appear as nodal singular points which are in general reached along the integral curves only after an infinitely long time. If we want to construct $\mathfrak{S}_n$ by all solutions of the characteristic equations starting from the Koebe points, we have to apply Poincaré's general theory on singular points of systems of differential equations. It is obvious that the whole surface $\mathfrak{S}_n$ may be swept out by characteristics if a small neighborhood of each Koebe point on $\mathfrak{S}_n$ is known.

A Koebe point ceases to be a true nodal point if the corresponding B_k are such that

$$(6.11) \qquad \sum_{k=-n}^{n} k^2 B_k \kappa^k = 0.$$

In this case the equation $Q = 0$ has at the point $z = 1/\kappa$ a root of at least fourth order and we see easily that it is possible to prolong a slit from the tip of the straight-line slit corresponding to the function (6.6′) and the prolonged slit will not be a straight continuation of the rectilinear one. In general if Q has on $|z| = 1$ other roots than the root $1/\kappa = -1$, we may start from the Koebe point and reach other points on $\mathfrak{S}_n$ in a finite time. All these points are characterized by the property that the corresponding schlicht functions $f(z)$ map the unit circle upon the w-plane with a slit boundary which has a rectilinear component at infinity.

The above reasoning may be carried over to the more general functions (6.7). If we start at a point belonging to one of these functions, we travel along special boundary manifolds which are characterized by the property that the Γ-structure in the neighborhood of ∞ lies on rays. These special manifolds are characterized by the statement that any point of them can be reached in finite time starting from a star-like schlicht function. However, if we consider any subset Σ of $\mathfrak{S}_n$ which contains $\mathfrak{S}_n^{\blacktriangle}$ in its interior, then all points of $\mathfrak{S}_n$ may be reached in finite time from points of Σ. A manifold Σ may be constructed by adding to it all points $a^{\blacktriangle}$ obtained from it by small perturbations. In this way we obtain a parameterization of $\mathfrak{S}_n$.

7. A different parameterization has been used in [12], [13] with the aid of curves which will shortly be described. These curves have an interesting duality relationship with the characteristic curves described above.

In order to understand this duality, let us begin with some interpretive remarks.

Let

$$(7.1) \qquad w = g(z, t) = e^t(z + a_2(t)z^2 + \cdots)$$

map $|z| < 1$ onto the w-plane minus a subcontinuum γ_t of Γ as described above, and define (as in formula (6.2))

$$(7.2) \qquad h(z, t) = g^{-1}\{g(z, 0), t\} = e^{-t}(z + b_2(t)z^2 + \cdots),$$

that is, $g\{h(z, t), t\} = g(z, 0)$. The function $h(z, t)$ maps $|z| < 1$ into the interior of the unit circle minus a slit structure χ_t. χ_t is the image by $g^{-1}(w, t)$ of that portion of the slit system of $g(z, 0)$ which has been destroyed at the time t. When $t = 0$, $h(z, 0) = z$ and as $t \to +\infty$,

$$(7.2') \qquad e^t h(z, t) \to g(z, 0).$$

Let us identify two points on $|z| = 1$ which map by $g(z, 0)$ into the same point of the plane but which lie on opposite edges of a slit. Identified points are considered to be equivalent in the sense that together they form a point of a new space. We then say that $g(z, 0)$ realizes this identification. It is clear that $h(z, t)$ realizes only a part of the identification but realizes more and more of it as t increases until in the limit it realizes all of it.

Thus the functions $g(z, t)$ correspond to pieces of one and the same slit, whereas the functions $h(z, t)$ correspond to pieces of one and the same identification of points on $|z| = 1$. In one case the slit is fixed and the identification on $|z| = 1$ is varying, in the other the identification is fixed and the slit is varying.

Along a characteristic we have

$$(7.3) \qquad da_{k+1}/dt = ka_{k+1} + 2\sum_{\nu=1}^{k} (k + 1 - \nu)a_{k+1-\nu}\kappa^\nu.$$

This implies that g and h satisfy the differential equations

$$(7.4) \qquad \frac{\partial g}{\partial t} = z\frac{\partial g}{\partial z}\frac{1 + \kappa z}{1 - \kappa z},$$

$$(7.4') \qquad \frac{\partial h}{\partial t} = -h\frac{1 + \kappa h}{1 - \kappa h}.$$

Here $1/\kappa$ maps by $g(z, t)$ into the tip of the slit being destroyed at time t. In terms of h, $1/\kappa$ is the point on the unit circumference in the h-plane at which the slit is being recreated at time t.

Let the c_k ($k = 1, 2, \cdots, n$) be defined by (5.24) with $\mu = 1$ and write

$$(7.5) \qquad \Pi_n(w) = \sum_{k=1}^{n} c_k/w^k.$$

We have seen that $w = g(z, t)$ satisfies the differential equation

$$(7.6) \qquad \left(\frac{z}{w}\frac{dw}{dz}\right)^2 \Pi_n(w) = Q_n(z, t),$$

where $Q_n(z, t) = \sum_{-n}^{n} B_k(t)z^{-k}$. Along a characteristic the c_k are constant and the B_k are varying, and we have

$$(7.7) \qquad \frac{\partial \Pi}{\partial t} = 0, \qquad \frac{\partial Q}{\partial t} = z \frac{\partial Q}{\partial z} \frac{1 + \kappa z}{1 - \kappa z} + \frac{4\kappa z}{(1 - \kappa z)^2} Q.$$

The first of these equations expresses the fact that $\Pi(w)$ is an integral of the characteristic equations.

Now consider the dual system of curves which have the same relation to $h(z, t)$ as the characteristics have to $g(z, t)$. Let a be a point of $\mathfrak{S}_n$ and let $f(z) = g(z, 0)$ be the function belonging to it. Then $w = f(z)$ satisfies an equation

$$(7.8) \qquad \left(\frac{z}{w}\frac{dw}{dz}\right)^2 \Pi(w) = Q(z)$$

where $Q(z) = Q(z, 0)$. Writing

$$(7.9) \qquad Z = \int (Q(z))^{\frac{1}{2}} z^{-1} \, dz, \qquad |\, dZ \,|^2 = |\, Q(z)z^{-2} \,| \,|\, dz \,|^2,$$

this defines a metric on $|\, z \,| = 1$. It is clear from (7.8) that in the identification of points on $|\, z \,| = 1$ realized by $f(z)$ the metric (7.9) is invariant.

Conversely, let $Q(z) = \sum_{-n}^{n} B_k z^{-k}$, $B_{-k} = B_k^*$, be given, where $Q(z) \geq 0$ on $|\, z \,| = 1$ with at least one zero there. Let $2m$ arcs of $|\, z \,| = 1$ ($m \geq 1$) be identified in pairs such that: (i) the unit circle $|\, z \,| \leq 1$ with boundary arcs identified in this way is equivalent to the sphere; (ii) if two identified arcs abut, the common end-point is a zero of Q with at most one exception; (iii) in the identification of any pair of arcs the metric (7.9) is preserved. Then it can be shown that there is a schlicht function

$$(7.10) \qquad f(z) = z + \cdots$$

which realizes the identification and that there is a corresponding $\Pi(w)$,

$$(7.11) \qquad \Pi(w) = \sum_{k=1}^{n} c_k / w^k$$

such that $f(z)$ satisfies the corresponding differential equation (7.8). If $Q(z)$ has N zeros on $|\, z \,| = 1$, the identification can be chosen in ∞^{N-1} ways.

Let $g(z, t)$ map $|\, z \,| < 1$ onto a subcontinuum γ_t of the structure Γ satisfying

$$(7.12) \qquad \left(\frac{1}{w}\frac{dw}{d\tau}\right)^2 \Pi(w) + 1 = 0.$$

We suppose, for simplicity, that only one branch is being annihilated at any given instant t in which case g will satisfy (7.9).

Now $h(z, t)$ has realized only a part of the identification on $|\, z \,| = 1$ at time t, but at a pair of points on $|\, z \,| = 1$ which have been identified we see that

$$(7.13) \qquad \left(\frac{z}{w}\frac{dw}{dz}\right)^2 \Big/ Q(z) = -\frac{1}{w^2} (dw)^2 \Big/ \left|\frac{Q(z)}{z^2}\right| \,|\, dz \,|^2$$

has the same value. This is a consequence of the fact that the metric (7.9) is preserved in the identification. Thus the expression (7.13) considered as a function of $w = h(z, t)$ is single-valued across the slits in the circle $| w | = | h | < 1$. Let

$$(7.14) \qquad \Pi(w, t) = \sum_{k=-n}^{n} c_k(t) w^{-k} \qquad (c_{-k} = c_k^*, \, c_0 \text{ real})$$

be chosen such that it vanishes with the proper orders at points in $| w | = | h | < 1$ which are images of zeros of $Q(z)$ in $| z | < 1$. At points on $| z | = 1$ which map into points on $| w | = | h | = 1$ where slits enter, $(dw/dz)^2$ will be infinite but these infinities can be cancelled by making $\Pi(w, t)$ have appropriate zeros at these points. Points on $| z | = 1$ where $Q(z)$ vanishes either map into tips of slits in the h-plane or they correspond to side points of slits. In either case these zeros of $Q(z)$ are cancelled by corresponding zeros of $(dw/dz)^2 = (dh/dz)^2$ or by making Π vanish with suitable multiplicity. Thus $\Pi(w, t)$ must vanish at points corresponding to some but not all of the zeros of $Q(z)$ and in addition at certain points where $(dw/dz)^2$ becomes infinite. Although the expression (7.13) may become infinite like $(z - z_0)^{-p/q}$, p, q integers, we remark that when expressed as a function of w it becomes infinite like a negative integer power of $(w - w_0)$. It also turns out that we never need more zeros than are provided by (7.14) to achieve this cancelling. The expression

$$(7.15) \qquad \left[\left(\frac{z}{w} \frac{dw}{dz} \right)^2 \Big/ Q(z) \right] \Pi(w, t)$$

is real on $| w | = | h | = 1$ and finite and single-valued in $| w | = | h | < 1$. By the reflection principle it is finite, single-valued and regular everywhere and so must equal a constant. Without loss of generality we may take this constant to be unity. Then

$$(7.16) \qquad \left(\frac{z}{w} \frac{dw}{dz} \right)^2 \Pi(w, t) = Q(z).$$

Taking logarithms of both sides of (7.16) and differentiating with respect to t (this argument is similar to the one given in [17] to obtain a differential equation for $Q(z, t)$), we have

$$\frac{d}{dt} \log \left\{ \left(\frac{z}{w} \frac{dw}{dz} \right)^2 \Pi(w, t) \right\} = 0$$

and so

$$d\{\log \Pi(w, t)\}/dt = -2\partial \log (w'/w)/\partial t = 4\kappa w/(1 - \kappa w)^2$$

by (7.4′). By (7.4′) we have also

$$\frac{d}{dt} \left\{ \Pi(w, t) \right\} = -w \frac{\partial \Pi}{\partial w} \frac{1 + \kappa w}{1 - \kappa w} + \frac{\partial \Pi}{\partial t},$$

and so

(7.17)
$$\frac{\partial \Pi}{\partial t} = w \frac{\partial \Pi}{\partial w} \frac{1 + \kappa w}{1 - \kappa w} + \frac{4\kappa w}{(1 - \kappa w)^2} \Pi$$

which is of the same form as the second equation in (7.7). Thus the dual of (7.7) is

(7.7′)
$$\frac{\partial \Pi}{\partial t} = w \frac{\partial \Pi}{\partial w} \frac{1 + \kappa w}{1 - \kappa w} + \frac{4\kappa w}{(1 - \kappa w)^2} \Pi, \qquad \frac{\partial Q}{\partial t} = 0.$$

As $t \to \infty$, $e^t h(z, t)$ tends to the boundary function $f(z) = g(z, 0)$ corresponding to $Q(z)$ and the chosen identification, and

(7.18)
$$\Pi(h(z, t), t) \to \Pi(f(z))$$

where $\Pi(w)$ is defined by (7.11).

The other formulas for these curves may be obtained using (7.7′). (Numerical calculations concerning $\mathfrak{S}_3$ based on the method of §7 have been carried out under a project at Stanford University sponsored by the Office of Naval Research (see [12]).)

8. Along a characteristic we have seen that $g(z, t)$ satisfies

(8.1)
$$\frac{\partial g}{\partial t} = z \frac{\partial g}{\partial z} \frac{1 + \kappa z}{1 - \kappa z}.$$

If $G[g(z, t)]$ is any function of $g(z, t)$, we observe that we also have

(8.1′)
$$\frac{\partial G}{\partial t} = z \frac{\partial G}{\partial z} \frac{1 + \kappa z}{1 - \kappa z}$$

since the term G' cancels. In particular, we may take

(8.2)
$$G[g(z, t)] = \frac{g(z, t)^\mu}{1 - xg(z, t)} = e^{\mu t} \sum_{k=0}^{\infty} (a_{k+\mu}^{(\mu)} + S_{k+\mu}^{(\mu)}(xe^t)) z^{k+\mu}.$$

Thus

(8.3)
$$\frac{d(a_{k+\mu}^{(\mu)} + S_{k+\mu}^{(\mu)}(xe^t))}{dt} = k(a_{k+\mu}^{(\mu)} + S_{k+\mu}^{(\mu)}(xe^t))$$
$$+ 2 \sum_{\nu=1}^{k} (k + \mu - \nu)(a_{k+\mu-\nu}^{(\mu)} + S_{k+\mu-\nu}^{(\mu)}(xe^t)) \kappa^\nu.$$

By (5.8) we obtain for $S_{k+\mu}^{(\mu)}(xe^t)$ the equation

(8.4)
$$\frac{dS_{k+\mu}^{(\mu)}(xe^t)}{dt} = kS_{k+\mu}^{(\mu)}(xe^t) + 2 \sum_{\nu=1}^{k} (k + \mu - \nu) S_{k+\mu-\nu}^{(\mu)}(xe^t) \kappa^\nu$$

which is of exactly the same form as (5.8).

The $(k + 1)$-st Faber polynomial T_{k+1} of $f(z, t) = e^{-t}g(z, t)$ has been defined in [16] by the formula

$$\log (1 - xf) - \log (f/z) = - \sum_{k=0}^{\infty} \frac{1}{k + 1} T_{k+1}(x)z^{k+1},$$

that is,

$$(8.5) \qquad \log [1 - xe^{-t}g(z, t)] - \log \frac{e^{-t}g(z, t)}{z} = - \sum_{k=0}^{\infty} \frac{1}{k + 1} T_{k+1}(x)z^{k+1}$$

or

$$G[g(z, t)] = \log [1 - xg(z, t)] - \log g(z, t)$$

$$(8.5')$$

$$= t - \log z - \sum_{k=0}^{\infty} \frac{1}{k + 1} T_{k+1}(xe^t)z^{k+1}.$$

By (8.1') we obtain

$$(8.6) \qquad \frac{dT_{k+1}(xe^t)}{dt} = (k + 1)T_{k+1}(xe^t) + 2(k + 1) \sum_{\nu=1}^{k+1} T_{k+1-\nu}(xe^t)\kappa^{\nu}.$$

The curves originally considered by Löwner [6] do not lie on the boundary of $\mathfrak{B}_n$ but in its interior. In fact, they extend from the origin $a_{k+1} = 0$ $(k = 1, 2, \cdots, n)$ to some other interior point of $\mathfrak{B}_n$. To obtain these curves let

$$(8.7) \qquad g(z, t) = e^{t-T}(z + a_2(t)z^2 + \cdots)$$

where $g(z, t) = z$ and $g(z, 0)$ is a function (no longer unique) belonging to the interior point a of $\mathfrak{B}_n$. The function $g(z, t)$ satisfies the same equation (8.1). If we wish to obtain Löwner formulas for the coefficients $a_{k+\mu}^{(\mu)}$ we take

$$(8.8) \qquad G[g(z, t)] = g(z, t)^{\mu} = e^{\mu t} \sum_{k=0}^{\infty} a_{k+\mu}^{(\mu)} z^{k+\mu}$$

and this function satisfies (8.1'). Equating coefficients of like powers of z in the differential equation we obtain

$$(8.9) \qquad da_{k+\mu}^{(\mu)}/dt = ka_{k+\mu}^{(\mu)} + 2 \sum_{\nu=1}^{k} (k + \mu - \nu)a_{k+\mu-\nu}^{(\mu)}\kappa^{\nu}$$

which agrees with (5.8). However the terminal conditions are different and we have now

$$(8.9') \qquad a_{\nu+\mu}^{(\mu)}(T) = 0 \qquad \qquad (\nu = 1, 2, \cdots).$$

Integrating (8.9), we have

$$(8.10) \qquad a_{k+\mu}^{(\mu)}(t) = -2e^{kt} \int_t^T e^{-k\tau} \sum_{\nu=1}^{k} (k + \mu - \nu)a_{k+\mu-\nu}^{(\mu)}\kappa^{\nu} \, d\tau.$$

Thus

$$a_\mu^{(\mu)} = 1, \qquad a_{1+\mu}^{(\mu)} = -2\mu e^t \int_t^T e^{-\tau} \kappa(\tau) \, d\tau,$$

$$a_{2+\mu}^{(\mu)} = -2\mu e^{2t} \int_t^T e^{-2\tau_1} \kappa(\tau_1)^2 \, d\tau_1$$

$$(8.11) \qquad\qquad + 4\mu(\mu + 1)e^{2t} \int_{\tau_1=t}^T \int_{\tau_2=\tau_1}^T e^{-\tau_1-\tau_2} \kappa(\tau_1) \kappa(\tau_2) \, d\tau_1 \, d\tau_2 ,$$

$$\cdots \quad \cdots \quad \cdots \quad \cdots ,$$

$$a_{k+\mu}^{(\mu)} = e^{kt} \sum (-1)^m \Gamma_{\alpha_1 \alpha_2 \cdots \alpha_m} \int_{\tau_1=t}^T \cdots \int_{\tau_m=\tau_{m-1}}^T e^{-\Sigma^m_1 \alpha_\nu \tau_\nu} \prod_1^m \kappa(\tau_\nu)$$

$$\cdot \, d\tau_1 \cdots d\tau_m ,$$

where

$$(8.11') \qquad \begin{aligned} \Gamma_{\alpha_1 \alpha_2 \cdots \alpha_m} &= 2^m(k + \mu - \alpha_1)(k + \mu - \alpha_1 - \alpha_2) \\ &\qquad \cdots (k + \mu - \alpha_1 - \alpha_2 - \cdots - \alpha_m) \end{aligned}$$

and the summation is over all sets of integers α_1, α_2, $\cdots$, α_m for which $\alpha_1 + \alpha_2 + \cdots + \alpha_m = k$ and $\alpha_i \geq 1$. m can be any integer from 1 to k, and each order of integers is counted except those which result by interchanging equal integers.

We observe that

$$(8.12) \qquad \begin{aligned} \lim_{\mu \to \infty} \frac{a_{k+\mu}^{(\mu)}}{\mu^k} &= (-1)^k e^{kt} 2^k \int_{\tau_1=t}^T \cdots \int_{\tau_k=\tau_{k-1}}^T e^{-\Sigma^k_1 \tau_\nu} \prod_1^k \kappa(\tau_\nu) \, d\tau_1 \cdots d\tau_k \\ &= (-1)^k \frac{2^k}{k!} e^{kt} \left(\int_{\tau_1=t}^T e^{-\tau_1} \kappa(\tau_1) \, d\tau_1 \right)^k, \end{aligned}$$

$$(8.12') \qquad \lim_{\mu \to -\infty} \frac{a_{k+\mu}^{(\mu)}}{\mu^k} = \frac{2^k}{k!} \left(\int_{\tau_1=t}^T e^{-\tau_1} \kappa(\tau_1) \, d\tau_1 \right)^k.$$

Taking $t = 0$, $T = \infty$ and writing

$$(8.13) \qquad l_k = k! 2^{-k} \lim_{\mu \to \infty} a_{k+\mu}^{(\mu)} \mu^{-k},$$

we obtain the limit variability region $\mathfrak{B}_n^{(\infty)}$ of the point $(l_1, l_2, \cdots, l_n)$. Points on the boundary of $\mathfrak{B}_n^{(\infty)}$ correspond to $\kappa = e^{i\theta}$, where θ is a real constant independent of τ. Hence the only function belonging to the boundary of $\mathfrak{B}_n^{(\infty)}$ is the Koebe function

$$(8.14) \qquad f(z) = z/(1 + e^{i\theta}z)^2.$$

In $\mathfrak{B}_n^{(\infty)}$ we have $l_k = l_1^k$ where l_1 is subject only to the restriction that $|\, l_1| \leq 1$. Thus $\mathfrak{B}_n^{(\infty)}$ is a 2-dimensional surface bounded by the spherical gauge-curve

$$(8.15) \qquad e^{i\theta}, e^{2i\theta}, \cdots, e^{ni\theta} \qquad\qquad (\theta \text{ real}).$$

Hence for large μ the structure of $\mathfrak{B}_n^{(\mu)}$ is rather trivial. Since $|\, a_{n+\mu}^{(\mu)}|$ always has a local maximum at points corresponding to functions (8.14) and since $\mathfrak{B}_n^{(\mu)}$ depends continuously on μ, we easily conclude that for all sufficiently large μ, $|\, a_{n+\mu}^{(\mu)}|$ has no other local maxima. This raises the interesting question whether this statement remains true for small μ—in particular for $\mu = 1$.

It is easily shown that

$$(8.16) \qquad |\, a_{2+\mu}^{(\mu)}| \leq \begin{cases} |\, \mu\,| \,(1 + 2e^{2(\mu-1)/(\mu+1)}) & (|\, \mu\,| \leq 1), \\[2mm] \mu(2\mu + 1) & (|\, \mu\,| > 1). \end{cases}$$

Finally, let $\varphi(w, t)$ be the function inverse to $w = h(z, t)$ where $h(z, t) = g^{-1}(g(z, 0), t)$. The terminal conditions for $\varphi(w, t)$ are

$$(8.17) \qquad \varphi(w, 0) = w, \qquad \varphi(w, T) = g^{-1}(w, 0) = e^{T}(z + \cdots),$$

and $\varphi(w, t)$ satisfies the equation

$$(8.18) \qquad \frac{\partial \varphi}{\partial t} = w \frac{\partial \varphi}{\partial w} \frac{1 + \kappa w}{1 - \kappa w}.$$

Writing

$$(8.19) \qquad \varphi(w, t) = e^{t}(w + \beta_2 w^2 + \cdots),$$

we see that $\{\varphi(w, t)\}^{\mu}$ also satisfies (8.18) and so

$$(8.20) \qquad d\beta_{k+\mu}^{(\mu)}/dt = k\beta_{k+\mu}^{(\mu)} + 2 \sum_{\nu=1}^{k} (k + \mu - \nu)\beta_{k+\mu-\nu}^{(\mu)}\kappa^{\nu},$$

where $\beta_{k+\mu}^{(\mu)}(0) = 0$ for $k > 0$. Integrating (8.20) we obtain formulas which differ from (8.11) only in the absence of a minus sign:

$$(8.21) \qquad \beta_{k+\mu}^{(\mu)} = e^{kt} \sum \Gamma_{\alpha_1 \alpha_2 \cdots \alpha_m} \int_{\tau_1 = t}^{T}$$

$$\cdots \int_{\tau_m = \tau_{m-1}}^{T} e^{-\sum^{m}_1 \alpha_r \tau_r} \prod_{1}^{m} \kappa(\tau_r)^{\alpha_r} \, d\tau_1 \cdots d\tau_m.$$

We obtain in this case the estimate

$$(8.22) \qquad |\beta_{2+\mu}^{(\mu)}| \leq \begin{cases} |\mu|(1 + 2e^{(\mu+3)/(\mu+1)}) & (|\mu+2| \leq 1), \\ \mu(2\mu+3) & (|\mu+2| > 1). \end{cases}$$

9. The above methods yield information about the structure of the coefficient domains in the large. Such information is not only interesting in itself, but may lead to the solution of specific extremal problems such as, for example, the determination of the exact bound for the n-th coefficient a_n of schlicht functions.

It was shown in [15] that any function $w = f(z)$ which maximizes $|a_{n+1}|$ maps $|z| < 1$ onto the w-plane minus slits which have no finite critical points. Thus, if $\mathfrak{T}_n$ is that portion of $\mathfrak{S}_n$ defined by functions $w = f(z)$ whose equations (2.11) have the property that $P_n(w)$ does not vanish at finite points of the slits of f, we know that the maximum of $|a_{n+1}|$ occurs on $\mathfrak{T}_n$. Variational methods have frequently been applied to the problem of maximizing $|a_{n+1}|$. In addition to first order terms, higher order terms in the variation may also be considered. It turns out that the second order terms do not yield any essentially new information concerning the maximizing functions. The reason is that if a function satisfies a differential equation obtained from the first order terms, the second order terms lead to a real decrease in value of the coefficient considered. This result, at first glance disappointing, becomes of value if applied to all points of $\mathfrak{T}_n$ and leads to certain convexity relations. In fact, we have only to consider linear forms $L_n = \mathrm{Re}\{\sum_{k=1}^{n} \lambda_{k+1}a_{k+1}\}$ which are maximal at points of $\mathfrak{T}_n$, and we find that the second order terms decrease L_n. The second order terms involve $P_n(w)$, and the decrease is a consequence of the property that $P_n(w)$ does not vanish.

In this way a knowledge of the structure of $\mathfrak{B}_n$ in the large may throw light on the existence of other possible local maxima and thus lead to a solution of the problem. A more detailed discussion of this convexity will be taken up in a subsequent paper.

REFERENCES

1. J. BASILEWITSCH, *Zum Koeffizientenproblem der schlichten Funktionen*, Matematiceski Sbornik (Recueil Mathématique) (N. S.), vol. 1(1936), pp. 211–228.
2. G. GOLUSIN, *Method of variations in the theory of conform representation. I*, Matematiceski Sbornik (Recueil Mathématique) (N. S.), vol. 19(1946), pp. 203–236.
3. G. GOLUSIN, *Method of variations in the theory of conform representation. II*, Matematiceski Sbornik (Recueil Mathématique) (N. S.), vol. 21(1947), pp. 83–117.
4. G. GOLUSIN, *Method of variations in the theory of conform representation. III*, Matematiceski Sbornik (Recueil Mathématique) (N. S.), vol. 21(1947), pp. 119–132.
5. GASTON JULIA, *Sur une équation aux dérivées fonctionnelles liée à la représentation conforme*, Annales Scientifiques de l'École Normale Supérieure (3), vol. 39(1922), pp. 1–28.
6. KARL LÖWNER, *Untersuchungen über schlichte konforme Abbildungen des Einheitskreises. I*, Mathematische Annalen, vol. 89(1923), pp. 103–121.

7. E. PESCHL, *Zur Theorie der schlichten Funktionen*, Journal für die reine und angewandte Mathematik, vol. 176(1937), pp. 61–94.
8. A. C. SCHAEFFER AND D. C. SPENCER, *The coefficients of schlicht functions*, this Journal, vol. 10(1943), pp. 611–635.
9. A. C. SCHAEFFER AND D. C. SPENCER, *The coefficients of schlicht functions. II*, this Journal, vol. 12(1945), pp. 107–125.
10. A. C. SCHAEFFER AND D. C. SPENCER, *The coefficients of schlicht functions. III*, Proceedings of the National Academy of Sciences of the United States of America, vol. 32(1946), pp. 111–116.
11. A. C. SCHAEFFER AND D. C. SPENCER, *A variational method in conformal mapping*, this Journal, vol. 14(1947), pp. 949–966.
12. A. C. SCHAEFFER AND D. C. SPENCER, *The coefficients of schlicht functions, IV*, Proceedings of the National Academy of Sciences of the United States of America, vol. 35(1949), pp. 143–150.
13. A. C. SCHAEFFER AND D. C. SPENCER, *Coefficient Regions for Schlicht Functions*, American Mathematical Society Colloquium Publications (to appear).
14. MENAHEM SCHIFFER, *A method of variation within the family of simple functions*, Proceedings of the London Mathematical Society (2), vol. 44(1938), pp. 432–449.
15. MENAHEM SCHIFFER, *On the coefficients of simple functions*, Proceedings of the London Mathematical Society (2), vol. 44(1938), pp. 450–452.
16. MENAHEM SCHIFFER, *Variation of the Green function and theory of the p-valued functions*, American Journal of Mathematics, vol. 65(1943), pp. 341–360.
17. MENAHEM SCHIFFER, *Sur l'équation différentielle de M. Löwner*, Comptes Rendus de l'Académie des Sciences, Paris, vol. 221(1945), pp. 369–371.
18. MENAHEM SCHIFFER, *Hadamard's formula and variation of domain-functions*, American Journal of Mathematics, vol. 68(1946), pp. 417–448.
19. D. C. SPENCER, *Some problems in conformal mapping*, Bulletin of the American Mathematical Society, vol. 53(1947), pp. 417–439.
20. OSWALD TEICHMÜLLER, *Ungleichungen zwischen den Koeffizienten schlichter Funktionen*, Sitzungsberichte der preussischen Akademie der Wissenschaften, Physikalisch-mathematische Klasse, 1938, pp. 363–375.

PURDUE UNIVERSITY,
HARVARD UNIVERSITY,
AND
STANFORD UNIVERSITY.

Commentary on

[28] A. C. Schaeffer, M. Schiffer and D. C. Spencer, *The coef"cient regions of schlicht functions*, Duke Math. J. **16** (1949), 493 527.

While colleagues at Stanford in the 1940s, Albert Schaeffer and Donald Spencer produced a series of papers on coefficient regions for schlicht (= univalent) functions, culminating in their well-known book [SS2]. Although it appeared earlier, the paper [28] is actually a sequel to the book. Largely expository, it provides new insights into the description of coefficient regions as developed in [SS2]. In particular, it highlights a remarkable connection with Loewner's theory [L] and Schiffer's paper [16].

In the notation of [SS2], V_n denotes the nth coefficient region for the class S. In other words, V_n is the subset of $\mathbb{C}^{n-1}$ consisting of points $(a_2, a_3, \ldots, a_n)$ that correspond to initial coefficients of some function $f(z) = z + a_2 z^2 + \cdots + a_n z^n + \ldots$ analytic and univalent in the unit disk. The "coefficient problem" in its most general form is to describe the region V_n. A theoretical description of the boundary ∂V_n was developed in [SS2], and the purpose of [28] is to clarify and elucidate the main lines of argument. The authors begin by showing how a suitable version of Schiffer's method of interior variation, as introduced in [13], can be used to derive a partial differential equation for the surface ∂V_n. (Schaeffer and Spencer had previously based a derivation on a more general variational method given in [SS1].) The characteristic curves of this partial differential equation are then found to be governed by Loewner's equation, and the method of [16] effectively leads to an explicit form for the characteristics.

General features of the regions V_n are described in [SS2] (see also [D]). Bieberbach's result of 1916 shows that V_2 is the closed disk $|a_2| \leq 2$. For $n = 3$, Schaeffer and Spencer [SS2]

integrated the differential equation to give a complete parametric description of ∂V_3 in terms of elementary functions. Because that description was complicated, it was difficult to visualize the surface; and so the authors commissioned the construction of two physical models of 3-dimensional cross sections of the 4-dimensional region V_3, one for a_2 real and the other for a_3 real. Color plates of these models appear as a frontispiece of their book [SS2]. Their paper [SS3] summarizes the parametric description for a general mathematical readership and gives further information about construction of the models.

By the time I met Don Spencer (at Princeton in 1968), he had left function theory for differential geometry, but he told me that his conversion had been motivated by efforts to describe the coefficient regions for univalent functions. He also told me a favorite story about his journey back to Stanford with the two models, which had been constructed at a special factory in Indiana. When he went to the railroad station, he explained to the ticket agent that he would need a compartment because he would be traveling with two models. "Aren't you lucky," came the instant reply.

References

[D] Peter L. Duren, *Univalent Functions*, Springer-Verlag, 1983.

[L] Karl Löwner (Charles Loewner), *Untersuchungen über schlichte konforme Abbildungen des Einheitskreises, I*, Math. Ann. **89** (1923), 103 121.

[SS1] A. C. Schaeffer and D. C. Spencer, *A variational method in conformal mapping*, Duke Math. J. **14** (1947), 949 966.

[SS2] A. C. Schaeffer and D. C. Spencer, *Coef"cient Regions for Schlicht Functions*, American Mathematical Society, 1950.

[SS3] A. C. Schaeffer and D. C. Spencer, *Models illustrating the third coef"cient region for schlicht functions*, Scripta Math. **16** (1950), 67 71 (2 plates).

PETER DUREN

[29] (with P. R. Garabedian) On existence theorems of potential theory and conformal mapping

[29] (with P. R. Garabedian) On existence theorems of potential theory and conformal mapping. *Ann. of Math.* **52** (1950), 164–187.

ANNALS OF MATHEMATICS
Vol. 52, No. 1, July, 1950

ON EXISTENCE THEOREMS OF POTENTIAL THEORY AND CONFORMAL MAPPING

By P. R. GARABEDIAN[1] AND M. SCHIFFER

(Received June 20, 1949)

1. Introduction

The object of the present paper is to give a new set of proofs of some fundamental existence theorems of conformal mapping and potential theory by using the notion of a kernel function introduced by Bergman [2, 3, 4]. The proofs which we present here are a more or less natural outcome of the recent researches done in this theory [5, 6, 7, 11, 12]. Our new attack upon the existence problem has on the one hand the advantage of generality, since it is applicable in the case of partial differential equations of elliptic type and since for functions of several complex variables the theory of the kernel function is well developed, while on the other hand the approach is simple and elementary in comparison with, say, the method using the Dirichlet principle. For potential theory and conformal mapping in multiply-connected plane domains, our proof is probably as simple as any known, and even for simply-connected domains the method has the advantage that it is based on a general theory rather than upon special artifices. A further point which we gain is that the kernel functions associated with the various norms of function theory and elliptic partial differential equations are well adapted to numerical computation, in view of their relation to orthogonal functions, so that we obtain at one and the same time an existence proof and a computational algorithm. The most serious drawback in our method is, perhaps, that we must make assumptions upon the smoothness of the boundary of the domains we consider, so that the general case is reached only after a topological approximation argument is given.

From the broader point of view, our treatment is important in that it yields a unified attack upon various existence problems. An extensive class of existence theorems of function theory of one or more variables and of partial differential equations of elliptic type can be developed using one basic procedure. This procedure can be outlined as follows. We set up the reproducing kernel function in a given function space by solving a suitable extremal problem, or alternatively by means of orthogonal functions, and we investigate the scalar product of the kernel function with what may be described as the fundamental singularity associated with this space. A local argument is used to study the behavior of this scalar product as the infinity of the fundamental singularity crosses the boundary of the domain of definition of the functions in our class, and the desired existence theorems follow from the geometric properties thus obtained. The local investigations depend in a general way upon a knowledge of our problem in simple domains, such as, for example, the unit circle.

---

[1] National Research Fellow.

164

The proofs may also be viewed from the standpoint of the variational method. Here we consider the fundamental singularity with an infinity outside the domain, which belongs to our function space, as a simple type of variation function for the extremal problem which determines the kernel function. This point of view is illustrated best in the last section, where the notion of a kernel function is much less in evidence.

Another aspect to which we should draw attention is that in this paper we obtain existence theorems using each of the better known compact classes of functions. Thus in Sections 1 and 2 we use normal families, while in Section 3 we use square-integrable functions forming a Hilbert space, and in the last section we use functions of bounded variation. For these three classes, the compactness follows, respectively, from Vitali's theorem, from the Riesz-Fischer theorem, and from Helly's theorem.

In this paper we do not try to present a complete discussion of the many cases in which our fundamental idea yields the existence proofs, but we rather restrict ourselves to the most important examples. It is significant, however, that there are very many function spaces and norms in these spaces which lead to highly interesting existence proofs which may easily suggest themselves to the reader. We shall, in particular, consider only plane regions, although the restriction to two dimensions is by no means essential. In this way it is hoped that our method will be made more easily understandable and will in time reach its broadest realm of application. Thus we present our material in four sections. In the first we show the existence of the conformal mappings of a multiply-connected plane domain upon the well-known parallel slit domains obtained often from the Dirichlet principle. We base our considerations in this section upon Bergman's original kernel function. In the second section, we use the kernel function of the space of real harmonic functions normed by the energy integral. We prove the existence of Green's function and thus open the way to the solution of the Dirichlet problem for harmonic functions, and we give indications of the generalization to the study of partial differential equations of elliptic type. Our method in this section is the simplest of all and should be of considerable interest because of the facility with which it allows one to present in completeness a fundamental part of potential theory. In the third section, we replace the area orthogonality by an orthogonalization over the boundary of the domain. We obtain in this way a set of important canonical conformal mappings whose existence has heretofore been quite difficult to prove for multiply-connected domains. The second new proof of Riemann's mapping theorem which we obtain with this boundary norm is again closely tied in with the theory of Hilbert space. In the final section, we leave Hilbert space and base our considerations instead upon functions of bounded variation and a length integral which is not a quadratic form. However, we are lead back by several remarkable identities to the very results which we found in Section 3.

For a better insight into the consequences of the identities and results which we derive here from basic elementary principles, the reader should refer to the recent work on kernel functions [5, 6, 7, 8, 11, 12]. It is to be hoped that the independent basis upon which we are now able to put this theory will help to place it among the fundamental tools of function theory, potential theory and partial differential equations.[2]

2. Proof by orthogonalization over the domain

Let D be a finite domain of the z-plane bounded by n simple analytic curves C. We shall denote by $C_1, \cdots, C_n$ the components of C and we shall denote by $B_1, \cdots, B_n$ the n domains of the z-plane which are complementary to D and are bounded, respectively, by $C_1, \cdots, C_n$. Let $\mathfrak{L}^2$ denote the class of functions $f(z)$ which are analytic in D and have there a finite Lebesgue integral

$$\iint_D |f'(z)|^2 \, d\tau, \qquad f'(z) = \frac{df(z)}{dz},$$

where $d\tau$ denotes the area element of D at the point z. Then it is possible to introduce in $\mathfrak{L}^2$ a complete system of functions $\{\varphi_\nu(z)\}$ which are orthonormal in the sense

$$\iint_D \phi'_\nu(z)\overline{\varphi'_\mu(z)} \, d\tau = \delta_{\nu\mu},$$

where $\delta_{\nu\mu}$ denotes, as usual, the Kronecker delta. The completeness of the system $\{\varphi_\nu(z)\}$ is taken to mean that each function $f(z) \,\epsilon\, \mathfrak{L}^2$ has a representation

$$f'(z) = \sum_{\nu=1}^{\infty} a_\nu \varphi'_\nu(z), \qquad a_\nu = \iint_D f'(z)\overline{\varphi'_\nu(z)} \, d\tau,$$

which is uniformly convergent in every closed subdomain of D.

It is not difficult to show that the kernel function

$$K(z, t) = \sum_{\nu=1}^{\infty} \varphi'_\nu(z)\overline{\varphi'_\nu(t)}$$

is uniformly convergent in every closed subdomain of D and that the function

$$M(z, t) = \sum_{\nu=1}^{\infty} \varphi_\nu(z)\overline{\varphi'_\nu(t)}, \qquad \frac{\partial M(z, t)}{\partial z} = K(z, t),$$

is in $\mathfrak{L}^2$. The reproducing property

$$(1) \qquad\qquad f'(t) = \iint_D \overline{K(z, t)}f'(z) \, d\tau$$

of $K(z, t)$ is then a simple consequence of Parseval's identity. For detailed discussion and proofs of these results, see [3].

We shall now show directly the existence of the reproducing kernel function

---

[1] (*Added in proof*) In a recent paper "Anwendung orthogonaler Systeme auf gewisse funktionentheoretische Extremal-und Abbildungsprobleme," Ann. Acad. Sci. Fennicae AI 59, (1949), Olli Lehto also used the method of the kernel function in order to prove the existence of the canonical slit mappings. This method is equivalent to the reasoning of our first section.

$K(z, t)$ by using the fact that $M(z, t)K(t, t)^{-1}$ is the function of class $\mathfrak{L}^2$ which yields the minimum of the norm

$$\iint_D |f'(z)|^2 \, d\tau$$

under the condition $f'(t) = 1$. It is for the sake of the completeness of this paper that we carry out this proof.

By the theory of normal families of analytic functions, there exists an extremal function $f_0(z) \in \mathfrak{L}^2$ which solves the above minimum problem. Consider now any function $f(z) \in \mathfrak{L}^2$ and set

$$\varphi(z) = f'(t)f_0(z) - f(z).$$

Clearly $\varphi'(t) = 0$, and hence $f_0(z) + \lambda\varphi(z)$ is a competing function for our minimum problem for every complex number λ. It follows that

$$\iint_D |f_0'(z)|^2 \, d\tau \leqq \iint_D |f_0'(z) + \lambda\varphi'(z)|^2 \, d\tau$$

$$= \iint_D |f_0'(z)|^2 \, d\tau + 2\text{Re}\left\{\lambda \iint_D \overline{f_0'(z)}\phi'(z) \, d\tau\right\} + |\lambda|^2 \iint_D |\varphi'(z)|^2 \, d\tau.$$

We conclude from the arbitrariness of λ that

$$\iint_D \overline{f_0'(z}\, \varphi'(z) \, d\tau = 0,$$

whence

$$f'(t) \iint_D |f_0'(z)|^2 \, d\tau = \iint_D \overline{f_0'(z)}f'(z) \, d\tau.$$

Thus the function

$$f_0'(z) \left\{\iint_D |f_0'(z)|^2 \, d\tau\right\}^{-1}$$

has the reproducing property (1) and we can define $K(z, t)$ by means of this expression. We obtain readily as a corollary that $K(z, t)$ is unique and depends only upon the domain D and not, for example, upon the particular choice of a system $\{\varphi_r(z)\}$.

We now apply (1) to the special function

$$f'(z) = \frac{1}{(z - w)^2},$$

where w is any point exterior to D. We obtain the relation

$$\iint_D \frac{\overline{K(z, t)} \, d\tau}{(z - w)^2} = \frac{1}{(t - w)^2}.$$

Integration with respect to w yields

(2) $$I(w, t) = \iint_D \frac{\overline{K(z, t)}d\tau}{z - w} = \frac{1}{t - w} + b_\mu,$$

for $w \in B_\mu$, $\mu = 1, \cdots, n$.

We now investigate the integral

$$I(w, t) = \iint_D \frac{\overline{K(z, t)}\, d\tau}{z - w}$$

for values of w which lie in D. This scalar product of the kernel function with the fundamental singularity $(z - w)^{-1}$ is obviously convergent, since $K(z, t)$ is continuous in D, and setting

$$w = u + iv, \qquad \frac{\partial}{\partial w} = \frac{1}{2}\left(\frac{\partial}{\partial u} - i\frac{\partial}{\partial v}\right), \qquad \frac{\partial}{\partial \overline{w}} = \frac{1}{2}\left(\frac{\partial}{\partial u} + i\frac{\partial}{\partial v}\right),$$

we obtain the expression

$$I(w, t) = 2\frac{\partial}{\partial w} \iint_D \overline{K(z, t)} \log\frac{1}{|z - w|}\, d\tau.$$

It follows that

$$\frac{\partial I(w, t)}{\partial \overline{w}} = 2\frac{\partial^2}{\partial w \partial \overline{w}} \iint_D \overline{K(z, t)} \log\frac{1}{|z - w|}\, d\tau.$$

But

$$\iint_D \overline{K(z, t)} \log\frac{1}{|z - w|}\, d\tau$$

is merely the potential of the distribution of mass with density $\overline{K(z, t)}$ over D, and $4(\partial^2/\partial w \partial \overline{w})$ is the Laplace operator with respect to the real variables u and v. Hence Poisson's equation yields (cf. [10])

(3) $$\frac{\partial I(w, t)}{\partial \overline{w}} = -\pi\overline{K(w, t)}, \qquad\qquad w \in D.$$

We can integrate the differential equation (3) for $I(w, t)$ directly. We find that there is a function $\lambda(w, t)$, analytic for $w \in D$, such that

(4) $$I(w, t) = -\pi\overline{M(w, t)} + \lambda(w, t), \qquad\qquad w \in D.$$

We may view $\lambda(w, t)$ as a constant with respect to the variable $\overline{w}$, because of the Cauchy-Riemann equations.

Our next step is to show that the integral $I(w, t)$ is continuous across the boundary C of D, and therefore in the entire w-plane. It is here that we make use of the analyticity of C, and we use quite strongly the assumption that the curves C are non-selfintersecting. To be quite precise, we only use from the analyticity of C the fact that there is a positive number $\rho > 0$ such that the closed circles of radius ρ tangent to C at each point $w_0 \in C$ lie the one entirely interior to D and the other entirely exterior to D, except for the single point w_0 of tangency lying on C. We could thus prove the existence theorems under somewhat more general hypotheses. We denote the interior tangent circle of radius ρ at w_0 by Γ and we denote its boundary by γ.

A first remark on the continuity of $I(w, t)$ across C is that from the explicit relation (2) it is evident that $I(w, t)$ is uniformly continuous in the exterior of D. Thus it suffices in order to prove the continuity of $I(w, t)$ to show that if w_1 and w_2 are inverse points with respect to the circle γ through w_0, then $I(w_2, t) - I(w_1, t)$ tends to zero as w_1 and w_2 approach w_0. It is even sufficient to assume that w_1 and w_2 both lie on the normal to γ at w_0 in this limit process.

Now we can write

$$I(w, t) = \iint_{D-\Gamma} \frac{\overline{K(z, t)}\, d\tau}{z - w} + \iint_{\Gamma} \frac{\overline{K(z, t)}\, d\tau}{z - w} = I_1(w, t) + I_2(w, t),$$

and it is clear that $I_1(w_2, t) - I_1(w_1, t)$ tends to zero as w_1 and w_2 tend to w_0 by Schwarz's inequality and the convergence of the integral

$$\iint_{D-\Gamma} \frac{d\tau}{|z - w_0|^2}.$$

The convergence of this integral follows easily from the fact that both tangent circles of radius ρ at w_0 do not intersect C.

Furthermore, it is seen from simple transformations that we can assume Γ to be the unit circle and w_0 to be the point $z = 1$. We have then merely to show that

$$\lim_{s \to 1} \iint_{|z| < 1} \overline{K(z, t)} \left(\frac{1}{z - s} - \frac{1}{z - s^{-1}} \right) d\tau = 0, \qquad s < 1,$$

and we shall indeed prove that this integral vanishes identically. In the proof it is permissible to integrate by parts by virtue of the regularity of $K(z, t)$ on γ, except at w_0, and by the fact that the function $M(z, t)$ of class $\mathfrak{L}^2$ can grow at most logarithmically as $z \to w_0$.

Thus we have, if Σ is a small circle of radius ϵ about $z = s$ and σ is its boundary,

$$\iint_{|z| < 1} \overline{K(z, t)} \left(\frac{1}{z - s} - \frac{1}{z - s^{-1}} \right) d\tau = \frac{1}{2i} \oint_{|z| = 1} \overline{M(z, t)} \left(\frac{1}{z - s} - \frac{1}{z - s^{-1}} \right) dz$$

$$+ \iint_{\Sigma} \overline{K(z, t)} \frac{d\tau}{z - s} - \frac{1}{2i} \oint_{\sigma} \overline{M(z, t)} \frac{dz}{z - s}$$

$$= \frac{1}{2i} \oint_{|z| = 1} M(z, t) \left(\frac{1}{z^{-1} - s} - \frac{1}{z^{-1} - s^{-1}} \right) \frac{dz}{z^2} - \pi \overline{M(s, t)}$$

$$= \overline{\pi(M(0, t) - M(0, t) + M(s, t))} - \pi \overline{M(s, t)} = 0.$$

The area integral over Σ vanishes, by a simple Fourier argument. Thus the continuity of $I(w, t)$ in the entire w-plane is proved.

We now proceed to a comparison of the formulas (2) and (4) for the integral $I(w, t)$. Since $I(w, t)$ is continuous across C, we must have

$$\lim_{w \to C_\mu} \left\{ -\pi \overline{M(w, t)} + \lambda(w, t) - \frac{1}{t - w} - b_\mu \right\} = 0,$$

or, writing

$$(5) \qquad N(w, t) = \frac{1}{\pi} \left(\frac{1}{w - t} + \lambda(w, t) \right),$$

we obtain

$$(6) \qquad \lim_{w \to C_\mu} \{ N(w, t) - \overline{M(w, t)} \} = \frac{b_\mu}{\pi}, \qquad \mu = 1, \cdots, n.$$

The formula (6) is in essence the statement that both (2) and (4) hold on C.

We can finally set up the functions

$$\varphi(w) = N(w, t) + M(w, t),$$

$$\psi(w) = N(w, t) - M(w, t).$$

It is apparent from (6) that

$$\text{Im}\{\varphi(w)\} = \text{const.}, \qquad \text{Re}\{\psi(w)\} = \text{const.},$$

for values of w on each component C_μ of C. For example,

$$\lim_{w \to C_\mu} \text{Im}\{\varphi(w)\} = \lim_{w \to C_\mu} \frac{1}{2i} \{ N(w, t) + M(w, t) - \overline{N(w, t)} - \overline{M(w, t)} \}$$

$$= \frac{1}{2\pi i} \{ b_\mu - \overline{b_\mu} \}.$$

Thus, by the Schwarz principle of reflection, $\varphi(w)$ and $\psi(w)$ are analytic on C and, as is easily seen from the argument principle and the fact that $\varphi(w)$ and $\psi(w)$ each have one simple pole at $w = t$ in D, $\varphi(w)$ maps D conformally upon the schlicht φ-plane cut along n horizontal rectilinear slits, while $\psi(w)$ maps D conformally upon the schlicht ψ-plane cut along n vertical rectilinear slits.

Thus we have proved the existence of the classical slit mappings. The proof may be summarized by saying that we analyze the scalar product in $\mathfrak{L}^2$ of the fundamental singularity with the kernel function, using our knowledge of the kernel function in the unit circle to do this.

Finally, we note that by setting $L(w, t) = \partial N(w, t)/\partial w$, we can bring (6) into the more elegant form

$$(7) \qquad L(w, t)\, dw = \overline{K(w, t)\, dw},$$

where dw is measured along C. Such conjugate pairs of differentials are characteristic of the kernel function theory, and we shall encounter them again in Section 4.

3. Proof using the Dirichlet integral

We now take up the investigation of the existence proofs of potential theory. We confine ourselves to the case of two variables, although the argument is quite general and goes through in precisely the same fashion in any number of dimensions. It should be pointed out also that our method applies equally well to a large class of partial differential equations of elliptic type associated with positive definite energy integrals. It is hardly to the point at present, however, to go into a discussion of this very general case.

We continue to make use of the notations of Section 1. We base our considerations upon the Dirichlet integral

$$D(\varphi, \psi) = \iint_D \left\{ \frac{\partial \varphi}{\partial x} \frac{\partial \psi}{\partial x} + \frac{\partial \varphi}{\partial y} \frac{\partial \psi}{\partial y} \right\} d\tau$$

of a pair of harmonic functions $\varphi(z)$ and $\psi(z)$ in D, $z = x + iy$. It is convenient to define the class $\mathfrak{M}^2$ of real-valued harmonic functions $\varphi(z)$ in D which possess a finite Dirichlet integral $D(\varphi, \varphi)$. We fix two points t_1 and t_2 in D and we pose the following extremal problem. We seek a function $\varphi_0(z)$ in $\mathfrak{M}^2$ which minimizes the Dirichlet integral $D(\varphi, \varphi)$ among all functions $\varphi(z) \in \mathfrak{M}^2$ which satisfy the condition

$$\varphi(t_2) - \varphi(t_1) = 1.$$

An extremal function of this type exists in $\mathfrak{M}^2$ by the theory of normal families. Furthermore, if $\varphi_1(z) \in \mathfrak{M}^2$ satisfies the relation

$$\varphi_1(t_2) - \varphi_1(t_1) = 0,$$

then

$$D(\varphi_0, \varphi_0) \leqq D(\varphi_0 + \lambda \varphi_1, \varphi_0 + \lambda \varphi_1)$$

$$= D(\varphi_0, \varphi_0) + 2\lambda D(\varphi_0, \varphi_1) + \lambda^2 D(\varphi_1, \varphi_1)$$

for any real number λ. Consequently

$$D(\varphi_0, \varphi_1) = 0,$$

and from this identity it is easy to conclude that for any function $\varphi(z) \in \mathfrak{M}^2$ we have

$$[\varphi(t_2) - \varphi(t_1)]D(\varphi_0, \varphi_0) = D(\varphi_0, \varphi).$$

Indeed, it suffices to put $\varphi_1(z) = [\varphi(t_2) - \varphi(t_1)]\varphi_0(z) - \varphi(z)$ in the above argument. Thus the kernel function

(8) $$H(z; t_1, t_2) = \varphi_0(z)D(\varphi_0, \varphi_0)^{-1}$$

has the characteristic reproducing property

(9) $$\varphi(t_2) - \varphi(t_1) = D(H(z; t_1, t_2), \varphi(z))$$

for any $\varphi(z)$ of class $\mathfrak{M}^2$. Note that (8) only determines $H(z; t_1, t_2)$ up to an additive constant.

We remark in passing that it is possible to introduce the harmonic kernel function $H(z; t_1, t_2)$ equally well by using the theory of orthogonal functions. To do this one introduces a complete orthonormal system $\{\psi_\nu(z)\}$ in the subclass $\mathfrak{M}_0^2$ of $\mathfrak{M}^2$ of harmonic functions $\psi(z)$ which vanish for $z = t_1$. The orthonormality condition is taken to mean that

$$D(\psi_\nu, \psi_\mu) = \delta_{\nu\mu}.$$

One then sets

(10)
$$H(z; t_1, t_2) = \sum_{\nu=1}^{\infty} \psi_\nu(z)\psi_\nu(t_2),$$

and it is easy to show that this definition of the harmonic kernel function coincides with the definition (8), except for an additive constant which is not determined in (8).

We now consider, as in Section 1, the scalar product

$$J(w) = D\left(H(z; t_1, t_2), \log \frac{1}{|z - w|}\right)$$

between the kernel function $H(z; t_1, t_2)$ and the fundamental singularity $\log 1/|z - w|$. For values of w in the exterior B of D we have, by the reproducing property (9) of the kernel function,

(11)
$$J(w) = \log \frac{|w - t_1|}{|w - t_2|}.$$

On the other hand, for values of w inside D it is easy to show by an integration by parts that $J(w)$ is harmonic. For let Σ be any small circle about the point $w \in D$ which lies entirely inside D, and let σ be the boundary of Σ. Then Green's formula yields readily for $J(w)$ the expression

$$J(w) = \iint_{D-\Sigma} \left\{ \frac{\partial H}{\partial x} \frac{\partial}{\partial x} \log \frac{1}{|z - w|} + \frac{\partial H}{\partial y} \frac{\partial}{\partial y} \log \frac{1}{|z - w|} \right\} d\tau$$
$$- \oint_\sigma \frac{\partial H}{\partial n} \log \frac{1}{|z - w|} ds,$$

where n denotes the inner normal and s the arc length of the circle σ. But it is now evident that $J(w)$ is harmonic, since we can differentiate under the integral signs on the right.

We wish next to show that $J(w)$ is continuous across C. Our assumptions on C and our notation will be the same as in Section 1. To prove continuity at a point $w_0 \in C$, it is again sufficient to show that

$$J(w_2) - J(w_1)$$

tends to zero as w_1 and w_2 tend to w_0, where w_1 and w_2 lie on the normal to C at

w_0 and are inverse points in the circle γ through w_0 which is tangent to C and has an interior Γ contained in D. As before, this simplification of the continuity proof results from the uniform continuity of the explicit expression (11) which we found for $J(w)$ when $w \in B$.

We can write

$$J(w) = \iint_{D-\Gamma} \left\{ \frac{\partial H}{\partial x} \frac{\partial}{\partial x} \log \frac{1}{|z-w|} + \frac{\partial H}{\partial y} \frac{\partial}{\partial y} \log \frac{1}{|z-w|} \right\} d\tau$$

$$+ \iint_{\Gamma} \left\{ \frac{\partial H}{\partial x} \frac{\partial}{\partial x} \log \frac{1}{|z-w|} + \frac{\partial H}{\partial y} \frac{\partial}{\partial y} \log \frac{1}{|z-w|} \right\} d\tau$$

$$= J_1(w) + J_2(w),$$

and it is obvious from Schwarz's inequality that the first integral, $J_1(w)$, is continuous across C, as in Section 1. Thus it remains to treat $J_2(w)$. Of course, there is no loss of generality if we assume that Γ is the unit circle and that $z = w_0$ is the point $z = 1$. Also, we can write $w_1 = r$, $w_2 = r^{-1}$, and we can assume that $r < 1$. We have merely to show that

$$(12) \qquad \lim_{r \to 1} \{J_2(r) - J_2(r^{-1})\} = 0.$$

But, as in Section 1, we remark that it is permissible to use integration by parts upon the integrals $J_2(r)$ and $J_2(r^{-1})$, since $H(z; t_1, t_2)$ is regular on γ except at $z = 1$, and since $H(z; t_1, t_2)$ is in $\mathfrak{M}^2$ and therefore has a gradient which grows at most like $(1 - |z|)^{-1}$ as $z \to 1$. We obtain

$$J_2(r) - J_2(r^{-1}) = -\oint_{|z|=1} \frac{\partial H}{\partial n} \log \frac{|z - r^{-1}|}{|z - r|} \, ds = -\oint_{|z|=1} \frac{\partial H}{\partial n} \log \frac{1}{r} \, ds = 0$$

for all values of r, where n and s denote now the inner normal and arc length on $|z| = 1$, respectively. Thus (12) certainly holds, and we have proved that $J(w)$ is continuous across C. It is worth remarking that this step is really a consequence of the fact that

$$\log \left| \frac{r(z - r^{-1})}{z - r} \right|$$

is the Green's function of $|z| < 1$ with logarithmic pole at $z = r$, a fact which is useful in generalizing this proof.

Since $J(w)$ is continuous across C and since we have the relation (11) for $w \in B$, we find that $J(w)$ is a harmonic function in D which assumes the boundary values

$$\log \left| \frac{w - t_1}{w - t_2} \right|$$

continuously on C. Thus the function

$$G(w; t_1, t_2) = \log \left| \frac{w - t_1}{w - t_2} \right| - J(w)$$

assumes, continuously, the boundary values zero on C. It is in particular, therefore, harmonic on C, by Schwarz's principle of reflection. Thus we have shown in a very simple way the existence in D of the difference of the Green's function with pole at $z = t_2$ and the Green's function with pole at $z = t_1$.

In order to obtain the Green's function $G(z; t_2)$ with pole at $z = t_2$ alone we have merely to allow t_1 to tend to the boundary C of D. It is quite elementary to see that this leads in the limit to the Green's function $G(z; t_2)$ since, for example, when t_1 tends to $w_0 = 1$ along the inner normal to C, we find that the function

$$G(z; t_1, t_2) - \log \left| \frac{z - t_1}{1 - t_1 z} \right|,$$

regular at t_1, tends to the same limit and has boundary values which tend to zero on C.

Thus we have, in fact, proved the existence of Green's function using the harmonic kernel function $H(z; t_1, t_2)$. It is not very difficult now to construct the Neumann's function $N(z; t_1, t_2)$ with positive logarithmic pole at $z = t_2$, negative logarithmic pole at $z = t_1$ and zero normal derivatives on C. We then have the well-known identity

$$H(z; t_1, t_2) = \frac{1}{2\pi} \{ N(z; t_1, t_2) - G(z; t_1, t_2) \}$$

relating the Green's function and Neumann's function to the kernel function. This identity is, of course, a consequence of the fact that the regular expression on the right has the characteristic reproducing property (9), as can be seen immediately from Green's formula, and it is the basis upon which the existence of Neumann's function is to be placed.

A final remark is that the existence of Robin's functions can be proved by a similar procedure, using, however, a different norm (see [6, 12]).

4. Proof by use of the boundary norm

We now proceed to considerations which could be formulated in terms of a complete system $\{\psi_\nu(z)\}$ of analytic functions $\psi_\nu(z)$ in D which are orthonormal in the sense that

$$\oint_C \psi_\nu(z) \overline{\psi_\mu(z)} \, ds = \delta_{\nu\mu}, \qquad\qquad ds = |dz|,$$

and the corresponding kernel function [13]

$$(13) \qquad\qquad k^*(z, t) = \sum_{\nu=1}^{\infty} \psi_\nu(z) \overline{\psi_\nu(t)}.$$

However, we shall present the material from a somewhat different point of view which proves to be more convenient.

Let s be the arc length along C and let $z(s)$ be the parametric representation of C. We introduce the class L^2 of all complex-valued functions $\mu(s)$ defined on C which have a finite Lebesgue integral

$$(14) \qquad \|\mu\|^2 = \oint_C |\mu(s)|^2 \, ds.$$

The class L^2 is a Hilbert space with the scalar product

$$(\mu, \nu) = \oint_C \mu(s)\overline{\nu(s)} \, ds,$$

defined for any pair of elements $\mu(s)$ and $\nu(s)$ in L^2. We consider the linear subspace Λ^2 of L^2 consisting of those functions $\mu(s) \, \epsilon \, L^2$ which satisfy the identity

$$(15) \qquad \oint_C \frac{\mu(s) \, dz}{z - w} = 0, \qquad z = z(s), \qquad w \, \epsilon \, B,$$

where $B = \sum_{\nu=1}^{n} B_\nu$ is the exterior of D. It is clear that Λ^2 is closed in L^2, since each function

$$\frac{z'(s)}{z(s) - w}, \qquad w \, \epsilon \, B,$$

is in L^2. The class Λ^2 can be thought of as representing the class of boundary values of functions which are analytic in D. This formulation of the analyticity property is fundamental for the work of this section and the next, and in modified form it proves to be a useful tool for further applications in function theory and partial differential equations.

We select any element $M(s) \, \epsilon \, L^2$, and we ask for the function $\mu(s)$ in Λ^2 which is nearest to $M(s)$ in the sense that

$$\| M(s) - \mu(s) \|^2 = \oint_C | M(s) - \mu(s) |^2 \, ds = \text{minimum}.$$

A function $\mu(s) \, \epsilon \, \Lambda^2$ with this extremal property exists in view of the fact that Λ^2 is closed. Now let $\nu_1(s)$ be any fucntion of Λ^2 and let λ be any complex number. We obtain from the extremal character of $\mu(s)$ the inequality

$$\oint_C | M(s) - \mu(s) |^2 \, ds \leqq \oint_C | M(s) - \mu(s) - \lambda\nu_1(s) |^2 \, ds$$

$$= \oint_C | M(s) - \mu(s) |^2 \, ds - 2 \, \text{Re} \left\{ \lambda \oint_C \overline{(M(s) - \mu(s))}\nu_1(s) \, ds \right\}$$

$$+ | \lambda |^2 \oint_C | \nu_1(s) |^2 \, ds.$$

It follows immediately from the arbitrariness of λ that

$$(16) \qquad \oint_C \overline{(M(s) - \mu(s))}\nu_1(s) \, ds = 0, \qquad \nu_1(s) \, \epsilon \, \Lambda^2.$$

We can specialize this identity to the case where $\nu_1(s)$ has the form

$$\nu_1(s) = \frac{1}{z(s) - w}, \qquad w \in B$$

since $\nu_1(s) \in \Lambda^2$ by Cauchy's theorem. We thus obtain the relation

$$\oint_c \frac{\overline{(M(s) - \mu(s))z'(s)}\, dz}{z - w} = 0, \qquad w \in B.$$

We conclude from the definition (15) that $\overline{\{M(s) - \mu(s)\}\, z'(s)}$ is in the class Λ^2. Denoting this element of Λ^2 by $\nu(s)$, we have

$$\overline{\{M(s) - \mu(s)\}z'(s)} = \nu(s),$$

or

(17) $$M(s) = \mu(s) + \overline{\nu(s)\, z'(s)}, \qquad \mu(s),\ \nu(s) \in \Lambda^2.$$

Thus we have shown that every function $M(s) \in L^2$ has a representation (17) in terms of functions of the class Λ^2.

We choose for $M(s)$ in (17) the particular function

$$M(s) = \frac{1}{2\pi i} \frac{1}{z(s) - t},$$

where t is a fixed point of D. The importance of this function stems from the fact that $(z - t)^{-1}$ is the fundamental singularity function in D. We have then from (17)

(18) $$\frac{1}{2\pi i} \frac{1}{z(s) - t} = \rho(z, t) + \overline{k(z, t)\, z'(s)},$$

where $\rho(z(s), t)$ and $k(z(s), t)$ are in the class Λ^2. For points $w \in D$ we can now set up the pair of analytic functions

$$k(w, t) = \frac{1}{2\pi i} \oint_c \frac{k(z, t)\, dz}{z - w},$$

$$\rho(w, t) = \frac{1}{2\pi i} \oint_c \frac{\rho(z, t)\, dz}{z - w},$$

and we set

(19) $$l(w, t) = \frac{1}{2\pi i} \frac{1}{w - t} - \rho(w, t).$$

Clearly, the function

$$l(z(s), t) = \frac{1}{2\pi i} \frac{1}{z(s) - t} - \rho(z(s), t)$$

is in the class L^2. The reader would suspect that the analytic functions $k(w, t)$ and $l(w, t)$ have boundary values which are given by $k(z(s), t)$ and $l(z(s), t)$, and we shall proceed to prove this fact.

We first notice that by Schwarz's inequality

$$| k(w, t) |^2 \leq \frac{1}{4\pi^2} \| k \|^2 \oint_C \frac{ds}{| z(s) - w |^2} = 0 \left(\frac{1}{r} \right),$$

where r is the distance of the point $w \in D$ to the boundary C of D. The same result holds for $l(w, t)$ when w is bounded away from t. Thus we have

(20) $\qquad\qquad | k(w, t) | = 0(r^{-\frac{1}{2}}), \qquad | l(w, t) | = 0(r^{-\frac{1}{2}}).$

We next make a few supplementary remarks about functions defined by Cauchy integrals. Let $\sigma(z)$ be any integrable function defined on C,

$$\oint_C | \sigma(z) | \, ds < \infty,$$

which satisfies the relation

$$\oint_C \frac{\sigma(z) \, dz}{z - w} = 0, \qquad\qquad w \in B.$$

We prove that the indefinite integral $F(w)$ of the analytic function

$$f(w) = \frac{1}{2\pi i} \oint_C \frac{\sigma(z) \, dz}{z - w}$$

in D is continuous in $D + C$ and has, up to an additive constant, the boundary values $\int^z \sigma(z) \, dz$. It is sufficient to consider the case when

$$\oint_{C_\nu} \sigma(z) \, dz = 0, \qquad\qquad \nu = 1, \cdots, n,$$

since every integrable function $\sigma(z)$ differs from such a "derivative" by a continuous expression of the form

$$\sum_{\nu=1}^n \frac{\alpha_\nu}{z(s) - w_\nu}, \qquad\qquad w_\nu \in B_\nu, \nu = 1, \cdots, n.$$

We can define the indefinite integral

$$\omega(z) = \int^z \sigma(z) \, dz$$

on each component C_ν of C, and we set

$$E(w) = \frac{1}{2\pi i} \oint_C \frac{\omega(z) \, dz}{z - w}.$$

Since $\omega(z)$ is absolutely continuous, we can integrate the expression

$$E'(w) = \frac{1}{2\pi i} \oint_C \frac{\omega(z) \, dz}{(z - w)^2}$$

by parts, and we find

$$E'(w) = \frac{1}{2\pi i} \oint_C \frac{\sigma(z)\, dz}{z - w} = f(w).$$

Thus the indefinite integral $F(w)$ of $f(w)$ is given by the expression $E(w)$ above.

Now let z_0 and z_1 be points on one component of C and let w_0 and w_1 be points of D near z_0 and z_1, respectively, while w_0' and w_1' are to denote points in B near z_0 and z_1. The points z_0 and z_1 are joined by an arc C' of C, and the points w_0, w_1 and w_0', w_1' are joined pair-wise by arcs C_1' and C_2' in D and B, respectively, which are near the arc C' joining z_0, z_1. We have from the definition of $F(w)$

$$(21) \qquad F(w_1) - F(w_0) = \int_{w_0}^{w_1} f(z)\, dz = \frac{1}{2\pi i} \oint_C \log\left(\frac{z - w_0}{z - w_1}\right) \sigma(z)\, dz,$$

while from the definition of $\sigma(z)$ we find

$$(22) \qquad \frac{1}{2\pi i} \oint_C \log\left(\frac{z - w_0'}{z - w_1'}\right) \sigma(z)\, dz = 0.$$

Here we understand integrations between the points w_0, w_1 and w_0', w_1' to be taken over the arcs C_1' and C_2', respectively. Thus we obtain in the limit as $w_0 \to z_0$, $w_0' \to z_0$ and $w_1 \to z_1$, $w_1' \to z_1$ and as C_1' and C_2' tend to C' the relation

$$(23) \qquad \begin{aligned} \lim \{F(w_1) - F(w_0)\} &= \lim \frac{1}{2\pi i} \oint_C \log\left(\frac{z - w_0}{z - w_1} \frac{z - w_1'}{z - w_0'}\right) \sigma(z)\, dz \\ &= \int_{z_0}^{z_1} \sigma(z)\, dz = \omega(z_1) - \omega(z_0). \end{aligned}$$

Indeed, the reader will easily verify that the branch of

$$\log\left(\frac{z - w_0}{z - w_1} \frac{z - w_1'}{z - w_0'}\right)$$

defined by C_1' and C_2' tends to $2\pi i$ on C' and tends to zero elsewhere on C. Thus in order to show that $F(w)$ has, up to an additive constant, the boundary values $\omega(z)$, it is sufficient to show that there is a sequence of points $w_0^{(n)} \, \epsilon \, D$ tending to $z_0 \, \epsilon \, C$ such that

$$\lim_{n \to \infty} F(w_0^{(n)})$$

exists. For this, again, it is sufficient to show that $F(w_0)$ is bounded, by the Bolzano-Weierstrass theorem.

To prove that $F(w_0)$ is bounded, we refer back to the formulas (21) and (22). If we fix w_1, w_1', we have the estimate

$$| F(w_0) | \leq | F(w_1) | + \oint_C | \sigma(z) | \left| \log\left(\frac{z - w_0}{z - w_1} \frac{z - w_1'}{z - w_0'}\right) \right| ds.$$

Now we have merely to select $w_0' \, \epsilon \, B$ in dependence upon w_0 in such a way that $| (z - w_0)/(z - w_0') |$ is uniformly bounded away from zero for $z \, \epsilon \, C$. This is easily done, and, in fact, it suffices when w_0 tends towards C to take w_0' to lie on the normal to C which passes through w_0 and symmetric to w_0 with respect to the tangent. Thus $F(w_0)$ is bounded in D, and it is evident that $F(w)$ has, up to an additive constant, the boundary values $\omega(z)$ and is continuous in $D + C$.

Now let

$$k^+(z, t) = \int^z k(z, t) \, dz,$$

$$\rho^+(z, t) = \int^z \rho(z, t) \, dz,$$

$$l^+(z, t) = \int^z l(z, t) \, dz$$

on C. The indefinite integrals $k^+(w, t)$, $\rho^+(w, t)$ and $l^+(w, t)$ of the functions $k(w, t)$, $\rho(w, t)$ and $l(w, t)$ have, continuously, the boundary values $k^+(z, t)$, $\rho^+(z, t)$ and $l^+(z, t)$.

Let γ be any small arc of C, and note that γ is by assumption the image of a segment (a, b) of the real axis by the analytic transformation $z(s)$. The map $z(s)$ carries a one-sided neighborhood U of (a, b) into a small, simply-connected subdomain D^* of D bounded by γ and, as we may obviously assume, an analytic arc γ^* in D. We have by Cauchy's theorem

$$\oint_{\gamma+\gamma^*} k^+(z, t) \, d\left(\frac{z'(s)^{-\frac{1}{2}}}{z-w}\right) = 0,$$

$$\oint_{\gamma+\gamma^*} l^+(z, t) \, d\left(\frac{z'(s)^{-\frac{1}{2}}}{z-w}\right) = 0,$$

for each point w exterior to D^*, by the analyticity of $z'(s)$ throughout the region U and therefore throughout the region D^*. Integration by parts yields the identities

$$\oint_{\gamma+\gamma^*} \frac{k(z, t)z'(s)^{-\frac{1}{2}} \, dz}{z-w} = 0,$$

$$\oint_{\gamma+\gamma^*} \frac{l(z, t)z'(s)^{-\frac{1}{2}} \, dz}{z-w} = 0.$$

This step is justified by the absolute integrability of $k(z, t)$ and $l(z, t)$ over γ and γ^*. Indeed, the analytic functions $k(z, t)$ and $l(z, t)$ are integrable over γ^* by the estimate (20) of their growth.

Applying now our remarks, originally made for D and the function $f(w)$, to the domain D^* and the functions

$$\frac{d}{dw} \alpha(w) = \frac{1}{2\pi i} \oint_{\gamma+\gamma^*} \frac{k(z, t)z'(s)^{-\frac{1}{2}} \, dz}{z-w},$$

$$\frac{d}{dw} \beta(w) = \frac{1}{2\pi i} \oint_{\gamma+\gamma^*} \frac{l(z, t)z'(s)^{-\frac{1}{2}} \, dz}{z-w},$$

analytic in D^*, we see from the absolute integrability of $k(z, t)z'(s)^{-\frac{1}{2}}$ and $l(z, t)z'(s)^{-\frac{1}{2}}$ over $\gamma + \gamma^*$ that $\alpha(w)$ and $\beta(w)$ assume continuously on γ the boundary values

$$(24) \qquad \alpha(z) = \int^z k(z, t)z'(s)^{\frac{1}{2}} ds,$$

$$(25) \qquad \beta(z) = \int^z l(z, t)z'(s)^{\frac{1}{2}} ds.$$

But from (18) and (19) we see that

$$(26) \qquad l(z, t) z'(s)^{\frac{1}{2}} = \overline{k(z, t) z'(s)^{\frac{1}{2}}}$$

almost everywhere on C. Thus

$$\alpha(z) = \overline{\beta(z)} + \text{const.}$$

everywhere on γ. It follows that

$$\alpha(w) + \beta(w),$$
$$i(\alpha(w) - \beta(w)),$$

have constant imaginary part on γ. Schwarz's principle of reflection then tells us that $\alpha(w)$ and $\beta(w)$ are analytic on γ, and differentiation of (24) and (25) shows that $l(z, t)$ and $k(z, t)$ are also analytic on C. We conclude immediately that the functions $k(z, t)$ and $l(z, t)$, defined by the Cauchy integral formula throughout D, are analytic in $D + C$, except for the simple pole of $l(z, t)$ at $z = t$, and that they satisfy the boundary relation

$$(27) \qquad l(z, t) = \overline{k(z, t)z'(s)}$$

everywhere on C.

We have now merely to make a few manipulations with the important pair of functions $l(z, t)$ and $k(z, t)$, whose existence we have just demonstrated. First, let us consider the scalar product

$$\oint_C \overline{k(z, t)}\varphi(z)\, ds$$

in Λ^2 of $k(z, t)$ with any function $\varphi(z)$ analytic in D and continuous in $D + C$. We have from (27)

$$\oint_C \overline{k(z, t)}\varphi(z)\, ds = \oint_C \varphi(z)\overline{k(z, t)z'}\, dz$$

$$= \oint_C \varphi(z)l(z, t)\, dz$$

$$= \varphi(t).$$

Thus $k(z, t)$ has the characteristic reproducing property

$$(28) \qquad \varphi(t) = \oint_C \overline{k(z, t)}\varphi(z)\, ds$$

of a kernel function. It is not difficult to show that $k(z, t)$ is actually identical with the function $k^*(z, t)$ defined by the series (13). Minimum properties and identities follow in the usual fashion.

Another point of interest is that the differentials $k(z, t)^2 \, dz$ and $l(z, t)^2 \, dz$ satisfy on C the identity

$$l(z, t)^2 \, dz = \overline{k(z, t)^2 \, dz},$$

an immediate consequence of the fundamental property (27). Note the similarity of this relation with (7). It is easy to conclude now that the indefinite integrals of the sum and difference of these two differentials yield multiple-valued slit mappings in D. Also, it can be shown that $[4\pi l(z, t)^2 - L(z, t)] \, dz$ and $[4\pi k(z, t)^2 - K(z, t)] \, dz$ are differentials of the first kind in D. In fact, this is a consequence of the regularity of both expressions and the combined identity

$$[4\pi l(z, t)^2 - L(z, t)] \, dz = \overline{[4\pi k(z, t)^2 - K(z, t)] \, dz}.$$

More interesting is the important function

$$F(z) = \frac{k(z, t)}{l(z, t)}.$$

For $z \in C$ we have

$$|F(z)| = \left| \frac{k(z, t)}{\overline{k(z, t) z'}} \right| = 1.$$

By the argument principle, it is found that $k(z, t)$ has $n - 1$ zeros in D and that $l(z, t)$ has no zeros, but precisely one pole. For by (27) it is clear that $lk z' \geqq 0$ on C, so that l and k together have $n - 1$ zeros, whereas also

$$(e^{i\alpha/2} l - e^{-i\alpha/2} k)^2 z' \leqq 0$$

for each real number α, so that the variation of arg $(e^{\frac{1}{2}(i\alpha)} l - e^{-\frac{1}{2}(i\alpha)} k)$ over each C_ν is $\pm \pi$. Thus $e^{\frac{1}{2}(i\alpha)} l - e^{-\frac{1}{2}(i\alpha)} k)$ vanishes on each C_ν, and the function $F(z)$ must assume each value of modulus one at least once on each C_ν. But if now $F(z)$ had $p > 0$ poles and $n - p > 0$ zeros, which would be the only alternative possibility, we should arrive at a contradiction by elementary topology. Thus $F(z)$ is regular and maps D conformally and $(1, n)$ upon the interior of the unit circle.

Let $\mathfrak{E}$ be the class of analytic functions $G(z)$ in D which are bounded in modulus by 1,

$$|G(z)| \leqq 1, \qquad\qquad z \in D.$$

We have the remarkable inequality [1, 7]

$$|G'(t)| \leqq |F'(t)| = 2\pi k(t, t), \qquad\qquad G(z) \in \mathfrak{E}.$$

For

$$|G'(t)| = \left| 2\pi i \oint_C l(z, t)^2 G(z) \, dz \right| \leqq 2\pi \oint_C |l(z, t)^2| \, ds$$

$$= 2\pi \oint_C l(z, t)k(z, t)z' \, ds = 2\pi \oint_C l(z, t)^2 \frac{k(z, t)}{l(z, t)} \, dz$$

$$= 2\pi \oint_C l(z, t)^2 F(z) \, dz = \frac{F'(t)}{i}.$$

Obviously, we have the equality only for the functions $e^{i\theta}F(z)$, θ real.

Thus our procedure leads to the existence of the circle mapping $F(z)$ which is associated with Schwarz's lemma. It is to be noted that the existence of this function lies somewhat deeper than the existence of the slit mappings $\varphi(w)$ and $\psi(w)$ in multiply-connected domains, and therefore it is not too surprising that the present section is more difficult than the preceding ones. Of course, for $n = 1$, $F(z)$ is just the function found in the elementary Koebe proof of the Riemann mapping theorem.

5. Method of minimum length

We give a final existence proof which depends neither upon the theory of normal families nor upon the completeness of Hilbert space, but rather upon the properties of functions of bounded variation [9].

Let Ω be the class of functions $f(z)$ defined on C which are of bounded variation there, and which may have periods about the C_ν, but which satisfy the following condition. For a fixed point $t \, \epsilon \, D$ and any point $w \, \epsilon \, B$ we require

$$(29) \qquad \oint_C \frac{df(z)}{z - w} = \frac{1}{(t - w)^2}.$$

The class Ω is not empty, since $1/2\pi i \, (z - t)^{-1}$ is in the class. It follows from Helly's theorem that there exists in Ω a function, which we shall again call $f(z)$, for which the total variation

$$\oint_C | df(z) |$$

is a minimum.

We remark that $f(z)$ is continuous on C. For if this function possessed a jump at a point $w_0 \, \epsilon \, C$, we should have the absurd relation

$$\frac{1}{(t - w_0)^2} = \lim_{w \to w_0} \oint_C \frac{df(z)}{z - w} = \infty, \qquad\qquad w \, \epsilon \, B.$$

Thus we obtain easily by considerations of the type in Section 4 that the multiple-valued function

$$f(w) = \frac{1}{2\pi i} \oint_C \log\left(\frac{z - w^*}{z - w}\right) df(z) + \frac{1}{2\pi i} \frac{1}{w - t},$$

where w^* is an arbitrary fixed point of D which makes the logarithm single-valued on C, is continuous in $D + C$ and has, up to additive constants, the boundary values $f(z)$. For we have in the notation of the formula (23), and because of (29), the relation

$$\lim\{f(w_1) - f(w_0)\} = \lim \frac{1}{2\pi i} \oint_C \log\left(\frac{z - w_0}{z - w_1}\right) df(z) - \lim \frac{1}{2\pi i} \int_{w_0}^{w_1} \frac{dz}{(z - t)^2}$$

$$= \lim \frac{1}{2\pi i} \oint_C \log\left(\frac{z - w_0}{z - w_1} \frac{z - w_1'}{z - w_0'}\right) df(z)$$

$$+ \lim \frac{1}{2\pi i}\left[\frac{1}{w_1 - t} - \frac{1}{w_0 - t} - \frac{1}{w_1' - t} + \frac{1}{w_0' - t}\right]$$

$$= \int_{z_0}^{z_1} df(z).$$

Also, the argument of Section 4 can be applied to the function $f(w)$ in order to show that $f(w_0)$ is bounded in D, and the Bolzano-Weierstrass theorem then yields a sequence $w_0^{(n)} \to z_0$ such that $\lim_{n \to \infty} f(w_0^{(n)})$ exists. Thus we have defined an analytic function with boundary values $f(z)$ and a simple pole of residue $1/2\pi i$ at the point of normalization $w = t$.

Now for any function $\varphi(z)$ which is analytic in $D + C$ with $\varphi(t) = \varphi'(t) = 0$, and for any complex number λ, we have by the residue theorem

$$\oint_C \frac{(1 + \lambda\varphi(z))\, df(z)}{z - w} = -\oint_{C+C_0} f(z)\, d\left(\frac{1 + \lambda\varphi(z)}{z - w}\right)$$

$$= \frac{1}{(t - w)^2},$$

where C_0 represents a set of two sided cuts in D which reduce D to a simply-connected domain. Therefore the function $\int^z (1 + \lambda\varphi(z))\, df(z)$ is in the class Ω, and the minimum property of $f(z)$ implies that

$$\oint_C |df(z)| \leq \oint_C |1 + \lambda\varphi(z)|\, |df(z)|.$$

It follows from the arbitrariness of λ that

$$\oint_C \varphi(z)\, |df(z)| = 0$$

for each function $\varphi(z)$ with $\varphi(t) = \varphi'(t) = 0$. In particular, setting

$$\varphi(z) = \frac{1}{z - w} - \frac{1}{t - w} - \frac{t - z}{(t - w)^2}, \qquad\qquad w \in B,$$

we have

$$\oint_C \frac{|df(z)|}{z - w} = \frac{1}{t - w} \oint_C |df(z)| + \frac{1}{(t - w)^2} \oint_C (t - z)\, |df(z)|, \qquad w \in B,$$

whence, upon integration by parts and integration with respect to w we find

$$\oint_c \frac{\mu(z)\,dz}{z-w} - \sum_{\nu=1}^n \log(z_\nu - w) \oint_{C_\nu} |df(z)| = \frac{1}{t-w} \oint_c (t-z)|df(z)|$$
$$- \log(w-t) \oint_c |df(z)| + A_\nu, \qquad w \epsilon B_\nu,$$

where

$$\mu(z) = \int_{z_\nu}^z |df(z)| \quad \text{on} \quad C_\nu, \qquad z_\nu \epsilon C_\nu, \nu = 1, \cdots, n.$$

Since $\mu(z)$ is continuous and of bounded variation on C, the integral

$$\frac{1}{2\pi i} \oint_c \frac{\mu(z)\,dz}{z-w}$$

has a jump of $\mu(z)$ when w crosses C at the point $z \epsilon C$. This is once more a consequence of an identity of the form (23). Thus the function

(30)
$$p(w) = \frac{1}{2\pi} \log(w-t) \oint_c |df(z)| + \frac{1}{2\pi} \frac{1}{w-t} \oint_c (t-z)|df(z)|$$
$$+ \frac{1}{2\pi} \oint_c \frac{\mu(z)\,dz}{z-w} - \frac{1}{2\pi} \sum_{\nu=1}^n \log(z_\nu - w) \oint_{C_\nu} |df(z)|,$$

defined for $w \epsilon D$, has, up to additive constants, the boundary values $i\mu(z)$ on C and is continuous in $D + C$, although, of course, it is multiple-valued in D. Note that this function is independent of the choice of the points $z_\nu \epsilon C_\nu$, and thus the continuity does not break down at these points.

We now conclude by the Schwarz principle of reflection that the function $p(w)$ is analytic in $D + C$, since $\operatorname{Re}\{p(w)\} = \text{const.}$ on each C_ν. We obtain this result using only the theory of the Riemann-Stieltjes integral. Now, however, we must introduce Lebesgue integrals in the proof. Since $\mu(z) = \operatorname{Im}\{p(z)\}$ is analytic on C, it is clear that $f(z)$ satisfies a Lipschitz condition and is absolutely continuous on C. Therefore the derivative $f'(z)z'(s)$ exists almost everywhere on C and

$$\oint_c |df(z)| = \oint_c |f'(z)|\,ds.$$

Also, $|f'(x)| = |p'(x)|$ almost everywhere on C, and $f'(z)$ vanishes on C at most on a set of measure zero.

Now the integral

$$\oint_c |df(z)| = \oint_c |f'(z)|\,ds$$

clearly represents the length of the curves upon which $f(z)$ maps C, and thus our minimum problem in the class Ω becomes a minimum length problem in conformal mapping. Indeed, the multiple-valued function $f(z)$ maps D upon a Riemann surface with shortest boundary in the class of multiple-valued conformal mappings with a simple pole at $z = t$ of fixed residue $1/2\pi i$. This can also be expressed by saying that $f'(z)$ is that analytic function in D with fixed double pole at $z = t$ with coefficient $-(1/2\pi i)$ and with zero residue which yields the smallest value of the length integral

$$\oint_C |f'(z)| \, ds.$$

In order to study $f'(z)$ in more detail, we proceed to make straightforward variations of $f'(z)$. If $\varphi(z)$ is analytic in $D + C$, then in view of the minimum property of $f'(z)$

$$\oint_C |f'(z)| \, ds \leq \oint_C |f'(z) + \lambda\varphi(z)| \, ds$$

for any complex number λ, and hence we find, by taking the derivative with respect to λ, first for λ real and then for λ imaginary,

$$\oint_C \varphi(z) \frac{|f'(z)|}{f'(z)} \, ds = 0.$$

It follows, in particular, that

$$\oint_C \frac{1}{z - w} \frac{|f'(z)|}{f'(z)} \, ds = 0, \qquad\qquad w \in B.$$

Therefore, by the usual argument, the analytic function

$$S(w) = \frac{1}{2\pi i} \oint_C \log\left(\frac{z - w^*}{z - w}\right) \frac{|f'(z)|}{f'(z)} \overline{z'(s)} \, dz, \qquad w^* \in D \text{ fixed, } w \in D,$$

has, up to additive constants, the boundary values

$$\int^z \frac{|f'(z)|}{f'(z)} \overline{z'(s)} \, dz.$$

Thus if D^* denotes a small neighborhood in D of C bounded by an arc γ of C and an analytic arc γ^* in D, we have by Cauchy's theorem

$$\oint_{\gamma + \gamma^*} S(z) \, d\left(\frac{p'(z)}{z - w}\right) = 0$$

for every point w exterior to D^*. Hence, upon integration by parts, we obtain, by reasoning analogous to that in Section 4,

$$\oint_{\gamma + \gamma^*} \frac{p'(z)}{z - w} S'(z) \, dz = 0, \qquad\qquad w \text{ exterior to } D^*,$$

where we take $S'(z) = (|f'(z)|)/(f'(z)) \overline{z'(s)}$ on γ. Thus the function

$$T(w) = \frac{1}{i} \frac{1}{2\pi i} \oint_{\gamma + \gamma^*} \log\left(\frac{z - w_1^*}{z - w}\right) p'(z) S'(z) \, dz, \qquad w, w_1^* \in D^*,$$

is analytic in D^* and has, except for an additive constant, the boundary values

$$\frac{1}{i}\int^z p'(z)S'(z)\ dz = \frac{1}{i}\int^z p'(z)\ \frac{\overline{f'(z)}}{|f'(z)|}\ \overline{z'(s)}\ dz$$

$$= \int^z \overline{f'(z)}\ \overline{dz} = \overline{f(z)}$$

on γ. Thus by the Schwarz principle of reflection, $T(z) + f(z)$ and $T(z) - f(z)$ are analytic on γ, and it follows that $f(z)$ is analytic on C.

We now find easily that the analytic function

$$E(w) = S'(w) = \frac{1}{2\pi i}\oint_C \frac{1}{z-w}\ \frac{|f'(z)|}{f'(z)}\ ds$$

$$= \frac{1}{2\pi i}\oint_C \frac{1}{z-w}\ \frac{1}{i}\ \frac{dp(z)}{df(z)}\ dz$$

has the boundary values

$$\frac{|f'(z)|}{f'(z)z'(s)} = \frac{1}{i}\ \frac{dp(z)}{df(z)}$$

of unit modulus on C. Returning to the class $\mathfrak{E}$ of Section 4, we have for arbitrary $G(z) \in \mathfrak{E}$

$$|G'(t)| = \left|\oint_C G(z)f'(z)\ dz\right|$$

$$\leq \oint_C |df(z)|,$$

with equality only for the functions $e^{i\theta}E(z)$, θ real, which are, of course, in $\mathfrak{E}$. Indeed, we have on C

$$E(z)f'(z)\ dz = \frac{dp(z)}{i} = |df(z)|.$$

Thus the function $E(z)$ is identical with the extremal function $F(z)$ of Section 4, except for a constant factor,

$$E(z) = i\ F(z).$$

In particular, we find that $E(t) = 0$, as is also obvious from the extremal character of $E(z)$ in $\mathfrak{E}$, and thus

$$p'(z) = i\ E(z)f'(z)$$

has only a simple pole at $z = t$. By means of the functions $k(z, t)$ and $l(z, t)$ which were used in the definition of $F(z)$ we can now easily formulate the interesting identities

$$f'(z) = -2\pi i\ l(z, t)^2,$$

$$p'(z) = 2\pi i\ l(z, t)k(z, t),$$

relating the functions of the present section to those of Section 4. Thus we have found two more or less similar proofs of the existence of these functions, based on two different compact classes of functions encountered in the study of functions of a real variable.

We remark, finally, that the method of the present section can also be applied readily to the class of single-valued functions of bounded variation on C. We can thus construct the extremal function for our minimum length problem in the class of single-valued mappings. The new derivation is considerably simpler than that used previously for this problem in multiply-connected domains.

In this paper we have made the assumption that the boundary C of D is analytic in order to present our method with the least complication. This assumption can often be weakened to mere smoothness of these boundary curves, as is the case, for example, for the slit mappings φ and ψ in Section 2, the Green's function in Section 3 and the function $p(z)$ of the present section. In the more general situation, we should have to use, instead of Schwarz's principle of reflection, the results of Lindelöf on the conformal mapping of smooth curves, if we desired to study the boundary behaviour of the kernel functions and the derivatives of the mapping functions we consider.

STANFORD UNIVERSITY

BIBLIOGRAPHY

1. AHLFORS, L. V., *Bounded analytic functions*, Duke Math. J., vol. 14 (1947), p. 1–11.
2. BERGMAN, S., *Ueber die Entwicklung der harmonischen Funktionen der Ebene und des Raumes nach Orthogonalfunktionen*, Math Ann., vol. 86 (1922), pp. 238–271.
3. BERGMAN, S., Partial differential equations, Brown University Publications, Providence, 1941.
4. BERGMAN, S., *Sur les fonctions orthogonales de plusieurs variables complexes*, Mémorial des Sciences mathématiques, vol. 106, Paris, 1947.
5. BERGMAN, S. and SCHIFFER, M., *A representation of Green's and Neumann's functions in the theory of partial differential equations of second order*, Duke Math. J., vol. 14 (1947), pp. 609–638.
6. BERGMAN, S., and SCHIFFER, M., *Kernel functions in the theory of partial differential equations of elliptic type*, Duke Math. J., vol. 15 (1948), pp. 535–566.
7. GARABEDIAN, P. R., *Schwarz's lemma and the Szegö kernel function*, Trans. Amer. Math. Soc., vol. 67 (1949), pp. 1–35.
8. GARABEDIAN, P. R., *Distortion of length in conformal mapping*, Duke Math. J., vol. 16 (1049), pp. 439–459.
9. RIESZ, F., *Ueber Potenzreihen mit vorgeschriebenen Anfangsgliedern*, Acta Math., vol. 42 (1920), pp. 145–171.
10. SCHIFFER, M., *The span of multiply connected domains*, Duke Math. J., vol. 10 (1943), pp. 209–216.
11. SCHIFFER, M., *The kernel function of an orthonormal system*, Duke Math. J., vol. 13 (1946), pp. 529–540.
12. SCHIFFER, M., *On various types of orthogonalization*, Duke Math J., to appear shortly.
13. SZEGÖ, G., *Ueber orthogonale Polynome, die zu einer gegebenen Kurve der komplexen Ebene gehoeren*, Math. Zeit., vol. 9 (1921), pp. 218–270.

Commentary on

[29] P.R. Garabedian and M. Schiffer, *On existence theorems of potential theory and conformal mapping*, Ann. of Math. (2) **52** (1950), 164–187.

This is the second Garabedian–Schiffer paper, following [26] by just a year in date of publication. The setting is again multiply connected plane domains, and the questions addressed are some of the most fundamental existence theorems in conformal mapping and potential theory. Specifically, in this one paper, the authors present new proofs of the existence of the mapping onto a parallel slit plane, of Green's function for Laplace's equation, and of the extremal function for a generalization of Schwarz's lemma. The arguments make thorough use of the Bergman kernel and its relatives and provide a common approach to the different problems. Moreover, the existence of the kernel itself, also independently shown here as the solution of an extremal problem, is fairly considered to be less deep than previous proofs of the existence results that the paper aims for, a point made by the authors. If the kernel played a supporting role in [26], here it takes center stage.

Let D be a bounded, multiply connected domain whose boundary C consists of n analytic Jordan curves. For each of the problems the authors consider in the domain D, it is essential to analyze boundary behavior and continuity across the boundary of functions related to kernel functions. Ultimately, the desired properties are consequences of the boundary behavior.

Take first the example of the parallel slit mappings. In the notation of the paper, let $K(z,t)$, $z,t \in D$ be the Bergman kernel[1] of D and introduce the function

$$I(w,t) = \iint_D \frac{\overline{K(z,t)}}{z-w} \, dx dy, \quad z = x+iy,$$

which the authors regard as the inner product of the kernel function with the "fundamental singularity" $1/(z-w)$. They examine $I(w,t)$ first for w

in the exterior of D, where they show $I(w,t) = 1/(t-w) + \text{const.}$; then for $w \in D$, where they show $I(w,t) = -\pi\overline{M(w,t)} + \lambda(w,t)$ with $\lambda(w,t)$ analytic in w and $(\partial/\partial z)M(z,t) = K(z,t)$; and finally for $w \to C$, where they show that $I(w,t)$ is continuous as w crosses the boundary. Let $\pi N(w,t) = 1/(w-t) + \lambda(w,t)$; then, from the identities for $I(w,t)$ together with the continuity, it is straightforward to show that

$$\varphi(w) = N(w,t) + M(w,t),$$

$$\psi(w) = N(w,t) - M(w,t)$$

are the conformal mappings of D onto a horizontal-slit plane and a vertical-slit plane, respectively. Each function has its pole at t.

The analyticity assumption on the boundary curves is used at the end of the argument for an appeal to the Schwarz reflection principle and earlier for the continuity proof essentially to guarantee a uniform upper bound on the curvatures of the boundary curves. The latter allows the authors to localize and prove continuity at a boundary point by working in two circles tangent at the boundary point, one interior and one exterior to the domain, whose radii are uniformly bounded below. This argument itself is interesting, and additional details are given in [Br], Chap. IX.

At almost the same time, and independently, a proof of the existence of the parallel slit mappings via the kernel function was given by Lehto in his thesis [Le]. When D is simply connected, one obtains a new proof of the Riemann mapping theorem, albeit with restrictions on smoothness of the boundary. Garabedian revisited this in his later paper [G4].

As for the existence of Green's function, in [19] Schiffer had forged a link between the Bergman kernel and Green's function by deriving the famous formula (in the notation of [19])

$$K(z,\bar\zeta) = -\frac{2}{\pi} \frac{\partial^2 g(z;\zeta)}{\partial z \partial \bar\zeta}.$$

However, this starts by assuming the existence of Green's function, $g(z;\zeta)$. The proof of the existence of Green's function in the present paper

---

[1] The relevant L^2 space consists of the analytic functions in D for which $\iint_D |f'|^2 dxdy < \infty$.

follows generally along the same lines as for the parallel slit mapping but using a kernel tailored to harmonic functions.

The authors introduce a harmonic version of the Bergman kernel $H(z;t_1,t_2)$, depending on two parameters $t_1, t_2 \in D$, which reproduces the difference $u(t_1) - u(t_2)$ of values of a harmonic function u. They then analyze the inner product

$$J(w) = \mathfrak{D}(H(z;t_1,t_2), -\log|z - w|)$$

of H with the harmonic fundamental singularity $-\log|z - w|$. Here the inner product is given by the Dirichlet integral $\mathfrak{D}(u,v) = \iint_D \nabla u \cdot \nabla v\, dx dy$. Then $J(w)$ is harmonic in D and assumes the boundary values $\log|w - t_1| - \log|w - t_2|$ continuously, again a key point. From this it is a short step to obtain Green's function with singularity at t_1 (or t_2). One can also produce Neumann's function.

By this method, the (harmonic) kernel function provides an alternative to Dirichlet's principle as a basis for the existence of solutions to the boundary value problems of potential theory. On the one hand, Dirichlet's principle became, with Weyl [W], the "method of orthogonal projection" and was ever after firmly tied to Hilbert space.[2] On the other hand, at about the same time as [29], Aronszajn's landmark paper [Ar] appeared, giving a Hilbert space setting for reproducing kernels in a manner that depends on orthogonal projection. This led Lax [L1] to present the Garabedian Schiffer results in terms of orthogonal projection in Hilbert space with no mention of the kernel function. Read this as an attractive companion paper to [29]; it is also covered in Bergman and Schiffer's book [51]. Furthering the functional analysis framework, Lax [L2] established the existence of Green's function via the Hahn Banach theorem.

The final two sections of [29] treat an extremal problem formulated and solved in [A]: for fixed $t \in D$, maximize $|G'(t)|$ over all analytic functions G in D with $|G(z)| < 1$. The

extremal, now called the Ahlfors function, is n to 1 in D and has modulus equal to 1 on the boundary of D. The authors' first method employs the Szego kernel, an approach initiated by Garabedian in his thesis [G1]. Instead of working with the Szego kernel as defined through an expansion in orthonormal functions on the boundary, Garabedian and Schiffer obtain it as the solution to an L^2 minimization problem on the boundary, essentially projecting from all of $L^2(\partial D)$ to the boundary values of functions that are analytic in D. This gives an orthogonal decomposition of $L^2(\partial D)$, also discussed in [31], and has proved to be a very influential point of view. In keeping with their treatment of the earlier problems, the authors apply this decomposition to the fundamental singularity $1/(z - t)$ on the boundary and then define two associated analytic kernels $k(w,t)$ and $l(w,t)$ in D. (Here $k(w,t)$ is the analytic extension of the Szego kernel.) The extremal function for Schwarz's lemma is found to be $F(z) = k(z,t)/l(z,t)$, a stunning result. The second method of getting to the extremal, in the last section of the paper, involves a minimum length problem for the image of the boundary.

Questions on kernels and boundary values have been studied in different contexts and from different points of view, notably in several complex variables. It is not unreasonable to trace these efforts to the ideas and techniques in [29]. For a bridge from the classical to the modern, see Bell's excellent monograph [B] and the references there to several of his important papers.

The relative ease and similar approaches by which these various existence results could be established in the classical setting led the authors to anticipate wider applications. As they state in the introduction, and as quoted in [H], "Our new attack ... has the advantage of generality, since it is applicable in the case of partial differential equations of elliptic type and since for functions of several complex variables the theory of the kernel function is well developed ... ." Then also: "It is to be hoped that the independent basis upon which we are now able to put this theory will help to place it among the fundamental tools of function theory, potential theory and partial

---

[2]Near the end of his paper Weyl says: "The method of orthogonal projection in Hilbert space is a pleasant variant of the Dirichlet principle of minimization."

differential equations." There have been any number of successes, but also some roadblocks.

In [G2, G3], Garabedian took up the challenge of using kernel functions and Green's functions in several complex variables, particularly to study existence theorems for the system of Cauchy Riemann equations. The formal similarities are very appealing, but the technical difficulties were considerable. One could say that these efforts served to illuminate some fundamental differences between the single and several variable cases. In particular, they led to Garabedian and Spencer's paper [GS], which is credited with introducing the $\bar{\partial}$-Neumann problem and which in turn led to new techniques in partial differential equations, pursued over several decades. The tale of the trials and triumphs can be found in Hörmander's history [H].

References

[A] Lars V. Ahlfors, *Bounded analytic functions*, Duke Math. J. **14** (1947) 1 11.

[Ar] N. Aronszajn. *Theory of reproducing kernels*, Trans. Amer. Math. Soc. **68** (1950), 337 404.

[B] Steven R. Bell, *The Cauchy Transform, Potential Theory, and Conformal Mapping*, CRC Press, 1992.

[Br] Stefan Bergman, *The Kernel Function and Conformal Mapping*, second revised edition, American Mathematical Society, 1970.

[G1] P.R. Garabedian, *Schwarz s lemma and the Szego kernel function*, Trans. Amer. Math. Soc. **67** (1949), 1 35.

[G2] P.R. Garabedian, *A new formalism for functions of several complex variables*, J. Analyse Math. **1** (1951), 59 80.

[G3] P.R. Garabedian, *A Green s function in the theory of functions of several complex variables*, Ann. of Math. (2) **55** (1952), 19 33.

[G4] P.R. Garabedian, *Univalent functions and the Riemann mapping theorem*, Proc. Amer. Math. Soc. **61** (1976), 242 244.

[GS] P.R. Garabedian and D.C. Spencer, *Complex boundary value problems*, Trans. Amer. Math. Soc. **73** (1952), 223 242.

[H] Lars Hörmander, *A history of existence theorems for the Cauchy-Riemann complex in L^2 spaces*, J. Geom. Anal. **13** (2003), 329 357.

[L1] Peter D. Lax, *A remark on the method of orthogonal projections*, Comm. Pure Appl. Math. **4** (1951), 457 464.

[L2] Peter D. Lax, *On the existence of Green s function*, Proc. Amer. Math. Soc. **3** (1952), 526 531.

[Le] Olli Lehto, *Anwendung orthogonaler Systeme auf gewisse funktionentheoretische Extremal- und Abbildungsprobleme*, Ann. Acad. Sci. Fenn. Ser. A I Math.-Phys. **59** (1949).

[W] Hermann Weyl, *The method of orthogonal projection in potential theory*, Duke Math. J. **7** (1940), 411 444.

BRAD OSGOOD

[35] (with S. Bergman) Kernel functions and conformal mapping

[35] (with S. Bergman) Kernel functions and conformal mapping. *Compositio Math.* **8** (1951), 205–249.

Kernel functions and conformal mapping [1])

by

S. Bergman and M. Schiffer

To our teacher, Erhard Schmidt,
on the occasion of his 75th birthday

Introduction.

The concept of a kernel function has found increasing application in the theory of functions which satisfy certain linear differential equations in a fixed domain. It has permitted a unified treatment of different important theories in analysis.

A particular role is played by the reproducing kernel of the class of functions considered which can easily be constructed by means of a complete orthonormal system in this class and is, on the other hand, closely related to such important domain functions as Green's and Neumann's functions. This kernel was originally introduced in the study of pseudo-conformal mapping by means of pairs of analytic functions of two complex variables (Bergman[1]). Its usefulness for the classical theory of analytic functions of one complex variable was soon realized (Bergman[2]) and, finally, its connection established with Green's function and the canonical map functions (Schiffer[3]). Its role was also studied from a general point of view by stressing its reproducing property in a linear function space with hermitian metric. Most of these results were extended to the theory of partial differential equations of elliptic type and a new approach to the boundary value problems was obtained (Bergman—Schiffer [1][2][3]). The dependence of the kernel functions upon the basic metric and its significance were investigated (Garabedian [1], Schiffer [5]).

At this occasion it became clear that certain other kernels should also be taken into consideration which are closely related to the reproducing kernel but have important properties of their

---

[1]) Work done at Harvard University, Cambridge, Mass. U.S.A., under Navy Contract N5 ori 76—16, NR 043—046. The present paper was accepted on April 11, 1949, for publication in the Duke Mathematical Journal. For technical reasons it was transferred to this journal.

own. It is the purpose of this paper to study such an additional kernel in great detail for the theory of analytic functions of one complex variable. We show how this new kernel leads to important inequalities for the reproducing kernel which is still considered as the fundamental one. From these inequalities numerous estimates for the coefficients of univalent functions in a given domain are derived and it is shown that Grunsky's necessary and sufficient conditions for univalence (Grunsky [1]) are an immediate consequence of the theory of these two kernels.

Since our new kernel does not have the reproducing property we are naturally led to study those functions which are reproduced by it except for a constant factor. This introduces a homogeneous integral equation the eigen values and eigen functions of which are to be determined. It appears that this integral equation is closely related to the classical one used in the treatment of boundary value problems by integral equations; thus, a connection is established between these different approaches to the theory of conformal mapping.

Using formal identities between the two kernels we establish a quickly convergent series for the reproducing kernel which seems to us of great importance for the numerical side of the theory of mapping of multiply-connected domains upon canonical domains.

Finally, we establish variation formulas which show the dependance of some of the quantities discussed upon the domain of definition if the latter varies. All our formulas show a great symmetry and simplicity which seems to justify the introduction of the new concepts.

1. Generalities and notations.

We consider in the complex z-plane a finite domain B which is bounded by n closed analytic curves $C_\nu (\nu = 1, 2, \ldots n)$; we denote the boundary $\sum\limits_{\nu=1}^{n} C_\nu$ of B by C. If a complex-valued function $F(x, y)$ is differentiable in both arguments for every point $x + iy = z \, \epsilon \, B$, we can define two complex differential operators on F:

$$(1.1) \quad \frac{\partial F}{\partial z} = \tfrac{1}{2}\left(\frac{\partial F}{\partial x} - i\frac{\partial F}{\partial y}\right), \quad \frac{\partial F}{\partial z^\dagger} = \tfrac{1}{2}\left(\frac{\partial F}{\partial x} + i\frac{\partial F}{\partial y}\right), \quad z^\dagger = x - iy.$$

Analytic functions $f(z)$ are characterized by the Cauchy-Riemann condition

$$(1.2) \qquad\qquad\qquad \frac{\partial f}{\partial z^\dagger} = 0$$

while anti-analytic functions $f(z^\dagger)$ satisfy correspondingly

$$(1.2a) \qquad\qquad\qquad \frac{\partial f}{\partial z} = 0.$$

Let $f(z)$ and $g(z)$ be analytic in the closed region $B + C$; by means of (1.2) and (1.2a) we may establish the following simple rules on integration by parts:

$$(1.3) \qquad \iint_B [f'(z)]^\dagger g(z) d\tau = \frac{1}{2i} \int_C [f(z)]^\dagger g(z) dz, \quad d\tau = dx dy,$$

$$(1.4) \qquad \iint_B f'(z)[g(z)]^\dagger d\tau = -\frac{1}{2i} \int_C f(z)[g(z) dz]^\dagger.$$

Here and in the following the contour integration on C will be understood to be in the positive sense with respect to B.

We shall denote the complement of $B + C$ in the z-plane by $\tilde{B}$, and $\tilde{B}_\nu$ will be that component of B which is bounded by C_ν.

We assume that C is given in a parametric form $z(s)$ where s is the length parameter on C; thus $z' = \dfrac{dz}{ds}$ represents in each point of C the tangential unit vector. C has at each point $z(s)$ a normal and we denote by $\dfrac{\partial}{\partial n_s}$ the differential operator in the direction of the interior normal with respect to B.

In the following we shall often write $a(z)^\dagger$ instead of $[a(z)]^\dagger$.

2. Green's function.

Green's function $g(z, \zeta)$ of B is defined in the usual way by its three fundamental properties:

(a) $g(z, \zeta)$ is harmonic in z, for $\zeta \in B$ fixed, except for $z = \zeta$.

(b) $g(z, \zeta) + \log|z - \zeta|$ is harmonic in the neighbourhood of $z = \zeta$.

(c) $g(z, \zeta) \equiv 0$ for $z \in C$ and $\zeta \in B$.

The symmetry of $g(z, \zeta)$ in z and ζ follows easily from these properties; our assumptions on the analyticity of C ensure the harmonicity of Green's function even on the boundary C of B as the two argument points z and ζ stav apart.

From the identity

(2.1) $$g(z(s), \zeta) \equiv 0, \qquad z(s) \, \epsilon \, C, \qquad \zeta \, \epsilon \, B,$$

we derive by differentiation with respect to s:

(2.2) $$z' \frac{\partial g}{\partial z} + (z')^\dagger \frac{\partial g}{\partial z^\dagger} = 2\Re \left\{ z' \frac{\partial g}{\partial z} \right\} \equiv 0$$

i.e.

(2.2a) $$z'(s)\frac{\partial}{\partial z} g(z(s), \zeta) = i \cdot \mathscr{F}(\xi, \eta; s), \qquad \zeta = \xi + i\eta,$$

where $\mathscr{F}$ is a real-valued function of its arguments.

We now define the two functions

(2.3) $$K(z, \zeta^\dagger) = -\frac{2}{\pi} \frac{\partial^2 g(z, \zeta)}{\partial z \partial \zeta^\dagger}, \qquad L(z, \zeta) = -\frac{2}{\pi} \frac{\partial^2 g(z, \zeta)}{\partial z \partial \zeta}.$$

They are both analytic in their arguments which easily follows from the harmonicity of Green's function. From (2.2a) and (2.8) we conclude

(2.4) $$z'(s) L(z(s), \zeta) = -[z'(s)K(z(s), \zeta^\dagger)]^\dagger, \quad z(s) \, \epsilon \, C, \zeta \, \epsilon \, B.$$

The function $K(z, \zeta^\dagger)$ is for fixed $\zeta \, \epsilon \, B$ regular in the closed region $B + C$; the logarithmic pole of Green's function has been destroyed by the particular process of differentiation leading to K. The function $L(z, \zeta)$, however, has a double-pole for $z = \zeta$ and may be written in the form

(2.5) $$L(z, \zeta) = \frac{1}{\pi(z - \zeta)^2} - l(z, \zeta)$$

where $l(z, \zeta)$ is, for $\zeta \, \epsilon \, B$, regular for $z \, \epsilon \, B + C$.

We further notice the symmetry relations which follows from the definitions:

(2.6) $$[K(z, \zeta^\dagger)]^\dagger = K(\zeta, z^\dagger), \quad L(z, \zeta) = L(\zeta, z), \quad l(z, \zeta) = l(\zeta, z).$$

For instance in the case of the unit-circle $|z| < 1$ we have

(2.7) $$g(z, \zeta) = \log \left| \frac{1 - \zeta^\dagger z}{z - \zeta} \right|, \quad K(z, \zeta^\dagger) = \frac{1}{\pi(1 - \zeta^\dagger z)^2}$$
$$L(z, \zeta) = \frac{1}{\pi(z - \zeta)^2}, \quad l(z, \zeta) = 0.$$

The functions K and L play a central role in the theory of logarithmic potential and conformal mapping and it is the principal aim of this paper to investigate their properties and to

show their applications. The following result (Schiffer [3]) illustrates their importance:

Let $\mathfrak{L}^2$ be the class of all functions $f(z)$ which are analytic in B
and for which the Lebesgue integral

$$(2.8) \qquad \iint_B |f(z)|^2 \, d\tau < \infty.$$

For each $f(z) \,\epsilon\, \mathfrak{L}^2$ we have the identities

$$(2.9) \qquad \iint_B K(z, \zeta^\dagger) \, f(\zeta) d\tau_\zeta = f(z)$$

and

$$(2.10) \qquad \iint_B L(z, \zeta)^\dagger f(\zeta) d\tau_\zeta = 0.$$

Both integrals are to be understood in the Lebesgue sense, and
the improper integral in (2.10) is the limit of an integral over
the domain $B_{\varepsilon, z}$ which is obtained from B by elimination of a
circle around z with radius ε.

3. The kernel functions.

We shall call $K(z, \zeta^\dagger)$ and $l(z, \zeta)$ the kernel functions of the
first and second kind with respect to the class $\mathfrak{L}^2$, since they
appear as kernels of certain integral operators applied to the
class; $K(z, \zeta^\dagger)$ might also be called the reproducing kernel of the
class because of (2.9). The significance of $l(z, \zeta)$ follows from
the identity

$$(3.1) \qquad \iint_B l(z, \zeta)^\dagger f(\zeta) d\tau_\zeta = \frac{1}{\pi} \iint_B [(z - \zeta)^{-2}]^\dagger f(\zeta) d\tau_\zeta$$

which is a consequence of (2.5) and (2.10). We see that $l(z, \zeta)$
is a kernel of the class $\mathfrak{L}^2$ which has on each function $f(z) \,\epsilon\, \mathfrak{L}^2$ the
same effect as the important but singular kernel $[\pi(z - \zeta)^2]^{-1}$.
Numerous applications of this fact will be given in the following.

Let $w = \varphi(z)$ map the domain B univalently upon a domain B_1
with analytic boundary C_1. If $\omega = \varphi(\zeta)$ and $g_1(w; \omega)$ is Green's
function with respect to B_1, we have the well-known identity

$$(3.2) \qquad g_1(w, \omega) = g_1(\varphi(z), \; \varphi(\zeta)) = g(z, \zeta).$$

Differentiating with respect to z and ζ and denoting by K_1, L_1
and l_1 the kernels with respect to B_1 which correspond to K,
L and l, we find in view of (2.3):

$$(3.3) \qquad K_1(w, \omega^\dagger) \, \varphi'(z)[\varphi'(\zeta)]^\dagger = K(z, \zeta^\dagger)$$

and

(3.4) $$L_1(w, \omega)\, \varphi'(z)\, \varphi'(\zeta) = L(z, \zeta).$$

Hence, in view of definition (2.5)

(3.5) $$l_1(w, \omega)\varphi'(z)\varphi'(\zeta) = l(z, \zeta) + \frac{1}{\pi}\left\{ \frac{\varphi'(z)\varphi'(\zeta)}{(\varphi(z) - \varphi(\zeta))^2} - \frac{1}{(z-\zeta)^2} \right\}.$$

This formula is better understood if we introduce the expression

(3.6) $$\Phi(z, \zeta) = \frac{1}{\pi} \log \frac{\varphi(z) - \varphi(\zeta)}{z - \zeta}$$

which is analytic in the closed region $B + C$ because of the univalence of $\varphi(z)$. Then we may write instead of (3.5):

(3.5a) $$l_1(w, \omega)\varphi'(z)\varphi'(\zeta) = l(z, \zeta) + \frac{\partial^2\Phi}{\partial z\, \partial\zeta}.$$

We notice the formal identity

(3.7) $$\frac{\partial^2\Phi}{\partial z\, \partial\zeta}\Big|_{z=\zeta} = -\frac{1}{6\pi}\left[\frac{d^2}{dz^2} \log \frac{d\varphi}{dz} - \frac{1}{2}\left(\frac{d}{dz} \log \frac{d\varphi}{dz} \right)^2 \right] = -\frac{1}{6\pi}\{\varphi, z\}$$

where $\{\varphi, z\}$ denotes the well-known differential parameter of Schwarz. This shows the interest of the function $\Phi(z, \zeta)$ of two complex variables in connection with the conformal mapping produced by $\varphi(z)$.

We now make the following application of the transformation formulas (3.3) and (3.5a). Ley B_1 be a simply-connected domain in the w-plane and let $w = \varphi(z)$ be the map of the unit-circle $|z| < 1$ upon B_1. Since $\varphi(z)$ is still analytic on $|z| = 1$ we see from (2.7) that

(3.8)
$$K_1(w, \omega^\dagger) = (\pi\varphi'(\zeta)^\dagger\varphi'(z)(1 - \zeta^\dagger z)^2)^{-1}$$
$$l_1(w, \omega) = (\varphi'(z)\varphi'(\zeta))^{-1}\frac{\partial^2\Phi}{\partial z\, \partial\zeta}$$

are still analytic on the boundary C_1 of B_1 if w and ω are separated. If $w = \omega$, however, K_1 becomes strongly infinite while l_1 remains regular even then. Thus, in the case of a simply-connected domain $l(z, \zeta)$ is regular in both arguments in the closed region $B + C$.

We now want to extend this result to the case of an arbitrary finite connectivity. We choose one boundary curve C_ν, say C_1, and consider the complement of the domain $\bar{B}_1$. This domain contains our initial domain B as subdomain; let $g_1(z, \zeta)$ be its Green's function. $g_1(z, \zeta)$ vanishes on C_1 and is still harmonic on the curve itself. If $g(z, \zeta)$ is again Green's function of B, the term $g_1(z, \zeta) - g(z, \zeta)$ is regular harmonic in B and may therefore be expressed by its boundary values and $g(z, \zeta)$. In fact, we have

$$
\begin{aligned}
g_1(z, \zeta) - g(z, \zeta) &= \frac{1}{2\pi} \int_C g_1(z, t) \frac{\partial g(t, \zeta)}{\partial n_t} ds_t = \\
&= \frac{1}{2\pi} \sum_{\nu \neq 1} \int_{C_\nu} g_1(z, t) \frac{\partial g(t, \zeta)}{\partial n_t} ds_t
\end{aligned}
$$
(3.9)

where $\dfrac{\partial}{\partial n}$ denotes differentiation in the direction of the interior normal. We notice that the integration in (3.9) runs over all boundary components of B except for C_1; the point $t \in C$ therefore never lies on C_1.

From (3.9) and (2.3), (2.5) we easily deduce

$$
(3.10) \qquad l_1(z, \zeta) - l(z, \zeta) = \frac{1}{\pi^2} \sum_{\nu \neq 1} \int_{C_\nu} \frac{\partial g_1(z, t)}{\partial z} \frac{\partial^2 g(t, \zeta)}{\partial n_t \partial \zeta} ds_t.
$$

Now, $l_1(z, \zeta)$ is regular even on C_1 since it is the l-kernel of a simply-connected domain. In (3.10) the right-hand integral is regular on C_1 since t does not run over this particular curve. Hence we proved the following theorem:

The function $l(z, \zeta)$ is regular analytic in the closed region $B + C$.

This property of the l-kernel will be of of great use for the general theory; it is one of the main reasons for the importance of this kernel. The K-kernel with its reproducing proporty and its simple definition (2.3) attracted the interest much earlier than the l-kernel; it has not, however, the property of regularity in the closed region $B + C$ and its infinity on the boundary was of some difficulty in its theory. By establishing a simple relationship between the two kernel functions we will be able to overcome this difficulty and to remove the infinity of the kernel function by addition of an elementary function.

4. Identities and inequalities for the kernel functions.

The functions of the class $\mathfrak{L}^2$ form a linear space Λ and we may introduce into this space a hermitian metric based on the scalar product between two elements f and g:

$$
(4.1) \qquad\qquad (f, g^\dagger) = \iint_B f(z)\, (g(z))^\dagger d\tau_z.
$$

Then it is important to determine the various scalar products between kernel functions.

From the reproducing property (2.9) of the K-kernel and the symmetry laws (2.6), we deduce immediately

$$(4.2) \quad \iint_B K(z, \zeta^\dagger)[K(z, w^\dagger)]^\dagger d\tau_z = \iint_B K(w, z^\dagger)K(z, \zeta^\dagger)d\tau_z = K(w, \zeta^\dagger)$$

and

$$(4.3) \quad \iint_B l(z, w)[K(z, \zeta^\dagger)]^\dagger d\tau_z = \iint_B K(\zeta, z^\dagger) l(z, w)d\tau_z = l(w, \zeta).$$

It is a little more difficult to determine the scalar products between l-kernels. Using the identity (3.1), we find

$$(4.4) \quad \iint_B l(z, \zeta)^\dagger l(z, w)d\tau_z = \frac{1}{\pi} \iint_B l(z, w)[(z - \zeta)^{-2}]^\dagger d\tau_z.$$

By integration by parts of the type (1.3), we transform this into

$$(4.5) \quad \iint_B l(z, \zeta)^\dagger l(z, w)d\tau_z = -\frac{1}{2\pi i} \int_C l(z, w) \frac{dz}{(z - \zeta)^\dagger}.$$

For $z \in C$, we have by (2.5) and (2.4)

$$(4.6) \quad l(z, w)dz = \frac{1}{\pi} \frac{dz}{(z - w)^2} - L(z, w)dz = \frac{1}{\pi} \frac{dz}{(z - w)^2} + [K(z, w^\dagger)dz]^\dagger.$$

Hence, (4.5) obtains the form

$$(4.7) \quad \iint_B l(z, \zeta)^\dagger l(z, w)d\tau_z = \left[\frac{1}{2\pi i} \int_C \frac{K(z, w^\dagger)}{z - \zeta} dz \right]^\dagger - \frac{1}{2\pi^2 i} \int_C \frac{dz}{(z - w)^2(z - \zeta)^\dagger}.$$

The first right-hand integral may be computed by the residue theorem; the second integral may be transformed into an integral over the complement $\overline{B}$ of B by means of (1.3). Finally, we arrive at the identity:

$$(4.8) \quad \iint_B l(z, \zeta)^\dagger l(z, w)d\tau_z = K(w, \zeta^\dagger) - \Gamma(w, \zeta^\dagger)$$

with

$$(4.9) \quad \Gamma(w, \zeta^\dagger) = \frac{1}{\pi^2} \iint_{\overline{B}} \frac{d\tau_z}{(z - w)^2[(z - \zeta)^2]^\dagger}.$$

Hence the scalar product between two l-kernels leads to a K-kernel and a Γ-term. The characteristic property of the latter is that it can be computed by elementary integration over the exterior of the considered domain B. It does not depend on the solution of a boundary value problem in harmonic functions as do the K- and l-kernels. We shall call expressions of this type geometric quantities and consider a problem in harmonic functions solved if it can be reduced to the computation of such terms. The geometric quantities are elementary ones as compared with the function-theoretic terms involving Green's function.

We now make the following natural application of our identities: We choose $r + s$ points $\zeta_1, \zeta_2, \ldots \zeta_r, \eta_1, \ldots \eta_s$ in B and $r + s$ arbitrary constants $\alpha_1 \ldots \alpha_r, \beta_1 \ldots \beta_s$. We start from the obvious inequality

$$(4.10) \quad \iint_B \Big| \sum_{\nu=1}^r \alpha_\nu^\dagger K(z, \zeta_\nu^\dagger) + \lambda \sum_{\mu=1}^s \beta_\mu l(z, \eta_\mu) \Big|^2 d\tau \geqq 0, \; \lambda \text{ real},$$

and compute the left-hand integral by means of the identities (4.2), (4.3) and (4.8). We obtain

$$(4.11) \quad \sum_{\nu, \mu=1}^r \alpha_\nu \alpha_\mu^\dagger K(\zeta_\nu, \zeta_\mu^\dagger) + 2\lambda \Re \Big\{ \sum_{\nu=1}^r \sum_{\mu=1}^s \alpha_\nu \beta_\mu l(\zeta_\nu, \eta_\mu) \Big\} +$$
$$+ \lambda^2 \sum_{\nu, \mu=1}^s \beta_\nu \beta_\mu^\dagger [K(\eta_\nu, \eta_\mu^\dagger) - \Gamma(\eta_\nu, \eta_\mu^\dagger)] \geqq 0.$$

For $\lambda = 0$ we obtain the well-known inequality

$$(4.11a) \quad \sum_{\nu, \mu=1}^r \alpha_\nu \alpha_\mu^\dagger K(\zeta_\nu, \zeta_\mu^\dagger) \geqq 0$$

which is often expressed by the statement that $K(z, \zeta^\dagger)$ is a definite kernel. This property is characteristic for any kernel which has the reproducing property with respect to a certain Hilbert space, as has been stressed in the abstract theory of such kernels (Aronszajn [1]); the proof in the general case is also based on the fact that the norm of every element in a Hilbert space is non-negative.

We conclude further from (4.11) the inequality

$$(4.12) \quad \sum_{\nu, \mu=1}^s \beta_\nu \beta_\mu^\dagger K(\eta_\nu, \eta_\mu^\dagger) \geqq \sum_{\nu, \mu=1}^s \beta_\nu \beta_\mu^\dagger \Gamma(\eta_\nu, \eta_\mu^\dagger).$$

This is a real improvement of (4.11a) since the kernel $\Gamma(z, \zeta^\dagger)$ is a positive-definite kernel, too. In fact, we may write

$$(4.13) \quad \sum_{\nu, \mu=1}^s \beta_\nu \beta_\mu^\dagger \Gamma(\eta_\nu, \eta_\mu^\dagger) = \frac{1}{\pi^2} \iint_B \Big| \sum_{\nu=1}^s \frac{\beta_\nu}{(z - \eta_\nu)^2} \Big|^2 d\tau_z,$$

which proves our assertion. By means of (4.12) we can estimate the hermitian forms connected with the kernel function in terms of geometric expressions.

Finally, we obtain from (4.11) the discriminant inequality

$$(4.14) \quad \Re\left\{ \sum_{\nu=1}^{r} \sum_{\mu=1}^{s} \alpha_\nu \beta_\mu l(\zeta_\nu, \eta_\mu) \right\}^2 \leqq$$

$$\leqq \sum_{\mu,\nu=1}^{r} \alpha_\nu \alpha_\mu^\dagger K(\zeta_\nu, \zeta_\mu^\dagger) \cdot \sum_{\nu,\mu=1}^{s} \beta_\nu \beta_\mu^\dagger [K(\eta_\nu, \eta_\mu^\dagger) - \Gamma(\eta_\nu, \eta_\mu^\dagger)].$$

If we replace in this inequality each β_μ by $\beta_\mu e^{i\sigma}$, the right-hand side remains unchanged while the left-hand side varies. The best possible inequality thus obtained is

$$(4.14a) \quad \left| \sum_{\nu=1}^{r} \sum_{\mu=1}^{s} \alpha_\nu \beta_\mu l(\zeta_\nu, \eta_\mu) \right|^2 \leqq$$

$$\leqq \sum_{\mu,\nu=1}^{r} \alpha_\nu \alpha_\mu^\dagger K(\zeta_\nu, \zeta_\mu^\dagger) \cdot \sum_{\nu,\mu=1}^{s} \beta_\nu \beta_\mu^\dagger [K(\eta_\nu, \eta_\mu^\dagger) - \Gamma(\eta_\nu, \eta_\mu^\dagger)].$$

Because of the definite character of $\Gamma(z, \zeta^\dagger)$ this inequality implies

$$(4.14b) \quad \left| \sum_{\nu=1}^{r} \sum_{\mu=1}^{s} \alpha_\nu \beta_\mu l(\zeta_\nu, \eta_\mu) \right|^2 \leqq$$

$$\leqq \sum_{\mu,\nu=1}^{r} \alpha_\nu \alpha_\mu^\dagger K(\zeta_\nu, \zeta_\mu^\dagger) \cdot \sum_{\nu,\mu=1}^{s} \beta_\nu \beta_\mu^\dagger K(\eta_\nu, \eta_\mu^\dagger).$$

If we finally choose $r = s$, $\alpha_\nu = \beta_\nu$, $\zeta_\nu = \eta_\nu$, we arrive at

$$(4.14c) \quad \left| \sum_{\mu,\nu=1}^{r} \alpha_\nu \alpha_\mu l(\zeta_\nu, \zeta_\mu) \right| \leqq \sum_{\mu,\nu=1}^{r} \alpha_\nu \alpha_\mu^\dagger K(\zeta_\nu, \zeta_\mu^\dagger).$$

Another very important consequence of (4.8) is the identity

$$(4.15) \quad K(z, z^\dagger) - \Gamma(z, z^\dagger) = \iint_B |l(z, \zeta)|^2 d\tau_\zeta \geqq 0.$$

Since $l(z, \zeta)$ is regular and analytic in the closed region $B + C$, we conclude from (4.15) that $K(z, z^\dagger) - \Gamma(z, z^\dagger)$ is bounded in $B + C$. This shows that the geometric quantity $\Gamma(z, z^\dagger)$ has at the boundary C the same asymptotic behavior as $K(z, z^\dagger)$ and that their difference behaves quite regular. At the same time, this elementary term provides at each interior point z a lower bound for $K(z, z^\dagger)$.

We have further the important theorem:

The hermitian kernel $K(z, \zeta^\dagger) - \Gamma(z, \zeta^\dagger)$ is regular in the closed region $B + C$.

The irregular behavior of $K(z, \zeta^\dagger)$ on C led to the phenomenon that the homogeneous integral equation

$$(4.16) \quad \varphi(z) = \lambda \iint_B K(z, \zeta^\dagger)\, \varphi(\zeta) d\tau_\zeta$$

had the value $\lambda = 1$ as eigenvalue of infinite order so that each analytic function $\varphi(z)$ was a corresponding eigenfunction. The classical theory for regular hermitian kernels is, however, applicable to the regularized kernel $K(z, \zeta^\dagger) - \Gamma(z, \zeta^\dagger)$ and we shall later study its eigenvalues and eigenfunctions. The close relation between the two hermitean kernels $K(z, \zeta^\dagger)$ and $\Gamma(z, \zeta^\dagger)$ is illustrated by the easily established fact that $K(z, \zeta^\dagger)[\Gamma(z, \zeta^\dagger)]^{-1}$ is invariant with respect to linear transformations of B.

We mention further the special instance of (4.14a)

$$(4.17) \qquad |l(z, \zeta)|^2 \leqq K(z, z^\dagger) \cdot [K(\zeta, \zeta^\dagger) - \Gamma(\zeta, \zeta^\dagger)]$$

which implies

$$(4.17a) \qquad\qquad |l(z, z)| \leqq K(z, z^\dagger).$$

5. The l-transforms

The l-kernel transforms every analytic function $f(z)$ of the class $\mathfrak{L}^2$ into a new analytic function $T\,f(z)$ by means of the operation

$$(5.1) \qquad\qquad T f(z) = \iint\limits_{B} l(z, \zeta)\, f(\zeta)^\dagger\, d\tau_\zeta.$$

We call $T\,f$ the l-transform of f and want to study the class of all these transforms. Using Schwarz' inequality and (4.15), we find

$$(5.2) \qquad | T f(z) |^2 \leqq [K(z, z^\dagger) - \Gamma(z, z^\dagger)] \iint\limits_{B} | f(\zeta) |^2 d\tau_\zeta$$

while the same reasoning applied to (2.9) yields

$$(5.2a) \qquad\qquad | f(z) |^2 \leqq K(z, z^\dagger) \iint\limits_{B} | f(\zeta) |^2 d\tau_\zeta.$$

We see from (5.2) that the class of all l-transforms of $\mathfrak{L}^2$ forms a proper subclass of $\mathfrak{L}^2$ which contains only bounded functions. One easily sees that all l-transforms are analytic in the closed region $B + C$.

Because of the fundamental property (3.1) of the l-kernel we may express the l-transform of $f(z)$ by means of the improper integral

$$(5.3) \qquad\qquad T f(z) = \frac{1}{\pi} \iint\limits_{B} f(\zeta)^\dagger (\zeta - z)^{-2}\, d\tau_\zeta.$$

This representation has the advantage of possessing an elementary kernel and of admitting simple transformations. Applying for example the integration rule (1.4) we obtain

$$(5.4) \qquad\qquad T f(z) = \frac{1}{2\pi i} \int\limits_{C} (\zeta - z)^{-1} f[(\zeta) d\zeta]^\dagger$$

in the case that $f(z)$ is continuous in the closed region $B + C$.
Further we way use the representation (5.8) in order to define
the transform $\mathrm{T} f(z)$ in the whole complex z-plane. In each
domain $\overline{B}_\nu$ the function $\mathrm{T} f(z)$ then represents an analytic function.

The different analytic functions $\mathrm{T} f(z)$ defined in the domains
B and $\overline{B}_\nu$ do not form a continuous function in the whole z-plane.
In order to study their behavior on C let us assume that $f(z)$ is
continuous in $B + C$, so that the representation (5.4) holds.
According to a classical theorem by Plemelj (Plemelj [1]) the
function $\mathrm{T} f(z)$ has a saltus of the value

(5.5) $\Delta(\mathrm{T} f(z)) = - [f(z) \cdot z'^2]^\dagger$

if we cross at the point $z \in C$ from B into the complementary
region $\overline{B}$.

Let us illustrate these formulas by the following example.
We have

(5.6) $\mathrm{T} K(z, w^\dagger) = \iint\limits_B K(\zeta, w^\dagger)^\dagger l(\zeta, z) d\tau_\zeta = l(z, w)$ for $z \in B$

and

(5.6a) $\mathrm{T} K(z, w^\dagger) = \frac{1}{\pi} \iint\limits_B K(\zeta, w^\dagger)^\dagger (\zeta - z)^{-2} d\tau_\zeta = \frac{1}{\pi(z-w)^2}$ for $z \in \overline{B}$.

The latter result follows from the fact that $(\zeta - z)^{-2}$ for $z \in \overline{B}$
belongs to the class $\mathfrak{L}^2$ with respect to ζ and that, therefore, the
reproducing property of the kernel function may be applied
here. The saltus condition (5.5) takes the form

(5.7) $- [K(z, w^\dagger)z'^2]^\dagger = \frac{1}{\pi(z-w)^2} - l(z, w) = L(z, w)$

which is just the important boundary relation (2.4).

At this point we notice that (5.6) may also be written in
the form

(5.6b) $l(z, w) = \frac{1}{\pi} \iint\limits_B K(\zeta, w^\dagger)^\dagger (\zeta - z)^{-2} d\tau_\zeta$ for $z \in B$

which shows that $l(z, w)$ may be computed elementarily, once
the K-kernel has been determined.

Similarly, we have

(5.8) $\mathrm{T} l(z, w) = \iint\limits_B l(\zeta, w)^\dagger l(\zeta, z) d\tau_\zeta = K(z, w^\dagger) - \Gamma(z, w^\dagger)$ for $z \in B$

and

(5.8a) $\mathrm{T} l(z, w) = \frac{1}{\pi} \iint\limits_B l(\zeta, w)^\dagger (\zeta - z)^{-2} d\tau_\zeta =$

$= \frac{1}{\pi^2} \iint\limits_B \lceil (\zeta - w)^{-2}) \rceil^\dagger (\zeta - z)^{-2} d\tau_\zeta, \quad z \in \overline{B}.$

One can show that in this example, too, the saltus condition (5.5) leads to the boundary relation (2.4) between the two kernels. If we bring (5.8) into the form

$$(5.8b) \quad K(z, w^\dagger) = \Gamma(z, w^\dagger) + \frac{1}{\pi} \iint_B l(\zeta, z)[(\zeta - w)^{-2}]^\dagger \, d\tau_\zeta, \quad z \in B,$$

we find that the K-kernel can be expressed by elementary computations in terms of the l-kernel. One sees that it is sufficient to find a construction for either kernel and that the other is then easily obtained.

Now let $f(z)$ and $g(z)$ be a pair of functions of the class $\mathfrak{L}^2$. Defining $\mathrm{T}\,f$ and $\mathrm{T}\,g$ by (5.1), we can easily compute by means of (4.8) the scalar product:

$$(5.9) \quad \iint_B \mathrm{T}f \cdot (\mathrm{T}g)^\dagger d\tau = \iint_B \iint_B [K(\zeta, \eta^\dagger) - \Gamma(\zeta, \eta^\dagger)]f(\zeta)^\dagger g(\eta) d\tau_\zeta d\tau_\eta.$$

Using the reproducing property (2.9) of the K-kernel and the definitions (4.9), (5.3), we may bring (5.9) into the elegant form

$$(5.10) \quad \iint_{B+\bar{B}} \mathrm{T}f \cdot (\mathrm{T}g)^\dagger d\tau = \iint_B f^\dagger g \, d\tau.$$

This result suggests the following concepts. Just as the metric in the space Λ was based on the scalar product (4.1), we may base a metric in the linear space Λ_T of transforms on the metric

$$(5.11) \quad [\mathrm{T}f, \ (\mathrm{T}g)^\dagger] = \iint_{B+\bar{B}} \mathrm{T}f \cdot (\mathrm{T}g)^\dagger \, d\tau.$$

In fact, the transforms being defined in the whole complex plane, it is natural to integrate in their scalar products over this whole region. In this notation, we now may express (5.10) in the form

$$(5.10a) \quad [\mathrm{T}f, \ (\mathrm{T}g)^\dagger] = (f^\dagger, g).$$

The close relation between the linear spaces Λ and Λ_T becomes more evident by the following inversion formulas: Given a function $\mathrm{T}f \in \Lambda_\mathrm{T}$, we want to find its generating function $f \in \Lambda$. For this purpose, we determine

$$(5.12) \quad \mathrm{T}[\mathrm{T}f(z)] = \iint_B l(\zeta, z)[\iint_B l(w, \zeta) \, f(w)^\dagger d\tau_w]^\dagger d\tau_\zeta, \quad z \in B.$$

Using (4.8) and (5.3), we have

$$(5.13) \quad \frac{1}{\pi} \iint_B [\mathrm{T}f(\zeta)]^\dagger \frac{d\tau_\zeta}{(\zeta - z)^2} = \iint_B [K(z, \zeta^\dagger) - \Gamma(z, \zeta^\dagger)]f(\zeta) \, d\tau_\zeta.$$

By virtue of (2.9) and (4.9), this may be written in the form

$$(5.14) \qquad f(z) = \frac{1}{\pi} \iint_{B+\bar{B}} [\mathbf{T} f(\zeta)]^\dagger \frac{d\tau_\zeta}{(\zeta - z)^2}, \qquad z \in B.$$

This formula shows the great symmetry existing between the spaces Λ and Λ_T; the corresponding elements transform into each other by an integral operation with the same kernel, extended in each case over the proper domain of definition only.

The meaning of (5.14) becomes very clear if we transform the domain integral into a contour integral along C by means of the integration rule (1.4). We arrive at

$$(5.14a) \quad \frac{1}{\pi} \iint_{B+\bar{B}} [\mathbf{T} f(\zeta)]^\dagger \frac{d\tau_\zeta}{(\zeta - z)^2} = \frac{1}{2\pi i} \int_C - [\varDelta (\mathbf{T} f(\zeta))]^\dagger \frac{d\zeta^\dagger}{\zeta - z}$$

where $\varDelta(\mathbf{T} f)$ is the discontinuity of $\mathbf{T} f$ on C as given by (5.5). Thus, (5.14) is nothing but the identity

$$(5.15) \qquad f(z) = \frac{1}{2\pi i} \int_C \frac{f(\zeta)}{\zeta - z} d\zeta$$

in the case that $f(z)$ is continuous in the closed region $B \in C$. This transformation shows also clearly that the value of the right-hand integral in (5.14) has the value zero for $z \in \bar{B}$.

Each function $\mathbf{T} f \in \Lambda_T$ is also a function of Λ and has a norm $(\mathbf{T} f, (\mathbf{T} f)^\dagger)$. We may compare its norm in Λ_T with that in Λ and find by (5.11)

$$(5.16) \quad [\mathbf{T} f, (\mathbf{T} f)^\dagger] = (\mathbf{T} f, (\mathbf{T} f)^\dagger) + \iint_{\bar{B}} |\mathbf{T} f|^2 d\tau \geq (\mathbf{T} f, (\mathbf{T} f)^\dagger).$$

There arises the question under what circumstances equality might hold in (5.16). It is obvious that in this case necessarily

$$(5.17) \qquad \mathbf{T} f(z) \equiv 0 \quad \text{for} \quad z \in \bar{B}.$$

From (5.14) we then conclude that

$$(5.18) \qquad f(z) = \mathbf{T} (\mathbf{T} f(z)), \quad \text{for} \quad z \in B,$$

i.e. $f(z)$ belongs also to the space Λ_T and is, therefore, analytic in $B + C$. Hence we may apply the saltus condition (5.5) to $\mathbf{T} f(z)$ and since $\mathbf{T} f(z)$ vanishes in $\bar{B}$, we simply obtain for the limit of $\mathbf{T} f$ an interior approach to $z \in C$:

$$(5.19) \qquad \mathbf{T} f(z) = - [f(z) z'^2]^\dagger, \quad z \in C.$$

We write this result in the more symmetric forms

(5.19a) $(Tf(z) + f(z))z' = - [(Tf(z) + f(z)z']^\dagger_s,\quad z \in C$

(5.19b) $i(Tf(z) - f(z))z' = - [i(Tf(z) - f(z))z']^\dagger,\quad z \in C.$

We introduce two real harmonic functions $\Omega_1(x, y)$ and $\Omega_2(x, y)$ such that

(5.20) $Tf(z) + f(z) = \dfrac{\partial \Omega_1}{\partial z},\quad i(Tf(z) - f(z)) = \dfrac{\partial \Omega_2}{\partial z}.$

The formulas (5.19a) and (5.19b) then simply state that on C

(5.21) $\dfrac{d\Omega_1}{ds} = 0,\quad \dfrac{d\Omega_2}{ds} = 0.$

Hence Ω_1 and Ω_2 are two real harmonic functions in B which are constant on each boundary curve C_ν. Therefore, they may be linearly composed of the harmonic measure functions

(5.22), $\omega_\nu(x, y) = \dfrac{1}{2\pi} \displaystyle\int_{C_\nu} \dfrac{\partial g(z, \zeta)}{\partial n_\zeta} ds_\zeta,\quad z = x + iy,$

which have the value 1 on C_ν and 0 on the rest of C. We introduce the analytic functions

(5.23) $w'_\nu(z) = 2\dfrac{\partial}{\partial z} \omega_\nu(x, y),\quad w'_\nu(z) \equiv \dfrac{dw_\nu(z)}{dz}$

which clearly satisfy the boundary conditions

(5.23') $w'_\nu(z)z' = - [w'_\nu(z)z']^\dagger.$

Then it follows from our considerations above that

(5.24) $f(z) = \displaystyle\sum_{\nu=1}^{n} a_\nu w'_\nu(z)$

with complex coefficients a_ν.

If inversely $f(z)$ has the form (5.24), it is analytic in $B + C$ and we may apply the formula (5.4). Because of (5.23a) this leads to

(5.25) $Tf(z) = \dfrac{1}{2\pi i} \displaystyle\int_C \sum_{\nu=1}^{n} a^\dagger_\nu [w'_\nu(\zeta)\zeta']^\dagger (\zeta - z)^{-1} ds = - \dfrac{1}{2\pi i} \int_C a^\dagger_\nu w'_\nu(\zeta) \dfrac{d\zeta}{\zeta - z}$

i.e.

(5.25a) $Tf(z) = - \displaystyle\sum_{\nu=1}^{n} a^\dagger_\nu w'_\nu(z)$ for $z \in B$

(5.25b) $Tf(z) = 0$ for $z \in \widetilde{B}.$

We see that in this case (5.17) and (5.18) indeed are fulfilled and that, therefore the equality sign in (5.16) holds. Since the harmonic measures have non-vanishing derivatives only in multiply-connected domains B, we see that equality in (5,16) is impossible in simply-connected domains, except for identically vanishing $f(z)$.

We notice finally that each function

$$(5.26) \qquad f(z) = i \sum_{\nu=1}^{n} a_\nu \overline{w}_\nu'(z), \qquad a_\nu \text{ real}$$

satisfies the condition

$$(5.27) \qquad f(z) = Tf(z)$$

and that every solution of (5.27) must necessarily have the form (5.26).

6. The eigen functions of the l-kernel

It is natural to ask for those functions in Λ which coincide with their l-transforms except for a numerical factor, i.e. which satisfy the integral equation

$$(6.1) \qquad \varphi_\nu(z) = \lambda_\nu \iint_B \varphi_\nu(\zeta)^\dagger \, l(\zeta, z) d\tau_\zeta.$$

Every multiple of $\varphi_\nu(z)$ will have the same property, since we may put (6.1) into the form

$$(6.1\text{a}) \qquad a\varphi_\nu(z) = \lambda_\nu a(a^{-1})^\dagger \iint_B [a\varphi_\nu(\zeta)]^\dagger l(\zeta, z) d\tau_\zeta.$$

We use this fact in order to put normalizing restrictions on the functions $\varphi_\nu(z)$ which we will consider. It is sufficient to deal with functions $\varphi_\nu(z)$ for which

$$(6.2) \qquad \iint_B |\varphi_\nu(z)|^2 \, d\tau_z = 1$$

and for which the corresponding λ_ν satisfies the condition

$$(6.3) \qquad \lambda_\nu \geqq 0.$$

We shall call such a function $\varphi_\nu(z)$ an eigen function and the corresponding value λ_ν an eigen value of the kernel l.

At the end of the preceding section we saw that the value 1 is an eigen value of the l-kernel in each domain of connectivity $n \geqq 1$ and that it belongs to the $n-1$ linearly independent eigen functions $i\,w_1'(z), \ldots i\,w_{n-1}'(z)$. In such a case we say that the eigen value $\lambda_\nu = 1$ is of degeneracy $n-2$.

The study of the integral equation (6.1) may easily be reduced to the classical theory of integral equations with hermitian definite kernels. In fact, iterating the integral equation (6.1) we obtain

$$(6.4) \qquad \varphi_\nu(z) = \lambda_\nu \iint_B [\lambda_\nu \iint_B \varphi_\nu(w)^\dagger l(w, \zeta) d\tau_w]^\dagger l(\zeta, z) d\tau_\zeta,$$

which leads because of (4.8) to the new integral equation

$$(6.5) \qquad \varphi_\nu(z) = \lambda_\nu^2 \iint_B [K(z, \zeta^\dagger) - \Gamma(z, \zeta^\dagger)] \varphi_\nu(\zeta) d\tau_\zeta.$$

Hence every eigen function of (6.1) is also a solution of the simpler integral equation (6.5) and to each eigen value λ_ν of (6.1) corresponds an eigen value λ_ν^2 of (6.5). Now we shall show that the converse of this statement is also true and derive from this fact the existence of solutions of (6.1).

The kernel $K(z, \zeta^\dagger) - \Gamma(z, \zeta^\dagger)$ is hermitian, regular in $B + C$ and positive definite, since we have for an arbitrary continuous function $\mu(z)$ in B in view of (4.8):

$$(6.6) \qquad \iint_B \iint_B [K(z, \zeta^\dagger) - \Gamma(z, \zeta^\dagger)] \, \mu(z)^\dagger \mu(\zeta) d\tau_z d\tau_\zeta =$$

$$\iint_B \left| \iint_B l(z, w) \mu(z)^\dagger d\tau_z \right|^2 d\tau_w \geqq 0.$$

Thus, the existence theorems for such kernels become applicable. We conclude:

a)   There exists a sequence of positive eigen values λ_ν^2 for the kernel $K(z, \zeta^\dagger) - \Gamma(z, \zeta^\dagger)$.

b)   The corresponding eigen functions $\psi_\nu(z)$ are analytic in the closed region $B + C$.

c)   We have the orthogonality relation for two eigen functions ψ_ν and ψ_μ which belong to different eigen values λ_ν^2, λ_μ^2:

$$(6.7) \qquad \iint_B \psi_\nu \psi_\mu^\dagger \, d\tau = 0 \quad \text{if} \quad \lambda_\nu^2 \neq \lambda_\mu^2.$$

d)   The eigen functions $\psi_\nu^{(\varrho)}(z)$ ($\varrho = 1, 2, \ldots m$) which belong to an eigen value λ_ν^2 of degeneracy $m - 1$ may be supposed orthonormalized, i.e.

$$(6.7a) \qquad \iint_B \psi_\nu^{(\varrho)} [\psi_\nu(\sigma)]^\dagger \, d\tau = \delta_{\varrho\sigma}.$$

However, this condition fixes the $\psi_\nu^{(\varrho)}$, only up to a unitary transformation.

To a set $\psi_\nu^{(\varrho)}(z)$ of m eigen functions belonging to the eigen value λ_ν^2 we introduce a new set of functions by the definition

$$(6.8) \qquad \Psi_\nu^{(\varrho)}(z) = \lambda_\nu \iint\limits_B \psi_\nu^{(\varrho)}(\zeta)^\dagger l(\zeta, z)\, d\tau_\zeta, \qquad \lambda_\nu > 0.$$

In view of the integral equation (6.5) satisfied by $\psi_\nu^{(\varrho)}$, we have also

$$(6.8a) \qquad \psi_\nu^{(\varrho)}(z) = \lambda_\nu \iint\limits_B \Psi_\nu^{(\varrho)}(\zeta)\, l(\zeta, z) d\tau_\zeta.$$

One sees easily from the definition (6.8) that the $\Psi_\nu^{(\varrho)}(z)$ form an orthonormalized set of eigen functions for the same eigen value λ_ν^2 with respect to the integral equation (6.5). Therefore, there exists a unitary matrix $U = (u_{\varrho\sigma})$ such that

$$(6.9) \qquad \Psi_\nu^{(\varrho)}(z) = \sum_{\sigma=1}^{m} u_{\varrho\sigma}\, \psi_\nu^{(\sigma)}(z).$$

Introducing this representation for $\Psi_\nu^{(\sigma)}(z)$ into (6.8a) we obtain

$$(6.9a) \qquad \psi_\nu^{(\varrho)}(z) = \sum_{\sigma=1}^{m} u_{\varrho\sigma}^\dagger\, \Psi_\nu^{(\sigma)}(z),$$

which gives the matrix formula

$$(6.10) \qquad U \cdot U^\dagger = E, \qquad E = \text{unit matrix},$$

for the unitary matrix U. Because of the unitary property of U this is equivalent to the symmetry of U.

Now, it is well-known that every symmetric unitary matrix U may be expressed by means of a unitary matrix V in the form

$$(6.11) \qquad U = VV', \qquad V' = \text{transposed matrix of } V,$$

where V is only determined up to a real-orthogonal matrix factor.

If we introduce another orthonormal system $\varphi_\nu^{(\varrho)}$ of eigen functions for λ_ν^2 which is obtained from the system $\psi_\nu^{(\varrho)}$ by means of a unitary matrix W, we see easily that their corresponding functions $\Phi_\nu^{(\varrho)}$, obtained by a transformation (6.8), evolve from the $\Psi_\nu^{(\varrho)}$ by means of the unitary matrix $W^\dagger$. One concludes then immediately from (6.9) that the eigen functions $\Phi_\nu^{(\varrho)}(z)$ and $\varphi_\nu^{(\varrho)}(z)$ are interrelated by a linear transformation with the unitary matrix $W^\dagger U W^{-1}$. If we now choose the arbitrary unitary matrix W by the condition

$$(6.12) \qquad W = V'$$

one sees that we have the identity

$$(6.18) \qquad \Phi_\nu^{(\varrho)}(z) = \lambda_\nu \iint\limits_B \varphi_\nu^{(\varrho)}(\zeta)^\dagger\, l(\zeta, z) d\tau_\zeta = \varphi_\nu^{(\varrho)}(z).$$

Hence we have proved that *each eigen value λ_ν^2 of the integral equation (6.5) leads to an eigen value $\lambda_\nu > 0$ of the integral equation (6.1). An appropriate complete set of orthonormalized eigen functions of (6.5) can be chosen such that it is simultaneously a complete set of orthonormal eigen functions with respect to (6.1).*

The complete equivalence of the two integral equations (6.1) and (6.5) is, therefore, proved.

The set of eigen functions $\{\varphi_\nu(z)\}$ may be a complete orthonormal set with respect to the function space Λ, i.e. every $\mathfrak{L}^2$-integrable function $f(z)$ in B may be expressed by the Fourier development

$$(6.14) \qquad f(z) = \sum_{\nu=1}^{\infty} a_\nu \varphi_\nu(z), \qquad a_\nu = \iint_B f(z) \varphi_\nu(z)^\dagger d\tau_z$$

which converges uniformly in every closed subdomain of B. In case of incompleteness there exist functions $f(z) \in \Lambda$ which do not vanish identically and are orthogonal to all eigen functions $\varphi_\nu(z)$. Hence these functions are, also orthogonal to the kernel $l(z, \zeta)$ and may be considered as eigen functions of (6.1) and (6.5) to the eigen value $\lambda = \infty$. We may then complete our system $\{\varphi_\nu(z)\}$ by addition of further eigen functions to this eigen value. We will not exclude in this paper the possibility of the eigen value $\lambda = \infty$ and may, therefore, always assume a complete orthonormal system of eigen functions $\{\varphi_\nu(z)\}$. We arrange the eigen functions in such order that the corresponding eigen values λ_ν form a non-decreasing sequence.

We now may express every function $f \in \Lambda$ in terms of these eigen functions. In view of the integral equation (6.1) it is now exceedingly simple to express the l-transform Tf of f. In fact, we have in view of definition (5.1), (6.14) and (6.1):

$$(6.15) \qquad Tf(z) = \sum_{\nu=1}^{\infty} \frac{1}{\lambda_\nu} a_\nu^\dagger \varphi(z).$$

From (5.10a) and (5.16), we have the inequality

$$(6.16) \qquad (Tf, (Tf)^\dagger) \leqq (f, f^\dagger)$$

which may be expressed in terms of the Fourier coefficients a_ν as follows:

$$(6.16a) \qquad \sum_{\nu=1}^{\infty} \frac{1}{\lambda_\nu^2} |a_\nu|^2 \leqq \sum_{\nu=1}^{\infty} |a_\nu|^2.$$

As we noticed already in the beginning of this section we know $n - 1$ linearly independent eigen functions $i w_\nu'(z)$ to the eigen value $\lambda = 1$. Let us orthonormalize these $n - 1$ functions in the way described before and we obtain the first $n - 1$ eigen functions $\varphi_\nu(z)$. For every function which is linearly independent of these initial eigen functions the inequality (6.16a) must be a proper one. Hence we conclude:

All eigen values λ_ν of $l(z, \zeta)$ are $\geqq 1$ and only the derivatives of the harmonic measures belong to the eigen value 1.

Let us now consider an eigen function $\varphi_\nu(z)$ with $\nu \geqq n$. This function is orthogonal to all functions $w_i'(z)$, $i = 1, 2, \ldots n$, i.e.

$$(6.17) \qquad \iint\limits_B \varphi_\nu(z) w_1'(z)^\dagger d\tau_z = 0.$$

Using the definition (5.23) of $w_i'(z)$ we obtain by integration by parts

$$(6.17a) \qquad \int\limits_C \varphi_\nu(z) \omega_i(x, y) dz = 0.$$

Since $\omega_i(x, y) = 0$ on C except for the boundary component C_i where $\omega_i(x, y) = 1$, we may write instead of (6.17a)

$$(6.18) \qquad \int\limits_{C_i} \varphi_\nu(z) dz = 0.$$

In other words: The eigen functions $\varphi_\nu(z)$ belonging to the eigen values $\lambda_\nu > 1$ possess single-valued integrals $\Phi_\nu(z)$ in B.

The subspace Λ_s of Λ, consisting of all functions with single-valued integrals, may also be defined as consisting of all functions $f(z) \in \Lambda$ which are orthogonal to all $w_i'(z)$; this is shown by the same reasoning which leads from (6.17) to (6.18). Hence it is evident that the functions $\varphi_\nu(z)$ which belong to the eigen values $\lambda_\nu > 1$ form a complete orthonormal system for this subspace Λ_s.

Finally we apply the orthonormal system $\{\varphi_\nu(z)\}$ in order to express our kernel functions in Fourier form. From the reproducing property (2.9) we obtain for the K-kernel the typical form, valid in every complete orthonormal system:

$$(6.19) \qquad K(z, \zeta^\dagger) = \sum_{\nu=1}^\infty \varphi_\nu(z) \varphi_\nu(\zeta)^\dagger.$$

Using the integral equation (6.1) in order to compute the Fourier coefficients of the l-kernel, we obtain the development:

$$(6.20) \qquad l(z, \zeta) = \sum_{\nu=1}^\infty \frac{1}{\lambda_\nu} \varphi_\nu(z) \varphi_\nu(\zeta).$$

From (6.20) and (4.8) we conclude next

$$(6.21) \qquad K(z, \zeta^\dagger) - \Gamma(z, \zeta^\dagger) = \sum_{\nu=1}^{\infty} \frac{1}{\lambda_\nu^2} \varphi_\nu(z) \varphi_\nu(\zeta)^\dagger$$

a result which also follows immediately from the general theory of positive definite kernels. From (6.19) and (6.21) we derive further

$$(6.22) \qquad \Gamma(z, \zeta^\dagger) = \sum_{\nu=1}^{\infty} \left(1 - \frac{1}{\lambda_\nu^2} \right) \varphi_\nu(z) \varphi_\nu(\zeta)^\dagger.$$

This shows that the eigen functions $\varphi_\nu(z)$ may also be considered as belonging to the purely geometric kernel $\Gamma(z, \zeta^\dagger)$ with the eigen value $(1 - \lambda_\nu^{-2})$.

The positive definite character of $\Gamma(z, \zeta^\dagger)$ and the development (6.22) provide a new proof for the fact that all λ_ν are greater or equal to one.

7. Discussion of the eigen functions

The significance of the eigen functions of the integral equation (6.1) and their connection with a classical problem of potential theory become clear by the following considerations. In view of (5.4) we may write the integral equation (6.1) in the form

$$(7.1) \qquad \varphi_\nu(z) = \frac{\lambda_\nu}{2\pi i} \int_C (\zeta - z)^{-1} [\varphi_\nu(\zeta) d\zeta]^\dagger.$$

From Cauchy's theorem we have, on the other hand, immediately

$$(7.2) \qquad \lambda_\nu \varphi_\nu(z) = \frac{\lambda_\nu}{2\pi i} \int_C (\zeta - z)^{-1} \varphi_\nu(\zeta) d\zeta.$$

Adding these two equations and introducing the harmonic real function $h_\nu(x, y)$ for which

$$(7.3) \qquad \varphi_\nu(z) = \frac{\partial}{\partial z} h_\nu(x, y)$$

we obtain

$$(7.4) \quad \varphi_\nu(z) = \frac{\lambda_\nu}{2\pi i (1 + \lambda_\nu)} \int_C \frac{d h_\nu(\xi, \eta)}{\zeta - z} = \frac{\lambda_\nu}{2\pi i (1 + \lambda_\nu)} \int_C \frac{h_\nu(\xi, \eta)}{(\zeta - z)^2} d\zeta.$$

Integrating this equation, we arrive at the integral equation for $h_\nu(x, y)$

$$(7.5) \quad h_\nu(x, y) = \frac{\lambda_\nu}{\pi (1 + \lambda_\nu)} \int_C h_\nu(\xi, \eta) \, \Re \left\{ \frac{\zeta'}{i(\zeta - z)} \right\} ds_\zeta + \text{const.}$$

Now it is well known that

(7.6) $$\frac{\partial}{\partial n_\zeta} \log \frac{1}{|\zeta - z|} = \Re\left\{\frac{\zeta'}{i(\zeta - z)}\right\}.$$

Thus, (7.5) obtains the form

(7.7) $$h_\nu(x, y) = \frac{\lambda_\nu}{\pi(1 + \lambda_\nu)} \int_C h_\nu(\xi, \eta) \frac{\partial}{\partial n_\zeta}\left(\log \frac{1}{|\zeta - z|}\right) ds_\zeta + \text{const.}$$

The integral equation (7.7) for $h_\nu(x, y)$ contains an arbitrary constant of integration which is evident since the definition (7.3) of h_ν determines this function only up to an additive constant.

Let us now suppose that we know a solution $h_\nu(x, y)$ of (7.7). The function $h_\nu(x, y) + a$ will also be a solution of the same integral equation because of the well-known fact that for $z \in B$ we have the identity

(7.8) $$\frac{1}{2\pi} \int_C \frac{\partial}{\partial n_\zeta}\left(\log \frac{1}{|\zeta - z|}\right) ds_\zeta = 1.$$

Thus, we conclude from (7.7):

(7.7a) $$h_\nu(x, y) + a = \frac{\lambda_\nu}{\pi(1 + \lambda_\nu)} \int_C [h_\nu(\xi, \eta) + a] \frac{\partial}{\partial n_\zeta}\left(\log \frac{1}{|\zeta - z|}\right) ds_\zeta +$$
$$+ a\frac{1 - \lambda_\nu}{1 + \lambda_\nu} + \text{const.}$$

We see that for $\lambda_\nu > 1$ we are able to introduce such a constant a into our function $h_\nu(x, y)$ that it satisfies the simpler integral equation

(7.9) $$h_\nu(x, y) = \frac{\lambda_\nu}{\pi(1 + \lambda_\nu)} \int_C h_\nu(\xi, \eta) \frac{\partial}{\partial n_\zeta}\left(\log \frac{1}{|\zeta - z|}\right) ds_\zeta, \quad z \in B.$$

We may derive from (7.9) an integral equation for the function $h_\nu(s) = h_\nu(x(s), y(s))$ considered as a function of the arc length on C. Using the discontinuity behavior of the dipole potential on the charged line C, we find

(7.10) $$h_\nu(s_z) = \frac{\lambda_\nu}{\pi} \int_C h_\nu(s_\zeta) \frac{\partial}{\partial n_\zeta}\left(\log \frac{1}{|\zeta - z|}\right) ds_\zeta.$$

This result gives a clear understanding of the significance of the eigen functions $\varphi_\nu(z)$. The inhomogeneous integral equation

(7.11) $$f(s_z) = \varphi(s_z) - \frac{\lambda}{\pi} \int_C \varphi(s_\zeta) \frac{\partial}{\partial n_\zeta}\left(\log \frac{1}{|\zeta - z|}\right) ds_\zeta,$$

plays a central role in the boundary value problem of harmonic functions if treated with the Fredholm theory. Now we see that our eigen functions $\varphi_\nu(z)$ are closely connected with the eigen functions of the corresponding homogeneous integral equation (7.10). Their importance for general theory thus becomes obvious.

To illustrate the theory we shall now determine the eigen functions $\varphi_\nu(z)$ and' their corresponding eigen values λ_ν for a few simple domains.

Let the domain B be mapped by the linear transformation

$$(7.12) \qquad\qquad w = \varphi(z) = \frac{\alpha z + \beta}{\gamma z + \delta}$$

into a new domain B_1. The function $\Phi(z, \zeta)$, defined in (3.6), is in this case

$$(7.12a) \quad \Phi(z, \zeta) = \frac{1}{\pi}\{\log(\alpha\delta - \beta\gamma) - \log(\gamma z + \delta) - \log(\gamma\zeta + \delta)\}$$

whence

$$(7.13) \qquad\qquad \frac{\partial^2 \Phi}{\partial z \partial \zeta} = 0.$$

Hence the transformation formula (3.5a) for the l-kernel now takes the simple form

$$(7.14) \qquad\qquad l_1(w, \omega)\varphi'(z)\varphi'(\zeta) = l(z, \zeta),$$

and in view of (6.20) we have the series development

$$(7.15) \qquad l_1(w, \omega) = \sum_{\nu=1}^{\infty} \frac{1}{\lambda_\nu}[\varphi_\nu(z)\varphi'(z)^{-1}] \cdot [\varphi_\nu(\zeta)\varphi'(\zeta)^{-1}].$$

Now it is easily verified that the functions

$$(7.16) \qquad\qquad \psi_\nu(w) = \varphi_\nu(z(w))\varphi'(z(w))^{-1}$$

form a complete orthonormal set of analytic functions in B_1. Hence

$$(7.15a) \qquad\qquad l_1(w, \omega) = \sum_{\nu=1}^{\infty} \frac{1}{\lambda_\nu} \psi_\nu(w)\, \psi_\nu(\omega)$$

and this clearly shows that the $\psi_\nu(w)$ are the eigen functions in B_1 with the same eigen values λ_ν. We proved, therefore:

If a domain B is mapped into a domain B_1 by a linear transformation (7.12) the eigen functions of both domains are related by (7.16) and the eigen values are the same.

For the case of the unit circle we have $l(z, \zeta) = 0$; hence all eigen values are infinite and because of the previous theorem this is true for every circle. We may now also consider domains B

which contain the point at infinity since we may always transform such domains into finite domains by linear transformation.

Consider the domain B_1 obtained by mapping the exterior of the unit circle by means of

$$(7.17) \qquad w = \varphi(z) = z + \frac{a}{z}, \qquad a < 1.$$

B_1 is the exterior of an ellipse with the principal axes $(1 + a)$ and $(1 - a)$. Using (3.8), we have immediately a formula for the l-kernel of this simply-connected domain:

$$(7.18) \quad l_1(w, \omega) = (\varphi'(z)\varphi'(\zeta))^{-1} \cdot \frac{-a}{\pi(z\zeta - a)^2} = \sum_{\nu=1}^{\infty} a^{\nu} \psi_{\nu}(w)\psi_{\nu}(\omega)$$

with

$$(7.18a) \qquad \psi_{\nu}(w) = i\sqrt{\frac{\nu}{\pi}} \cdot z^{-(\nu+1)}\varphi'(z)^{-1}, \qquad w = \varphi(z)$$

Since the functions $i\sqrt{\dfrac{\nu}{\pi}}z^{-(\nu+1)}$ $(\nu = 1, 2, \ldots)$ form a complete orthonormal system in $|z| > 1$ the $\psi_{\nu}(w)$ do the same in B_1. The representation of the l-kernel shows that the $\psi_{\nu}(w)$ are the eigen functions of the exterior of the ellipse and we have in this case:

$$(7.19) \qquad \lambda_{\nu} = a^{-\nu}.$$

The domain B_1 has an interesting extremum property with respect to the eigen values λ_{ν}. Consider an arbitrary simply-connected domain B which is bounded by a closed analytic curve C and contains the point at infinity. One shows by a linear transformation that even for such a domain B the development (6.22) for $\Gamma(z, \zeta^{\dagger})$ is valid; it is also obvious that $\lambda_1 > 1$ because of the simple connectivity of B. Hence we derive from (6.22) the inequality

$$(7.20) \quad \Gamma(z, z^{\dagger}) \geq \left(1 - \frac{1}{\lambda_1^2}\right)\sum_{\nu=1}^{\infty}|\varphi_{\nu}(z)|^2 = \left(1 - \frac{1}{\lambda_1^2}\right)K(z, z^{\dagger})$$

This inequality assumes a simple meaning if we let $z \to \infty$. Let

$$(7.21) \qquad z = d(\zeta + c_0 + c_1\zeta^{-1} + \ldots)$$

be the function which maps the domain $|\zeta| > 1$ upon B. The constant d is called the mapping radius of B and plays a considerable role in the conformal geometry of B. One easily verifies the limit relations, which follow from (3.8), (7.21) and (4.9):

$$(7.22) \qquad \lim_{z \to \infty}|z|^4 K(z, z^{\dagger}) = \frac{1}{\pi}d^2$$

and

(7.23) $$\lim_{z \to \infty} |z|^4 \, \Gamma(z, z^\dagger) = \frac{1}{\pi^2} A$$

where A is the area of the finite complement $\overleftarrow{B}$ of B.

From (7.20), (7.22) and (7.23) we obtain the inequality

(7.24) $$A \geqq \left(1 - \frac{1}{\lambda_1^2}\right) \pi d^2.$$

It is well known that between the mapping radius d and the area A of $\overleftarrow{B}$ the following inequality holds:

(7.25) $$\pi d^2 \geqq A.$$

Hence we can transform (7.24) into

(7.26) $$\lambda_1^2 \leqq \frac{\pi d^2}{\pi d^2 - A}.$$

We see that the lowest eigen value λ_1 provides an upper bound for the excess of πd^2 over A. In the case of a circle we have $\lambda_1 = \infty$ and $\pi d^2 = A$.

Now it is interesting that the inequality (7.26) is an equality for all ellipses; these may therefore be considered as the extremum domains with respect to (7.26). In fact we have in the case of the ellipse $B_1 : d = 1$, $A = \pi(1 - a^2)$ and $\lambda_1^2 = a^{-2}$ which shows that equality holds in (7.26).

Another interesting result may be obtained for the eigen functions $\varphi_\nu(z)$ of a simply-connected domain B. In this case the whole plane is divided into two complementary domains B and $\overleftarrow{B}$ and let $\varphi_\nu(z)$, λ_ν and $\overleftarrow{\varphi}_\nu(z)$, $\overleftarrow{\lambda}_\nu$ denote the corresponding eigen functions and eigen values. The eigen functions $\varphi_\nu(z)$ of B have an l-transform $T\varphi_\nu$ which is defined in B and in $\overleftarrow{B}$. In view of the integral equation we have in B

(7.27) $$T\varphi_\nu(z) = \frac{1}{\lambda_\nu} \, \varphi_\nu(z) \qquad \text{for } z \, \epsilon \, B.$$

Hence we may write the identity (5.10) in the form

(7.28) $$\iint_{\overleftarrow{B}} T\varphi_\nu \cdot (T\varphi_\mu)^\dagger d\tau + \frac{1}{\lambda_\nu \lambda_\mu} \iint_B \varphi_\nu \, \varphi_\mu^\dagger \, d\tau = \iint_B \varphi_\nu^\dagger \varphi_\mu \, d\tau.$$

Because of the orthogonality relations between the eigen functions we obtain

(7.29) $$\iint_{\overleftarrow{B}} T\varphi_\nu (T\varphi_\mu)^\dagger \, d\tau = \left(1 - \frac{1}{\lambda_\nu^2}\right) \delta_{\nu\mu}.$$

Hence the transforms of the eigen functions $\varphi_\nu(z)$ of B create an orthonormal system of analytic functions in $\widetilde{B}$:

$$(7.30) \qquad \psi_\nu(z) = i\left(1 - \frac{1}{\lambda_\nu^2}\right)^{-\frac{1}{2}} T\varphi_\nu(z) \qquad \text{for } z \,\epsilon\, \widetilde{B}.$$

Because of (7.27) and the saltus condition (5.5) for $T\varphi_\nu$, we find for the boundary value of $\psi_\nu(z)$ at a point $z \,\epsilon\, C$:

$$(7.31) \qquad \psi_\nu(z) = i\left(1 - \frac{1}{\lambda_\nu^2}\right)^{-\frac{1}{2}}\left[\frac{1}{\lambda_\nu}\varphi_\nu(z) - (\varphi_\nu(z)z'^2)^\dagger\right], \qquad z \,\epsilon\, C.$$

We have, therefore, integrating along C in the positive sense with respect to $\widetilde{B}$:

$$(7.32) \qquad \frac{1}{2\pi i}\int_C \frac{[\psi_\nu(\zeta)\,d\zeta]^\dagger}{\zeta - z} = -i\left(1 - \frac{1}{\lambda_\nu^2}\right)^{-\frac{1}{2}}\left\{\frac{1}{\lambda_\nu}\cdot\frac{1}{2\pi i}\int_C \frac{[\varphi_\nu(\zeta)d\zeta]^\dagger}{\zeta - z} - \right.$$
$$\left. - \frac{1}{2\pi i}\int_C \frac{\varphi_\nu(\zeta)\,d\zeta}{\zeta - z}\right\}, \qquad z \,\epsilon\, \widetilde{B}.$$

The last right-hand integral vanishes because of Cauchy's theorem, and using the definition (7.30) of $\psi_\nu(z)$ we find by means of (5.4):

$$(7.33) \qquad \psi_\nu(z) = \frac{\lambda_\nu}{2\pi i}\int_C \frac{[\psi_\nu(\zeta)\,d\zeta]^\dagger}{\zeta - z} = \lambda_\nu\, T\psi_\nu(z).$$

This proves that the functions $\psi_\nu(z)$ are the eigen functions $\widetilde{\varphi}_\nu$ of $\widetilde{B}$ and that the sequences λ_ν and $\widetilde{\lambda}_\nu$ are identical. Hence

Two complementary simply-connected domains have the same set of eigen values. This is a very useful result since the determination of the eigen values of one domain may be much easier than those of the other. We see for example that the interior of an ellipse with principal axes $(1 + a)$ and $(1 - a)$ has the eigen values $a^{-\nu}$; while the mapping function of a circle on the exterior of this ellipse is an elementary function, the map of the circle upon the interior of the ellipse is given by quite complicated elliptic functions. The importance of our result for the conformal geometry of domains becomes quite obvious from this example.

We understand the last result better if we notice that the λ_ν are the eigen values of the integral equation (7.10) which is defined on C alone and does not indicate which adjacent domain of C is to be considered. For the same reason the treatment by integral equations of the boundary value problem for harmonic functions leads to a simultaneous solution for the so-called interior and exterior problems.

If the eigen functions $\varphi_\nu(z)$ of a simply-connected domain B are known, we may easily construct the kernel functions and by their aid map B upon the unit-circle. At the same time, we can compute by elementary integrations the functions $\overline{\varphi}_\nu(z)$ of the complementary domain $\widetilde{B}$ and determine its kernel functions, too. Hence knowledge of the $\varphi_\nu(z)$ permits at the same time the conformal mapping of the two complementary domains upon the unit circle. We have the formulas for the kernel functions $\widetilde{K}(z, \zeta^\dagger)$ and $\overline{l}(z, \zeta)$:

$$(7.34) \qquad \widetilde{K}(z, \zeta^\dagger) = \sum_{\nu=1}^{\infty} (1 - \lambda_\nu^{-2})^{-1} \cdot T\varphi_\nu(z)[T\varphi_\nu(\zeta)]^\dagger,$$

$$(7.35) \qquad \overline{l}(z, \zeta) = -\sum_{\nu=1}^{\infty} (\lambda_\nu - \lambda_\nu^{-1})^{-1} T\varphi_\nu(z) \cdot T\varphi_\nu(\zeta).$$

We notice also the elegant formula:

$$(7.36) \qquad \widetilde{\Gamma}(z, \zeta^\dagger) = \sum_{\nu=1}^{\infty} T\varphi_\nu(z)[T\varphi_\nu(\zeta)]^\dagger$$

which follows easily from (6.22) and (7.30).

8. The space Λ_s and its kernel functions.

Let

$$(8.1) \qquad F(z; u, v) = \log \frac{z - u}{z - v} + \text{regular terms}$$

and

$$(8.2) \qquad G(z; u, v) = \log \frac{z - u}{z - v} + \text{regular terms}$$

be the logarithms of two univalent functions in B which map this domain on the whole complex plane slit along concentric circular arcs around the origin or along rectilinear slits directed towards the origin, respectively. The points $u, v \in B$ shall correspond to the origin and the point at infinity after the mapping. The functions F and G are determined by this description up to an additive constant. On each contour C_ν of B, we have

$$(8.3) \qquad F(z; u, v) = a_\nu + iK_\nu(s), \quad z \in C_\nu,$$

$$(8.4) \qquad G(z; u, v) = l_\nu(s) + ib_\nu, \quad z \in C_\nu$$

where a_ν and b_ν are constants and $k_\nu(s)$, $l_\nu(s)$ real-valued functions of the arc length s.

Let us define two functions

$$(8.5) \qquad P(z; u, v) = \frac{1}{2\pi}\{F(z; u, v) - G(z; u, v)\}$$

and

$$(8.6) \quad Q(z; u, v) = \frac{1}{2\pi}\{F(z; u, v) + G(z; u, v)\} = \frac{1}{\pi}\log\frac{z-u}{z-v} - q(z; u, v).$$

They are both analytic in B, except for two simple logarithmic poles of Q at u and v. On each C_ν one has because of (8.3) and (8.4)

$$(8.7) \qquad P(z; u, v) = - Q(z; u, v)^\dagger + c_\nu, \qquad z \in C_\nu.$$

The functions $P'(z; u, v)$ and $q'(z; u, v)$ (where the dash denotes the differentiation with respect to the first argument) are both of the class Λ_s and we want to develop them in Fourier series with respect to the complete orthonormal system $\varphi_\nu(z)$ with $\lambda_\nu > 1$, i.e. $\nu \geq n$. We have by virtue of (1.4)

$$(8.8) \qquad \iint\limits_B P'(z; u, v)\varphi_\nu(z)^\dagger d\tau_z = -\frac{1}{2i}\int\limits_C P(z; u, v)[\varphi_\nu(z)dz]^\dagger.$$

Using (6.18) and (8.7) this may also be written in the form

$$(8.9) \qquad \frac{1}{2i}\int\limits_C [Q(z; u, v)\varphi_\nu(z)dz]^\dagger = \left[\frac{1}{2i}\int\limits_C Q'(z; u, v)\Phi_\nu(z)dz\right]^\dagger$$

where $\Phi_\nu(z)$ denotes again the single-valued integral of $\varphi_\nu(z)$. From (8.6) and the residue theorem we finally find

$$(8.10) \qquad \iint\limits_B P'(z; u, v)\varphi_\nu(z)^\dagger d\tau_z = [\Phi_\nu(u) - \Phi_\nu(v)]^\dagger$$

whence the Fourier series

$$(8.11) \qquad P'(z; u, v) = \sum_{\nu=n}^{\infty} \varphi_\nu(z)[\Phi_\nu(u) - \Phi_\nu(v)]^\dagger$$

and integrating this identity between z and ζ, we finally obtain:

$$(8.12) \quad P(z; u, v) - P(\zeta; u, v) = \sum_{\nu=n}^{\infty} [\Phi_\nu(z) - \Phi(\zeta)] \cdot [\Phi_\nu(u) - \Phi_\nu(v)]^\dagger.$$

It should be noticed that in this derivation no use was made of the integral equation satisfied by the $\varphi_\nu(z)$ so that the representation (8.12) will hold for each complete orthonormal system $\varphi_\nu(z)$ with single-valued integrals $\Phi_\nu(z)$.

Next, we compute

$$(8.13) \qquad \iint\limits_B q'(z; u, v)\varphi_\nu(z)^\dagger d\tau_z = -\frac{1}{2i}\int\limits_C q(z; u, v)[\varphi_\nu(z)dz]^\dagger.$$

Using the definition (8.6) of q and the boundary relation (8.7), we obtain

$$(8.14) \quad \iint_B q'(z; u, v)\varphi_\nu(z)^\dagger d\tau_z =$$
$$-\frac{1}{2\pi i} \int_C \log \frac{z-u}{z-v} [\varphi_\nu(z)dz_\nu]^\dagger + \left[\frac{1}{2i} \int_C P(z; u, v)\varphi_\nu(z)dz\right]^\dagger.$$

The last integral vanishes by Cauchy's theorem, while the integral equation (7.1) yields by integration between u and v:

$$(8.15) \qquad \Phi_\nu(u) - \Phi_\nu(v) = \frac{-\lambda_\nu}{2\pi i} \int_C \log \frac{z-u}{z-v} [\varphi_\nu(z)dz]^\dagger.$$

Thus, we finally arrive at the Fourier series

$$(8.16) \qquad q'(z; u, v) = \sum_{\nu=n}^\infty \frac{1}{\lambda_\nu} \varphi_\nu(z)(\Phi_\nu(u) - \Phi_\nu(v)).$$

Integrating again between z and ζ, we obtain at last

$$(8.17) \quad q(z; u, v) - q(\zeta; u, v) = \sum_{\nu=n}^\infty \frac{1}{\lambda_\nu}(\Phi_\nu(z) - \Phi_\nu(\zeta))(\Phi_\nu(u) - \Phi_\nu(v)).$$

Most of the important domain functions, as for example Green's and Neumann's functions of B and many others may easily be expressed in terms of P and Q (Garabedian—Schiffer [1]). The formulas (8.12) and (8.17) show the simple construction of these functions in terms of the $\varphi_\nu(z)$.

Let further

$$(8.18) \qquad f_0(z, u) = \frac{1}{z-u} + \text{regular terms},$$

$$g_0(z, u) = \frac{1}{z-u} + \text{regular terms}$$

be univalent in B, mapping the domain upon the whole plane slit along straight segments parallel to the real and the imaginary axis, respectively. The point $z = u$ obviously corresponds to infinity.

Using the well-known relations between f_0, g_0 on the one hand and P, Q on the other, it is possible to show that

$$(8.19) \quad \frac{1}{2\pi}[f_0'(z, u) - g_0'(z, u)] = \sum_{\nu=n}^\infty \varphi_\nu(z)\varphi_\nu(u)^\dagger, \qquad f_0'(z, u) = \frac{df_0(z, u)}{dz},$$

and

$$(8.20) \quad \frac{1}{2\pi}[f_0'(z, u) + g_0'(z, u)] + \frac{1}{\pi(z-u)^2} = \sum_{\nu=n}^\infty \frac{1}{\lambda_\nu} \varphi_\nu(z)\varphi_\nu(u).$$

These results could also be obtained directly by applying to f_0 and g_0 the same reasoning as we did before to F and G.

We see that the kernel functions of the Λ_s-space

$$(8.21) \qquad K_s(z, \zeta^\dagger) = \sum_{\nu=n}^\infty \varphi_\nu(z)\varphi_\nu(\zeta)^\dagger$$

and

$$(8.22) \qquad l_s(z, \zeta) = \sum_{\nu=n}^\infty \frac{1}{\lambda_\nu} \varphi_\nu(z) \varphi_\nu(\zeta)$$

have an important geometric significance.

It is of interest to notice that K_s and l_s lead to the same algebra under scalar multiplication as did K and l. In fact, we clearly have:

$$(8.23) \qquad \iint_B K_s(z, \zeta^\dagger)K_s(\zeta, w^\dagger)d\tau_\zeta = K_s(z, w^\dagger),$$

$$(8.24) \qquad \iint_B K_s(z, \zeta^\dagger)\, l_s(\zeta, w)d\tau_\zeta = l_s(z, w),$$

and in view of (8.22) and (6.22):

$$(8.25) \qquad \iint_B l_s(z, \zeta)\, l_s(\zeta, w)^\dagger d\tau_\zeta = K_s(z, w^\dagger) - \Gamma(z, w^\dagger).$$

In fact, in the series development (6.22) for $\Gamma(z, w^\dagger)$ the first $n-1$ eigen functions do not appear. Thus, all inequalities which we deduced in section 4 for the kernel functions K and l of the space Λ remain valid if we replace those kernels by K_s and l_s. It is also easily verified that K_s and l_s behave under conformal transformations just as K and l and that analogous formulas to (3.3) and (3.5) hold for them.

Since the developments (6.19) and (6.20) may be expressed in the form

$$(8.26) \quad K(z, \zeta^\dagger) = -\frac{2}{\pi} \frac{\partial^2 g(z, \zeta)}{\partial z\, \partial \zeta^\dagger} = \sum_{i,k=1}^{n-1} p_{ik} w_i'(z) w_k'(\zeta)^\dagger + K_s(z, \zeta^\dagger)$$

and

$$(8.27) \quad l(z, \zeta) = \frac{1}{\pi(z-\zeta)^2} + \frac{2}{\pi} \frac{\partial^2 g(z, \zeta)}{\partial z\, \partial \zeta} = -\sum_{i,k=1}^{n-1} p_{ik} w_i'(z) w_k'(\zeta) + l_s(z, \zeta)$$

with real coefficients p_{ik}, we have

$$(8.28) \quad K_s(z, \zeta^\dagger) = -\frac{2}{\pi} \frac{\partial^2 G(z, \zeta)}{\partial z\, \partial \zeta^\dagger}, \quad l_s(z, \zeta) = \frac{1}{\pi(z, \zeta)^2} + \frac{2}{\pi} \frac{\partial^2 G(z, \zeta)}{\partial z\, \partial \zeta}$$

with

$$(8.29) \qquad G(z, \zeta) = g(z, \zeta) + 2\pi \sum_{i, k = 1}^{n-1} p_{ik} \omega_i(z) \omega_k(\zeta).$$

It is easily seen that $G(z, \zeta)$ is the real part of the logarithm of a univalent function which maps the domain B upon the exterior of a circle which is slit along concentric circular arcs. The point ζ corresponds in this map to infinity.

The fact that in the series development for $K_s(z, \zeta^\dagger)$ only eigen functions occur with $\lambda_\nu > 1$ leads to the following important application. We define the ϱ-th iterated Γ-kernel by the formula

$$(8.30) \qquad \Gamma^{(\varrho)}(z, \zeta^\dagger) = \sum_{\nu = n}^{\infty} \left(1 - \frac{1}{\lambda_\nu^2} \right)^{\varrho} \varphi_\nu(z) \varphi_\nu(\zeta)^\dagger.$$

Obviously, we have

$$(8.31) \quad \Gamma^{(\varrho+1)}(z, \zeta^\dagger) = \iint_B \Gamma^{(\varrho)}(z, w^\dagger) \, \Gamma(w, \zeta^\dagger) d\tau_w, \ \Gamma^{(1)}(z, \zeta^\dagger) = \Gamma(z, \zeta^\dagger).$$

Hence all kernels $\Gamma^{(\varrho)}(z, \zeta^\dagger)$ are geometric integrals and may be computed elementarily.

Next we consider the kernels

$$(8.32) \quad \Delta_\mu(z, \zeta^\dagger) = \sum_{\varrho = 0}^{\mu} (-1)^\varrho \binom{\mu}{\varrho} \Gamma^{(\varrho+1)}(z, \zeta^\dagger) =$$

$$= \sum_{\nu = n}^{\infty} \left(1 - \frac{1}{\lambda_\nu^2} \right) \frac{1}{\lambda_\nu^{2\mu}} \varphi_\nu(z) \varphi_\nu(\zeta)^\dagger.$$

The $\Delta_\mu(z, \zeta^\dagger)$ are also elementary expressions being linear combinations of the $\Gamma^{(\varrho)}$-kernels. It is obvious that they are positive-definite kernels.

Finally, we construct the sum

$$(8.33) \qquad \sum_{\mu = 0}^{\infty} \Delta_\mu(z, \zeta^\dagger) = \sum_{\nu = n}^{\infty} \varphi_\nu(z) \varphi_\nu(\zeta)^\dagger = K_s(z, \zeta^\dagger).$$

We see that we can express the kernel $K_s(z, \zeta^\dagger)$ as an infinite sum of elementary integrals. It is of particular interest that

$$(8.34) \qquad K_s(z, z^\dagger) = \sum_{\mu = 0}^{\infty} \Delta_\mu(z, z^\dagger)$$

appears as a sum of positive terms. This leads to an infinity of inequalities for $K_s(z, z^\dagger)$. Since $\Delta_0(z, z^\dagger) = \Gamma(z, z^\dagger)$, we see that the inequalities

$$(8.35) \qquad K_s(z, z^\dagger) \geqq \Gamma(z, z^\dagger)$$

is only the first in a series of improving inequalities for the kernel function.

From (8.32) and (6.22) we obtain the inequality

$$(8.36) \qquad \Delta_\mu(z, z^\dagger) \leqq \frac{1}{\lambda_n^{2\mu}} \Gamma(z, z^\dagger);$$

by Schwarz' inequality we have on the other hand:

$$(8.37) \quad |\Delta_\mu(z, \zeta^\dagger)|^2 \leqq \dot\Delta_\mu(z, z^\dagger) \Delta_\mu(\zeta, \zeta^\dagger) \leqq \frac{1}{\lambda_n^{4\mu}} \Gamma(z, z^\dagger) \Gamma(\zeta, \zeta^\dagger).$$

Hence we see that the series development (8.33) for the K_s-kernel converges geometrically. It seems that it leads to a very useful numerical method in conformal mapping.

Each domain B can be mapped into a canonical domain which plays a distinguished role with respect to the kernel functions K_s and l_s. It is well known that the function

$$(8.38) \quad h(z, u) = \tfrac{1}{2}[f_0(z, u) + g_0(z, u)] = \frac{1}{z - u} + \text{regular terms}$$

maps the domain B univalently upon a domain B_1 which solves the following extremum problem: Among all domains which are obtained from B by a conformal map with a pole of residue 1 at $z = u$, B_1 possesses a complement $\tilde{B}_1$ with maximal area A_μ (Schiffer [1]). B_1 is a canonical domain for all domains B which can be mapped into each other by means of a univalent function $\varphi(z)$ with the normalization

$$(8.39) \qquad \qquad \varphi(u) = u, \qquad \varphi'(u) = 1.$$

In fact, let $w = \varphi(z)$ map our original domain B into a new domain B_2. Let $f_{02}(w, u)$, $g_{02}(w, u)$ and $h_2(w, u)$ denote analogous univalent functions with respect to B_2 as were $f_0(z, u)$, $g_0(z, u)$ and $h(z, u)$ with respect to B. We clearly may put

$$(8.40) \qquad f_{02}(\varphi(z), u) = f_0(z, u), \quad g_{02}(\varphi(z), u) = g_0(z, u)$$

and hence

$$(8.41) \qquad \qquad h_2(\varphi(z), u) = h(z, u).$$

Hence the same procedure (8.38) leads to the same canonical domain B_1 for all domains B_2 which are equivalent to B by means of a function (8.39).

The function

$$(8.42) \qquad \qquad w = H(z, u) = u + h(z, u)^{-1}$$

has clearly the normalization (8.39) and maps B upon a canonical domain D which is more suitable for our purposes. The transition from D to B_1 is given by the linear transformation

$$(8.43) \qquad h_D(w, u) = (w - u)^{-1}.$$

On the other hand, we have because of (8.38), (8.20) and (8.22) for each domain B the identity

$$(8.44) \qquad \overset{.}{h}'(z, u) = \pi l_s(z, u) - \frac{1}{(z - u)^2}.$$

Thus, in the particular case of the domain D we conclude from (8.43) and (8.44)

$$(8.45) \qquad l_s(w, u) \equiv 0, \qquad w \in D.$$

From (8.25) we easily obtain

$$(8.46) \qquad K_s(w, u^\dagger) = \Gamma(w, u^\dagger).$$

This shows that in the case of the canonical domain D the series development (8.38) for $K_s(w, u^\dagger)$ may be stopped after the first term.

From the series development (8.22) and (8.45) we conclude that in the case of the domain D all eigen function $\varphi_\nu(w)$ which do not belong to the eigen value ∞ vanish at the distinguished point u. Since $\Gamma(u, u^\dagger) \neq 0$ we conclude also that in the case of the canonical domain D there exists at least one eigen value ∞.

We now define the following concept: Let R be a domain in the z-plane which does not contain the point u. We call the expression

$$(8.47) \qquad A_u(R) = \iint_R \frac{d\tau_t}{|t - u|^4}$$

the area of R with respect to u. Clearly, $A_u(R)$ represents the area of the image of R under the linear transformation $\dfrac{1}{z - u}$.

From (8.46) we conclude

$$(8.48) \qquad K_s(u, u^\dagger) = \Gamma(u, u^\dagger) = \frac{1}{\pi^2} A_u(\overset{\frown}{D}).$$

By definition of D, we clearly have $A_u(\overset{\frown}{D}) = A_u$. Because of the behavior of $K_s(u, u^\dagger)$ under conformal transformation, its value is the same for all equivalent domains obtained from each other by means of a function (8.39). From (6.22), we obtain on the other hand the inequality

$$(8.49) \qquad \Gamma(u, u^\dagger) \geqq \left(1 - \frac{1}{\lambda_n^2}\right) K(u, u^\dagger) = \frac{1}{\pi^2}\left(1 - \frac{1}{\lambda_n^2}\right) A_u.$$

Clearly, $\Gamma(u, u^\dagger) = \dfrac{1}{\pi^2} A_u(\bar{B})$; hence we proved the inequality

(8.50) $$A_u(\bar{B}) \geqq \left(1 - \frac{1}{\lambda_u^2}\right) A_u.$$

Because of the extremum property of B_1, mentioned above, we always have $A_\mu(\bar{B}) \leqq A_\mu$ and hence (8.50) leads to

(8.51) $$\lambda_n^2 \leqq \frac{A_u}{A_u - A_u(\bar{B})}.$$

A somewhat different approach is necessary if the point $u = \infty$ lies in B. In this case the class of univalent functions with the normalization

(8.39a) $$\varphi(\infty) = \infty, \qquad \varphi'(\infty) = 1$$

must be considered. Let B_1 be that domain obtained from B by means of a function (8.39a) which has a complement B_1 with the largest possible area A. This maximum area is related to the

span S of B by means of the identity $A = \dfrac{\pi}{2} S$. The span plays

a role in various problems of conformal mapping of multiply-connected domains (Schiffer [1]). One shows easily that (8.51) tends for $u \to \infty$ to the inequality

(8.52) $$\lambda_n^2 \leqq \frac{A}{A - A(\bar{B})} = \frac{\pi s}{\pi s - 2A(\bar{B})}$$

where $A(\bar{B})$ is the area of the complement $\bar{B}$ of B. This is the generalization of (7.26) to the case of multiply-connected domains.

9. Applications to the theory of univalent functions:

We proved in section 4 inequalities of the type (4.14a) between the kernel functions K and l; exactly the same inequalities can be derived from (8.23—8.25) for the kernel functions K_s and l_s of the class Λ_s. Since the functions K and l show a very different behavior under conformal transformation these inequalities represent also important inequalities for the univalent mapping functions in the domain.

Let $w = \varphi(z)$ be univalent and analytic in the closed region $B + C$. It maps B upon a domain B_1 of the same type as B and we have, therefore, between the kernels K_1 and l_1 the following inequalities in analogy to (4.14):

(9.1) $$\left| \sum_{\nu, \mu=1}^{r} \alpha_\nu \alpha_\mu l_1(\omega_\nu, \omega_\mu) \right| \leqq \sum_{\nu, \mu=1}^{r} \alpha_\nu \alpha_\mu^\dagger K_1(\omega_\nu, \omega_\mu^\dagger)$$

for every choice of the complex numbers α_ν and points $\omega_\nu \in B_1$. Now let ζ_ν be the point in B corresponding to ω_ν, i.e. $\omega_\nu = \varphi(\zeta_\nu)$. Using the transformation formulas (8.3) and (8.5a) for the kernel functions, we obtain

$$(9.2) \quad \left| \sum_{\nu,\,\mu=1}^{r} \alpha_\nu \alpha_\mu [l(\zeta_\nu, \zeta_\mu) + U(\zeta_\nu, \zeta_\mu)] \right| \leqq \sum_{\nu,\,\mu=1}^{r} \alpha_\nu \alpha_\mu^\dagger K(\zeta_\nu, \zeta_\mu^\dagger)$$

with

$$(9.3) \qquad\qquad U(z, \zeta) = \frac{\partial^2 \Phi(z, \zeta)}{\partial z\, \partial \zeta}.$$

If we assume our domain B and its kernel functions well-known and fixed, we have in (9.2) an important condition on all univalent functions in B.

We illustrate our result by considering special cases of (9.2). Let B be the unit circle $|z| < 1$. In this case K and l are given by (2.7). Hence we have the inequality

$$(9.2a) \quad \left| \sum_{\nu,\,\mu=1}^{r} \alpha_\nu \alpha_\mu U(\zeta_\nu, \zeta_\mu) \right| \leqq \frac{1}{\pi} \sum_{\nu,\,\mu=1}^{r} \alpha_\nu \alpha_\mu^\dagger \frac{1}{(1 - \zeta_\mu^\dagger \zeta_\nu)^2}.$$

Specializing further to $r = 1$, we obtain the interesting necessary condition for univalence in the unit circle, expressed in terms of Schwarz' differential parameter (8.7) For a similar sufficient condition see (Nehari [1]):

$$(9.2b) \qquad\qquad \{\varphi, z\} \leqq \frac{6}{(1 - |z|^2)^2}.$$

We may generalize this result to the case of multiple connectivity by use of (9.2) for $r = 1$:

$$(9.2c) \qquad\qquad \left| l(z, z) - \frac{1}{6\pi} \{\varphi, z\} \right| \leqq K(z, z^\dagger).$$

Let Σ be a closed rectifiable curve in B and $p(s)$ a complex-valued function of the length parameter on Σ; we obtain from (9.2) by a limit process

$$(9.4) \quad \left| \int_{z \in \Sigma} \int_{\zeta \in \Sigma} [l(z, \zeta) + U(z, \zeta)] p(s_z) p(s_\zeta) ds_z\, ds_\zeta \right| \leqq \int_{z \in \Sigma} \int_{\zeta \in \Sigma} K(z, \zeta^\dagger) p(s_z) p(s_\zeta)^\dagger ds_z\, ds_\zeta$$

Let us now assume that the analytic functions l, U, K are developed into power series of their variables around the origin 0 which we suppose in B:

$$(9.5) \qquad U(z, \zeta) = \sum_{\mu,\,\nu=0}^{\infty} c_{\mu\nu} z^\mu \zeta^\nu, \qquad l(z, \zeta) = \sum_{\mu,\,\nu=0}^{\infty} l_{\mu\nu} z^\mu \zeta^\nu,$$

$$K(z, \zeta^\dagger) = \sum_{\mu,\,\nu=0}^{\infty} k_{\mu\nu} z^\mu (\zeta^\nu)^\dagger.$$

Because of the symmetries of these three functions, we find for their coefficients

(9.5a) $c_{\mu\nu} = c_{\nu\mu}$, $l_{\mu\nu} = l_{\nu\mu}$, $k_{\mu\nu} = k_{\nu\mu}^{\dagger}$.

We now choose for Σ a curve in the common domain of convergence of all three developments (9.5) which surrounds the origin. Let

(9.6) $q(z) = \dfrac{1}{2\pi i} \sum\limits_{\nu=0}^{N} \alpha_{\nu} z^{-(\nu+1)}$, $p(s) = q(z(s))\, z'(s)$.

We clearly have

(9.7) $\displaystyle\iint\limits_{z,\,\zeta \in \Sigma} [l(z,\zeta) + U(z,\zeta)]p(s_z)p(s_\zeta)ds_z\, ds_\zeta = \sum\limits_{\mu,\,\nu=0}^{N}(c_{\mu\nu} + l_{\mu\nu})\alpha_\mu \alpha_\nu$

and

(9.8) $\displaystyle\iint\limits_{z,\,\zeta \in \Sigma} K(z,\zeta^{\dagger})p(s_z)p(s_\zeta)^{\dagger}ds_z\, ds_\zeta = \sum\limits_{\mu,\,\nu=0}^{N} k_{\mu\nu}\, \alpha_\mu \alpha_\nu^{\dagger}$.

Hence the inequality (9.4) leads to relations between certain quadratic and hermitian forms which are connected with the coefficient matrices of the kernel functions:

(9.9) $\left| \sum\limits_{\mu,\,\nu=0}^{N}(c_{\overline{\mu\nu}} + l_{\mu\nu})\alpha_\mu \alpha_\nu \right| \leqq \sum\limits_{\mu,\,\nu=0}^{N} k_{\mu\nu}\, \alpha_\mu \alpha_\nu^{\dagger}$.

Similar inequalities are obtained for the coefficient matrices of the kernels in the Λ_e-space. These inequalities were first discovered by Grunsky (Grunsky [1]) who gave them a somewhat different formulation. Compare also (Schiffer [4]).

Grunsky showed also that when the necessary conditions (9.9) are satisfied for every N and every choice of the α_ν, the univalence of the function $\varphi(z)$ considered is ensured; i.e. all conditions (9.9) are a sufficient condition for univalence. We want to give here a new and shorter proof for this fact which is based on an important result concerning the kernel function. We first announce the following theorem:

Let $V(z,\zeta)$ be symmetric and analytic in both arguments in a neighbourhood of the origin; let

(9.10) $V(z,\zeta) = \sum\limits_{m,\,n=0}^{\infty} d_{mn} z^m \zeta^n$, $K(z,\zeta^{\dagger}) = \sum\limits_{m,\,n=0}^{\infty} k_{mn} z^m (\zeta^n)^{\dagger}$

be the series for V and the K-kernel around the origin. If for every complex vector $\alpha_0, \alpha_1, \ldots \alpha_N$

(9.11) $\left| \sum\limits_{m,\,n=0}^{N} d_{mn} \alpha_m \alpha_n \right| \leqq \sum\limits_{m,\,n=0}^{N} k_{mn} \alpha_m \alpha_n^{\dagger}$

then $V(z,\zeta)$ is analytic in the whole domain B.

In order to prove this theorem we introduce a complete set of orthonormal functions $\chi_\nu(z)$ which is very useful in dealing with power series developments. Each $\chi_\nu(z)$ has around the origin the series development

$$(9.12) \qquad \chi_\nu(z) = \sum_{\mu=\nu}^{\infty} \beta_{\nu\mu} z^\mu, \qquad \nu = 0, 1, 2, \ldots$$

The condition that the matrix $(\beta_{\nu\mu})$ be triangular determines the set $\chi_\nu(z)$ in a unique way (Bergman [2]) Since the K-kernel can be expressed in the forms

$$(9.13) \qquad K(z, \zeta^\dagger) = \sum_{\nu=0}^{\infty} \chi_\nu(z)\chi_\nu(\zeta)^\dagger = \sum_{\mu,\nu=0}^{\infty} k_{\mu\nu} z^\mu (\zeta^\nu)^\dagger$$

we conclude from (9.12) the identities

$$(9.14) \qquad k_{\mu\nu} = \sum_{\varrho=0}^{\infty} \beta_{\varrho\mu} \beta_{\varrho\nu}^\dagger.$$

Since $\beta_{\varrho\mu} = 0$ for $\varrho > \mu$, the matrix $(k_{\mu\nu})$ consists of finite combinations of β-terms.

The relation (9.12) between the z^μ and $\chi_\nu(z)$ can easily be inverted:

$$(9.15) \qquad z^\nu = \sum_{\mu=\nu}^{\infty} b_{\nu\mu} \chi_\mu(z), \qquad \nu = 0, 1, \ldots$$

The matrix $(b_{\nu\mu})$ is of the same triangular form as its inverse $(\beta_{\nu\mu})$. Introducing (9.15) into (9.13) and comparing the coefficients of $\chi_\mu(z)\chi_\nu(\zeta)^\dagger$, we find

$$(9.16) \qquad \delta_{\mu\nu} = \sum_{\varrho,\sigma}^{\infty} k_{\varrho\sigma} b_{\varrho\mu} b_{\sigma\nu}^\dagger;$$

again, all sums (9.16) are only of finite range.

We rearrange formally the series (9.10) for $V(z, \zeta)$ by means of (9.15) and obtain

$$(9.17) \qquad \sum_{m,n=0}^{\infty} d_{mn} z^m \zeta^n = \sum_{\mu,\nu=0}^{\infty} t_{\mu\nu} \chi_\mu(z) \chi_\nu(\zeta), \qquad t_{\mu\nu} = \sum_{m,n=0}^{\infty} d_{mn} b_{m\mu} b_{n\nu}.$$

We do not know if and where the second sum (9.17) converges. But the series for $t_{\mu\nu}$ are finite expressions and well-defined.

Introduce an arbitrary complex vector $a_\nu (\nu = 0, \ldots, N)$ and let

$$(9.18) \qquad \alpha_\nu = \sum_{\mu=0}^{N} b_{\nu\mu} a_\mu.$$

Consider now the expression

$$(9.19) \qquad \sum_{\mu,\nu=0}^{N} t_{\mu\nu} a_\mu a_\nu = \sum_{m,n=0}^{N} d_{mn} \alpha_m \alpha_n.$$

Using the assumption (9.11) of our theorem, we find by virtue of (9.16) and (9.18)

$$(9.20) \qquad \left| \sum_{\mu, \nu=0}^{N} t_{\mu\nu} a_{\mu} a_{\nu} \right| \leq \sum_{\nu=0}^{N} |a_{\nu}|^2$$

for arbitrary choice of the complex vector a_{ν}.

From (9.20) we derive easily that for any two vectors a_{ν} and a'_{ν}

$$(9.20a) \qquad \left| \sum_{\mu, \nu=1}^{N} t_{\mu\nu} a_{\mu} a'_{\nu} \right| \leq \sum_{\nu=0}^{N} |a_{\nu}|^2 + \sum_{\nu=0}^{N} |a'_{\nu}|^2$$

holds.

Now let B' be a closed subdomain of B. In this domain the kernel function $K(z, z^{\dagger})$ is uniformly bounded, say by the constant M. Hence, in view of (9.13), (9.20a) leads to the inequality

$$(9.21) \qquad \left| \sum_{\mu, \nu=0}^{N} t_{\mu\nu} \chi_{\mu}(z) \chi_{\nu}(\zeta) \right| \leq 2M$$

for arbitrary choice of z and ζ in B'. Hence the functions

$$(9.22) \qquad V_N(z, \zeta) = \sum_{\mu, \nu=1}^{N} t_{\mu\nu} \chi_{\mu}(z) \chi_{\nu}(\zeta)$$

are uniformly bounded in B' and form a normal family there. Therefore we can select subsequences of our set which converge uniformly in each closed subdomain of B'. But in view of (9.17) and (9.22) the limit functions will always coincide with $V(z, \zeta)$ in the neighbourhood of the origin. Hence the whole sequence $V_N(z, \zeta)$ possesses the same limit and converges uniformly in each closed subdomain of B'. The limit function is the analytic continuation of the power series $V(z, \zeta)$ over the whole domain. Hence our theorem is proved.

The application of this result to the theory of univalent functions is immediate. Grunsky's conditions (9.9) guarantee the regularity of the function $U(z, \zeta)$ in B, which shows that $\Phi(z, \zeta)$ is regular is B and that except for $z = \zeta$ we never have $\varphi(z) = \varphi(\zeta)$. This is just the univalence property required. It is remarkable how closely the proof of necessity and sufficiency of (9.9) is connected with the kernel functions.

Finally we want to study the extremum problem, for which functions the Grunsky inequalities (9.9) may become equalities. Since these inequalities have been derived from the more general inequalities (4.14c) it will be sufficient to determine these domains B for which equality can hold in (4.14c) under an appropriate choice of points ζ_{ν} and constants α_{ν}. If we go back in the derivation of these inequalities we see that they can only become precise if the corresponding non-negative integral (4.10) vanishes, i.e. if there exists a real constant λ such that

$$(9.29) \qquad \sum_{\nu=1}^{r} \alpha_\nu^\dagger K(z, \zeta_\nu^\dagger) + \lambda \sum_{\nu=1}^{r} \alpha_\nu l(z, \zeta_\nu) \equiv 0 \quad \text{for } z \in B.$$

In proceeding from (4.14a) to (4.14c) we furthermore neglected the term

$$(9.30) \qquad \sum_{\nu,\mu=1}^{r} \alpha_\nu \alpha_\mu^\dagger l'(\zeta_\nu, \zeta_\mu^\dagger) = \frac{1}{\pi^2} \iint_{\widetilde{B}} \left| \sum_{\nu=1}^{r} \frac{\alpha_\nu}{z - \zeta_\nu} \right|^2 d\tau_z.$$

This integral can only vanish if the area of $\widetilde{B}$ is zero, i.e. if B is a slit domain. It is true that we developed our theory only for domains B which are bounded by closed analytic curves; at this stage, however, the consideration of more general domains becomes inevitable. Using the continuity of K and l in dependence of their domain of definition B it can be shown that the identities (4.2), (4.3) and (4.8) hold in the most general case. For slit domains, the term Γ is to be taken as zero in (4.8).

We multiply (9.29) with $l(w, s)^\dagger$ and integrate the identity over all $z \in B$. Using (4.3) and (4.8), we obtain

$$(9.29a) \qquad \sum_{\nu=1}^{r} \alpha_\nu^\dagger l(w, \zeta_\nu)^\dagger + \lambda \sum_{\nu=1}^{r} \alpha_\nu K(w, \zeta_\nu^\dagger)^\dagger \equiv 0 \quad \text{for } w \in B.$$

From (9.29) and (9.29a) we conclude the identity

$$(9.31) \qquad \sum_{\nu=1}^{r} [\alpha_\nu^\dagger K(z, \zeta_\nu^\dagger) + \alpha_\nu l(z, \zeta_\nu)] \equiv 0.$$

In view of (2.5) this may also be written as

$$(9.31a) \qquad \sum_{\nu=1}^{r} [\alpha_\nu^\dagger K(z, \zeta_\nu^\dagger) - \alpha_\nu L(z, \zeta_\nu)] + \frac{1}{\pi} \sum_{\nu=1}^{r} \frac{\alpha_\nu}{(z - \zeta_\nu)^2} \equiv 0.$$

We want to study this expression at the boundary C of B; however, this boundary may be a very complicated one and we prefer, therefore, to map B upon an auxiliary domain B_1 with smooth boundary C_1. Let $z = f(w)$ give the map of B_1 into B. We multiply (9.31a) with $f'(w)$ and have by virtue of (3.3) and (3.4)

$$(9.32) \quad \sum_{\nu=1}^{r} [A_\nu^\dagger K_1(w, \omega_\nu^\dagger) - A_\nu L_1(w, \omega_\nu)] + \frac{1}{\pi} \sum_{\nu=1}^{r} \frac{\alpha_\nu f'(w)}{(f(w) - \zeta_\nu)^2} \equiv 0, \quad w \in B_1,$$

where $\zeta_\nu = f(\omega_\nu)$ and $A_\nu = \alpha_\nu [f'(\omega_\nu)^{-1}]$. Let $w'(s)$ be the tangent vector at the point $w \in C_1$; multiplying (9.32) with $w'(s)$ and using the boundary relation (2.4) between K and L, we obtain

$$(9.33) \qquad \frac{d}{ds} \sum_{\nu=1}^{r} \frac{\alpha_\nu}{f(w(s)) - \zeta_\nu} = \text{real}, \quad w(s) \in C_1.$$

Integrating with respect to s, we find at last

$$(9.34) \qquad \Im \left\{ \sum_{\nu=1}^{r} \frac{\alpha_\nu}{z - \zeta_\nu} \right\} = \text{const. on } C.$$

This proves that C consists of analytic slits with the algebraic equation (9.34).

It is evident that the same treatment leads to the extremum domains in the particular case of Grunsky's coefficient inequalities (9.9).

10. Variation formulas for the eigen functions $\varphi_\nu(z)$.

We now want to study the dependance of some of the important domain functions on the varying domain B. Let $\nu(s)$ be continuous on C and $\varepsilon\nu(s)$ denote the shift of each boundary point $z(s) \in C$ along the interior normal direction at this point. This defines a deformation of the boundary $C \equiv z(s)$ into a new curve system C^* with the parametric representation

$$(10.1) \qquad z^*(s) = z(s) + iz'(s) . \varepsilon\nu(s) = z(s) + iz'(s)\delta n(s).$$

C^* is the boundary of a new domain B^* which differs very little from B for small ε. We may choose the deformation function $\nu(s)$ in such a way that C^* is a system of closed analytic curves.

Let $g^*(z, \zeta)$ be Green's function of B^*; according to a classical formula by Hadamard (Hadamard [1], Lévy [1]), we have

$$(10.2) \qquad g^*(z, \zeta) = g(z, \zeta) - \frac{1}{2\pi} \int_C \frac{\partial g(z, t)}{\partial n_t} \frac{\partial g(t, \zeta)}{\partial n_t} \delta n(t) ds_t + \varepsilon^2 \gamma_\varepsilon(z, \zeta)$$

where $\gamma_\varepsilon(z, \zeta)$ is bounded and harmonic in each closed subdomain of B. Using the definitions (2.8) and (2.5) for the kernels K and l, we obtain from (10.2) by differentiation the following formulas for their first order variations:

$$(10.3) \qquad \delta K(z, \zeta^\dagger) = \frac{1}{\pi^2} \int_C \frac{\partial^2 g(z, t)}{\partial z \, \partial n_t} \frac{\partial^2 g(t, \zeta)}{\partial n_t \, \partial \zeta^\dagger} \partial n_t \, ds_t$$

and

$$(10.3a) \qquad \delta l(z, \zeta) = -\frac{1}{\pi^2} \int_C \frac{\partial^2 g(z, t)}{\partial z \, \partial n_t} \frac{\partial^2 g(t, \zeta)}{\partial n_t \, \partial \zeta} \partial n_t \, ds_t.$$

Further it is easily seen that

$$(10.4) \qquad \frac{\partial g(z, t)}{\partial n_t} = \frac{2}{i} \frac{\partial g(z, t)}{\partial t} t' = -\frac{2}{i} \frac{\partial g(z, t)}{\partial t^\dagger} (t')^\dagger.$$

Using this identity in (10.3) and (10.8a) we find

(10.5) $\delta K(z, \zeta^{\dagger}) = \int\limits_C K(z, t^{\dagger}) K(t, \zeta^{\dagger}) \delta n_t \, ds_t = \int\limits_C L(z, t) L(t, \zeta)^{\dagger} \delta n_t \, ds_t$

(10.5a) $\delta l(z, \zeta) = \int\limits_C L(z, t) L(t, \zeta) t'^2 \delta n_t \, ds_t = \int\limits_C K(z, t^{\dagger}) K(\zeta, t^{\dagger}) (t'^2)^{\dagger} \delta n_t \, ds$

$= - \int\limits_C L(z, t) K(\zeta, t^{\dagger}) \delta n_t \, ds_t = - \int\limits_C K(z, t^{\dagger}) L(t, \zeta) \delta n_t \, ds_t.$

Let now $\zeta_k (\nu = 1, 2, \ldots, r)$ be an arbitrary set of points of B and $\alpha_\nu (\nu = 1, 2, \ldots, r)$ a set of complex numbers. We have by virtue of (10.5) and (10.5a)

(10.6) $\delta\left\{ \sum\limits_{\nu, \mu=1}^r \alpha_\nu \alpha_\mu^{\dagger} K(\zeta_\nu, \zeta_\mu^{\dagger}) \right\} = \int\limits_C |\sum\limits_{\mu=1}^r \alpha_\mu^{\dagger} K(t, \zeta_\mu^{\dagger})|^2 \delta n_t \, ds_t$

(10.6a) $\delta\left\{ \sum\limits_{\nu, \mu=1}^r \alpha_\nu \alpha_\mu l(\zeta_\nu, \zeta_\mu) \right\} = \int\limits_C \left(\sum\limits_{\mu=1}^r \alpha_\mu K(\zeta_\mu, t^{\dagger}) \right)^2 (t'^2)^{\dagger} \delta n_t \, ds_t.$

If the domain B decreases under variation, $\delta n_t \geqq 0$ on C and we see from (10.6) and (10.6a) that the expressions

(10.7) $\sum\limits_{\nu, \mu=1}^r \alpha_\nu \alpha_\mu^{\dagger} K(\zeta_\nu, \zeta_\mu^{\dagger}) \pm |\sum\limits_{\nu, \mu=1}^r \alpha_\nu \alpha_\mu l(\zeta_\nu, \zeta_\mu)|$

increase with decreasing domain. The terms of the inequality (4.14c) have, therefore, the following behavior; if the domain decreases, the bigger term increases quicker than the smaller term and the inequality becomes continually stronger.

We now want to determine the Fourier coefficients of the functions δK and δl with respect to a given orthonormal system in B. For this purpose, we have to prove the identity (2.9) for the case that the analytic function $f(z)$ is continuous in the closed region $B + C$ and for $z \epsilon C$. In this case, we may apply the boundary condition (2.4) and we obtain

(10.8) $\iint\limits_B K(z, \zeta^{\dagger}) f(\zeta) d\tau_\zeta = - [z'^2 \iint\limits_B L(z, \zeta) f(\zeta)^{\dagger} d\tau_\zeta]^{\dagger}, \quad z \epsilon C.$

Because of (2.5), we have

(10.9) $\iint\limits_B L(z, \zeta) f(\zeta)^{\dagger} d\tau_\zeta = \frac{1}{\pi} \iint\limits_B \frac{f(\zeta)^{\dagger}}{(z - \zeta)^2} d\tau_\zeta - \iint\limits_B l(z, \zeta) f(\zeta)^{\dagger} d\tau_\zeta.$

Using the definitions (5.1) and (5.3), the regularity of $l(z, \zeta)$ in the closed region $B + C$ and elementary proporties of improper integrals, we have

(10.10) $\dfrac{1}{\pi}\iint\limits_{B} f(\zeta)^{\dagger}(z-\zeta)^{-2}d\tau_{\zeta} = \lim\limits_{w\to z} Tf(w),$ $w\,\epsilon\,\bar{B},$

(10.10a) $\iint\limits_{B} f(\zeta)^{\dagger}\,l(z,\zeta)d\tau_{\zeta} = \lim\limits_{w\to z} Tf(w),$ $w\,\epsilon\,B$

Because of the saltus condition (5.5) we find therefore

(10.11) $\iint\limits_{B} L(z,\zeta)\,f(\zeta)^{\dagger}d\tau_{\zeta} = -\,[f(z)z'^{2}]^{\dagger},$ $z\,\epsilon\,C$

and hence from (10.8):

(10.12) $\iint\limits_{B} K(z,\zeta^{\dagger})f(\zeta)d\tau_{\zeta} = f(z),$ $z\,\epsilon\,C$

Hence the identity (2.9) has been extended to the closed region $B+C$.

The eigen functions $\varphi_{\nu}(z)$ are continuous in the closed region $B+C$ and form a complete orthonormal system in B. Using (10.5), (10.5a) and (10.12), we compute the following identities:

(10.13) $\iint\limits_{B}\iint\limits_{B}\int \delta K(z,\zeta^{\dagger})\varphi_{\nu}(z)^{\dagger}\varphi_{\mu}(\zeta)d\tau_{z}d\tau_{\zeta} = \int\limits_{C}\varphi_{\nu}(t)^{\dagger}\varphi_{\mu}(t)\,\delta n_{t}ds_{t}$

(10.13a) $\iint\limits_{B}\iint\limits_{B}\int \delta l(z,\zeta)\varphi_{\nu}(z)^{\dagger}\varphi_{\mu}(\zeta)d^{\dagger}\tau_{z}d\tau_{\zeta} = \left[\int\limits_{C}\varphi_{\nu}(t)\varphi_{\mu}(t)t'^{2}\delta n_{t}ds_{t}\right]^{\dagger}.$

From the definition of the eigen functions $\varphi_{\nu}(z)$ as solutions of the integral equation (6.5) it can easily be shown that each $\varphi_{\nu}(z)$ varies continuously with the domain B, if it does not belong to a degenerate eigenvalue λ_{ν}; its first order variation is of the class $\mathfrak{L}^{2}$ in B and we put:

(10.14) $\delta\varphi_{\nu}(z) = \sum\limits_{\mu=1}^{\infty} v_{\nu\mu}\varphi_{\mu}(z).$

We denote further the first order variation of the non-degenerate eigen value λ_{ν} by $\delta\lambda_{\nu}$.

In view of (6.19) and (6.20), the notation (10.14) leads to the formulas:

(10.15) $\delta K(z,\zeta^{\dagger}) = \sum\limits_{\nu,\mu=1}^{\infty} v_{\nu\mu}^{\dagger}\varphi_{\nu}(z)\varphi_{\mu}(\zeta)^{\dagger} + \sum\limits_{\nu,\mu=1}^{\infty} v_{\nu\mu}\varphi_{\mu}(z)\varphi_{\nu}(\zeta)^{\dagger},$

(10.16a) $\delta l(z,\zeta) = \sum\limits_{\nu,\mu=1}^{\infty}\dfrac{1}{\lambda_{\nu}} v_{\nu\mu}\varphi_{\nu}(z)\varphi_{\mu}(\zeta) +$

$+ \sum\limits_{\nu,\mu=1}^{\infty}\dfrac{1}{\lambda_{\nu}} v_{\nu\mu}\varphi_{\mu}(z)\varphi_{\nu}(\zeta) - \sum\limits_{\nu=1}^{\infty}\dfrac{\delta\lambda_{\nu}}{\lambda_{\nu}^{2}}\varphi_{\nu}(z)\varphi_{\nu}(\zeta).$

Introducing these expressions into (10.13) and (10.13a) respectively and using the orthonormality of the system $\{\varphi_\nu(z)\}$, we find the following equations for $v_{\nu\mu}$ and $\delta\lambda_\nu$:

$$(10.16) \qquad v^\dagger_{\nu\mu} + v_{\mu\nu} = \int\limits_C \varphi_\nu(t)^\dagger \varphi_\mu(t)\, \delta n_t\, ds_t$$

$$(10.16a) \quad \frac{1}{\lambda_\nu} v_{\nu\mu} + \frac{1}{\lambda_\nu} v_{\mu\nu} - \frac{\delta\lambda_\nu}{\lambda_\nu^2}\, \delta_{\nu\mu} = \left[\int\limits_C \varphi_\nu(t)\varphi_\mu(t) t'^2 \delta n_t\, ds_t\right]^\dagger.$$

These equations determine completely the variation of an eigen function $\varphi_\nu(z)$ and its corresponding eigen value λ_ν provided that λ_ν is non-degenerate.

From (10.16) and (10.16a) we conclude

$$(10.17) \quad -\frac{\delta\lambda_\nu}{\lambda_\nu^2} = \delta\left(\frac{1}{\lambda_\nu}\right) = \int\limits_C \left(\Re\{\varphi_\nu(t)^2 t'^2\} - \frac{1}{\lambda_\nu} |\varphi_\nu(t)|^2\right) dn_t\, \delta s.$$

We may write

$$(10.18) \quad \Re\{\varphi_\nu(t)^2 t'^2\} = \frac{1}{2}\left(\lambda_\nu + \frac{1}{\lambda_\nu}\right) |\varphi_\nu(t)|^2 - \frac{\lambda_\nu}{2}\left|\frac{\varphi_\nu(t)}{\lambda_\nu} - (\varphi_\nu(t)t'^2)^\dagger\right|^2.$$

If we introduce in the complementary domain $\widetilde{B}$ the function $\psi_\nu(z)$ defined by (7.30) we may express the variation formula (10.17) by means of (10.18) and (7.31) in the form

$$(10.19) \quad \delta\left(\frac{1}{\lambda_\nu}\right) = \frac{1}{2}\left(\lambda_\nu - \frac{1}{\lambda_\nu}\right)\int\limits_C (|\varphi_\nu(t)|^2 - |\psi_\nu(t)|^2)\delta n_t\, ds_t.$$

Since we proved in section 7 that in the case of a simply-connected domain B the function $\psi_\nu(z)$ is an eigen function of the complementary domain $\widetilde{B}$ with the same eigen value λ_ν, the great symmetry of (10.19) is obvious.

Further interesting formulas appear when the type of variation (10.1) is specialized. The following kind of variation has been of great use in the general theory (Schiffer [2]); let z_0 be an arbitrary fixed point in B. Let the boundary C be subjected to the variation

$$(10.20) \qquad \delta z = \frac{a\varrho^2}{z - z_0}, \qquad 0 < \varrho,\ z_0 \in B,$$

which is for ϱ small enough of the type (10.1). One sees immediately that the normal shift of a point $t \in C$ is given by the formula

$$(10.21) \qquad \delta n = \Re\left\{\frac{1}{it'}\frac{a\varrho^2}{t - z_0}\right\}.$$

Under this particular variation the formula (10.17) may be transformed into

$$(10.22) \quad \delta\left(\frac{1}{\lambda_\nu}\right) = \Re\left\{\frac{a\varrho^2}{2i}\int_C \{(\varphi_\nu t')^2 + [(\varphi_\nu t')^2]^\dagger - \frac{2}{\lambda_\nu}|\varphi_\nu|^2\}\frac{dt}{t-z_0}\right\}.$$

A slight rearrangement of this formula and the residue theorem yield

$$(10.23) \quad \delta\left(\frac{1}{\lambda_\nu}\right) = \Re\{\pi a\varrho^2\left(1-\frac{1}{\lambda_\nu^2}\right)\varphi_\nu(z_0)^2\} +$$

$$+ \Re\left\{\frac{a\varrho^2}{2i}\int_C \left(\frac{\varphi_\nu}{\lambda_\nu}-(\varphi_\nu t'^2)^\dagger\right)^2\frac{dt}{t-z_0}\right\}.$$

Now we remember that the function $T\varphi_\nu$ is regular in each complementary domain $\widehat{B}_\varrho$ and has the boundary values $\frac{\varphi_\nu(t)}{\lambda_\nu} - (\varphi_\nu \cdot t'^2)^\dagger$. Hence the integrand of the integral (10.23) is regular inside each boundary curve C_ϱ and the integral vanishes. Therefore, finally:

$$(10.24) \qquad \delta\left(\frac{1}{\lambda_\nu}\right) = \Re\{a\varrho^2\pi\left(1-\frac{1}{\lambda_\nu^2}\right)\varphi_\nu(z_0)^2\}.$$

This is a variation formula of the „interior" type where all boundary integrals have been eliminated.

A similar result is obtained if the point z_0 in the variation (10.20) is chosen in a complementary domain $\widehat{B}_\sigma$. One finds easily by the same considerations

$$(10.25) \qquad \delta\left(\frac{1}{\lambda_\nu}\right) = \Re\{a\varrho^2\pi\left(1-\frac{1}{\lambda_\nu^2}\right)\psi_\nu(z_0)^2\}$$

where the function $\psi_\nu(z_0)$ is connected with the l-transform $T\varphi_\nu$ by (7.30).

It is not difficult to determine the variation of the eigen functions $\varphi_\nu(z)$ under a variation (10.20). The corresponding formulas for the kernels K and l have been given in (Schiffer [8]). Formulas of this type are of particular use if one extends the definition of the functions and functionals considered to domains of the most general type; in this case, the boundary C may be so involved that a description (10.1) of the domain variation becomes impossible.

BIBLIOGRAPHY

N. ARONSZAJN

[1] La théorie des noyaux reproduisants et ses applications, I, Proc. Cambridge Phil. Soc., vol. 39 (1943), pp. 133—153.

S. BERGMAN

[1] Sur les fonctions orthogonales de plusieurs variables complexes avec les applications a la théorie des fonctions analytiques, Interscience Publishers, New York, 1941, Memorial des Sciences Mathématiques, Fascicules 106 and 108, Paris 1947—1948.

[2] Partial differential equations, Advanced topics, Brown University, Providence 1941.

S. BERGMAN—M. SCHIFFER

[1] A representation of Green's and Neumann's functions in the theory of partial differential equations of second order. Duke Math. J., vol. 14 (1947), pp. 609—638.

[2] On Green's and Neumann's functions in the theory of partial differential equations, Bull. Am. Math. Soc. vol. 53 (1947), pp. 1141—1151.

[3] Kernel functions in the theory of partial differential equations of elliptic type, Duke Math. J., vol. 15 (1948), pp. 535—566.

P. R. GARABEDIAN

[1] Schwarz' lemma and the Szegö kernel functions, Transactions Am. Math. Soc., vol. 67 (1949) pp. 1—35.

P. R. GARABEDIAN—M. SCHIFFER

[1] Identities in the theory of conformal mapping, Transactions Am. Math. Soc., vol. 65 (1949), pp. 187—238,

H. GRUNSKY

[1] Koeffizientenbedingungen für schlicht abbildende meromorphe Funktionen, Math. Zeitschrift, vol. 45 (1939), pp. 29—61.

J. HADAMARD

[1] Leçons sur le calcul des variations, Paris, 1910.

P. LÉVY

[1] Leçons d'Analyse fonctionelle, Paris, 1922.

Z. NEHARI

[1] The Schwarzian derivative and schlicht functions, Bull. Am. Math. Soc., vol. 55 (1949), pp. 545—551.

J. PLEMELJ

[1] Ein Ergänzungssatz zur Cauchy'schen Integraldarstellung analytischer Funktionen, Randwerte betreffend, Monatsh. Math. Phys. 19 (1908), pp. 205—210.

M. SCHIFFER

[1] The span of multiply-connected domains, Duke Math. J., vol. 10 (1943), pp. 209—216.

[2] Hadamard's formula and variation of domain functions, Amer. J. Math., vol. 68 (1946), pp. 417—448.

[3] The kernel function of an orthonormal system, Duke Math. J., vol. 13 (1946), pp. 529—540.

[4] Faber polynomials in the theory of univalent functions, Bull. Am. Math. Soc., vol. 54 (1948). pp. 503—517.

[5] On different types of orthogonalization, Duke Math. J. vol. 17 (1950).

(Received July 20th 1950).

Commentary on

[35] S. Bergman and M. Schiffer, *Kernel functions and conformal mapping*, Compositio Math. **8** (1951), 205 249.

In this well-known and widely cited paper, the authors return to their study of the Bergman reproducing kernel $K(z,\overline{\zeta})$ in the Hilbert space of square integrable analytic functions on a bounded plane domain B whose boundary C consists of finitely many real analytic closed curves. The kernel $K(z,\overline{\zeta})$ becomes strongly unbounded whenever z and ζ tend to the same boundary point in C, so the standard theory of integral equations does not apply. To circumvent this difficulty, the authors consider the *regularized* (Hermitian) kernel $K(z,\overline{\zeta}) - \Gamma(z,\overline{\zeta})$, where

$$\Gamma(z,\overline{\zeta}) = \frac{1}{\pi^2} \iint_{\mathbb{C}\backslash B} \frac{du\,dv}{(w-z)^2(\overline{w}-\overline{\zeta})^2},$$

$$w = u + iv,$$

which is shown to be regular in the closed region $B \cup C$ and thus amenable to study via the classical theory.

The Hermitian operator $K - \Gamma$ is positive definite and hence has infinitely many eigenfunctions. These turn out to coincide with the eigenfunctions associated with the regular part $\ell(z,\zeta)$ of the analytic kernel

$$L(z,\zeta) = -\frac{2}{\pi}\frac{\partial^2 g}{\partial z\,\partial \zeta} = \frac{1}{\pi(z-\zeta)^2} - \ell(z,\zeta),$$

introduced by Schiffer in [19], with the eigenvalues λ_n of $K - \Gamma$ being related to the eigenvalues μ_n of the ℓ-kernel via the relation $\lambda_n = \mu_n^2$. (Actually, as was customary for that period, λ_n and μ_n stand for the *reciprocals* of the eigenvalues as defined in current terminology.) The authors show that $\lambda_n \geq 1$ and exploit this fact in devising a procedure for numerical calculation of the Bergman kernel, which is shown to converge geometrically.

The paper concludes with estimates of the eigenvalues λ_n in terms of geometric characteristics of the domain (area, mapping radius, etc.), applications to the theory of univalent functions, and the derivation of variational formulas for the eigenfunctions associated with the kernel $K(z,\overline{\zeta}) - \Gamma(z,\overline{\zeta})$. For further studies concerned with the operators L and Γ, see [DP].

This paper can be viewed as the first step in a wider program to study properties of the Fredholm eigenvalues of a plane domain, subsequently elaborated in [62, 68, 78, 82, 103, 106, 125]. More recent applications of the results of [35] to the study of Fredholm eigenvalues appear in [KPS, Sh1, Sh2, PP], as well as the references cited there.

References

[DP] Philip Davis and Henry Pollak, *On the analytic continuation of mapping functions*, Trans. Amer. Math. Soc. **87** (1958), 198 225.

[KPS] Dmitry Khavinson, Mihai Putinar, and Harold S. Shapiro, *Poincaré s variational problem in potential theory*, Arch. Rational Mech. Anal. **185** (2007), 143 184.

[PP] Karl-Mikael Perfekt and Mihai Putinar, *Spectral bounds for the Neumann-Poincaré operator on planar domains with corners*, arXiv:1209.3918.

[Sh1] Yuliang Shen, *Generalized Fourier coef"cients of a quasisymmetric homeomorphism and the Fredholm eigenvalue*, J. Analyse Math. **112** (2010), 33 48.

[Sh2] Yuliang Shen, *Fredholm eigenvalue for a quasicircle and Grunsky functionals*, Ann. Acad. Sci. Fenn. Math. **35** (2010), 581 593.

DMITRY KHAVINSON

Corrigendum. Brad Osgood has informed the editors of a misprint in the authors' celebrated expression for the Schwarzian derivative $\{\varphi, z\}$. In the formula (3.7) on page 210, the minus signs preceding the two expressions on the right should be deleted. A correct version of an equivalent formula appears as equation (A2.47) on page 275 of [34].

Editors Note. Stefan Bergman and Max Schiffer dedicated this paper to their teacher in Berlin, Erhard Schmidt. As Schiffer makes clear in his obituary article [124], he first became acquainted with Bergman, who was 16 years his senior, during his student years in Berlin. Later they overlapped for a time at Harvard, before moving together to Stanford in 1952. Here, in Schiffer's words [124], is the remarkable story of Bergman's discovery of the kernel function:

Shortly after his arrival at Berlin, Bergman participated in Schmidt's seminar and was charged to give a lecture on development of arbitrary functions with finite square integral in terms of an orthogonal set. As he told me, he misunderstood the task and instead of dealing with real functions over a real interval, he attacked the problem for analytic functions over a complex domain. He found the task hard but attacked it courageously and carried it through. This was the genesis of his famous theory of orthogonal functions and the kernel function. A first fruit was his thesis which gave him the Doctor's degree in 1922.

[40] Variational methods in the theory of conformal mapping

[40] Variational methods in the theory of conformal mapping, in Proceedings of the International Congress of Mathematicians, Cambridge, Mass., 1950, *American Mathematical Society*, Providence, R.I., **2** (1952), 233–240.

Reprinted from Vol. II, Proceedings of the
International Congress of Mathematicians, 1950
Printed in U.S.A.

VARIATIONAL METHODS IN THE THEORY OF
CONFORMAL MAPPING

Menahem Schiffer

1. Let D be a domain in the complex z-plane which is bounded by n smooth curves C_ν ($\nu = 1, 2, \cdots, n$) which form the boundary C of D. A major problem in conformal mapping consists in determining univalent analytic functions $f(z)$ in D which map it on a canonical domain of specified type. The most important canonical maps can be easily expressed in terms of the Green's function of the domain. Green's function $g(z, \zeta)$ depends on its two argument points and on the domain D (or on its boundary curve system C). The latter dependence is of transcendental character and its investigation belongs to functional analysis as pointed out by Volterra and Hadamard. The theory of schlicht functions is also closely related to this field of research; for, every schlicht function, say, in the unit circle can be characterized by the domain upon which it maps and can easily be expressed in terms of the Green's function of the image domain. Thus, extremum problems with respect to schlicht functions can be reduced to extremum problems for Green's function and are incorporated into the theory of the latter.

A fundamental tool in the study of functionals are variation formulas which express the rate of change of the functional under given deformation of its domain. The value of such formulas for extremum problems, comparison formulas, etc. is evident. We shall discuss here various types of variation methods and compare their relative merits. Actual applications and illustrations by concrete problems will not be given here for lack of space.

2. The first and already very general variational formula for Green's function was given by Hadamard [5]. Let the domain D be deformed into D^* by subjecting every boundary point to a normal shift δn (counted positive along the interior normal vector n); the variation of the Green's function is that expressed by Hadamard's formula

$$(1) \qquad \delta g(z, \zeta) = -\frac{1}{2\pi} \int_C \frac{\partial g(z, t)}{\partial n_t} \frac{\partial g(t, \zeta)}{\partial n_t} \delta n_t \, ds_t .$$

Immediate applications of (1) are obvious; let $g(z, \zeta) = \log (1 / |z - \zeta|) + h(z, \zeta)$. We observe from (1) that the functionals $h(z, z) + h(\zeta, \zeta) - 2h(z, \zeta)$ and $\sum_{i,k=1}^N h(z_i, z_k)\alpha_i\alpha_k$ vary monotonically with the domain and useful inequalities and comparison formulas follow directly.

Let us next define for z and ζ on C the functional of the curve system C

$$(2) \qquad \mathfrak{G}(z, \zeta) = -\frac{1}{2\pi} \frac{\partial^2 g(z, \zeta)}{\partial n_z \, \partial n_\zeta}.$$

233

It is easily seen that $\mathfrak{G}$ is negative on C and that knowledge of it is sufficient to solve Dirichlet's problem with respect to D. It has the elegant variational formula

$$(3) \qquad \delta\mathfrak{G}(z, \zeta) = \int_C \mathfrak{G}(z, t)\mathfrak{G}(t, \zeta) \, \delta n_t \, ds_t.$$

The monotony of $\mathfrak{G}(z, \zeta)$ as a functional of D is obvious from (3) and the definiteness of $\mathfrak{G}$, and, hence, useful comparison formulas can be established for it. Formula (3) must, however, be used with caution since $\mathfrak{G}(z, \zeta)$ has an infinity for $z = \zeta$. A permissible deformation must leave arcs of C around z and ζ unchanged if (3) is to be used. Otherwise, corrective terms of rather complicated nature arise and the integral in (3), moreover, becomes improper.

One can avoid these difficulties by introducing the two complex functions [13]

$$(4) \quad K(z, \bar{\zeta}) = -\frac{2}{\pi} \frac{\partial^2 g(z, \zeta)}{\partial z \partial \bar{\zeta}}, \quad L(z, \zeta) = -\frac{2}{\pi} \frac{\partial^2 g(z, \zeta)}{\partial z \partial \zeta} = \frac{1}{\pi(z - \zeta)^2} - l(z, \zeta),$$

where $\partial/\partial z = (\partial/\partial x - i\partial/\partial y)/2$ and $\partial/\partial\bar{z} = (\partial/\partial x + i\partial/\partial y)/2$. Both functions are analytic in all their arguments; K is regular throughout D while L has a double pole at $z = \zeta$ as exhibited in (4). These two new functionals of D satisfy the following variational equations which are derived from (1) by differentiation and using the boundary behavior of Green's function:

$$(5) \quad \delta K(z, \bar{\zeta}) = \int_C K(z, \bar{t})K(t, \bar{\zeta}) \, \delta n_t \, ds_t, \quad \delta L(z, \zeta) = \int_C K(z, \bar{t})L(t, \zeta) \, \delta n_t \, ds_t.$$

3. The functions K and L to which we are led quite naturally by the formalism of the variational formulas play an important role in conformal mapping [3]. They satisfy the following equations on the boundary C of D:

$$(6) \quad \overline{z'K(z, \bar{\zeta})} = z'L(z, \zeta), \quad z \in C, \quad \zeta \in D, \quad z' = \text{tangent vector at } z \text{ to } C;$$

$$(6') \qquad z'K(z, \bar{\zeta})\bar{\zeta}' = z'L(z, \zeta)\zeta' = \mathfrak{G}(z, \zeta), \quad z \in C, \quad \zeta \in C.$$

The connection between the functional (2) and the two new functions K and L is thus established. $K(z, \bar{\zeta})$ satisfies the identity

$$(7) \qquad \iint_D K(z, \bar{\zeta})f(\zeta) \, d\tau_\zeta = f(z), \quad d\tau_\zeta = \text{area element in } \zeta,$$

for every analytic function $f(z)$ in D with $\iint_D |f|^2 \, d\tau < \infty$. From this reproducing property follows easily a representation of $K(z, \bar{\zeta})$ in terms of any complete orthonormal set $\{\varphi_\nu(z)\}$ in D:

$$(7') \qquad\qquad K(z, \bar{\zeta}) = \sum_{\nu=1}^{\infty} \varphi_\nu(z)\overline{\varphi_\nu(\zeta)}.$$

The kernel function $K(z, \bar{\zeta})$ was defined by Bergman [1] just by this formula and various applications to conformal mapping were given by him from this definition. (7') lends itself also easily to a generalization for the case of analytic functions of several complex variables. The kernel $L(z, \zeta)$ can be easily constructed in terms of $K(z, \bar{\zeta})$ by means of the identity

$$(8) \qquad L(z, \zeta) = \frac{1}{\pi(z - \zeta)^2} - \frac{1}{\pi} \iint_D K(z, \bar{t})(t - \zeta)^{-2} \, d\tau_t,$$

which shows that from $K(z, \bar{\zeta})$ ultimately Green's function can be obtained.

Let D_0 be a domain containing D and denote its corresponding kernels by K_0 and L_0. We observe that $L(z, \zeta) - L_0(z, \zeta)$ is regular analytic in D and we obtain by use of (6), after easy calculation, the identity, valid for any two points z and ζ in D [15]:

$$
\begin{aligned}
(9) \qquad & \iint_D (L(z, t) - L_0(z, t))\overline{(L(t, \zeta) - L_0(t, \zeta))} \, d\tau_t \\
& \qquad = K(z, \bar{\zeta}) - K_0(z, \bar{\zeta}) - \iint_{D_0 - D} L_0(z, t) \, \overline{L_0(t, \zeta)} \, d\tau_t.
\end{aligned}
$$

Suppose now that the boundary C_0 of D_0 lies in a Fréchet neighborhood of order ϵ to the corresponding curve system C. Then, we shall have $K - K_0 = O(\epsilon)$ and $L - L_0 = O(\epsilon)$. Introducing these estimates into (9), we arrive at

$$(10) \qquad \delta K_0(z, \bar{\zeta}) = \iint_{D_0 - D} L_0(z, t)\overline{L_0(t, \zeta)} \, d\tau_t + O(\epsilon^2)$$

which, in view of (6), can be easily brought into the form of the first formula (5). Thus, (9) represents a finite comparison formula which goes for small deformations into the Hadamard form. One can use (9) as the basic formula in a process of successive approximation for $K - K_0$ and express this difference as an infinite series of iterated integrals containing L_0 and K_0 only [15]. Formula (9) gives also rise to numerous inequalities; in particular, Grunsky's inequalities [4] for the coefficients of schlicht functions can be easily derived from it.

4. An essential assumption for the application of Hadamard's formula is the smoothness of the boundary C, since the normal of C plays a distinguished role in the formula. Thus, the formula is inapplicable for very many domains possessing a Green's function; this fact precludes the use of Hadamard's formula in extremum problems of conformal mapping. One is never sure, a priori, if the extremum domain in question permits the use of Hadamard's formula and one cannot compare it in this way with neighbor domains. One may, however, transform Hadamard's formula into such a form that this difficulty is removed. We introduce an infinitesimal deformation of the whole z-plane by the formula

$$(11) \qquad z^* = z + \frac{e^{i\varphi} \rho^2}{z - z_0} \qquad\qquad \rho > 0, \quad z_0 \in D.$$

The curve system C is deformed into a system C^* which is, for ρ sufficiently small, very near to C and of the same character. In fact, the map (11) is univalent and analytic outside of the circle $\gamma\colon |z - z_0| < \rho$. The domain D^* determined by the curve system C^* has the Green's function $g^*(z, \zeta)$. Let $D_{z_0,\rho}$ be the domain obtained from D by removing the circle γ; we remark that $g^*(z^*(z), \zeta^*(\zeta))$ is harmonic in $D_{z_0,\rho}$, except for $z = \zeta$, and vanishes on C. Hence, we may apply Green's identity to the functions $g(z, \zeta)$ and $g^*(z^*, \zeta^*)$ with respect to $D_{z_0,\rho}$; we easily obtain:

$$
(12) \qquad g^*(z^*, \zeta^*) = g(z, \zeta) - \frac{1}{2\pi}
$$
$$
\cdot \oint_\gamma \left\{ g^*(z^*, t^*) \frac{\partial}{\partial n_t} g(t, \zeta) - g(t, \zeta) \frac{\partial}{\partial n_t} g^*(z^*, t^*) \right\} ds_t .
$$

Thus, by means of Green's identity we expressed the difference between the two Green's functions in terms of an integral extended over a little circle γ entirely in D and removed from the dangerous boundary C. Hadamard's general formula is also an immediate consequence of Green's identity, but our special type of variation (11) allows the further transformation from (1) to (12). It is also obvious that a description of a variation in terms of normal shift on C is impossible for general curve systems and that a variation should rather be described by a rule determining the shift of all points in the neighborhood of C. (11) is a very special case of such a rule [cf. 8], but the most general variation of the described form can be approximated arbitrarily by superposition of elementary deformations (11).

Formula (12) can easily be evaluated by series developments in the neighborhood of the point z_0 and one arrives at [11; 12]:

$$
(13) \qquad g^*(z^*, \zeta^*) = g(z, \zeta) + \operatorname{Re} \left\{ 4e^{i\varphi} \rho^2 \frac{\partial g(z_0, z)}{\partial z_0} \frac{\partial g(z_0, \zeta)}{\partial z_0} \right\} + o(\rho^2)
$$

where $o(\rho^2)$ can be estimated uniformly in each closed subdomain Δ of $D_{z_0,\rho}$ with respect to all domains in a sufficiently small Fréchet neighborhood of D. So far, we have still to make the assumption that C is a smooth curve system in order to apply Green's identity. However, using the continuity of Green's function and all its derivatives in dependence of the domain D, we may extend the validity of (13) to the most general type of domains D bounded by n continua and possessing a Green's function. Thus, (13) is generally valid and may be used in extremum problems concerning Green's function, even if nothing is known about the nature of the boundary of the sought extremal domain. (13) has been used in the coefficient problem for schlicht and p-valued functions [11] and various related extremum problems for conformal functionals [2]. In most cases, one can show by using (13) that the extremum domain possesses piecewise analytic boundary curves C which satisfy certain differential equations [9]. A finer investigation of this domain and a study of higher order variations is then more conveniently carried out by Hadamard type variations which become permissible, once that the analytic character of the boundary has been proved. This remark explains why the type (11) of variation is general enough for most applications and why a greater generality in variational formulas of this type would often lead to unnecessary complications.

Consider the domain $D_{z_0,\rho}$ where the boundary $|z - z_0| = \rho$ has been removed by identifying points on it which have the same image points under the map (11). The domain thus obtained is topologically equivalent to D and D^*. Its Green's function $g^*(z, \zeta)$ is equal at each point to the right-hand side of (13). D^* is the realization of this Riemann manifold in the z-plane and (11) is the map which performs this realization. In this interpretation, we can easily extend the above type of interior variation to arbitrary domains on Riemann surfaces by changing their structure slightly through boundary correspondence along artificial holes and by realizing the domain again over the complex plane. Two further types of variation are also suggested by this point of view: (a) Hole punching: remove from D the domain $g(z, z_0) \geqq \log (1/\rho)$ which transforms D approximately into $D_{z_0,\rho}$. The Green's function of the new domain can again be computed by proper use of Green's identity. We obtain [15]:

$$(14) \quad g^*(z, \zeta) = g(z, \zeta) + (\log \rho)^{-1} g(z_0, z) g(z_0, \zeta)$$

$$- \operatorname{Re}\left\{4\rho^2 \frac{\partial g(z_0, z)}{\partial z_0} \frac{\partial g(z_0, \zeta)}{\partial \bar{z}_0}\right\} + o(\rho^2)$$

(b) Sewing on of handles: eliminate from D the two domains $G(z; z_0, z_1) > \log (1/\rho)$ and $G(z; z_0, z_1) < \log \rho$ with $G(z; z_0, z_1) = g(z, z_0) - g(z, z_1)$. Pairs w, ω of points on the newly created boundaries c and γ near z_0 and z_1 are identified if they satisfy

$$(15) \quad (c) \int_{w_0}^{w} \frac{\partial}{\partial n_t} G(t; z_0, z_1) \, ds_t = (\gamma) \int_{\omega_0}^{\omega} \frac{\partial}{\partial n_t} G(t; z_0, z_1) \, ds_t.$$

We obtain thus a new domain D^* with larger genus than D. Its corresponding Green's function is obtained again by means of Green's identity [15]:

$$g^*(z, \zeta) = g(z, \zeta) + \frac{1}{2} (\log \rho)^{-1} G(z; z_0, z_1) G(\zeta; z_0, z_1)$$
$$(16)$$
$$- \operatorname{Re}\left\{4e^{i\varphi} \rho^2 \left[\frac{\partial g(z_0, z)}{\partial z_0} \frac{\partial g(z_1, \zeta)}{\partial z_1} + \frac{\partial g(z_1, z)}{\partial z_1} \frac{\partial g(z_0, \zeta)}{\partial z_0}\right]\right\} + o(\rho^2).$$

Relation (15) establishes a relation $(w - z_0)(\omega - z_1) = \rho^2 e^{i\varphi} + o(\rho^2)$ between the identified points and the real constant φ in (16) is just defined by this relation. Variations (a) and (b) permit us to change connectivity and genus of the domain D and, together with the previous topology preserving variations, give a great freedom in changing the initial domain D.

5. Let $f(z)$ be schlicht in the unit circle E, $f(0) = 0$, and let it map E upon the simply-connected domain Δ with boundary curve Γ. If each point $\omega \in \Gamma$ is shifted along a vector $\delta\omega$ which points into the direction of the normal at ω and varies continuously along Γ, we obtain a domain Δ^* which is mapped by means of a schlicht function $f^*(z)$. Julia derived from Hadamard's formula (1) the following variational law [6]:

$$(17) \qquad \delta f(z) = f'(z) \cdot \frac{1}{2\pi i} \oint_\Gamma \frac{z}{\zeta^2} \frac{\zeta + z}{\zeta - z} \frac{\delta\omega \, d\omega}{f'(\zeta)^2}, \qquad \omega = f(\zeta).$$

This formula contains Loewner's differential equation [7] as a limit case, namely if we let $\delta\omega$ converge to zero everywhere on Γ except near one single point where it is made to grow beyond any limit. The great advantage of Loewner's form of variation is that it permits the introduction of a simple and natural parameter and thus reduces many problems of functional analysis to problems concerning ordinary differential equations.

Let us generalize Julia's formula to the following case: A domain D in the z-plane is given and the class of schlicht functions in D is to be studied. The variation δf of a map function under a Julia variation $\delta\omega$ of the image domain Δ is to be determined. We set up the formula

$$(18) \qquad \delta f(z) = f'(z) \cdot \frac{1}{2\pi i} \oint_\Gamma n(z, \zeta) \frac{\delta\omega \, d\omega}{f'(\zeta)^2}, \qquad \omega = f(\zeta),$$

where the integration is now to be extended over the whole boundary curve system Γ of Δ. In order to study $n(z, \zeta)$ suppose that $\delta\omega = 0$ along a subarc $\gamma \subset \Gamma$. Let $c \subset C$ be the corresponding arc on the boundary of D and let $z \to c$. Clearly, the image of z must lie on γ even after the variation and hence $\delta f(z)$ must have tangential direction. Thus $\delta f(z) \cdot f'(z)^{-1}$ must be a vector of tangential direction at $z \in c$ and denoting this vector by z', we easily recognize that

$$(19) \qquad \frac{n(z, \zeta)\zeta'^2}{z'} = \text{real}, \qquad z, \zeta \in C,$$

is the characteristic property of the Julia kernel $n(z, \zeta)$; i.e., $n(z, \zeta)$ must be a reciprocal differential in its first argument and a quadratic differential in the second. Besides, it must have a simple pole for $z = \zeta$ as is seen from (17). Let us assume for sake of simplicity that D has $n \geq 3$ boundary curves C_ν. There will exist two linearly independent harmonic functions $\omega_1(z)$ and $\omega_2(z)$ which are constant on each C_ν. By means of these functions, let us define

$$(20) \quad \Lambda(z, \zeta) = \begin{vmatrix} \dfrac{\partial\omega_1(z)}{\partial z} & \dfrac{\partial\omega_2(z)}{\partial z} \\[2ex] \dfrac{\partial\omega_1(\zeta)}{\partial\zeta} & \dfrac{\partial\omega_2(\zeta)}{\partial\zeta} \end{vmatrix}, \quad \Lambda(z, \bar{\zeta}) = \begin{vmatrix} \dfrac{\partial\omega_1(z)}{\partial z} & \dfrac{\partial\omega_2(z)}{\partial z} \\[2ex] \dfrac{\partial\omega_1(\zeta)}{\partial\bar{\zeta}} & \dfrac{\partial\omega_2(\zeta)}{\partial\bar{\zeta}} \end{vmatrix}, \quad T(z) = \begin{vmatrix} \dfrac{\partial\omega_1}{\partial z} & \dfrac{\partial\omega_2}{\partial z} \\[2ex] \dfrac{\partial^2\omega_1}{\partial z^2} & \dfrac{\partial^2\omega_2}{\partial z^2} \end{vmatrix}$$

and construct the kernels [14]

$$(21) \qquad n(z, \zeta) = \pi L(z, \zeta) \frac{\Lambda(z, \zeta)}{T(z)}, \qquad m(z, \bar{\zeta}) = \pi K(z, \bar{\zeta}) \frac{\Lambda(z, \bar{\zeta})}{T(z)}.$$

One easily verifies that $n(z, \zeta)$ has all the properties required above and that one has for ζ in D and z on C the relation:

$$(22) \qquad \frac{n(z, \zeta)}{z'} = \overline{\left(\frac{m(z, \bar{\zeta})}{z'} \right)}.$$

The functions n and m are not regular throughout D but have (besides the simple pole of n for $z = \zeta$) exactly $3n-6$ poles at the zeros of T. Thus, not every variation (18) is permissible but only such which make δf regular in $D + C$; this gives $3n-6$ conditions which are just the number of moduli determining the conformal type.

We are now able to attack the following important problem of conformal mapping. Two domains D and R are given; consider the family $\mathfrak{F}$ of all functions $f(z)$ in D which map D into a domain Δ which is schlicht relative to R and develop a calculus of variations for $\mathfrak{F}$. This problem includes the problem of schlicht functions within D ($R =$ complex plane), of bounded schlicht functions ($R =$ circle) and of p-valued functions in D ($R =$ Riemann surface with p sheets). We can define to every given $f(z) \in \mathfrak{F}$ a neighbor function of the same family by means of the kernels n and m of D and the corresponding kernels N and M of R. We have [14]:

$$(23) \qquad \begin{aligned} f^*(z) &= f(z) + \epsilon f'(z) \sum_{\nu=1}^{k} (r_\nu n(z, \zeta_\nu) + \bar{r}_\nu m(z, \zeta_\nu)) \\ &\quad - \epsilon \sum_{\nu=1}^{k} (P_\nu N(w, \omega_\nu) + \bar{P}_\nu M(w, \bar{\omega}_\nu)) + o(\epsilon), \qquad \epsilon \text{ real,} \end{aligned}$$

where $w = f(z)$, $\omega_\nu = f(\zeta_\nu)$, and $P_\nu = r_\nu f'(\zeta_\nu)^2$. One sees easily that the addition of the second right-hand term represents a tangential shift of each boundary point of Δ and that the third right-hand term has on the boundary of R again tangential direction. The last term $o(\epsilon)$ which is necessary in order to make $f^*(z)$ precisely univalent and its image domain Δ^* exactly lie in R can be estimated uniformly in each closed subdomain of D for any compact subclass of $\mathfrak{F}$. Thus, the variation formula (23) can be used in order to characterize the functions of $\mathfrak{F}$ solving significant extremum problems. It should be observed that we are not quite free in the choice of the values r_ν in (23). The formula has been constructed such that the right-hand side is regular at all points ζ_ν; however, it could become infinite at the fixed zeros of the kernels n, m and N, M which are independent of the ζ_ν. The r_ν have to be chosen such that these poles just cancel; since the number k of the arbitrary pole points ζ_ν is not bounded, one has ample possibility to keep this and finitely many other side conditions which one likes to impose on the variation. Numerous applications of this method to the coefficient problem for function classes $\mathfrak{F}$ are possible.

The coefficient problem for schlicht functions in a multiply-connected domain has previously been attacked by a different variational method [10]. The exterior of small pieces of the boundary in the image domain was mapped conformally, and, in view of the group property of schlicht conformal mappings, this led to a variation of the original map function. Differential equations for the boundary curves of the extremal domains considered were readily derived from this procedure. Comparing it with the above general method, we may say that as long as this boundary variation is applicable it is much easier to handle and more elegant. It breaks down, however, if many side conditions are to be observed and in this case one is obliged to use the heavier but more adaptable variation (23).

BIBLIOGRAPHY

1. S. BERGMAN, *Ueber die Entwicklung der harmonischen Funktionen der Ebene und des Raumes nach Orthogonalfunktionen*, Math. Ann. vol. 96 (1922) pp. 237–271.

2. G. GOLUSIN, *Method of variations in the theory of conformal mapping*, Rec. Math. (Mat. Sbornik) N. S. vol. 19 (1946) pp. 203–236, vol. 21 (1947) pp. 83–117, 119–132.

3. P. R. GARABEDIAN and M. SCHIFFER, *Identities in the theory of conformal mapping*, Trans. Amer. Math. Soc. vol. 65 (1949) pp. 187–238.

4. H. GRUNSKY, *Koeffizientenbedingungen fuer schlicht abbildende meromorphe Funktionen*, Math. Zeit. vol. 45 (1939) pp. 29–61.

5. J. HADAMARD, *Mémoire sur le problème d'analyse relatif à l'équilibre des plaques élastiques encastrées*, Mém. d. Sav. étr. vol. 33 (1908) pp. 1–128.

6. G. JULIA, *Sur une équation aux dérivées fonctionelles liées à la représentation conforme*, Ann. École Norm. (3) vol. 39 (1922) pp. 1–28.

7. K. LOEWNER, *Untersuchungen ueber schlichte konforme Abbildungen des Einheitskreises*, Math. Ann. vol. 89 (1923) pp. 103–121.

8. A. C. SCHAEFFER and D. C. SPENCER, *A variational method in conformal mapping*, Duke Math. J. vol. 14 (1947) pp. 949–966.

9. ———, *Coefficient regions for schlicht functions*, Amer. Math. Soc. Colloquium Publications vol. 35, New York, 1950.

10. M. SCHIFFER, *A method of variation within the family of simple functions*, Proc. London Math. Soc. (2) vol. 44 (1938) pp. 432–449.

11. ———, *Variation of the Green function and theory of p-valued functions*, Amer. J. Math. vol. 65 (1943) pp. 341–360.

12. ———, *Hadamard's formula and variation of domain functions*, Amer. J. Math. vol. 68 (1946) pp. 417–448.

13. ———, *The kernel function of an orthonormal system*, Duke Math. J. vol. 13 (1946) pp. 529–540.

14. M. SCHIFFER and D. C. SPENCER, *The coefficient problem for multiply-connected domains*, Ann. of Math. vol. 52 (1950) pp. 362–402.

15. ———, *Lectures on conformal mapping and extremal methods*, Princeton Lectures, 1949–1950.

HEBREW UNIVERSITY,
    JERUSALEM, ISRAEL.

Commentary on

[40] *Variational methods in the theory of conformal mapping*, Proceedings of the International Congress of Mathematicians, Cambridge, Mass., 1950; American Mathematical Society, Providence, R.I., 1952, Vol. 2, pp. 233 240.

[70] *Extremum problems and variational methods in conformal mapping*, Proceedings of the International Congress of Mathematicians, Edinburgh, 1958; Cambridge University Press, New York, 1960, pp. 211 231.

Max Schiffer was twice honored by invitations to address the International Congress of Mathematicians, at Cambridge (Massachusetts) in 1950 and at Edinburgh in 1958. The articles [40, 70] are written versions of his lectures. Both of the articles are devoted to the variational methods for which Schiffer was best known, but they are rather different in character. The paper [40] offers a broad discussion of variational methods and their relative merits, and it surveys a variety of applications. The paper [70], on the other hand, focuses more narrowly on applications to particular extremal problems in function theory such as coefficient problems, and it gives a fairly detailed account of technical advances in the use of variational methods. In this respect, [70] is somewhat dated, since the Bieberbach conjecture has now been proved (see commentaries on [5, 6, 72]) and the conjecture $|b_n| \leq \frac{2}{n+1}$ for functions of class Σ has been disproved for all $n \geq 3$ (see commentaries on [4, 59, 123]). The paper [70] also gives an extended description of Schiffer's method for obtaining a lower bound for the first nontrivial Fredholm eigenvalue of a simply connected domain with analytic boundary curve. Schiffer and others returned repeatedly to this problem (see Kühnau's commentary on [62, 68, 78, 125]). The lower bound is important numerically, because it allows an estimate on the rate at which the Neumann series converges to the solution of the classical Poincaré Fredholm integral equation associated with the solution of a Dirichlet problem. In his expository paper [78], Schiffer gives a clear and detailed account of this beautiful circle of ideas.

PETER DUREN

[44] (with P. R. Garabedian and H. Lewy) Axially symmetric cavitational flow

[44] (with P. R. Garabedian and H. Lewy) Axially symmetric cavitational flow. *Ann. of Math.* **56** (1952), 560–602.

ANNALS OF MATHEMATICS
Vol. 56, No. 3, November, 1952
Printed in U.S.A.

AXIALLY SYMMETRIC CAVITATIONAL FLOW[1]

By P. R. GARABEDIAN, H. LEWY, AND M. SCHIFFER

(Received May 26, 1952)

TABLE OF CONTENTS

1. Introduction

The mathematical theory of cavitational flow dates back to the early work of Helmholtz and Kirchhoff on free streamlines. There is an extensive literature on the subject for the case of the plane irrotational flow of an incompressible fluid. In addition to existence and uniqueness theorems, for plane cavitational flow many explicit examples are known and much qualitative information is available. However, for axially symmetric flow, there is almost no mathematical work on cavitation.

Cavitational flow occurs physically when an object, called projectile, travels so fast through water that the fluid pressure falls to vapor pressure, thus causing the water to boil and to form a cavity of steam following the object. When a projectile enters the water from the air, it is also possible for a pocket of air to follow, and thus the cavity in some cases consists partly of air and partly of steam. Typical of such a situation is that the projectile possesses axial symmetry; if the object moves in the direction of the axis of symmetry and if the fluid fills the entire space, it can be shown that the flow will be axially symmetric as well. A common procedure among the engineers in order to apply the highly developed mathematical theory of plane flow past free streamlines to the actual physical problem at hand has been to rotate the plane configuration about an axis of symmetry and estimate in this way the shape of the steam pocket and the drag on the object. Although obviously unsatisfactory to the mathematician, this approach has led to results which are in good agreement with experimental data [15].

The object of the present paper is to lay the foundation for an analysis and construction of axially symmetric cavitational flows. The development centers about the proof of existence and uniqueness of cavitational flow past a given

---

[1] This paper was written under the sponsorship of the Office of Naval Research.

rotationally symmetric body with prescribed cavitation constant, or with prescribed circle of separation. The method applies for various models of interest to the experimentalist, but it is carried out in complete detail only for the simplest case in which its main features can be clearly brought to light.

Our existence proof depends on a reformulation of the cavitational flow problem as a free boundary extremal problem in the calculus of variations. This variational formulation of the hydrodynamic question appears to have been given for the first time by Riabouchinsky [17]. A later discussion of the variational problem, which combined it with Dirichlet's principle, was given by Friedrichs [3], who deduced from it the local uniqueness of the flow from a nozzle. An existence theorem for plane cavitating flow based on the above variational theory has been given quite recently [1, 4, 5].

Basic for the present treatment of the variational problem is a lemma on the behavior of the virtual mass of an object under symmetrization which was first stated in [5] and later developed and generalized by Payne and Weinstein [14]. This lemma is an outgrowth of the original studies on symmetrization given by Pólya and Szegö [16]. It is used here in order to deduce the existence of a rectifiable solution of the extremal problem for cavity flow.

A preliminary analysis of the solution of the extremal problem is given in this paper by means of variational techniques which were developed originally for applications in the theory of conformal mapping [18], but which have already proved quite useful in studying physical problems [5, 19]. The main idea of the technique as it is used here is to make a variation of the independent variables by means of certain explicit conformal mappings. This method has occasionally been described as the method of interior variations because it involves mapping a boundary curve by a transformation which is regular on the curve, but has a pole in its interior.

One of the main difficulties encountered in generalizing the variational proof of existence of plane cavitating flow given in [5] to the case of axially symmetric flows is to show that the extremal free boundary which we obtain is an analytic curve. This difficulty has been overcome in the present paper by performing an analytic continuation of the steam function into the complex domain of the independent variables [9]; this allows us in turn to express the values of a certain complex analytic function on the free boundary in purely geometrical terms. The condition that an analytic function with these prescribed complex values on the free boundary does exist at all may be transformed into a functional equation connected with the geometry of the boundary. The solution of this functional equation yields the desired analyticity of the free boundary. Similar demonstrations of the analyticity of free boundaries for minimal surfaces and of free surfaces for plane flows in a gravity field have been given already [10, 11], and these earlier results were applied in a similar way to obtain the existence of plane cavitating flow in a gravity field [5].

Together with the existence of axially symmetric cavitational flow, we give a brief treatment of the uniqueness of the free boundary and the continuous dependence of the free surface on the point of separation and the cavitation parameter. Our results here are merely extensions of developments due to Lavrentieff [7] and Gilbarg [6], which were applied similarly in the existence proofs for plane free boundary flows given in [5]. The basic tool is a powerful comparison lemma which Lavrentieff [7] appears to have been the first to apply to cavities.

In summary, the principal new ideas appearing in this paper center about the formulation of the hydrodynamical problem as a problem in the calculus of variations, the analysis of this extremal problem by symmetrization and interior variations, the discussion of analyticity of the free surface by means of a functional equation developed in the complex domain, and the treatment of uniqueness by a comparison theorem parallel to Schwarz's lemma. The results of the paper are not only of use in current hydrodynamical questions, but also they may constitute a contribution to the study of free boundary extremal problems in the calculus of variations which arise in the theory of partial differential equations of elliptic type in two independent variables.

2. Formulation of the physical problem

Let there be given in the (x, y)-plane a simple curve C with the following properties. It is symmetric in the y-axis; it rises from a point $(-k_0, 0)$ on the negative x-axis monotonically to a point $(-k, h)$ in the second quadrant; then it remains horizontal up to the point (k, h); and thereafter is descends on a symmetric arc to the point $(k_0, 0)$ on the positive x-axis. The arc C can be represented by giving y as a positive non-increasing function of $|x|$.

We denote by B the object bounded by C and the x-axis. By L we denote a set of arcs outside B which lies in the half-strip $-k \leq x \leq k, y > 0$ and which bounds there, together with C, a set of regions W, which we term the cavity. We define the region D of the upper half-plane $y > 0$ as the complement of the closure of $B + W$ in the space $y > 0$, and we require it to be simply-connected.

In the three-dimensional space with cylindrical coordinates θ, y and x ($y =$ distance from x-axis) we introduce the axially symmetric domains B', W' and D' obtained by rotating B, W and D about the x-axis. These bodies are separated from one another by surfaces of revolution C' and L' generated by rotation of C and L about the x-axis. We shall be concerned here with the steady axially symmetric irrotational flow in D' of an incompressible fluid past the object $B' + W'$. We assume the constant density of the fluid to be $\rho = 1$. This flow has a velocity potential $\phi = \phi(x, y)$ and a stream function $\psi = \psi(x, y)$ which satisfy the generalized Cauchy-Riemann equations

$$(2.1) \qquad \frac{\partial \phi}{\partial x} = \frac{1}{y} \frac{\partial \psi}{\partial y}, \qquad \frac{\partial \phi}{\partial y} = -\frac{1}{y} \frac{\partial \psi}{\partial x}$$

in D and which fulfill on those portions of L and C which generate the surface of D' the equivalent boundary conditions

$$(2.2) \qquad \qquad \psi = 0$$

and

$$(2.3) \qquad \qquad \frac{\partial \phi}{\partial n} = 0,$$

where n denotes the outer normal of C and L with respect to the flow region D'.

Since ϕ and ψ do not depend on the angle θ, it will suffice in our discussions to treat the system of partial differential equations (2.1) in the plane region D rather than in the actual flow space D'. Thus we shall make very little explicit reference to the three-dimensional geometry of B', W', D', C', and L', and shall deal rather with a mathematical problem in two independent variables x and y.

We can eliminate ϕ in (2.1) to obtain for ψ the elliptic partial differential equation

$$(2.4) \qquad \frac{\partial^2 \psi}{\partial x^2} + \frac{\partial^2 \psi}{\partial y^2} - \frac{1}{y} \frac{\partial \psi}{\partial y} = 0,$$

while elimination of ψ yields

$$(2.5) \qquad \frac{\partial^2 \phi}{\partial x^2} + \frac{\partial^2 \phi}{\partial y^2} + \frac{1}{y} \frac{\partial \phi}{\partial y} = 0.$$

Equation (2.5) merely states that ϕ is an axially symmetric harmonic function in three-dimensional space. It is well known that any solution of (2.4) or (2.5) in a region of the half-plane $y > 0$ cannot assume its maximum or minimum value at an interior point of the region, unless it is identically constant. We shall refer to this statement as the maximum principle.

We assume that our flow past $B' + W'$ behaves at infinity like a uniform flow in the horizontal direction with velocity 1. Thus at large distances, ϕ and ψ have convergent expansions of the form

$$(2.6) \qquad \psi = \frac{y^2}{2} - \frac{ay^2}{r^3} + \cdots ,$$

$$(2.7) \qquad \phi = x + \frac{ax}{r^3} + \cdots ,$$

where $r^2 = x^2 + y^2$. Thus we imagine the fluid to be in motion with the object at rest.

The physical problem for which we have developed these preliminaries arises when the portion W' of the object $B' + W'$ consists of a cavity of steam and air generated by the rapid motion of the flow. While the shape of the body B', which we may describe as the projectile, is presumed known, the shape of the pocket of steam W' is not known. On the other hand, we may assume that the pressure p of the gas within W' is constant, since this gas in W' has an inertia which is negligible relative to that of the water in D', so that the kinetic energy of its motion has a negligible influence on the pressure in W' [15]. Thus on the surface L' between W' and D' the pressure p must be constant.

Since Bernoulli's law

$$(2.8) \qquad \tfrac{1}{2}(\nabla\phi)^2 + p = \text{const.}$$

holds throughout the flow, we deduce that the velocity $|\nabla\phi|$ of the fluid on the surface L' is constant. Thus on L we can write

$$(2.9) \qquad (\nabla\phi)^2 = \lambda,$$

or, equally well, by (2.1),

$$(2.10) \qquad\qquad (\nabla \psi)^2 = \lambda y^2.$$

We shall call the number λ the cavitation constant of the flow; it is a linear function of the difference between the pressure at infinity and the pressure on the free surface L'.

The essential difficulty in determining the flow just described lies in the fact that the shape of the cavity W, or, better, the shape of the curves L, is unknown. Thus we call L the free boundary of the flow. In order to find the free boundary, we must make use of the additional boundary condition (2.10).

The problem of constructing the free boundary L for a given fixed boundary C will be studied in the remainder of this paper. We shall find that for each value of the cavitation constant λ in a certain interval it is possible to construct a free surface L and a corresponding cavitational flow of the type described above. Alternatively, it will turn out that one can prescribe the point of separation of L from C.

3. The minimum problem

Our mathematical theory is based on an alternative formulation of the classical boundary value problem embodied in (2.2), (2.4), (2.6), and (2.10) as a variational problem [3, 5, 17]. We state the variational problem in this section and present a heuristic reasoning which indicates its relation to the boundary value problem.

A number of formal identities will prove to be quite useful. Let K denote the curve made up of arcs of C and L which bounds together with the x-axis the flow region D. If

$$\Psi = \frac{\alpha y^2}{r^3} + \cdots$$

is any solution of (2.4) in D which is regular and vanishes on the x-axis, then by Green's theorem we have

$$\int_K \Psi \frac{\partial \psi}{\partial n} \frac{1}{y} \, ds + \int_{S_r} \left\{ \Psi \frac{\partial \psi}{\partial n} - \psi \frac{\partial \Psi}{\partial n} \right\} \frac{ds}{y} = 0,$$

where s is the arc length and n is the outer normal along K and S_r with respect to the finite domain enclosed by K, S_r and the x-axis, and where S_r represents the arc of a large semi-circle of radius r in the upper half-plane. Letting r approach infinity, one finds in view of the normalization (2.6) of ψ and putting $y = r \sin \phi$

$$\lim_{r \to \infty} \int_{S_r} \left\{ \Psi \frac{\partial \psi}{\partial n} - \psi \frac{\partial \Psi}{\partial n} \right\} \frac{ds}{y} = \lim_{r \to \infty} \int_0^\pi \left[\frac{\alpha y^2}{r^3} \cdot y \sin \phi + \frac{1}{2} y^2 \cdot \frac{\alpha y^2}{r^4} \right] \frac{r d\phi}{y}$$

$$= \frac{3}{2} \alpha \int_0^\pi \sin^3 \phi \, d\phi = 2\alpha.$$

Thus

(3.1) $$2\alpha + \int_K \Psi \frac{\partial \psi}{\partial n} \frac{ds}{y} = 0.$$

We define the virtual mass M of the axially symmetric flow past $B + W$ by the formula

$$M = \iint_D \left(\nabla \psi - \nabla \frac{y^2}{2} \right)^2 \frac{dx \, dy}{y},$$

which is equivalent by (2.1) to

$$M = \iint_D (\nabla \phi - \nabla x)^2 \, y \, dx \, dy.$$

Clearly, $2\pi M$ represents the kinetic energy of the fluid motion induced if the body $B' + W'$ moves with velocity 1 against the fluid at rest. Since $\psi = 0$ on K, Green's theorem yields for this quantity the expression

$$M = -\frac{1}{2} \int_K y^2 \frac{\partial \psi}{\partial n} \frac{ds}{y} + \frac{1}{2} \int_K y^2 \frac{\partial y}{\partial n} \, ds.$$

The second integral on the right is proportional to the volume

$$2\pi V = \pi \int_K y^2 \, dx$$

enclosed by the surface of revolution generated by rotating K about the x-axis. This is merely the volume of $B' + W'$. Setting $\Psi = \psi - y^2/2$ in (3.1) and using the development (2.6), we obtain for the first integral on the right

$$\int_K y^2 \frac{\partial \psi}{\partial n} \frac{ds}{y} = -4a,$$

since ψ vanishes on K. Hence

(3.2) $$2a = M + V,$$

an identity which expresses the coefficient a in the expansion (2.6) in terms of the displaced mass and the virtual mass of the object $B + W$.

Let us now shift the curves L to a new position L^* by displacing them a small distance δn along their outer normals. This shift results in a corresponding variation into new domains D^* and K^* of D and K. In D^*, there is a varied axially symmetric flow with a stream function

$$\psi^* = \frac{y^2}{2} - \frac{a^* y^2}{r^3} + \cdots$$

and a virtual mass M^*, and K^* generates a surface of revolution with varied volume $2\pi V^*$. With accuracy of the first order in the magnitude of δn, we have immediately

$$(3.3) \qquad \delta V = V^* - V = -\int_L \delta n \, y \, ds.$$

On the other hand, setting $\Psi = \psi^* - \psi$ in (3.1), we find

$$2(a^* - a) = \int_K \psi^* \frac{\partial \psi}{\partial n} \frac{ds}{y}.$$

On K one has with accuracy of the first order in δn

$$\psi^* = -\frac{\partial \psi^*}{\partial n} \delta n = -\frac{\partial \psi}{\partial n} \delta n$$

by Taylor's theorem, since ψ^* vanishes on K^*. Thus the variation of the coefficient a is [19]

$$(3.4) \qquad \delta a = a^* - a = -\frac{1}{2} \int_L \left(\frac{\partial \psi}{\partial n} \right)^2 \delta n \frac{ds}{y}.$$

Combining (3.3) and (3.4) we obtain by (3.2) the formula for the variation of energy M

$$(3.5) \qquad \delta M = M^* - M = \int_L \left\{ 1 - \frac{1}{y^2} \left(\frac{\partial \psi}{\partial n} \right)^2 \right\} \delta n \, y \, ds.$$

If we write the additional boundary condition (2.10) which holds along the constant pressure free boundary for a cavitational flow in the form

$$(3.6) \qquad \frac{1}{y^2} \left(\frac{\partial \psi}{\partial n} \right)^2 = \lambda,$$

we see by (3.3) and (3.4) that the expression

$$2a - \lambda V$$

is stationary for shifts of the free boundary,

$$2\delta a - \lambda \delta V = 0.$$

Thus it is natural in order to show the existence of cavitational axially symmetric flows to attempt to choose the free curves L bounding W in such a way that

$$(3.7) \qquad 2a - \lambda V = \text{minimum}.$$

We shall show in the following, indeed, that for each suitable positive value of λ, the extremal problem (3.7) has a solution with L constrained to lie in the strip $-k \leq x \leq k$, $y > 0$ and outside B, and we shall prove that this solution yields a cavitational flow in which L appears as the free surface.

4. A lemma on symmetrization

Our treatment of the minimum problem (3.7) requires an analysis of the behavior of the virtual mass M of an arbitrary axially symmetric object under symmetrization in the y-axis and in the x-axis [5, 14].

To begin with, we shall suppose that the curve K bounding D is an analytic arc with at most a finite number of points of inflection. In the expression for M as an energy integral we have a factor y^{-1} which leads to difficulties for the symmetrization process. In order to remove it, we make the substitution $u = \psi y^{-1}$, and we obtain

$$M = \iint_D \left\{ y \left(\nabla u - \nabla \frac{y}{2} \right)^2 + \frac{1}{y} \left(u - \frac{y}{2} \right)^2 + \frac{\partial}{\partial y} \left(u - \frac{y}{2} \right)^2 \right\} dx \, dy$$

$$= \iint_D \left\{ y \left(\nabla u - \nabla \frac{y}{2} \right)^2 + \frac{1}{y} \left(u - \frac{y}{2} \right)^2 \right\} dx \, dy - \frac{1}{4} \int_K y^2 \, dx,$$

or, since $u - y/2 \to 0$ with $y \to \infty$,

$$(4.1) \qquad M = \iint_D \left\{ y \left(\frac{\partial u}{\partial x} \right)^2 + y \left(\frac{\partial u}{\partial y} - \frac{1}{2} \right)^2 + \frac{1}{y} \left(u - \frac{y}{2} \right)^2 \right\} dx \, dy - \tfrac{1}{2} V.$$

By the maximum principle, $\psi > 0$ in D and therefore also $u > 0$ there. Furthermore, by (2.4) we have

$$(4.2) \qquad \frac{\partial^2 u}{\partial x^2} + \frac{\partial^2 u}{\partial y^2} + \frac{1}{y} \frac{\partial u}{\partial y} = \frac{u}{y^2}.$$

It is convenient to define $u = 0$ in the region $B + W$ lying below K.

We now set $t = y^2/2$ and consider u as a function of x and t. We propose to symmetrize u in the x-axis, and to show that (4.1) does not increase under the symmetrization, whereas V is unchanged. To define the symmetrization, we note that for each fixed x and for each fixed positive number ρ, there is a finite odd number m of values of t, say $t_1 < t_2 < \cdots < t_m$, such that

$$(4.3) \qquad u(x, t_i) = \rho, \qquad\qquad i = 1, \cdots, m.$$

The number m is always finite because the surface $z = u(x, t)$ is piecewise analytic for all finite values of x and $t > 0$ and behaves at infinity like $(t/2)^{\frac{1}{2}}$. We set

$$(4.4) \qquad T = t_1 - t_2 + t_3 - \cdots + t_m.$$

It is easily seen that $T = T(\rho, x)$ is a continuous function of ρ for $0 < \rho < \infty$ and that it varies monotonically from a value $T_0(x) \geq 0$ to infinity if ρ increases from zero to infinity. In the domain D^* lying above the curve $t = T_0(x)$ we then define the symmetrized function $U(x, T)$ by the formula

$$(4.5) \qquad U(x, T) = \rho,$$

and we define $U(x, T) = 0$ elsewhere.

The surface $z = U(x, t)$ is the ordinary Steiner symmetrization of the surface $z = u(x, t)$ in the (x, z)-plane. The number $V = \int y^2 \, dx$ represents in the (x, t)-metric the area of the flat portion of the u-surface, that is, of the region $u(x, t) = 0$. The area of the corresponding plane section of the surface U has not been changed by the above process of symmetrization. We proceed to show that each of the three terms appearing in the integral (4.1) decreases, or, at least, does not increase, when we replace u by U and D by the region D^* in which $U > 0$.

For each fixed x we can replace u and U as functions of t and T by t and T as functions of z, the coordinate in the direction of u and U. The contribution of the first term in the integral (4.1) extended over the vertical line corresponding to this value of x alone is then of the form

$$\int y \left(\frac{\partial u}{\partial x}\right)^2 dy = \int \left(\frac{\partial u}{\partial x}\right)^2 dt = \int_0^\infty \sum_{i=1}^{m(z)} \left[\left(\frac{\partial t_i}{\partial x}\right)^2 \Big/ \left|\frac{\partial t_i}{\partial z}\right|\right] dz.$$

Here $m(z)$ is an odd integer which may change for a denumerable set of values $z_v(x)$ for which the curve $u(x, t) = z$ (x fixed) possesses a horizontal tangent. These exceptional points do not affect the value of the integral and may be disregarded. But by Schwarz's inequality,

$$\left(\sum_{i=1}^m (-1)^{i-1} \frac{\partial t_i}{\partial x}\right)^2 \leq \left(\sum_{i=1}^m \left[\left(\frac{\partial t_i}{\partial x}\right)^2 \Big/ \left|\frac{\partial t_i}{\partial z}\right|\right]\right)\left(\sum_{i=1}^m (-1)^{i-1} \frac{\partial t_i}{\partial z}\right),$$

since by the definition of the values $t_i(z)$

$$(-1)^{i-1} \frac{\partial t_i}{\partial z} \geq 0, \qquad\qquad i = 1, \cdots, m.$$

Hence by (4.4)

$$\int_0^\infty \left[\left(\frac{\partial T}{\partial x}\right)^2 \Big/ \frac{\partial T}{\partial z}\right] dz \leq \int_0^\infty \sum_{i=1}^m \left[\left(\frac{\partial t_i}{\partial x}\right)^2 \Big/ \left|\frac{\partial t_i}{\partial z}\right|\right] dz.$$

Integration of this inequality with respect to x yields finally, after a return to the original independent variables,

$$(4.6) \qquad\qquad \iint_{D^*} y \left(\frac{\partial U}{\partial x}\right)^2 dx \, dy \leq \iint_D y \left(\frac{\partial u}{\partial x}\right)^2 dx \, dy.$$

The later terms in the integrand of (4.1) can be integrated over any finite vertical segments, and since for large y, or t, the surface u is asymptotic to the cylinder $z = y/2$ and does not change under symmetrization, it will suffice to show that for these segments

$$\int y \left(\frac{\partial u}{\partial y}\right)^2 dy = \int \left(\frac{\partial u}{\partial t}\right)^2 t \, dt$$

is not increased by the symmetrization (4.3), (4.4), (4.5). We have

$$\int \left(\frac{\partial u}{\partial t}\right)^2 t \, dt = \int \sum_{i=1}^m \left[t_i \Big/ \left|\frac{\partial t_i}{\partial z}\right|\right] dz,$$

whence the inequality

$$\sum_{i=1}^{m} (-1)^{i-1} t_i \leq \left(\sum_{i=1}^{m} t_i \Big/ \left| \frac{\partial t_i}{\partial z} \right| \right) \left(\sum_{i=1}^{m} (-1)^{i-1} \frac{\partial t_i}{\partial z} \right),$$

trivial because $(-1)^{i-1} \frac{\partial t_i}{\partial z} = \left| \frac{\partial t_i}{\partial z} \right|$, yields

$$\int \left[T \Big/ \left| \frac{\partial T}{\partial z} \right| \right] dz \leq \int \sum_{i=1}^{m} \left[t_i \Big/ \left| \frac{\partial t_i}{\partial z} \right| \right] dz.$$

Integrating with respect to x and returning to x and y as independent variables, we find

(4.7)
$$\iint y \left(\frac{\partial U}{\partial y} \right)^2 dx \, dy \leq \iint y \left(\frac{\partial u}{\partial y} \right)^2 dx \, dy,$$

where the regions of integration are suitable bounded subsets of D^* and D outside of which $U \equiv u$.

We note that

(4.8)
$$\int \left\{ y \frac{\partial u}{\partial y} + u \right\} dy = \int \left\{ y \frac{\partial U}{\partial y} + U \right\} dy,$$

when the integrals are extended over any vertical lines from a position where u and U vanish up to a height beyond which $u \equiv U$. Also

(4.9)
$$\int \frac{u^2}{y} dy \geq \int \frac{U^2}{y} dy$$

over any such lines, since the area under the curve $z = u^2(y)$ is moved in the symmetrization away from the z-axis for each fixed x and is diminished. Integrating (4.8) and (4.9) with respect to x and combining with (4.6) and (4.7), we obtain

(4.10)
$$\iint_{D^*} \left\{ y \left(\frac{\partial U}{\partial x} \right)^2 + y \left(\frac{\partial U}{\partial y} - \frac{1}{2} \right)^2 + \frac{1}{y} \left(U - \frac{y}{2} \right)^2 \right\} dx \, dy$$

$$\leq \iint_{D} \left\{ y \left(\frac{\partial u}{\partial x} \right)^2 + y \left(\frac{\partial u}{\partial y} - \frac{1}{2} \right)^2 + \frac{1}{y} \left(u - \frac{y}{2} \right)^2 \right\} dx \, dy.$$

The object B lies in the domain where $U \equiv 0$, since the curve C is monotonic in each quadrant of the half-plane $y > 0$. The domain $U \equiv 0$ with B cut out will be defined to be the symmetrized cavity W^*. The volume V^* of $B + W^*$ is equal to V. We denote by

$$\psi^* = \frac{y^2}{2} - \frac{a^* y^2}{r^3} + \cdots$$

the stream function of the flow through the symmetrized region D^* and set $u^* = \psi^* y^{-1}$. The function $u^*(x, y)$ satisfies the differential equation (4.2) in the flow region D^*, vanishes on the boundary of D^* and behaves at infinity like $y/2$, that is, like U. Hence by Dirichlet's principle for (4.2) we have

$$
\begin{aligned}
(4.11) \quad & \iint_{D^*} \left\{ y \left(\frac{\partial u^*}{\partial x} \right)^2 + y \left(\frac{\partial u^*}{\partial y} - \frac{1}{2} \right)^2 + \frac{1}{y} \left(u^* - \frac{y}{2} \right)^2 \right\} dx\, dy \\
& \leqq \iint_{D^*} \left\{ y \left(\frac{\partial U}{\partial x} \right)^2 + y \left(\frac{\partial U}{\partial y} - \frac{1}{2} \right)^2 + \frac{1}{y} \left(U - \frac{y}{2} \right)^2 \right\} dx\, dy.
\end{aligned}
$$

Thus, by (4.10), the virtual mass M^* of the symmetrized object $B + W^*$ is smaller than the original virtual mass M,

$$
(4.12) \qquad\qquad M^* \leqq M.
$$

Therefore also

$$
(4.13) \qquad\qquad a^* \leqq a, \qquad V^* = V.
$$

The inequalities (4.12) and (4.13) can be derived for Steiner symmetrization of $B + W$ in the y-axis by a similar procedure. In this case, for each fixed y and positive ρ there are values $x_1, \cdots, x_m$, finite in number, such that

$$
(4.14) \qquad\qquad u(x_i, y) = \rho, \qquad\qquad i = 1, \cdots, m.
$$

We set

$$
(4.15) \qquad\qquad X = x_1 - x_2 + \cdots + x_m
$$

and define the symmetrized surface $z = U(X, y)$ by the formula

$$
(4.16) \qquad\qquad U(X, y) = \rho.
$$

Again we can use the coordinate z in the direction of u and U as independent variable, and we obtain in view of the trivial inequality $\sum \alpha_i \cdot \sum 1/\alpha_i \geqq 1$ valid for $\alpha_i > 0$,

$$
\begin{aligned}
(4.17) \quad & y \int \left(\frac{\partial U}{\partial x} \right)^2 dx = y \int \frac{\partial z}{\frac{\partial X}{\partial z}} = y \int \frac{dz}{\sum\limits_{i=1}^{m} \left| \frac{\partial x_i}{\partial z} \right|} \\
& \qquad\qquad\qquad \leqq \int \sum_{i=1}^{m} y \frac{dz}{\left| \frac{\partial x_i}{\partial z} \right|} = y \int \left(\frac{\partial u}{\partial x} \right)^2 dx,
\end{aligned}
$$

and

$$
(4.18) \qquad\qquad \frac{1}{y} \int \left(u - \frac{y}{2} \right)^2 dx = \frac{1}{y} \int \left(U - \frac{y}{2} \right)^2 dx.
$$

Since y is fixed in (4.14), (4.15) and (4.16), we see that the surface

$$
v(x, y) = u(x, y) - \frac{y}{2}
$$

is symmetrized by these formulas to yield the surface

$$V(X, y) = U(Y, y) - \frac{y}{2}.$$

We express now $X = \Omega(y, z)$, where $z = V(X, y)$. Similarly, we use $z = v(x, y)$ in order to express $x = w(y, z)$. Then, we find by Schwarz's inequality

$$
\begin{aligned}
y \int \left(\frac{\partial U}{\partial y} - \frac{1}{2}\right)^2 dx &= y \int \left(\frac{\partial V}{\partial y}\right)^2 dx = y \int \left[\left(\frac{\partial X}{\partial y}\right)^2 \bigg/ \left|\frac{\partial X}{\partial z}\right|\right] dz \\
&= y \int \left[\left(\sum_{i=1}^{m} (-1)^i \frac{\partial x_i}{\partial y}\right)^2 \bigg/ \sum_{i=1}^{m} \left|\frac{\partial x_i}{\partial z}\right|\right] dz \\
&\leq y \int \sum_{i=1}^{m} \left[\left(\frac{\partial x_i}{\partial y}\right)^2 \bigg/ \left|\frac{\partial x_i}{\partial z}\right|\right] dz = y \int \left(\frac{\partial v}{\partial y}\right)^2 dx \\
&= y \int \left(\frac{\partial u}{\partial y} - \frac{1}{2}\right)^2 dx.
\end{aligned}
$$

(4.19)

We integrate (4.17), (4.18) and (4.19) with respect to y and obtain (4.10) once more, with the function U and the domain D^* now interpreted as the symmetrizations of u and D in the y-axis. By Dirichlet's principle for (4.2) we find that (4.11) is again valid, and this yields (4.12) and (4.13) for symmetrization of $B + W$ in the y-axis, as was desired.

Thus far we assumed that the curve K bounding D is an analytic arc and has, therefore, finitely many inflection points. In order to remove this restriction, we will approximate a more general curve K from within D by analytic arcs and derive (4.13) by passage to the limit as the approximation becomes arbitrarily fine. The analytic arcs approximating K may be taken explicitly to be level curves of the Green's function for Laplace's equation in the domain consisting of D and its reflection in the x-axis.

In order to carry out the passage to the limit, we will use the following remarks. Let D_1 be a subdomain of D bounded by an arc K_1 and the two infinite segments of the x-axis. Let $\psi_1(x, y)$ be the stream function belonging to D_1 with the development

(4.20)
$$\psi_1(x, y) = \frac{y^2}{2} - \frac{a_1 y^2}{r^3} + \cdots$$

at infinity. Since $\psi_1(x, y)$ is a solution of (2.4), the difference function $\psi(x, y) - \psi_1(x, y)$ represents a solution of (2.4) which is finite and regular everywhere in D_1. This function vanishes at infinity and is non-negative on K_1; hence, we have by the maximum principle

(4.21)
$$\psi_1(x, y) \leq \psi(x, y) \qquad\qquad \text{in } D_1.$$

Since near infinity

(4.22)
$$\psi(x, y) - \psi_1(x, y) = \frac{(a_1 - a)y^2}{r^3} + \cdots$$

we can conclude from (4.21) also

$$(4.23) \qquad\qquad a \leqq a_1 .$$

Let now D_n be a sequence of such subdomains of D which converge increasingly to D. From the preceding remarks it follows that the corresponding stream functions $\psi_n(x, y)$ converge. Since for $n > k$ the maximum of $\psi - \psi_n$ in D_k is assumed on K_k and is $\leqq \psi$, and since ψ tends uniformly to zero on K_k as $k \to \infty$, it follows that for a fixed D_{k_0} the convergence of the ψ_n is uniform and toward ψ. Consequently, the coefficients a_n converge to a.

Let K be an arbitrary curve in the half-plane $y \geqq 0$ which bounds a flow region D. We can certainly find an analytic curve K_1 in D which determines a subregion $D_1 \subset D$ such that, for $\varepsilon > 0$ arbitrarily given, we have $a_1 < a + \varepsilon$. If we now symmetrize D and D_1 into two new regions D^* and D_1^*, it can easily be seen that $D_1^* \subset D^*$. Since we proved already that symmetrization decreases the coefficient a_1 of a domain with analytic boundary curves and because of (4.23), we have the chain of inequalities

$$(4.24) \qquad\qquad a^* \leqq a_1^* \leqq a_1 < a + \varepsilon.$$

Since this is true for every choice of $\varepsilon > 0$, we have

$$(4.13) \qquad\qquad a^* \leqq a$$

and have now proved that symmetrization diminishes the virtual mass for a general type of boundary curves K.

Let us recall finally some properties of the level curves $\psi = $ const. for a symmetrized body $B + W$. Let us suppose that the boundary curve K of $B + W$ is an analytic arc and that it is symmetrized in the y-axis. We clearly have $\psi(x, y) = \psi(-x, y)$ and hence $\partial\psi/\partial x = 0$ on the segment of the y-axis in D. On the x-axis in D we also have $\partial\psi/\partial x = 0$, while on the part of the curve K which lies in the quadrant $x > 0$, $y > 0$ we obviously have $\partial\psi/\partial x \geqq 0$. Hence, applying the minimum principle to the function $\partial\psi/\partial x$, which is a solution of (2.4), we find $\partial\psi/\partial x > 0$ in the whole common part of D and the first quadrant. The same inequality must also be valid if K is a curve of general type, since it can be approximated arbitrarily by symmetrized analytic curves and since $\partial\psi_n/\partial x \to \partial\psi/\partial x$ uniformly in each closed subdomain of the region considered. Thus, we find that the level curves $\psi = $ const. of an object $B + W$ symmetrized in the y-axis are themselves symmetrized in the y-axis.

If $B + W$ is symmetrized in the x-axis, we consider the function $(\partial\phi/\partial x) = (1/y)(\partial\psi/\partial y)$. This function is non-negative on the boundary of D, satisfies (2.5) and, hence, by the minimum principle we have $\partial\psi/\partial y > 0$ in D. This shows that now all level curves are symmetrized in the x-axis. Thus the level curves $\psi = $ const. for symmetrized bodies $B + W$ are monotonically rising in the second quadrant and monotonically descending in the first quadrant, a fact which is of considerable use in the applications.

5. Existence of an extremal configuration

For a given curve C of the type introduced in Section 1 enclosing a given object B, we seek to find curves L enclosing with C a shape W lying outside B, but lying within the strip $-k \leq x \leq k$, which possess the extremal property (3.7). We show in this section that rectifiable extremal curves of the desired type exist for each positive value of λ.

If, for a given λ, such an extremal curve could not be found, we would be able to select a sequence of curves L_n in the strip $-k \leq x \leq k$ and a sequence of flow regions D_n for which the corresponding expressions

$$2a_n - \lambda V_n$$

approach their greatest lower bound. We shall call such a sequence a minimal sequence. The object $B + W_n$ complementary to D_n can be assumed to be symmetrized in the x-axis and in the y-axis without loss of generality. For, if this were not the case, we could replace the minimal sequence $B + W_n$ by the sequence of objects obtained by symmetrizing them in the x-axis and y-axis. By Section 4, this would not increase the numbers a_n and V_n, and would thus leave us still with a minimal sequence. It is important to note here that symmetrization in the x-axis and y-axis, as described in Section 4, leaves the fixed object B invariant.

Symmetrized objects $B + W_n$ are intersected by each vertical or horizontal line in at most one segment. Hence the curves L_n and the related curves K_n bounding D_n are monotonic in each quadrant. If we consider a coordinate system σ, τ obtained by rotation of the (x, y)-plane through $45°$, we find that the representation

$$\tau = h_n(\sigma)$$

of the arc of K_n in the first quadrant $x > 0, y > 0$, satisfies the Lipschitz condition

$$| h_n(\sigma_2) - h_n(\sigma_1) | \leq | \sigma_2 - \sigma_1 |.$$

Thus the functions h_n are equicontinuous; we will show in the next paragraph that the curves K_n have bounded ordinates and, consequently, bounded values $h_n(\sigma)$. Hence we can select among them a uniformly convergent subsequence. Let us assume that the original sequence is uniformly convergent, with no loss of generality. It follows that the curves K_n and the regions D_n converge to a limit curve K and a limit domain D, and it follows that the object $B + W_n$ converges to a limiting object $B + W$.

We wish to show that W does not extend to infinity. To see this, we let

$$\psi_0 = \frac{y^2}{2} - \frac{a_0 y^2}{r^3} + \cdots$$

be the stream function for the flow past the unit disc $x = 0, 0 \leq y \leq 1$. This flow has a positive virtual mass and hence $a_0 > 0$. On the other hand, from

(3.4) we verify the well-known fact that the coefficient a_n decreases as the object $B + W_n$ diminishes. Thus if R_n denotes the ordinate of the highest point on K_n, we find that a_n is larger than the corresponding coefficient for the axially symmetric flow past a disc of radius R_n, since such a disc is contained within $B + W_n$. The stream function for the flow past the disc of radius R_n is easily derived from that for the unit disc, and is

$$R_n^2 \psi_0 \left(\frac{x}{R}, \frac{y}{R} \right) = \frac{y^2}{2} - \frac{R_n^3 a_0 y^2}{r^3} + \cdots ;$$

therefore

(5.1) $$a_n \geqq R_n^3 a_0 .$$

For the volume V_n we have obviously

(5.2) $$V_n \leqq 2k\pi R_n^2 ,$$

and hence a_n increases more rapidly than V_n as R_n tends to infinity. We conclude that R_n is bounded for a minimal sequence for (3.7), and hence W must be bounded.

Thus in order to show that for each positive λ there exists a set of monotonic extremal curves L in the strip $-k \leqq x \leqq k$ bounding a finite cavity W which is a solution of the minimum problem (3.7), we have only to prove that the stream functions ψ_n for the flows in D_n converge to the stream function ψ of the flow in D. Since the level curves of ψ_n descend in the first quadrant, $\partial \psi_n / \partial y > 0$ there. Hence by (2.4), ψ_n is subharmonic in the first quadrant. In the neighborhood of a point x_0, y_0 on K_n, ψ_n must therefore be dominated by the positive harmonic function

$$\text{Im } \{[i(x - x_0) - (y - y_0)]^{\frac{1}{2}}\},$$

since D_n does not intersect the vertical segment $0 \leqq y \leqq y_0$, $x = x_0$. A similar statement can be made concerning ψ, and we conclude by the maximum principle that if K_n has a maximal distance ε from K, then

$$|\psi - \psi_n| \leqq A\varepsilon^{\frac{1}{2}}$$

for a fixed positive number A. The desired convergence of ψ_n to ψ follows, and also we find that $a_n \to a$.

6. Interior variations

Once in possession of the rectifiable extremal curves L for (3.7) we can proceed to apply variational methods in order to show that they satisfy the constant pressure condition (2.10).

Let $z_0 = x_0 + iy_0$ be an interior point of an arc of L, and let $\rho > 0$ be so small that the circle

(6.1) $$|z - z_0| \leqq 2\rho$$

does not intersect B. Suppose that $F(z, \bar{z})$ is a suitably differentiable, complex-valued function of x and y which vanishes outside the circle (6.1). We make for sufficiently small complex values of the parameter ε the one-to-one transformation of coordinates

$$(6.2) \qquad z^* = z + \varepsilon F(z, \bar{z}).$$

This transforms L in the neighborhood of z_0 into a new free curve L^* and induces at the same time a variation

$$(6.3) \qquad \psi^{**}(x^*, y^*) = \psi(x, y)$$

of the function ψ, with $z = x + iy$, $z^* = x^* + iy^*$. The curve L^* bounds a varied flow region D^* with virtual mass M^*, and it bounds a varied object $B + W^*$ with volume V^*.

If ψ^* denotes the stream function of the normalized axially symmetric flow in D^*, then by Dirichlet's principle

$$M^* = \iint_{D^*} \left(\nabla \psi^* - \nabla \frac{y^2}{2} \right)^2 \frac{dx\, dy}{y}$$

$$\leqq \iint_{D^*} \left(\nabla \psi^{**} - \nabla \frac{y^2}{2} \right)^2 \frac{dx\, dy}{y},$$

where $\psi^{**} = \psi^{**}(x, y)$. But by (3.7) and (3.2)

$$M + V - \lambda V \leqq M^* + V^* - \lambda V^*,$$

whence

$$(6.4) \qquad (\lambda - 1) \left\{ \iint_{W^*} y\, dx\, dy - \iint_{W} y\, dx\, dy \right\}$$

$$\leqq \iint_{D^*} \left(\nabla \psi^{**} - \nabla \frac{y^2}{2} \right)^2 \frac{dx\, dy}{y} - \iint_{D} \left(\nabla \psi - \nabla \frac{y^2}{2} \right)^2 \frac{dx\, dy}{y}.$$

We introduce the symbols

$$\frac{\partial}{\partial z} = \frac{1}{2} \left(\frac{\partial}{\partial x} - i \frac{\partial}{\partial y} \right), \qquad \frac{\partial}{\partial \bar{z}} = \frac{1}{2} \left(\frac{\partial}{\partial x} + i \frac{\partial}{\partial y} \right)$$

and calculate the first order term in ε in the inequality (6.4), noting that the terms of order zero cancel. Since F vanishes outside the circle of radius 2ρ about z_0, the right-hand side of (6.4) can be written, after an integration by parts,

$$\iint_{\Omega^*} [(\nabla \psi^{**})^2 + y^2] \frac{dx\, dy}{y} - \iint_{\Omega} [(\nabla \psi)^2 + y^2] \frac{dx\, dy}{y},$$

where Ω is the intersection of D with the circle (6.1) and Ω^* is the image of Ω by the transformation (6.2). We find easily

$$\frac{\partial(x^*, y^*)}{\partial(x, y)} = 1 + 2 \operatorname{Re}\left\{\varepsilon \frac{\partial F}{\partial z}\right\} + 0(\varepsilon^2),$$

$$(\nabla \psi^{**})^2 = 4 \left|\frac{\partial \psi^{**}}{\partial z^*}\right|^2 = 4 \left|\frac{\partial \psi}{\partial z}\right|^2 - 8 \operatorname{Re}\left\{\varepsilon \frac{\partial F}{\partial z} \left|\frac{\partial \psi}{\partial z}\right|^2 + \varepsilon \frac{\partial F}{\partial \bar{z}} \left(\frac{\partial \psi}{\partial z}\right)^2\right\} + 0(\varepsilon^2)$$

$$y^* = y + \operatorname{Re}\left\{\frac{\varepsilon F}{i}\right\},$$

where $0(\varepsilon^2)$ denotes terms of the second order in ε, that is, terms such that $0(\varepsilon^2)/|\varepsilon^2|$ remains bounded as $\varepsilon \to 0$. Hence, if we replace in the integrals over D^* and W^* the variables x, y by $x^*(x, y)$ and $y^*(x, y)$ and integrate over D and W respectively, (6.4) can be written

$$(\lambda - 1) \operatorname{Re}\left\{\varepsilon \iint_W \left\{\frac{F}{i} + 2y \frac{\partial F}{\partial z}\right\} dx\, dy\right\}$$

$$\leqq \operatorname{Re}\left\{\varepsilon \iint_\Omega \left[\frac{F}{i} + 2y \frac{\partial F}{\partial z}\right] dx\, dy\right\} - 8 \operatorname{Re}\left\{\varepsilon \iint_\Omega \frac{\partial F}{\partial \bar{z}} \left(\frac{\partial \psi}{\partial z}\right)^2 \frac{dx\, dy}{y}\right\}$$

$$- 4 \operatorname{Re}\left\{\varepsilon \iint_\Omega \frac{F}{i} \left|\frac{\partial \psi}{\partial z}\right|^2 \frac{dx\, dy}{y^2}\right\} + 0(\varepsilon^2).$$

The circle (6.1) is transformed into itself by (6.2), and therefore the volume generated by rotating it about the x-axis is unchanged, whence

$$\iint_W \left\{\frac{F}{i} + 2y \frac{\partial F}{\partial z}\right\} dx\, dy + \iint_\Omega \left\{\frac{F}{i} + 2y \frac{\partial F}{\partial z}\right\} dx\, dy = 0.$$

Thus, finally,

$$(6.5) \quad \operatorname{Re}\left\{\varepsilon\lambda \iint_W \left[\frac{F}{i} + 2y \frac{\partial F}{\partial z}\right] dx\, dy + 4\varepsilon \iint_\Omega \frac{F}{i} \left|\frac{\partial \psi}{\partial z}\right|^2 \frac{dx\, dy}{y^2}\right.$$
$$\left. + 8\varepsilon \iint_\Omega \frac{\partial F}{\partial \bar{z}} \left(\frac{\partial \psi}{\partial z}\right)^2 \frac{dx\, dy}{y}\right\} + 0(\varepsilon^2) \leqq 0$$

for all sufficiently small ε.

Since ε is arbitrary in (6.5), we conclude in the usual way that the following variational identity holds

$$(6.6) \quad \lambda \iint_W \left[\frac{F}{i} + 2y \frac{\partial F}{\partial z}\right] dx\, dy + 4 \iint_\Omega \frac{F}{i} \frac{\partial \psi}{\partial z} \frac{\partial \psi}{\partial \bar{z}} \frac{dx\, dy}{y^2}$$
$$+ 8 \iint_\Omega \frac{\partial F}{\partial \bar{z}} \left(\frac{\partial \psi}{\partial z}\right)^2 \frac{dx\, dy}{y} = 0.$$

We now specialize (6.2) and (6.6) so as to obtain a more useful formula.

We let ω be a function of $|z - z_0|$ which is 1 for $|z - z_0| \leqq \rho$, which vanishes for $|z - z_0| \geqq 2\rho$, and which has continuous derivatives of all orders. We let t be a point with $|t - z_0| < \rho$, and for t in W we set

$$(6.7) \qquad F(z, \bar{z}) = \frac{1}{z - t}\,\omega.$$

For t in D, we denote by η a positive number so small that the circle $|z - t| \leqq \eta$ is contained in D and in the circle $|z - z_0| \leqq \rho$. We set

$$F(z, \bar{z}) = \frac{1}{z - t}\,\omega$$

for $|z - t| \geqq \eta$, as before, but for $|z - t| < \eta$, we define

$$(6.8) \qquad F(z, \bar{z}) = \frac{\bar{z} - \bar{t}}{\eta^2}.$$

The function F so defined is continuous, and (6.6) can be applied.

For t in W, we obtain by (6.6)

$$(6.9) \quad \lambda \iint_{W_0} \left[\frac{-i}{z - t} - \frac{2y}{(z - t)^2} \right] dx\,dy + 4 \iint_{\Omega_0} \frac{\partial \psi}{\partial z} \frac{\partial \psi}{\partial \bar{z}} \frac{dx\,dy}{iy^2(z - t)} = Q(t),$$

where $Q(t)$ is an analytic function in the circle $|t - z_0| < \rho$ which is defined by a definite integral over the ring $\rho \leqq |z - z_0| \leqq 2\rho$, where W_0 and Ω_0 are the intersections of W and D with $|z - z_0| < \rho$, and where the improper integral over W_0 is taken in the sense of the Cauchy principal value. The Cauchy principal value of a double improper integral of an integrand with point singularity at t is taken here to mean the limit of the same integral over the domain obtained by elimination of a circle about t with radius approaching zero. For t in D, we have

$$\lambda \iint_{W_0} \left[\frac{-i}{z - t} - \frac{2y}{(z - t)^2} \right] dx\,dy + 4 \iint_{\Omega_1} \frac{\partial \psi}{\partial z} \frac{\partial \psi}{\partial \bar{z}} \frac{dx\,dy}{iy^2(z - t)}$$

$$+ 4 \iint_{\Omega_2} \left\{ \frac{\bar{z} - \bar{t}}{i\eta^2} \frac{\partial \psi}{\partial z} \frac{\partial \psi}{\partial \bar{z}} \frac{1}{y^2} + \frac{2}{\eta^2} \left(\frac{\partial \psi}{\partial z} \right)^2 \frac{1}{y} \right\} dx\,dy = Q(t),$$

where Ω_2 is the circle $|z - t| \leqq \eta$ and where $\Omega_1 = \Omega_0 - \Omega_2$. Letting $\eta \to 0$, we find for t in D

$$(6.10) \quad \lambda \iint_{W_0} \left[\frac{-i}{z - t} - \frac{2y}{(z - t)^2} \right] dx\,dy + 4 \iint_{\Omega_0} \frac{\partial \psi}{\partial z} \frac{\partial \psi}{\partial \bar{z}} \frac{dx\,dy}{iy^2(z - t)}$$

$$+ 8\pi \frac{1}{y} \left(\frac{\partial \psi}{\partial z} \right)^2 \Big|_{z=t} = Q(t).$$

If l denotes the arc of L inside the circle $|z - z_0| \leqq \rho$, then by Green's theorem, for t in either Ω_0 or W_0,

$$\lambda \iint_{W_0} \left[\frac{-i}{z - t} - \frac{2y}{(z - t)^2} \right] dx\,dy = \lambda \iint_{W_0} \frac{\partial}{\partial z} \frac{2y}{z - t} dx\,dy = \frac{\lambda}{2i} \int_l \frac{2y\,d\bar{z}}{z - t} + Q_1(t),$$

where $Q_1(t)$ is an analytic function of t in the circle $|t - z_0| < \rho$ which is represented simply by the line integral on the right extended over arcs of $|z - z_0| = \rho$ bounding W_0, and where the integration over l is carried out in a direction such that D lies on the left. We denote by $\delta(t)$ the function which is 1 inside D and 0 outside D, and we set $Q_0(t) = iQ(t) - iQ_1(t)$, which is an analytical function of t throughout the circle $|t - z_0| < \rho$. The variational identities (6.9) and (6.10) can now be combined in the one formula

$$(6.11) \quad \lambda \int_l \frac{y\,d\bar{z}}{z - t} + 4 \iint_{\Omega_0} \frac{\partial\psi}{\partial z} \frac{\partial\psi}{\partial \bar{z}} \frac{dx\,dy}{y^2(z - t)} - \frac{16\pi}{t - \bar{t}} \left(\frac{\partial\psi}{\partial t}\right)^2 \delta(t) = Q_0(t),$$

valid for all complex values of t in $|t - z_0| < \rho$ which do not lie on l. This formula will be the basis of the following considerations.

7. Generalized boundary conditions

Our next objective is to obtain from (6.11) relations for the velocity $\partial\psi/\partial t$ as t approaches the arc l of the free boundary. In order to do this we must still rearrange the formula further by application of Green's theorem.

We will have to consider integrals of the form

$$(7.1) \quad \iint_{\Omega_0} \frac{\partial\psi}{\partial z} \frac{\partial\psi}{\partial \bar{z}} A(z, \bar{z}) \frac{dx\,dy}{z - t}, \qquad \iint_{\Omega_0} \frac{\partial\psi}{\partial z} \psi A(z, \bar{z}) \frac{dx\,dy}{z - t},$$

$$\iint_{\Omega_0} \frac{\partial\psi}{\partial \bar{z}} \psi A(z, \bar{z}) \frac{dx\,dy}{z - t}$$

with analytic coefficients $A(z, \bar{z})$ and to study their discontinuity character as the argument point t crosses the free boundary arc l. Since we may develop $A(z, \bar{z})$ into a power series in $z - t$ and $\bar{z} - \bar{t}$ near t and since integrals of the form

$$\iint_{\Omega_0} \frac{\partial\psi}{\partial z} \frac{\partial\psi}{\partial \bar{z}} \frac{\bar{z} - \bar{t}}{z - t}\,dx\,dy, \qquad \iint_{\Omega_0} \psi \frac{\partial\psi}{\partial z} \frac{\bar{z} - \bar{t}}{z - t}\,dx\,dy,$$

$$\iint_{\Omega_0} \psi \frac{\partial\psi}{\partial \bar{z}} \frac{\bar{z} - \bar{t}}{z - t}\,dx\,dy$$

are continuous by the Osgood-Lebesgue theorem as t crosses l, it is sufficient to study the character of the integrals (7.1) when $A(z, \bar{z})$ is replaced by the constant term $A(t, \bar{t})$ under the integral sign. We find easily

$$(7.1') \quad \iint_{\Omega_0} \frac{\partial\psi}{\partial \bar{z}} \psi \frac{dx\,dy}{z - t} = \frac{1}{2} \iint_{\Omega_0} \frac{\partial}{\partial \bar{z}} \left(\frac{\psi^2}{z - t}\right) dx\,dy = \frac{1}{4i} \int \frac{\psi^2}{z - t}\,dz - \frac{1}{2}\delta(t)\pi\psi(t)^2,$$

where the line integral is extended over the boundary arcs of Ω_0. Using (2.4), we have

$$(7.1'') \quad \iint_{\Omega_0} \frac{\partial\psi}{\partial z} \psi \frac{dx\,dy}{z - t} = \frac{1}{2i} \int \psi \frac{\partial\psi}{\partial z} \frac{\bar{z} - \bar{t}}{z - t}\,dz - \iint_{\Omega_0} \frac{\partial\psi}{\partial z} \frac{\partial\psi}{\partial \bar{z}} \frac{\bar{z} - \bar{t}}{z - t}\,dx\,dy$$

$$- \frac{1}{4} \iint_{\Omega_0} \psi \frac{\partial\psi}{\partial y} \frac{1}{y} \frac{\bar{z} - \bar{t}}{z - t}\,dx\,dy.$$

Finally, using again the differential equation (2.4) satisfied by ψ, we find

$$(7.1''') \qquad \iint_{\Omega_0} \frac{\partial \psi}{\partial z} \frac{\partial \psi}{\partial \bar{z}} \frac{dx\,dy}{z-t} = \frac{1}{2i} \int \psi \frac{\partial \psi}{\partial z} \frac{dz}{z-t} - \delta(t)\pi\psi(t) \frac{\partial \psi}{\partial t}$$

$$- \frac{1}{4} \iint_{\Omega_0} \psi \frac{\partial \psi}{\partial y} \cdot \frac{dx\,dy}{y(z-t)}.$$

Since

$$\frac{\partial \psi}{\partial y} = i\left[\frac{\partial \psi}{\partial z} - \frac{\partial \psi}{\partial \bar{z}} \right],$$

we may again reduce the integral on the right to the integrals (7.1') and (7.1'').

In all these transformations line integrals taken over the boundary of Ω_0 appear. In order to clarify their character, we make the following observations. On each level curve of ψ, the normal derivative of ψ is of one sign, since symmetrization showed the level curves of ψ to be monotonic in each quadrant. Hence

$$\int \left| \frac{\partial \psi}{\partial z} \right| |dz| = \int \left| \frac{\partial \psi}{\partial \bar{z}} \right| |dz|$$

can be estimated on each level curve in terms of the increment of ϕ, according to (2.1). Therefore the gradient of ψ has a uniformly bounded integral along such curves. Thus, if we evaluate the integrals (7.1) over a subregion of Ω_0 and let this subregion expand to fill Ω_0, we find that the line integrals occurring in the above transformations will approach zero for arcs tending to l, and they will as a consequence represent in the limit continuous functions in the circle $|t - z_0| < \rho$, since $\psi = 0$ on l.

Applying all these considerations to the particular integral which occurs as second term on the left in (6.11), we find

$$(7.2) \qquad 4 \iint_{\Omega_0} \frac{\partial \psi}{\partial z} \frac{\partial \psi}{\partial \bar{z}} \frac{dx\,dy}{y^2(z-t)} = \frac{16\pi\delta(t)}{(t-\bar{t})^2} \psi(t) \frac{\partial \psi}{\partial t} - \frac{4\pi\delta(t)}{(t-\bar{t})^3} \psi(t)^2 + Q_2(t),$$

where $Q_2(t)$ is a continuous function of t in the entire circle $|t - z_0| < \rho$. We can, therefore, replace (6.11) by the formula

$$(7.3) \qquad \lambda \int_l \frac{y\,d\bar{z}}{z-t} + \delta(t)\left\{ -\frac{16\pi}{t-\bar{t}} \left(\frac{\partial \psi}{\partial t}\right)^2 + \frac{16\pi}{(t-\bar{t})^2} \psi \frac{\partial \psi}{\partial t} - \frac{4\pi}{(t-\bar{t})^3} \psi^2 \right\} = q(t).$$

in which $q(t)$ is a continuous function of t in the circle $|t - z_0| < \rho$.

Since $\delta(t)$ vanishes for t in W and has the value 1 for t in D, the bracket in (7.3) must have a limit as t approaches l which is equal to the jump of the integral in (7.3) as t crosses l. We proceed to show that this jump exists almost everywhere along l.

On l, we define a function $A(z)$ by the relation

$$A(z) = \int_{z_0}^z y\,d\bar{z},$$

where the integration is to be carried out along l. We find that almost everywhere on l the derivative

$$(7.3') \qquad\qquad\qquad \dot{z} = \frac{\partial z}{\partial s}$$

exists and

$$(7.4) \qquad\qquad\qquad \frac{\partial A}{\partial s} = y\bar{\dot{z}},$$

since l consists of monotonic arcs. Let z_1 be a point on l satisfying $(7.3')$ and (7.4) and let $\dot{z}_1$ denote the tangent vector to l at that point. We know that l ascends monotonically in the second quadrant and descends monotonically in the first quadrant. Thus in each case points z on a line making an angle of $45°$ with the horizontal and intercepting l at $z_1 = x_1 + iy_1$, lie in Ω_0 for $y > y_1$ and in W_0 for $y < y_1$. Denote by t a point in Ω_0 which lies on such a line through z_1, and denote by

$$w = 2z_1 - t$$

the point in W_0 on the same line and at an equal distance

$$\varepsilon = |t - z_1|$$

from z_1. We wish to prove the jump condition

$$(7.5) \qquad \lim_{t \to z_1} \left\{ \int_l \frac{y\, d\bar{z}}{z - t} - \int_l \frac{y\, d\bar{z}}{z - w} \right\} = 2\pi i y_1 \bar{\dot{z}}_1^2,$$

where $z_1 = x_1 + iy_1$. By Cauchy's formula, we have immediately

$$\lim_{t \to z_1} \left\{ \int_l \frac{y_1 \bar{\dot{z}}_1^2\, dz}{z - t} - \int_l \frac{y_1 \bar{\dot{z}}_1^2\, dz}{z - w} \right\} = 2\pi i y_1 \bar{\dot{z}}_1^2.$$

Therefore an integration by parts shows that (7.5) is equivalent to

$$(7.6) \qquad \lim_{t \to z_1} \left\{ \int_l \left[\frac{A(z) - A(z_1)}{z - z_1} - y_1 \bar{\dot{z}}_1^2 \right] \frac{(w - t)(z - z_1)^2\, dz}{(z - t)^2 (z - w)^2} \right\} = 0.$$

We introduce a coordinate system σ, τ with its origin at z_1, and with the σ-axis inclined at $-45°$ with the x-axis if $x_1 > 0$, and at $+45°$ with the x-axis if $x_1 < 0$. If z and $\sigma + i\tau$ are corresponding points of l, and if $|\sigma| \leq 2\varepsilon$, then $|z - t| \geq 2^{-\frac{1}{2}}\varepsilon, |z - w| \geq 2^{-\frac{1}{2}}\varepsilon, |z - z_1| \leq 2^{\frac{1}{2}}\sigma$ and $|dz| \leq 2^{\frac{1}{2}}\, d\sigma$, while if $|\sigma| \geq \varepsilon^{\frac{1}{2}}$, then $|z - t| \leq \sigma, |z - w| \geq \sigma, |z - z_1| \leq 2^{\frac{1}{2}}\sigma$ and $|dz| \leq 2^{\frac{1}{2}}\, |d\sigma|$. Therefore the integral (7.6) is in absolute value smaller than a fixed constant times

$$\varepsilon^{-3} \int_{-2\varepsilon}^{2\varepsilon} \left| \frac{A(z) - A(z_1)}{z - z_1} - y_1 \bar{\dot{z}}_1^2 \right| \sigma^2\, d\sigma$$

$$(7.7) \qquad\qquad + \varepsilon \left\{ \int_{-\sqrt{\varepsilon}}^{-\varepsilon} + \int_{\varepsilon}^{\sqrt{\varepsilon}} \right\} \left| \frac{A(z) - A(z_1)}{z - z_1} - y_1 \bar{\dot{z}}_1^2 \right| \frac{d\sigma}{\sigma^2}$$

$$+ \varepsilon \int \left| \frac{A(z) - A(z_1)}{z - z_1} - y_1 \bar{\dot{z}}_1^2 \right| \frac{d\sigma}{\sigma^2} = I_1 + I_2 + I_3,$$

where the last integral I_3 is extended over the projection of l on the σ-axis outside the interval $-\sqrt{\varepsilon} \leqq \sigma \leqq \sqrt{\varepsilon}$. The first two integrals, I_1 and I_2, approach zero as $\varepsilon \to 0$ because

$$\lim_{z \to z_1} \left\{ \frac{A(z) - A(z_1)}{z - z_1} - y_1 \bar{z}_1^2 \right\} = 0$$

by (7.3') and (7.4). The last integral $I_3 \to 0$ as $\varepsilon \to 0$ because the difference quotient

$$\frac{A(z) - A(z_1)}{z - z_1}$$

is bounded. Hence (7.5) is proved and it follows from (7.3) and the continuity of $q(t)$ that

(7.8)
$$\lim_{z \to z_1} \left\{ -\frac{4}{y} \left(\frac{\partial \psi}{\partial z} \right)^2 + \frac{2}{iy^2} \psi \frac{\partial \psi}{\partial z} + \frac{1}{4y^3} \psi^2 \right\} = \lambda y_1 \bar{z}_1^2$$

for almost all z_1 on l, as z approaches z_1 from within D on a line inclined $45°$ to the x-axis.

The difference quotient

$$\frac{A(z) - A(z_1)}{z - z_1}$$

is bounded for all z_1 on l, even when (7.3') and (7.4) are not fulfilled at z_1. But then an estimate similar to that of the left side integral in (7.5) which leads to the integrals (7.7), shows in the general case that for each $\varepsilon > 0$ the bracket on the left in (7.8) is bounded when $|x| > \varepsilon$ and z is near L. This calculation is again based on (7.3), and the steps involve merely replacing $y_1 \bar{z}_1^2$ by 0 in previous formulas.

We conclude, since the bracket in (7.8) is a quadratic in $\partial \psi / \partial z$, that $\partial \psi / \partial z$ is bounded in Ω_0, if Ω_0 does not intersect the y-axis. Finally, since $\psi \to 0$ as $z \to l$, we deduce from (7.8) the generalized boundary condition

(7.9)
$$\lim \frac{2}{y} \frac{\partial \psi}{\partial z} = -i \sqrt{\lambda} \, \bar{z}$$

or

(7.10)
$$\lim 2 \frac{\partial \phi}{\partial z} = \sqrt{\lambda} \, \bar{z}$$

for z approaching almost all points on l along lines inclined at $45°$. This result, combined with the boundedness of the derivatives $\partial \psi / \partial x$ and $\partial \psi / \partial y$, is a substitute for the constant pressure condition (2.10) for the arc l of the free boundary.

The generalized boundary condition can be formulated also by stating that the normal derivative

(7.11)
$$-\frac{1}{y} \frac{\partial \psi}{\partial n} \to \sqrt{\lambda}$$

almost everywhere along any curve which approaches l in a suitable way, for example, on an infinitesimal translation of l. By integrating (7.11) and applying Lebesgue's convergence theorem to the formula

$$\phi = -\int \frac{\partial \psi}{\partial n} \frac{ds}{y},$$

we find that ϕ has on l the continuous boundary values

(7.12) $$\phi = \sqrt{\lambda}s.$$

8. Analyticity of the free boundary

On the free surface L we have by (2.2) and (7.12) the overdetermined analytic boundary condition

$$\phi + i\psi = s\sqrt{\lambda},$$

where s is arc length. We shall derive in this section from the known condition on the function $\phi + i\psi$ an analytic boundary condition, more complicated than the previous one, but which involves only an analytic function regular in the flow region. We shall be able to apply the new condition to obtain from its over-determined character the analyticity of the free boundary.

We set $z = x + iy$, $\bar{z} = x - iy$, $\zeta = \xi + i\eta$, $\bar{\zeta} = \xi - i\eta$, and we introduce the fundamental solution of (2.4),

$$S(z, \bar{z}; \zeta, \bar{\zeta}) = A(z, \bar{z}; \zeta, \bar{\zeta}) \log (z - \zeta)(\bar{z} - \bar{\zeta}) + B(z, \bar{z}; \zeta, \bar{\zeta}),$$

defined by the following properties. A and B are regular functions of x, y and ξ, η in the half-plane $y > 0$ or $\eta > 0$, and they are real for real values of x, y, ξ, η. S satisfies with respect to x, y the equation (2.4),

$$\frac{\partial^2 S}{\partial x^2} + \frac{\partial^2 S}{\partial y^2} = \frac{1}{y} \frac{\partial S}{\partial y},$$

and with respect to ξ, η it satisfies the adjoint equation

$$\frac{\partial^2 S}{\partial \xi^2} + \frac{1}{\eta} \frac{\partial S}{\partial \eta} + \frac{\partial^2 S}{\partial \eta^2} = \frac{S}{\eta^2}.$$

For $z = \zeta$ we have

$$A(z, \bar{z}; z, \bar{z}) \equiv 1,$$

and S is symmetric according to the rule

$$\sqrt{\frac{\zeta - \bar{\zeta}}{z - \bar{z}}} \, S(z, \bar{z}; \zeta, \bar{\zeta}) = \sqrt{\frac{z - \bar{z}}{\zeta - \bar{\zeta}}} \, S(\zeta, \bar{\zeta}; z, \bar{z}),$$

since the left side of equation (2.4) becomes self-adjoint if we multiply it by $y^{-\frac{1}{2}}$ and introduce $\psi y^{-\frac{1}{2}}$ as new dependent variable.

The function $A(z, \bar{z}; \zeta, \bar{\zeta})$ is merely the Riemann function for the hyperbolic equation

$$(8.1) \qquad \frac{\partial^2 \psi}{\partial z \partial \bar{z}} + \frac{1}{2(z - \bar{z})} \frac{\partial \psi}{\partial z} - \frac{1}{2(z - \bar{z})} \frac{\partial \psi}{\partial \bar{z}} = 0$$

equivalent to (2.4), and it has the explicit representation [2]

$$(8.2) \qquad A(z, \bar{z}; \zeta, \bar{\zeta}) = \frac{(z - \bar{\zeta})^{\frac{1}{2}}(\zeta - \bar{z})^{\frac{1}{2}}}{\zeta - \bar{\zeta}} F\left[-\tfrac{1}{2}, -\tfrac{1}{2}, 1, \frac{(z - \zeta)(\bar{z} - \bar{\zeta})}{(z - \bar{\zeta})(\bar{z} - \zeta)}\right],$$

where F is the hypergeometric series.

If Γ is any closed curve in D, we have by Green's formula

$$(8.3) \qquad \int_\Gamma \left\{ \psi \frac{\partial S}{\partial n} - S \frac{\partial \psi}{\partial n} \right\} \frac{ds}{y} = 0,$$

for points ζ outside the curve Γ. If we let one arc of Γ approach l, while restricting the remainder of Γ to lie in D but outside the circle $|z - z_0| < \rho$ enclosing l, we obtain from (8.3) by the boundary conditions (2.2) and (7.11)

$$(8.4) \qquad \int_l S(z, \bar{z}; \zeta, \bar{\zeta}) \, ds = P(\zeta, \bar{\zeta}),$$

for ζ in W_0, where $P(\zeta, \bar{\zeta})$ is a regular analytic function of ξ and η in W_0. This follows from an application of Green's theorem, and P merely stands for the integral (8.3) extended over the arcs of Γ lying in D after the limit process.

The identity (8.4) holds for real values of ξ, η in W_0, but by analytic continuation of S and P to complex values of ξ, η, we see that it holds also for any values of the independent complex numbers $\zeta = \xi + i\eta$ and $\bar{\zeta}$ replaced by $\zeta^* = \xi - i\eta$ with ζ and ζ^* in W_0. Indeed, the functions $A = A(z, \bar{z}; \zeta, \zeta^*)$ and $B(z, \bar{z}; \zeta, \zeta^*)$ are regular analytic functions of the independent complex numbers ζ and ζ^* in W_0, and S and P are defined in terms of A and B. Finally, we can let $\zeta^* \to z_0$ for each fixed ζ in W_0 to obtain there the relation

$$(8.5) \qquad \int_l S(z, \bar{z}; \zeta, z_0) \, ds = P(\zeta, z_0).$$

Since $S = A \log (z - \zeta)(\bar{z} - \bar{z}_0) + B$, we have to state the branch of the logarithm used in the formula. For this purpose, we draw an arbitrary curve from ζ to z_0 which lies in $|z - z_0| < \rho$ and which meets l only at z_0. Outside of this branch line the above logarithm is a single-valued function in the whole z-plane. We determine it in a unique way by requiring, for example, that its imaginary part at a given point $z_1 \, \varepsilon \, l$ lie between 0 and 2π (the latter value excluded).

For ζ in the intersection Ω_0 of D with the circle $|z - z_0| < \rho$, we define an analytic function $F(\zeta)$ by the formula

$$F(\zeta) = \int_l S(z, \bar{z}; \zeta, z_0) \, ds - P(\zeta, z_0).$$

We use again the above convention in the determination of S. Using (8.5), wish to find the boundary values of $F(\zeta)$ on l.

The function $P(\zeta, \bar{z}_0)$ and the integral

$$\int_l B(z, \bar{z}; \zeta, \bar{z}_0) \, ds$$

are continuous in ζ across l. On the other hand, the expression

$$\int_l A(z, \bar{z}; \zeta, \bar{z}_0) \log (z - \zeta)(\bar{z} - \bar{z}_0) \, ds$$

jumps by the integral along l

$$-2\pi i \int_{z_0}^t A(z, \bar{z}; t, \bar{z}_0) \, ds$$

as ζ crosses l into D at the point t, since A is continuous and the logarithm a determination which shifts by $2\pi i$ on the arc of l between z_0 and t. Here sign of ds has to be considered. It is determined by assuming that the arc le s increases if we proceed along the boundary curve with the region D on our Hence $F(\zeta)$ has on l the boundary values

$$(8.6) \qquad F(t) = -2\pi i \int_{z_0}^t A(z, \bar{z}; t, \bar{z}_0) \, ds,$$

where the integral is to be evaluated along l. This is the overdetermined ana boundary condition which we have been seeking. It gives the analytic func F explicitly along l as a geometric integral. While we may prescribe the re the imaginary part of an analytic function quite arbitrarily along any g curve l, it is not always possible to prescribe the whole analytic function a such an arc. Often such a boundary value problem is only solvable if the considered is analytic, and we shall show that the condition (8.6) will just to this consequence.

It is worth remarking, for the sake of the interested reader, that the ana function F represents essentially the analytic continuation of the stream func ψ into the four-dimensional, complex domain of two complex variables x y [9]. $F(t)(t - \bar{z}_0)$ is, up to a numerical factor, the stream function ψ or characteristic plane for (8.1) which passes through the point z_0 in the real don Thus (8.6) is nothing more than the explicit representation of ψ along characteristic in terms of the Riemann function A, and it could have been seen that such a representation would hold, since by (2.2) and (7.11) the Ca data for ψ are given explicitly along the free curve l.

For the applications, we wish to replace (8.6) by the condition obtained mally by differentiating it with respect to t. To do this, we differentiate (8.5) respect to ζ and for ζ in W_0 we obtain

$$(8.7) \qquad \int_l S_\zeta(z, \bar{z}; \zeta, \bar{z}_0) \, ds = P'(\zeta, \bar{z}_0),$$

where

$$S_\zeta = \frac{\partial S}{\partial \zeta} = \frac{A}{\zeta - z} + \frac{\partial A}{\partial \zeta} \log (z - \zeta)(\bar{z} - \bar{z}_0) + \frac{\partial B}{\partial \zeta}.$$

Similarly, in Ω_0

(8.8) $$F'(\zeta) = \int_l S_\zeta (z, \bar{z}; \zeta, \bar{z}_0) \, ds - P'(\zeta, \bar{z}_0).$$

The function $P'(\zeta, \bar{z}_0)$ and the integral

$$\int_l \frac{\partial B}{\partial \zeta} \, ds$$

are continuous in ζ across l, and the integral

$$\int_l \frac{\partial A}{\partial \zeta} \log (z - \zeta)(\bar{z} - \bar{z}_0) \, ds$$

jumps by

$$-2\pi i \int_{z_0}^t A_t(z, \bar{z}; t, \bar{z}_0) \, ds$$

when ζ crosses l at t, where $A_t = \partial A/\partial t$. On the other hand, we have by Green's theorem and (7.10), together with (2.5),

$$\int_l \frac{A(z, \bar{z}; \zeta, \bar{z}_0) \, ds}{z - \zeta} = \int_l \frac{A(z, \bar{z}; \zeta, \bar{z}_0) \bar{z} \, dz}{z - \zeta} = \frac{2}{\sqrt{\lambda}} \int_l \frac{A(z, \bar{z}; \zeta, \bar{z}_0)}{z - \zeta} \frac{\partial \phi}{\partial z} \, dz$$

(8.9) $$= \frac{4i}{\sqrt{\lambda}} \iint_{\Omega_0} \frac{\partial A}{\partial \bar{z}} \frac{\partial \phi}{\partial z} \frac{dx \, dy}{z - \zeta} - \frac{i}{\sqrt{\lambda}} \iint_{\Omega_0} \frac{A}{y} \frac{\partial \phi}{\partial y} \frac{dx \, dy}{z - \zeta}$$

$$+ \frac{4\pi i}{\sqrt{\lambda}} \delta(\zeta) A(\zeta, \bar{\zeta}; \zeta, \bar{z}_0) \frac{\partial \phi}{\partial \zeta} + P_1(\zeta),$$

where $P_1(\zeta)$ is a contour integral which is analytic for $| \zeta - z_0 | < \rho$. Therefore, by (7.10) and the boundedness of $\partial \phi / \partial z$,

$$\int_l \frac{A(z, \bar{z}; \zeta, \bar{z}_0) \, ds}{\zeta - z}$$

jumps at almost all points t on l by

$$-2\pi i A(t, \bar{t}; t, \bar{z}_0) \bar{t}.$$

By comparison of (8.7) and (8.8), we now conclude that for almost all t on l, $F'(t)$ has the boundary values

(8.10) $$F'(t) = -2\pi i A(t, \bar{t}; t, \bar{z}_0) \bar{t} - 2\pi i \int_{z_0}^t A_t(z, \bar{z}; t, \bar{z}_0) \, ds.$$

Furthermore, (8.9) shows that $F'(t)$ is bounded in Ω_0, because $\partial\phi/\partial t$ is bounded there, provided Ω_0 does not intersect the y-axis. Formula (8.10) is seen to be merely the derivative of (8.6).

We prefer to rewrite (8.10) in the form

$$(8.11) \quad \begin{cases} \bar{t} = -\dfrac{F'(t)}{2\pi i A(t,\bar{t};t,\bar{z}_0)} - \displaystyle\int_{z_0}^t \dfrac{A_t(z,\bar{z};t,\bar{z}_0)}{A(t,\bar{t};t,\bar{z}_0)}\,\bar{\bar{z}}\,dz, \\[2ex] \bar{t} = \bar{z}_0 + \displaystyle\int_{z_0}^t \bar{\bar{z}}^2\,dz, \end{cases}$$

valid for almost all t on l. We observe from (8.2) that

$$(8.2') \qquad\qquad A(t,\bar{t};t,\bar{z}_0) = \sqrt{\dfrac{t-\bar{t}}{t-\bar{z}_0}}$$

is an elementary function.

We introduce the system of two integral equations

$$(8.12) \quad \begin{cases} f(t) = -\dfrac{F'(t)}{2\pi i A(t,g(t);t,\bar{z}_0)} - \displaystyle\int_{z_0}^t \dfrac{A_t(z,g(z);t,\bar{z}_0)}{A(t,g(t);t,\bar{z}_0)}\,f(z)\,dz, \\[2ex] g(t) = \bar{z}_0 + \displaystyle\int_{z_0}^t f(z)^2\,dz, \end{cases}$$

for the determination of the unknown analytic functions $f(t)$ and $g(t)$ in $\Omega_0 + l$. These equations have a sense for analytic functions $f(t)$ and $g(t)$ in Ω_0, since by Cauchy's theorem the integrals on the right are independent of path in $\Omega_0 + l$. The objective now is to show that (8.12) has a unique solution f, g, since by (8.11) this solution would then have to agree with the known solution $\bar{\bar{t}}$, $\bar{t}$ on l. If this were the case, the analytic function $g(t)$ would have the continuous boundary values

$$g(t) = \bar{t}$$

on l, and thus we would have

$$\Phi = g(t) + t = \text{real}$$
$$\Psi = g(t) - t = \text{imaginary}$$

there. Denoting by $w(t)$ a conformal mapping of Ω_0 onto the upper half-plane, we check by the Schwarz principle of reflection that Φ and Ψ can be continued analytically across those segments of the real axis of the w-plane corresponding to l. As a result,

$$t(w) = \dfrac{\Phi - \Psi}{2}$$

is analytic on those segments and l consists of analytic arcs.

In the next section, we prove that (8.12) has a unique, analytic solution f, g

in a neighborhood of z_0 in $\Omega_0 + l$ by successive approximations, and we thus obtain the analyticity of the free boundary L.

The analyticity of ψ on the analytic free boundary L follows easily by the Cauchy-Kowalewski power series method. By this method, we can construct in the neighborhood of L a solution ψ_0 of (2.4) satisfying the boundary conditions (2.2) and (2.10) which is analytic on L. If we express $\psi - \psi_0$ as an integral along C and L using the elementary fundamental solution S, the integral along L will drop out, since $\psi - \psi_0 = 0$ and $\partial(\psi - \psi_0)/\partial n = 0$ there. Hence ψ_0 and $\psi - \psi_0$ are both analytic on L, and it follows that ψ is analytic there. An explicit formula for ψ which clearly exhibits its analytic character will be given in (12.3), providing an independent proof for this fact.

9. Solution of the functional equation

The functional equations (8.12) differ from the non-linear integral equations usually arrived at in the theory of ordinary differential equations by Picard's method only in that the independent variable t appears on the right, not only as an upper limit of integration, but also as an argument in the integrand and as an argument of g on the right. However, the standard procedure for discussing such problems can still be carried through, as we shall proceed to show [12].

Let us suppose at first that z_0 does not lie on the y-axis. We set $f_0 = 0$ and

$$g_0 = \bar{z}_0 + \int_{z_0}^{t} f_0(z)^2 \, dz = \bar{z}_0,$$

and we define sequences of analytic functions $f_n(t)$ and $g_n(t)$ recursively by the formulas

$$(9.1) \qquad f_{n+1}(t) = -\frac{F'(t)}{2\pi i A(t, g_n(t); t, \bar{z}_0)} - \int_{z_0}^{t} \frac{A_t(z, g_n(z); t, \bar{z}_0)}{A(t, g_n(t); t, \bar{z}_0)} f_n(z) \, dz,$$

$$(9.2) \qquad g_{n+1}(t) = \bar{z}_0 + \int_{z_0}^{t} f_{n+1}(z)^2 \, dz$$

wherever these integrals are defined. Let I denote an upper bound for

$$f_1(t) = -\frac{F'(t)}{2\pi i A(t, \bar{z}_0; t, \bar{z}_0)}$$

in a suitable neighborhood of $t = z_0$; it is possible to find such a number because $F'(t)$ is bounded near z_0 and $A(z_0, \bar{z}_0; z_0, \bar{z}_0) = 1$. It is readily verified that there exists a positive constant δ such that if

$$(9.2') \qquad |f_n| \leq I + \delta, \qquad |f_{n-1}| \leq I + \delta, \qquad |g_n - \bar{z}_0| \leq \delta,$$
$$|g_{n-1} - \bar{z}_0| \leq \delta, \qquad |z - z_0| \leq \delta, \qquad |t - z_0| \leq \delta,$$

then we have

$$|g_n(t) - g_{n-1}(t)| \leq N_1 \int_{z_0}^{t} |f_n(z) - f_{n-1}(z)| \, |dz|,$$

and therefore, by the analyticity of A,

$$\left| \frac{F'(t)}{2\pi i A(t, g_n(t); t, \bar{z}_0)} - \frac{F'(t)}{2\pi i A(t, g_{n-1}(t); t, \bar{z}_0)} \right| \leq N_2 \int_{z_0}^{t} |f_n - f_{n-1}| \, |dz|,$$

for suitable positive constants N_1 and N_2. Furthermore, for small enough $\delta > 0$, there are positive numbers N_3, N_4 and N_5 such that in addition

$$\left| \frac{A_t(z, g_n(z); t, \bar{z}_0)}{A(t, g_n(t); t, \bar{z}_0)} f_n(z) - \frac{A_t(z, g_{n-1}(z); t, \bar{z}_0)}{A(t, g_{n-1}(t); t, \bar{z}_0)} f_{n-1}(z) \right|$$

$$\leq N_3 \, |g_n(t) - g_{n-1}(t)| + N_4 \, |g_n(z) - g_{n-1}(z)| + N_5 \, |f_n(z) - f_{n-1}(z)|.$$

For paths from z_0 to t which have uniformly bounded length and lie in $\Omega_0 + l$, these Lipschitz conditions yield according to (9.1) the inequality

$$(9.3) \qquad |f_{n+1}(t) - f_n(t)| \leq N \int_{z_0}^{t} |f_n(z) - f_{n-1}(z)| \, |dz|,$$

for a suitably large positive value of N.

Let now Γ be any sufficiently short path from z_0 to t in $\Omega_0 + l$. Then the conditions (9.2') will be fulfilled for $n = 1$, and hence by (9.3)

$$|f_2(t) - f_1(t)| \leq N \int_{z_0}^{t} |f_1 - f_0| \, |dz| \leq INs,$$

where s is arc length from z_0 to t along Γ. For a short enough path Γ, (9.2') is fulfilled also for $n = 2$, and therefore

$$|f_3 - f_2| \leq N \int_{z_0}^{t} INs \, ds = IN^2 \frac{s^2}{2!}.$$

By the usual application of induction to (9.3), we obtain now in general

$$(9.4) \qquad |f_{n+1} - f_n| \leq I \frac{(Ns)^n}{n!}.$$

Therefore in a sufficiently small circle about z_0

$$|f_n(t)| \leq I e^{Ns},$$

$$|g_n(t) - \bar{z}_0| \leq I \frac{N_1}{N} \{e^{Ns} - 1\}.$$

Thus, we can find a uniform upper bound for the length of the path of integration Γ in order that the inequalities (9.2') be fulfilled for all values of n. This shows that for sufficiently short paths Γ the iteration (9.1), (9.2) is defined.

We deduce now from (9.4) that

$$\begin{cases} f(t) = f_1(t) + \displaystyle\sum_{n=1}^{\infty} \{f_{n+1}(t) - f_n(t)\}, \\ g(t) = g_0(t) + \displaystyle\sum_{n=0}^{\infty} \{g_{n+1}(t) - g_n(t)\} \end{cases}$$

converge uniformly in Ω_0 and uniformly almost everywhere along l to a solution of (8.12), provided that $|t - z_0|$ is sufficiently small, say, $|t - z_0| < \delta$. By (9.1), the iterated analytic functions $f_n(t)$, $g_n(t)$ are independent of the defining path of integration from z_0 to t, and therefore the same is true of the solution f, g. Weierstrass's convergence theorem shows by the uniform convergence of the above series that the solution is analytic near z_0 in Ω_0.

The solution is unique in the neighborhood of z_0. For if there were another bounded solution, $\bar{f}(t)$, $\bar{g}(t)$, of (8.12), then we could substitute f, f, g, g for f_{n-1}, f_n, g_{n-1}, g_n and $\bar{f}$, $\bar{f}$, $\bar{g}$, $\bar{g}$ for f_n, f_{n+1}, g_n, g_{n+1} in (9.1) and (9.2), in that order, and we could apply (9.3) to obtain

$$(9.3') \qquad \begin{cases} |\bar{f}(t) - f(t)| \leqq N \int_{z_0}^{t} |\bar{f} - f| \, |dz| \,, \\[2mm] |\bar{g}(t) - g(t)| \leqq N_1 \int_{z_0}^{t} |\bar{f} - f| \, |dz| \,, \end{cases}$$

for suitable large values of N and N_1. Consider now an arbitrary path of integration Γ from z_0 to t and let

$$m = \max \int_{z_0}^{\tau} |\bar{f} - f| \, |dz|$$

for τ on Γ between z_0 and t. Then by integration of (9.3') along Γ from z_0 to the point where the maximum is attained we find

$$m \leqq Nm \int_{z_0}^{t} |dz|.$$

This gives $m = 0$ along Γ and hence $\bar{f} = f$ almost everywhere, provided t is sufficiently near z_0. Consequently $\bar{g} = g$ also.

The function $g(t)$, as an integral of the bounded function f^2, must be continuous. By the uniqueness of the solution of (8.12) just shown, $g(t)$ must have on l near z_0 the boundary values $\bar{l}$, since by (8.11), $\bar{l}$ and $\bar{l}$ solve (8.12) along l. Thus by the argument given at the end of Section 8, l is near z_0 an analytic arc. Since z_0 can be taken as any interior point of the free boundary L not on the y-axis, this shows by the Heine-Borel theorem that the free boundary consists of analytic arcs.

The argument given thus far shows that the free boundary curve L is analytic in the neighborhood of each point z_0 not on the y-axis. In the proof, we have used the boundedness of the analytic function $F'(z)$ for values of z near L and outside a neighborhood of the y-axis. We next show that L is analytic in the neighborhood of its intersection with the y-axis, which we denote again by z_0. The result depends on showing, by means of the Phragmén-Lindelöf principle [13], that $F'(z)$ is bounded in D near z_0.

It is clear from (8.10) that $F'(z)$ is analytic and bounded on those arcs of L

for which $x \neq 0$, so that we may write $|F'(z)| \leq T_1$ there. We denote by $M(r)$ the quantity

$$(9.5) \qquad M(r) = \max_{\substack{z \in D \\ |z-z_0|=r}} |F'(z)|,$$

and we find by the Phragmén-Lindelöf principle that if $F'(z)$ is not bounded near z_0, then

$$(9.6) \qquad M(r) \geq T_2 \exp\{T_3 r^{-\frac{1}{2}}\},$$

where T_2 and T_3 are positive constants. By (8.8), $F'(z)$ differs in D from

$$\int_l \frac{A(t, \bar{t}; z, \bar{z}_0)\, ds}{z - t}$$

by a term which is bounded near z_0. On the other hand, there is a constant T_4 such that

$$\left| \int_l \frac{A(t, \bar{t}; z, \bar{z}_0)\, ds}{z - t} \right| \leq T_4 \int_l \frac{dx + |dy|}{|z - t|},$$

since l is monotonic on each side of the y-axis. Thus if ε denotes the shortest distance from z to l, we have

$$(9.7) \qquad |F'(z)| \leq \frac{T_5}{\varepsilon}$$

near z_0, for a suitable choice of T_5.

Let z be a point on $|z - z_0| = r$ such that $|F'(z)| = M(r)$. Then by (9.6) and (9.7),

$$(9.8) \qquad \varepsilon \leq T_6 \exp\{-T_3 r^{-\frac{1}{2}}\}, \qquad T_6 T_2 = T_5.$$

Thus if $|F'(z)|$ is unbounded near z_0, the points on $|z - z_0| = r$ where it is largest lie much closer to l than to the summit z_0.

We shall show that (9.8) is, in fact, impossible by estimating $F'(z)$ in terms of ε/r. Let z_1 be the point of l nearest to z on the line through z inclined at an angle of $45°$ with the x-axis, and let $w = 2z_1 - z$. We choose β to be a closed contour about z which consists of an open arc of l including z_0 and z_1 together with a curve in D whose shortest distance from z is of at least the order of magnitude r and which is composed of two level curves $\phi = $ const. and one level curve $\psi = $ const. We obtain in a manner similar to that leading from (8.3) to (8.8) the formula

$$F'(z) = -\frac{1}{\sqrt{\lambda}} \int_\beta \left\{ [S_z(t, \bar{t}; z, \bar{z}_0) - S_w(t, \bar{t}; w, \bar{z}_0)] \frac{\partial \psi}{\partial n} \frac{ds}{y} - \psi \frac{\partial}{\partial n}[S_z - S_w] \frac{ds}{y} \right\}.$$

This yields for $|F'(z)|$ the estimate

$$|F'(z)| \leq T_7 \int_\beta \frac{|w - z|\,|d\phi|}{|w - t|\,|z - t|} + T_8 \int_{\beta-l} \frac{|w - z|\,ds}{|w - t|^2\,|z - t|^2},$$

whence by the monotonicity of l in each quadrant

$$(9.9) \qquad |F'(z)| \leq T_9 \frac{\varepsilon}{r^4} + T_{10}.$$

The proof is analogous to the estimate performed in (7.7).

Since z was chosen so that $|F'(z)| = M(r)$, we have

$$M(r) \leq T_9 \frac{\varepsilon}{r^4} + T_{10}.$$

Combined with (9.6) and (9.8), this gives

$$T_2 \exp\{T_3 r^{-\frac{1}{2}}\} \leq T_6 T_9 r^{-4} \exp\{-T_3 r^{-\frac{1}{2}}\} + T_{10}.$$

For $r \to 0$, we arrive thus at a contradiction. Hence $F'(z)$ is, indeed, bounded in the neighborhood of z_0 on the y-axis, and therefore the previous discussion of the functional equation (8.12) shows the analyticity of L near z_0.

This completes the proof that the free boundary L is an analytic arc.

10. Uniqueness

While so far we have obtained the existence of axially symmetric cavitational flows using the minimum problem (3.7) and have shown the analyticity of the free boundary using (8.12), we wish now to show unique, continuous dependence of the flow region D on the cavitation constant λ. Basic for such a discussion will be the following lemma [5, 6, 7].

Let Ω be a region of the upper half-plane $y > 0$ bounded by a closed analytic curve Γ lying in $y > 0$. Corresponding to a positive solution of (2.4), let $u = \psi y^{-\frac{1}{2}}$ be a positive solution of

$$(10.1) \qquad \frac{\partial^2 u}{\partial x^2} + \frac{\partial^2 u}{\partial y^2} = \frac{3}{4y^2}\, u.$$

If u vanishes at a point t on Γ and has analytic boundary values on a sub-arc containing t, then at that point

$$(10.2) \qquad \frac{\partial u}{\partial n} < 0,$$

where n represents the outer normal of Γ.

For the proof, we map the domain Ω upon the upper half of the w-plane by a conformal mapping $w = f(z)$ which carries the point t into the origin. The differential equation (10.1) has in the w-plane the form

$$(10.1') \qquad \frac{\partial^2 u}{\partial \xi^2} + \frac{\partial^2 u}{\partial \eta^2} = p(\xi, \eta)u, \qquad w = \xi + i\eta,$$

where p is positive for $\eta > 0$ and analytic in both variables. The solution $u(\xi, \eta)$ is still analytic near the origin and has, therefore, a series development

$$u(\xi, \eta) = \sum_{\nu=1}^{\infty} P_\nu(\xi, \eta),$$

where $P_\nu(\xi, \eta)$ is a homogeneous polynomial in ξ, η of degree ν. In particular, because of the positive character of u in the upper half-plane, we have

$$P_1(\xi, \eta) = b\eta, \qquad\qquad b \geqq 0.$$

The statement of our lemma, that is, the inequality (10.2), is obviously equivalent to the inequality

(10.2′) $b > 0.$

Let us suppose that $b = 0$ and let n be the degree of the first non-vanishing polynomial $P_\nu(\xi, \eta)$ in the above development. From the differential equation (10.1′) follows

$$\frac{\partial^2 P_n}{\partial \xi^2} + \frac{\partial^2 P_n}{\partial \eta^2} = 0,$$

that is, $P_n(\xi, \eta)$ is a harmonic polynomial; it is homogeneous of degree $n \geqq 2$ and can be written in polar coordinates

$$P_n(\xi, \eta) = A \mid w \mid^n \cos (n\phi + \delta).$$

It changes sign along $2n$ rays through the origin forming equal angles π/n. Hence $P_n(\xi, \eta)$ becomes negative on rays through the origin with $\eta > 0$ and the same holds then also for $u(\xi, \eta)$ in view of the series development. Thus, the assumption $b = 0$ led to a contradiction. Inequality (10.2′) is, therefore, established and the statement of the lemma (10.2) follows.

The inequality (10.2) implies that any positive solution ψ of (2.4) in a region bounded by an analytic arc which vanishes at a point of that analytic arc must have there a negative outer normal derivative

(10.3) $$\frac{\partial \psi}{\partial n} < 0.$$

We utilize this lemma in order to describe the behavior of the free boundary at the point t of detachment. There are three possibilities; first, the free boundary may leave C at an interior point of its horizontal boundary, second, at the corner $(-k, h)$, and third, the flow domain is bounded by the arc of C over the interval $-k_0 \leqq x \leqq k$, a segment $h \leqq y < h_1$ of the vertical line $x = -k_0$ and a free streamline issuing from the point $(-k_0, h_1)$. In the first case, we will show that it is impossible to draw a segment issuing at t of positive slope which lies outside D. In the third case, it is impossible to draw a segment issuing at t which lies entirely in D.

In the first case, we make the observation, of interest in itself, that on the horizontal part of C between $(-k, h)$ and t

(10.4) $$\frac{1}{y^2} \left(\frac{\partial \psi}{\partial n} \right)^2 \geqq \lambda$$

holds. In fact, because of the minimum property of the flow region D, we must have

$$2\delta a - \lambda \delta V \geq 0$$

if we deform this horizontal piece by an upward normal shift ($\delta n < 0$). In view of the variational formulas (3.3) and (3.4), which are applicable in the case of the straight segment, we then obtain the inequality (10.4).

Now, denote by t' the first intersection of the hypothetical line through t of positive slope with the free boundary L of D. Let D' be the domain obtained from D by replacing the arc   of L by the segment tt' and let ψ' be its stream function. Since $D' \supset D$, we have $\psi' \geq \psi$ in D and, by (10.3), on the common horizontal boundary

$$\left(\frac{\partial \psi'}{\partial n}\right)^2 > \left(\frac{\partial \psi}{\partial n}\right)^2.$$

But it is easily seen that $\partial \psi'/\partial n$ tends to zero if we approach t along the horizontal. This fact contradicts (10.4) since obviously $\lambda > 0$, which proves our assertion.

Similarly, in the third case we use again the minimum property in order to show that

(10.4′) $$\frac{1}{y^2}\left(\frac{\partial \psi}{\partial n}\right)^2 \leq \lambda$$

holds on the vertical segment $x = -k$, $h < y < h_1$. Analogous reasoning then proves the assertion in the third case.

Note also that a free boundary cannot contain a horizontal segment because analytic continuation would show the extremal flow to be uniform parallel.

A next remark concerns the behavior of the stream function belonging to a domain D under a magnification from the origin as center. We denote the magnified domain by D^* and denote by α the factor of magnification. Denote by ψ and ψ^* the stream functions of the domains D and D^*, normalized at infinity to behave like $y^2/2$. We have

$$\psi^*(x^*, y^*) = \alpha^2 \psi\left(\frac{x^*}{\alpha}, \frac{y^*}{\alpha}\right)$$

and

(10.5) $$\frac{1}{y^*}\frac{\partial \psi^*}{\partial n} = \frac{1}{y}\frac{\partial \psi}{\partial n}$$

at corresponding points x^*, y^* and x, y on the boundaries of D^* and D, where $x^* = \alpha x$, $y^* = \alpha y$. Trivially, (10.5) also holds if magnification is combined with a translation parallel to the x-axis.

Next let D_1 and D_2 be two extremal flow regions for (3.7) with free boundaries L_1 and L_2 corresponding, respectively, to two values λ_1 and λ_2 of the cavitation parameter λ. Let us suppose that $D_1 \neq D_2$ and that we can magnify D_1 from the origin until it just includes D_2. We denote (as above) the magnification of D_1 by D_1^* and suppose that the factor of magnification is $\alpha < 1$. We want to prove the inequality

$$(10.6) \qquad\qquad \lambda_2 < \lambda_1 .$$

The free boundaries L_2 and L_1^* of D_2 and D_1^* must touch at some point t. This geometric lemma follows from the fact that these curves are monotonic arcs in each quadrant and that D_1 and D_2 belong to an identical fixed boundary C which consists within the strip $|x| \leq k$ of a horizontal segment. Let us suppose, at first, that t is an interior point of L_2 and L_1^* ; then the free boundaries are tangent at t. Since D_1^* includes D_2, the solution $\psi_1^* - \psi_2$ of (2.4) must be positive in D_2, by the maximum principle. Hence, by (10.3),

$$\frac{1}{y} \frac{\partial \psi_1^*}{\partial n} < \frac{1}{y} \frac{\partial \psi_2}{\partial n} \leqq 0$$

at $z = t$. By (10.5) this yields the asserted inequality (10.5$'$).

We need further considerations in order to deal with the case that t is not an interior point of both L_1^* and L_2. (a) t may be an interior point of L_1^*, but a point where L_2 detaches itself from the horizontal portion of C. (b) t may be an interior point of L_2 but a corner point of L_1^* which corresponds to a point of detachment of L_1 from a vertical segment. Finally, (c) t may be a point of detachment from a vertical segment for L_1^* and a point of detachment from a horizontal segment for L_2.

In case (a) L_1^* must necessarily have a horizontal tangent at t in view of the mentioned character of L_2 at its point of detachment.

Since L_1 is an analytic arc which is not a segment in the direction of the x-axis, such points with a horizontal tangent are isolated in each closed sub-arc of L_1^*. We magnify slightly more by a factor $\alpha_1 = \alpha(1 - \varepsilon)$ and shift the new region D_1^{**} to the left until its boundary L_1^{**} touches L_2 again. If ε is small enough, D_1^{**} will still contain D_2 and its free boundary L_1^{**} will now touch L_2 at a point near to the former contact point, and hence at a point with non-horizontal tangent. Thus, the contact will take place at an interior point of L_2, we will have tangency between L_1^{**} and L_2, and we derive again the inequality (10.6), as before. In case (b) L_2 would have to have a vertical tangent at t. Since the points with vertical tangent are isolated in the interior of L_2 we can again achieve by additional magnification and displacement in the negative x-direction that L_2 and L_1^* touch each other at interior points. Thus, the inequality (10.6) will be established in this case, too. The case (c) can easily be treated by the same method of additional magnification and translation as the preceding cases. This exhausts all possible configurations.

Hence, we have shown that if λ_1 and λ_2 are the cavitation parameters belonging to two domains D_1 and D_2 with free boundaries, and if D_1 can be magnified by a factor $\alpha < 1$ so that it contains D_2, then inequality (10.6) holds.

For small enough λ the extremal domain will reduce to the complement D_0 of B. In fact, let ψ_0 be the stream function for the flow past B. For any given λ with flow region $D \neq D_0$, we magnify D to include D_0, apply (10.3) and obtain

$$\left| \frac{1}{y} \frac{\partial \psi_0}{\partial n} \right| < \lambda^{\frac{1}{2}},$$

where the term on the left is to be evaluated at the intersection of C with the y-axis. This inequality shows that for sufficiently small λ the region D coincides with D_0, since by comparison with the stream function of a parallel flow in view of (10.3)

$$\left| \frac{1}{y} \frac{\partial \psi_0}{\partial n} \right| > 1.$$

This shows that if $D_2 = D_0$ and $D_1 \neq D_0$, then necessarily $\lambda_1 > \lambda_2$. For since D_0 is the extremum domain belonging to λ_2, we have by (3.3) and (3.4)

$$\frac{1}{y^2} \left(\frac{\partial \psi_0}{\partial n} \right)^2 \geq \lambda_2,$$

and as shown above

$$\lambda_1 > \frac{1}{y^2} \left(\frac{\partial \psi_0}{\partial n} \right)^2.$$

At the same time the preceding estimate on λ_2 shows that for sufficiently large values of λ the extremal domain cannot coincide with D_0 and hence has a free streamline.

As a first application of (10.6), we show that the cavitational flow solving the extremal problem (3.7) for a given λ is unique. If, indeed, there were two distinct flow regions D_1 and D_2 corresponding to the same value of λ, we could magnify at least one of them with $\alpha < 1$ until it just contained the other, and by (10.5) this would give the absurd inequality $\lambda < \lambda$.

A second application of (10.6) shows that the extremal flow region D for (3.7) diminishes steadily as λ increases, or, in other words, that the cavity W expands as λ increases. For the proof, suppose first that of two cavitational flow regions D_1 and D_2 neither one contains the other. Then each one could be magnified to just include the other with a factor $\alpha < 1$. Thus (10.6) shows that the parameters λ_1 and λ_2 for D_1 and D_2 would satisfy both of the inequalities

$$\lambda_1 < \lambda_2, \qquad \lambda_2 < \lambda_1,$$

a manifest contradiction. Thus one region, say D_2, includes the other, say D_1. We magnify D_1 to include D_2 and obtain (10.6), and this proves the desired monotonic dependence of D on λ.

Finally, we wish to show that D depends continuously on λ. Suppose that D_n is the flow region for a sequence λ_n of cavitation parameters which approach the limit λ. Since the flow region D is admissible for the extremal problem (3.7) corresponding to the value λ_n of the cavitation parameter, we have

$$2a - \lambda_n V \geqq 2a_n - \lambda_n V_n ,$$

in obvious notation. The volume V_n remains bounded, by (5.2), and therefore $\lambda_n \to \lambda$ implies

$$\varlimsup_{n \to \infty} \{2a_n - \lambda V_n\} \leqq 2a - \lambda V.$$

Hence the domains D_n are a minimal sequence for (3.7), and by Section 5 we can select a subsequence which converges to the unique solution D of (3.7). This means that D depends continuously on λ. Moreover, by a reasoning which will be detailed in the next section, it can also be shown that the free boundaries L_n of the domains D_n converge to the free boundary L of the limit domain D unless $D = D_0$.

Let λ^* be the upper limit of cavitation constants for which $D = D_0$; let λ_n be a sequence descending to λ^*. The domains D_n converge to D_0 and the limit of a sequence of free boundaries is a free boundary unless it reduces to a point. A nondegenerate segment of the horizontal part of C cannot be a free streamline. Hence, the limit of the free boundaries L_n is the point $(0, h)$ on the y-axis. Thus, when λ approaches λ^* from above the points of detachment of the free boundaries tend to the point $(0, h)$.

We can get an upper estimate on values of λ for which the extremal curve contains no vertical segment. We proceed as follows. Let γ_t be a smooth arc consisting of two vertical segments $x = t + \varepsilon$ and $x = t - \varepsilon, 0 \leqq y \leqq h$ joined at their upper end-points by the semi-circle S: $(x - t)^2 + (y - h)^2 = \varepsilon^2$, $y \geqq h$, with $0 < \varepsilon < k - |t|$. We consider the axially symmetric flow past the object belonging to γ_t and denote its stream function by ψ_t. Let $\mu(\varepsilon)$ be the minimum of the velocity of this flow on the semi-circle S.

For sufficiently large values of λ we can obviously find values of t and ε such that the curve γ_t lies outside of the flow region D. For such a value ε, taken fixed, we can take t so large that γ_t just touches the boundary of D. If the point of tangency lies on the free boundary L, then by (10.3)

$$\mu(\varepsilon) = \frac{1}{y}\left|\frac{\partial \psi_t}{\partial n}\right| \geqq \frac{1}{y}\left|\frac{\partial \psi}{\partial n}\right| = \lambda^{\frac{1}{2}}.$$

Thus, if $\lambda > \mu(\varepsilon)^2$ then γ_t cannot touch the free boundary, which must, therefore, detach itself from the vertical segments on the lines $|x| = k$ above $y = h$. This shows that for large λ, L must consist of two vertical segments on $x = \pm k$ together with a single free boundary arc joining the upper end-points of those segments.

We see that for small λ, L collapses onto C, while for large λ, L lies entirely above C. Since L depends continuously upon λ, we conclude that for at least one intermediate value of λ, the free boundary L consists of a single symmetric arc joining the intersection of C with $x = -k$ to the intersection of C with $x = +k$. Thus we obtain a model for an axially symmetric cavitational flow in which a prescribed nose (that portion of C lying in the half-plane $x \leqq -k$) is joined by a constant pressure free boundary L to a tail consisting of the reflection of the nose in the y-axis.

As a particular case of the above theory, we deduce the existence of a cavitational flow past two symmetric discs perpendicular to the x-axis of arbitrary size, with a pocket of steam between the discs enclosed by a free surface L. As the radius of the discs increases for fixed distance between them, the cavitation parameter λ increases, and the cavity expands.

11. The case of an infinite cavity

We have seen in the previous section that from a given nose generated by a monotonic curve we can obtain a cavitational flow with a free surface leading back to a tail which is the reflected image of the nose. While this model is satisfactory from the point of view of the experimentalist, it is nevertheless desirable to ask for a similar flow in which the tail is replaced by an infinite cavity. We shall find such a flow here by allowing the tail to tend towards infinity to the right and by performing a limit process on the finite cavity.

For the sake of simplicity, we assume here that the nose C is smooth enough so that the velocity

$$\frac{1}{y^2} \left(\frac{\partial \psi_0}{\partial n} \right)^2$$

is integrable along C after multiplication by the normal derivative of the Green's function of (2.5) for the exterior of a fixed object B.

Let D_n be a sequence of flow regions with the same nose C (changing slightly our earlier notation, we denote now by C the rising arc of the obstacle in the second quadrant), with free boundaries L_n, and with tails T_n which tend monotonically with n towards infinity to the right. We denote by λ_n the corresponding cavitation parameters. For $n < m$, we can translate D_n to the right and magnify it until it just includes D_m, and, indeed, this can be done in such a way that the free boundaries L_n and L_m just touch. Hence by (10.6) we have

(11.1) $\lambda_n > \lambda_m$,

and the numbers λ_m decrease to a limit λ.

In the coordinate system σ, τ obtained by a rotation of the usual system through 45°, we find, as in Section 5, that the representations $\tau = h_n(\sigma)$ of the monotonically ascending portions of the free curves L_n form an equicontinuous class of functions. We can select a subsequence of these functions which converges uniformly in every bounded region, and we assume without loss of generality that the original sequence has this property. Thus the curves L_n approach a monotonic limit curve L bounding a region D which is the limit of the flow regions D_n.

The stream functions ψ_n are equicontinuous in every closed subdomain of D, and hence a subsequence of them converges in D to a limit ψ which represents the stream function of an axially symmetric flow through D. We assume, again without loss of generality, that the original sequence ψ_n converges to ψ.

We wish to show that L, as the limit of free boundaries for the flows ψ_n, is a free boundary for the flow ψ. To show this, let z_0 be any interior point of L and let F be a suitably differentiable function which vanishes outside a circle $|z - z_0| \leq 2\rho$ so small that it does not intersect the lines $x = \pm k$, or $y = \pm h$. By (6.6) we obtain for each n the identity

$$
\text{(11.2)} \quad \lambda_n \iint_{W_n} \left[\frac{F}{i} + 2y \frac{\partial F}{\partial z} \right] dx\, dy + 4 \iint_{D_n} \frac{F}{i} \frac{\partial \psi_n}{\partial z} \frac{\partial \psi_n}{\partial \bar{z}} \frac{dx\, dy}{y^2}
$$
$$
+ 8 \iint_{D_n} \frac{\partial F}{\partial \bar{z}} \left(\frac{\partial \psi_n}{\partial z} \right)^2 \frac{dx\, dy}{y} = 0,
$$

where W_n is the cavity complementary to D_n. If we can show that $\partial \psi_n / \partial z$ is bounded for $|z - z_0| \leq 2\rho$ uniformly in n, then we shall be able to pass to the limit in (11.2) as $n \to \infty$ and $\psi_n \to \psi$ to obtain by Lebesgue's convergence theorem

$$
\text{(11.3)} \quad \lambda \iint_W \left[\frac{F}{i} + 2y \frac{\partial F}{\partial z} \right] dx\, dy + 4 \iint_D \frac{F}{i} \frac{\partial \psi}{\partial z} \frac{\partial \psi}{\partial \bar{z}} \frac{dx\, dy}{y^2}
$$
$$
+ 8 \iint_D \frac{\partial F}{\partial \bar{z}} \left(\frac{\partial \psi}{\partial z} \right)^2 \frac{dx\, dy}{y} = 0,
$$

where W is the infinite cavity for the flow ψ.

Let $\tilde{\psi}$ be the stream function for the axially symmetric flow past a fixed body B, normalized to behave like $y^2/2$ at infinity. Let ω be the harmonic function in 3-dimensional space which is regular outside the object generated by rotation of B about the x-axis (and vanishes, therefore, at infinity at least like r^{-1}) and assumes on that surface the boundary values

$$
\omega = \frac{1}{y^2} (\nabla \tilde{\psi})^2.
$$

Such ω exists since C has been assumed smooth for this purpose. The functions $(\nabla \phi_n)^2$ are subharmonic in the regions D_n' of space obtained by rotating the regions D_n about the x-axis, and they satisfy on the nose C

$$
(\nabla \phi_n)^2 \leq \frac{1}{y^2} (\nabla \tilde{\psi})^2 = \omega,
$$

by (10.3). On L_n,

$$
(\nabla \phi_n)^2 = \lambda_n \leq \lambda_1,
$$

and hence throughout the flow, since $(\nabla \phi_n)^2 = 1$ at infinity,

$$
(\nabla \phi_n)^2 \leq \omega + \lambda_1 + b_n + 1,
$$

by the maximum principle, where b_n is a term representing the estimate analogous to ω coming from the tail and tends to zero in the finite part of the plane.

The derivatives $\partial \psi_n / \partial z$ are thus uniformly bounded in the circle $| z - z_0 | \leqq 2\rho$, and (11.3) follows. The results of Sections 7, 8 and 9 now apply to L in the neighborhood of z_0, and we deduce that L is, indeed, an analytic, constant pressure, free boundary for the axially symmetric flow ψ.

We have thus obtained an infinite cavity W behind C as the limit of the finite cavities W_n. It is worth while to obtain a crude estimate on the geometrical nature of the infinite monotonic free streamline L (cf. [8]). Let us suppose that W contains a large circle with center on the x-axis, $| z - x_0 | < R$, whose circumference is tangent to L at $z = t$. Then if ψ_0 denotes the stream function of the flow past this circle, we get by (10.3)

$$\frac{1}{y^2} (\nabla \psi_0)^2 \geqq \lambda$$

at t. Explicit calculation gives

$$\frac{1}{y^2} (\nabla \psi_0)^2 = \frac{9}{4} \frac{y^2}{| z - x_0 |^2},$$

and hence if $t = t_1 + it_2$,

$$\frac{t_2^2}{R^2} \geqq \frac{4\lambda}{9}, \qquad \frac{(x_0 - t_1)^2}{R^2} \leqq 1 - \frac{4\lambda}{9},$$

and

$$\mu R \leqq R - (x_0 - t_1) \leqq t_1,$$

with

$$\mu = 1 - \left\{ 1 - \frac{4\lambda}{9} \right\}^{\frac{1}{2}} > 0.$$

But $t_2 \leqq R$, and therefore

(11.4) $$t_2 \leqq t_1/\mu.$$

This shows that L contains a sequence of points within the angle (11.4) which tend to infinity. In particular, the cavitational flow region D has points with arbitrarily large, positive abscissas. These results indicate that the cavity or wake, W recedes indefinitely from the nose C.

12. Summary and discussion of results

The principal result of Sections 4 through 10 is the existence of an axially symmetric cavitational flow leaving a prescribed nose surface generated by rotation of a monotonic curve about the x-axis and returning to a tail surface which is the reflection of the nose in a plane perpendicular to the x-axis. Section 4 shows in addition that the free boundary is generated by a curve which ascends to a maximum and then descends. The proof is based on the minimum virtual mass problem (3.7) and consists of five parts: demonstration of the existence of a minimum, application of symmetrization to get a smooth free boundary, interior variational analysis (6.11) leading to the boundary condition (7.9), proof of analyticity of the free boundary by means of the functional equations (8.12), and unique, continuous dependence on the cavitation constant.

In Section 11, for the same nose it is shown that there is a cavitational flow with an infinite wake rather than the tail of the Riabouchinsky model, which is of interest in view of the classical infinite wake theory of Helmholtz and Kirchhoff. This flow is obtained as a limiting case of the previous model.

The application of the integral equation (8.12) presented in Sections 8 and 9 has significance in its own right, since the analysis there shows that any constant pressure, free surface with rectifiable meridian curve for an axially symmetric flow with bounded velocity must be analytic. The method has quite general implications for overdetermined analytic boundary value problems in elliptic partial differential equations.

Further flow problems and free boundary extremal problems for linear elliptic partial differential equations in two independent variables can be treated by the methods developed in the present paper. We indicate here one of these applications.

One can carry through the existence proof for axially symmetric cavitational flow in the case where the flow' region D is bounded outside by a cylindrical tunnel, or pipe, $y = H$. The hypotheses on the object B and the cavity W are the same as before, except that in addition $B + W$ must lie always within the cylinder $0 \leqq y < H$. The line $y = H$ is a streamline, and the virtual mass M is defined exactly as before, except that D is now a region contained in the cylinder of radius H. The coefficient a no longer appears, but it suffices to replace (3.7) by the equally valid extremal problem

$$(12.1) \qquad\qquad M - (\lambda - 1)V = \text{minimum}.$$

Symmetrization, interior variations, the functional equation (8.12), and the comparison method for uniqueness can be carried through essentially in the same way as before, and the differences encountered for the new model are merely formal [5]. The flow described here occurs in actual experimental set-ups.

The above methods suggest essentially two lines of attack for calculating axially symmetric cavities. First, the minimum problem (3.7) and the uniqueness technique of Section 10 might allow one to approximate free streamline flows by a Ritz procedure.

Another, possibly more interesting, approach would start from the functional equations (8.12). Given any analytic curve L, we can calculate explicitly analytic functions $f(z)$ and $g(z)$ with the values $\bar{z}$ and $\bar{z}$ on L. Formulas (8.12) can then be used to calculate the analytic function $\lambda^{\frac{1}{2}}(z - \bar{z}_0)F(z)/8\pi i$, which merely represents the axially symmetric stream function ψ along a characteristic plane in the 4-dimensional, complex domain. Thus, for complex values of x, y with $z = x + iy$, $\bar{z}_0 = x - iy$, formulas (8.6) and (8.12) imply for z_0 on L

$$\psi(z, \bar{z}_0) = \lambda^{\frac{1}{2}} \frac{\bar{z}_0 - z}{4} \int_{z_0}^{z} A(\zeta, g(\zeta); z, \bar{z}_0) g'(\zeta)^{\frac{1}{2}} d\zeta,$$

with A defined by (8.2). Since $g(z_0) = \bar{z}_0$ on L, we obtain further

(12.2) $$\psi(z, g(z_0)) = \lambda^{\frac{1}{2}} \frac{g(z_0) - z}{4} \int_{z_0}^{z} A(\zeta, g(\zeta); z, g(z_0)) g'(\zeta)^{\frac{1}{2}} d\zeta.$$

This is an analytic identity in z_0 along L, and hence by analytic continuation it holds for all values of z_0 in a neighborhood of L. Let $w = g(z_0)$ in some region near L and let $z_0 = G(w)$ be the inverse of that analytic function. By (12.2), we find

$$\psi(z, w) = \lambda^{\frac{1}{2}} \frac{w - z}{4} \int_{G(w)}^{z} A(\zeta, g(\zeta); z, w) g'(\zeta)^{\frac{1}{2}} d\zeta.$$

In particular, for $w = \bar{z}$ we obtain in the real (x, y)-plane the real stream function

(12.3) $$\psi(z, \bar{z}) = \lambda^{\frac{1}{2}} \frac{\bar{z} - z}{4} \int_{G(\bar{z})}^{z} A(\zeta, g(\zeta); z, \bar{z}) g'(\zeta)^{\frac{1}{2}} d\zeta.$$

The stream function (12.3) yields an explicit example of an axially symmetric flow for which an arbitrary analytic curve L given by $\bar{z} = g(z)$ is a free streamline. This indirect method yields examples of cavitating flow, but it has the disadvantage that the fixed boundaries and singularities cannot be prescribed and may not be physically significant.

In order to get useful examples, one might try to use for the analytic curve L the exact free streamlines connected with the known two-dimensional cavitating flow patterns. In this case, the analytic functions f and g are easily computed in terms of the complex potential of the plane flow.

APPLIED MATHEMATICS AND STATISTICS LABORATORY
STANFORD UNIVERSITY

The following list contains references to papers known to us on axially symmetric flow plus papers from which we have borrowed ideas; we do not pretend to cover the extensive bibliography of plane cavity flow.

REFERENCES

1. BERGMAN, S. and SCHIFFER, M., Kernel functions in mathematical physics, Academic Press, New York, 1952.
2. DARBOUX, G., Leçons sur la théorie générale des surfaces, vol. 2, Paris, 1915.
3. FRIEDRICHS, K., Über ein Minimumproblem für Potentialströmungen mit freiem Rande, Math. Ann., vol. 109 (1933–34), pp. 60–82.
4. GARABEDIAN, P. R. and ROYDEN, H., A remark on cavitational flow, Proc. Nat. Acad. Sci. U. S. A., vol. 38 (1952), pp. 57–61.

5. GARABEDIAN, P. R. and SPENCER, D. C., *Extremal methods in cavitational flow*, Journal of Rational Mechanics and Analysis, vol. 1 (1952), pp. 359–409. Technical Report No. 3, Appl. Math. and Stat. Lab., Stanford Univ., 1951.

6. GILBARG, D., *Uniqueness of axially symmetric flows with free boundaries*, Journal of Rational Mechanics and Analysis, vol. 1 (1952), 309–320.

7. LAVRENTIEFF, M., *On certain properties of univalent functions and their application to wake theory* (in Russian), Rec. Math. (Math. Sbornik) N. S., vol. 46 (1938), pp. 391–458.

8. LEVINSON, N., *On the asymptotic shape of the cavity behind an axially symmetric nose moving through an ideal fluid*, Ann. of Math., vol. 47 (1946), pp. 704–730.

9. LEWY, H., *Neuer Beweis des analytischen Charakters der Lösungen elliptischer Differentialgleichungen*, Math. Ann., vol. 101 (1929), pp. 609–619.

10. LEWY, H., *A note on harmonic functions and a hydrodynamical application*, Proc. Amer. Math. Soc., vol. 3 (1952), pp. 111–113.

11. LEWY, H., *On steady free surface flow in a gravity field*, Comm. Pure Appl. Math., to appear.

12. LEWY, H., *A theory of terminals and the reflection laws of partial differential equations*, to appear.

13. NEVANLINNA, R., Eindeutige analytische Funktionen, Berlin, 1936.

14. PAYNE, L. E. and WEINSTEIN, A., *Capacity, virtual mass, and generalized symmetrization*, Pacific Journal of Mathematics, to appear.

15. PLESSET, M. S. and SHAFFER, P. A., *Drag in cavitating flow*, Reviews of Modern Physics, vol. 20 (1948), pp. 228–231.

16. PÓLYA, G. and SZEGÖ, G., Isoperimetric inequalities of mathematical physics, Princeton, 1951.

17. RIABOUCHINSKY, D., *Sur un problème de variation*, C. R. Acad. Sci. Paris, vol. 185 (1927), pp. 840–841.

18. SCHIFFER, M., *Variation of the Green function and the theory of the p-valued functions*, Amer. J. Math., vol. 65 (1943), pp. 341–360.

19. SCHIFFER, M. and SZEGÖ, G., *Virtual mass and polarization*, Trans. Amer. Math. Soc., vol. 67 (1949), pp. 130–205.

Commentary on

[44] P.R. Garabedian, H. Lewy, and M. Schiffer, *Axially symmetric cavitational "ow*, Ann. of Math. (2) **56** (1952), 560-602.

This well-known paper is concerned with axially symmetric cavitational flow in three dimensions. Such flow occurs when a body moves so quickly in water that the fluid pressure falls to vapor pressure and a cavity of steam develops behind the body. If the body is axially symmetric and moves in the direction of the axis, and if the fluid (and cavity) fill all of $\mathbb{R}^3$, then the flow is also symmetric about the axis. (For plane cavitational irrotational flow in two dimensions, there had been much earlier work.)

This basic paper is the first to provide existence of axially symmetric cavitational flow. The problem is treated as a variational one, in which the boundary of the cavity is a free a priori unknown boundary surface. The paper combines a number of basic arguments and techniques, including symmetrization and interior variations, i.e., a variation of the independent variables here via explicit conformal mappings.

It proves that the free boundary is analytic by extending the stream function analytically to complexified values of the independent variables. A functional equation is then obtained, in the complex domain, related to the geometry of the free boundary. Its solution yields the analyticity of the boundary. Uniqueness and continuous dependence on the "cavitation constant" are proved using a comparison lemma similar to Schwarz' lemma. In fact, this lemma follows easily from the Hopf boundary lemma for elliptic equations [H], which was also published in 1952.

References

[H] Eberhard Hopf, *A remark on linear elliptic differential equations of second order*, Proc. Amer. Math. Soc. **3** (1952), 791 793.

[N] Louis Nirenberg, *Comments on some of Hans Lewy s work*, Hans Lewy Selecta, Vol. 1, ed. D. Kinderlehrer, Birkhäuser, 2002, pp. xlv xlix.

LOUIS NIRENBERG
(EXCERPTED FROM [N])

[54] Variation of domain functionals

[54] Variation of domain functionals. *Bull. Amer. Math. Soc.* **60** (1954), 303–328.

VARIATION OF DOMAIN FUNCTIONALS[1]

M. SCHIFFER

1. **Introduction.** Let D be a domain in an m-dimensional space and $\phi[D]$ be a number which depends on the domain. We shall call $\phi[D]$ a domain functional. Insofar as D is also characterized by its boundary C or its complement $\tilde{D}$, we may sometimes consider the functional $\phi[D]$ as dependent on C or $\tilde{D}$ instead of D. The problem with which we are occupied is to establish a calculus for the functional with respect to changes of its variable; in other words, a calculus of variations for $\phi[D]$.

We give the following examples of domain functionals: The volume $V[D]$ of the domain and the surface area $A[C]$ of its boundary which are elementary geometric concepts. Let ∇^2 be the Laplace operator and consider the eigenvalue problem $\nabla^2 u + \lambda u = 0$ with the boundary condition $u = 0$ on C. The eigenvalues $\lambda_\nu[D]$ of this problem are often studied domain functionals. The corresponding eigenfunctions $u_\nu(x)$ taken at a fixed point x may be considered likewise as domain functionals. Finally, we mention the Green's function of D with respect to Laplace's equation $G(x, \xi)$. If we consider both its argument points x and ξ as fixed parameters, it becomes a functional of D.

The first domain functionals of nonelementary character to be studied seem to have been the eigenvalues $\lambda_\nu(D)$ of a plane domain [17] and the electrostatic capacity of a three-dimensional conductor [15]. Since both these quantities are closely related to the Green's function of the domains considered with respect to Laplace's equation, their theory can be reduced to that of the Green's function. Hadamard treated $G(x, \xi)$ from the general point of view of functional analysis and gave for the first time a systematic basis to the whole problem complex [7].

In order to study the dependence of a functional on its variable, we have, in the spirit of classical analysis, to investigate how the functional changes under an infinitesimal change of the domain. Hadamard assumed that the domain D is bounded by a closed smooth surface C. He deforms C into a surface C^* by pushing each point ζ on C by an amount δn in the direction of the exterior normal. Under

An address delivered before the Pasadena meeting of the Society on November 28, 1953, by invitation of the Committee to Select Hour Speakers for Far Western Sectional Meetings; received by the editors December 15, 1953.

[1] This work was done under a contract with the Office of Naval Research.

303

certain restrictive assumptions on δn, we can assert that C^* is also a smooth surface which bounds a domain D^* with a Green's function $G^*(x, \xi)$. If $\delta n > 0$, $G^*(x, \xi)$ is well-defined in D and, by Green's identity, we find

$$(1.1) \quad -\iint_C \left[G^*(\zeta, x) \frac{\partial G(\zeta, \xi)}{\partial n} - G(\zeta, \xi) \frac{\partial G^*(\zeta, x)}{\partial n} \right] d\sigma$$

$$= G^*(x, \xi) - G(x, \xi).$$

Now, $G(\zeta, \xi) = 0$ on C and

$$G^*(\zeta + \delta\zeta, \xi) = G^*(\zeta, \xi) + (\partial G^*/\partial n)\delta n + \cdots = 0.$$

Hence obviously

$$(1.2) \qquad \delta G(x, \xi) = \iint_C \frac{\partial G(\zeta, x)}{\partial n} \frac{\partial G(\zeta, \xi)}{\partial n} \delta n d\sigma.$$

This is Hadamard's variational formula for the Green's function.

Though the above derivation is purely formal, it can be justified if the surface C is regular enough and many interesting applications can be made. Since $\partial G/\partial n < 0$ on C, we recognize, for example, that $\delta G > 0$ if $\delta n > 0$, that is the monotony of the Green's function in dependence of the domain. Our formula would, at first, guarantee this monotonicity only for domains with sufficiently regular boundary. But it is easy to extend this result to the most general domains, since every domain can be approximated arbitrarily by domains whose boundary is even analytic.

Much deeper results can be derived by similar considerations; each time when we have a functional of D which is a combination of Green's functions and various partial derivatives of Green's functions such that the coefficient of $\delta n d\sigma$ in the variational formula is positive, we arrive at a monotonicity theorem for this functional.

Similar variational formulas can be easily established for the Green's functions of other partial differential equations of elliptic type, for higher orders, and for different boundary conditions. Hadamard studied, in particular, the Green's functions for the bi-harmonic equations which play a central role in the theory of elasticity of plates. But while the theory of boundary value problems in partial differential equations proceeded to more general domains, D, the variational method was inherently tied to a very restricted class of regular domains. Even very simple extremum problems for domain functionals could not be solved by variational methods, since one is not sure, a priori, that the sought extremum domain is regular enough to admit a variational formula. It is clear that the difficulty lies in the kinematics of our variation. Since we describe the transition from one domain to its neighbor by a normal shift, it is obvious that the normal at the varied points of C will play a role in the variational formula, and that we shall, a priori, exclude all domains D from the theory whose boundary C does not possess a normal at each point.

2. Theory of univalent functions. The theory of domain functionals was applied in a precise and successful manner at first in the theory of conformal mapping. Let Δ be a domain in the complex z-plane which contains the point at infinity. We call a domain D in the complex w-plane conformally equivalent to Δ if there exists in Δ a univalent analytic function which has near $z = \infty$ the development

$$(2.1) \qquad w = f(z) = z + a_0 + \frac{a_1}{z} + \cdots$$

and maps Δ onto D. Since D determines $f(z)$ in a unique way, we may consider all coefficients $a_\nu[D]$ as domain functionals of D and pose various extremum problems with respect to them. de Possel [16] asked for a domain D which is conformally equivalent to Δ and for which

$$(2.2) \qquad \operatorname{Re}\left\{a_1[D]\right\} = \max.$$

The existence of such a domain is insured by theorems on the normality of the family of univalent functions with the development (2.1) at infinity. There is only the question of characterizing the extremum function. de Possel showed that the extremum domain D consists of the whole w-plane slit along rectilinear segments parallel to the real axis. Since the existence of the extremum domain is sure, this constitutes an existence proof for a canonical conformal mapping. Since de Possel's proof many other canonical mappings have been derived in the same way. The preceding reasoning shows the great value of the variational method for the existence proof in complicated boundary value problems. The method is to be clearly distinguished from extremum methods as, for example, Dirichlet's principle. There we work with a fixed domain and a very wide class of admissible functions, while in our case we are dealing only with analytic univalent functions but varying domains.

Another classical problem in conformal mapping deals with the family of all functions

$$(2.3) \qquad f(z) = z + a_2 z^2 + \cdots + a_n z^n + \cdots$$

which are univalent in the unit circle $|z| < 1$. This family plays an important role in the general theory of conformal mapping and uniformization. It is a normal family which implies again that each sensible extremum problem has a solution, that is, at least one extremum function. The classical problem of the theory, as yet unsolved, is the question of max $|a_\nu|$ for the class $\mathfrak{F}$. Bieberbach's conjecture

$$(2.4) \qquad\qquad\qquad |a_\nu| \leqq \nu$$

which is proved for $\nu = 2, 3$ attracted the attention of many mathematicians because of its great simplicity.

Each function $f(z) \subset \mathfrak{F}$ determines a domain D in the complex plane which may be considered as the graphical representation of $f(z)$ and, conversely, $f(z)$ and its coefficients a_ν may be conceived as functionals of D. Let now $\phi(z)$ be the inverse function of $f(z)$ which maps D onto the unit circle; then

$$(2.5) \qquad\qquad -\frac{1}{2\pi} \log |\phi(z)| = G(z, 0)$$

is easily seen to be the harmonic Green's function of D with the source point at the origin. Thus, the coefficient problem for univalent functions may be considered as an extremum problem for the Green's function of a domain D. That this point of view is natural is seen from the following fact. The function $\phi(z)$ which is nearer to the Green's function than its inverse $f(z)$ also gives rise to a coefficient problem. This problem is simpler than the preceding one and was solved completely by Löwner in 1923 [11]. Thus, the problem closer to the Green's function problem seems for this reason to be easier to handle.

In his classical paper of 1923 Löwner considered a sequence of varying domains which depend on a parameter t as follows. Let Γ be a Jordan arc $z(t)$ which runs from $z(0)$ to infinity as t runs from zero to infinity. Let Γ_t be the subarc of Γ between $z(t)$ and infinity and let

$$(2.6) \qquad\qquad f(z, t) = \sum_{\nu=1}^{\infty} a_\nu(t) z^\nu$$

be the function which maps the unit circle onto the complex plane slit along the arc Γ_t. Then, $f(z, t)$ is differentiable in t and satisfies a simple and elegant first order partial differential equation in z and t. On the other hand, Julia [9] transformed Hadamard's variational formula for the Green's function into a variational formula for the corresponding mapping function $f(z)$ and Biernacki [1] showed that

Löwner's differential equation follows from Julia's formula by a proper passage to the limit. Thus, one of the most successful methods in the theory of univalent functions was tied to Hadamard's variational method.

In the special case of the Laplace equation in two variables there exists a simple device in order to extend Hadamard's formula to general domains and thus to make it applicable to extremum problems. This device is based on the fact that a harmonic function in two variables remains harmonic under conformal mapping; if $u(z)$ is harmonic in x, y $(z=x+iy)$, then $u[f(\zeta)]$ is harmonic in ξ, η $(\zeta=\xi+i\eta)$. Let now D be a domain in the z-plane with finitely many analytic boundary curves C; let Δ be a subdomain interior to D with finitely many analytic boundary curves Γ. Let $f(z)$ be regular analytic outside of Δ; then $z^*=z+\epsilon f(z)$ will be regular analytic on C and, for small enough ϵ, even univalent. Hence, the curve system C will be mapped onto a new curve system C^* which determines a new domain D^* with the Green's function $G^*(z;\zeta)$. Clearly

$$(2.7) \qquad g_\epsilon(z,\zeta) = G^*[z + \epsilon f(z), \zeta + \epsilon f(\zeta)]$$

is harmonic in $D-\Delta$ for $z \neq \zeta$ and vanishes for z or ζ on C. We assume that z, ζ lie in $D-\Delta$ and apply Green's identity:

$$(2.8) \qquad \begin{aligned} g_\epsilon(z,\zeta) &- G(z,\zeta) \\ &= \int_{C+\Gamma} \left[g_\epsilon(t,\zeta) \frac{\partial G(t,z)}{\partial n} - G(t,z) \frac{\partial g_\epsilon(t,\zeta)}{\partial n} \right] ds. \end{aligned}$$

Since $G(t,z)$ as well as $g(t,\zeta)$ vanishes on C, there remains only the integration over the curve Γ, interior to D.

Formula (2.8) was derived for a domain D with analytic boundary C; but now we may extend this formula to the most general case C by approximating C by analytic curves. Using theorems on the uniform convergence of the corresponding Green's functions in a closed subdomain of D, we can justify the above result for the general case.

The law of deformation

$$(2.9) \qquad z^* = z + \epsilon f(z)$$

defines to the most general boundary C of a domain D a new boundary C^* if C lies outside of Δ and if ϵ is small enough. For most applications it is sufficient to choose $f(z)$ in a very special way, namely,

$$(2.10) \qquad f(z) = \frac{1}{z - z_0}, \qquad\qquad z_0 \in D.$$

Using variations of this type, numerous extremum problems in the theory of univalent functions could be treated [6; 19; 20; 21]. One of the simplest and most important results states that the extremum functions of the coefficient problem are analytic on the periphery of the unit circle with the exception of finitely many points.

More precisely, a function $f(z)$ belonging to a maximal value of $|a_n|$ satisfies an ordinary differential equation

$$(2.11) \qquad f'(z)^2 P_n[f(z)^{-1}] = z^{-(n+1)} Q_n(z)$$

where P_n and Q_n are polynomials whose coefficients depend in a specified manner on the first n coefficients of $f(z)$ itself. It can be shown that the function

$$(2.12) \qquad f(z) = \frac{z}{(1-z)^2} = z + 2z^2 + \cdots + nz^n + \cdots,$$

which is the conjectured extremum function for all values of n, satisfies the above functional-differential equation.

Teichmüller has proved a converse of the above theorem [23]. Namely, suppose that a univalent function $f(z)$ satisfies a differential equation of the type (2.11) with any polynomial

$$(2.13) \qquad P_n(x) = \alpha_1 x + \alpha_2 x^2 + \cdots + \alpha_{n-1} x^{n-1}, \alpha_{n-1} = 1;$$

then $\mathrm{Re}\{a_n\}$ will be maximal for this function with respect to all functions of the family $\mathfrak{F}$ which have the same first $n-1$ coefficients as $f(z)$. This result shows that every univalent solution of (2.11) is in some sense an extremum function. However, in order to decide whether $f(z)$ is the extremum function of some specific extremum problem, more powerful methods seem to be needed.

It is natural to expect additional information from a theory of the second variation for the functionals considered. Since the vanishing of the first variation ensures already the analyticity of the boundary of the extremum domain, it seems sufficient to establish a formula for the second variation in the Hadamard kinematics. However, the computational difficulties in deriving the second variation for the Green's function are considerable; the formula was established only very recently and will be given in §4.

3. **General interior variations.** The variational kinematics used in the preceding section can be easily applied to much more general domain functionals. In particular, it permits us to deal with functionals which are connected with boundary value problems for self-adjoint elliptic differential equations. For the sake of simplicity, we

shall illustrate the method by treating Laplace's equation in three-space.

We define in the space with coordinates x_i ($i = 1, 2, 3$) a vector field $S_i(x)$ which is twice continuously differentiable and consider the transformation of the whole space

$$(3.1) \qquad x_i^* = x_i + \epsilon S_i(x)$$

which depends on the real parameter ϵ. If ϵ is small enough, this transformation will be univalent in a given sphere of radius R. Every domain D in this sphere is mapped topologically onto a domain D^* and we denote the corresponding harmonic Green's functions by $G(x, \xi)$ and $G^*(x, \xi)$, respectively. Our problem is to express $G^*(x, \xi)$ in terms of $G(x, \xi)$ and of the transformation vector field. We shall call the deformation of a domain by means of such a vector field an interior variation.

In order to study $G^*(x, \xi)$, we consider the function

$$(3.2) \qquad g(x, \xi; \epsilon) = G^*(x^*(x), \xi^*(\xi))$$

which is defined in the original domain D, is twice continuously differentiable there if $x \neq \xi$, and which vanishes if x or ξ lies on the boundary C of D. In the case of harmonic functions in the plane, we utilized a deformation vector field $S_i(x)$ of particular type, namely, we chose S_i so that $f(z) = S_1 + iS_2$ was an analytic function of $z = x_1 + ix_2$ outside of some subregion Δ. This particular choice had the advantage that $g(x, \xi; \epsilon)$ was still harmonic in D outside of Δ for arbitrary choice of ϵ. Now, the situation is more difficult; we cannot preserve the harmonicity of $G(x, \xi)$ under the deformation, but we are led to a new differential equation for $g(x, \xi; \epsilon)$ which expresses that $G^*(x^*, \xi^*)$ is harmonic in D^*. In fact, a simple calculation shows that Laplace's equation in D^* is translated into

$$(3.3) \qquad \sum_{i=1}^{3} \frac{\partial}{\partial x_i} \left(\sum_{k=1}^{3} A_{ik} \frac{\partial}{\partial x_k} g(x, \xi; \epsilon) \right) = 0$$

with

$$(3.3') \quad A_{ik} = A_{ki} = \theta \sum_{j=1}^{3} \frac{\partial x_i}{\partial x_j^*} \frac{\partial x_k}{\partial x_j^*}, \qquad \theta = \frac{\partial(x_1^*, x_2^*, x_3^*)}{\partial(x_1, x_2, x_3)}.$$

We observe that the coefficients of the new differential equation depend analytically on ϵ.

It is clear that $g(x, \xi; \epsilon)$ and $G^*(x, \xi)$ are equivalent functions and that the knowledge of one leads to the other in an elementary way.

We have expressed the Green's function $G^*(x, \xi)$ for the fixed Laplace equation and a variable domain D^* in terms of the Green's function $g(x, \xi; \epsilon)$ for a varying differential equation (3.3) but with respect to the original fixed domain D. This was achieved by transplanting $G^*(x, \xi)$ back into D by use of the inverse transformation of (3.1).

The dependence of the solution of a partial differential equation $L_\epsilon[u] = 0$ with fixed boundary value on the parameter ϵ occurring in the coefficients of the equation has been frequently investigated. Hilbert applied the parametrix method successfully for this problem [8]. It can be shown [3] that in our particular case the Green's function $g(x, \xi; \epsilon)$ depends analytically on the parameter ϵ.

We may develop g into a power series in ϵ and calculate without difficulty the coefficients of this development. These are clearly closely related to the variations of the respective orders of the Green's function $G(x, \xi)$. We find, for example:

$$(3.4) \qquad \frac{\partial}{\partial \epsilon} g(x, \xi; \epsilon) \bigg|_{\epsilon=0} = \iiint_D \sum_{i,k=1}^{3} T_{ik}(\eta; x, \xi) \frac{\partial S_i(\eta)}{\partial \eta_k} d\eta_1 d\eta_2 d\eta_3$$

with

$$(3.4') \qquad T_{ik}(\eta; x, \xi) = \frac{\partial G(\eta, x)}{\partial \eta_i} \frac{\partial G(\eta, \xi)}{\partial \eta_k} + \frac{\partial G(\eta, x)}{\partial \eta_k} \frac{\partial G(\eta, \xi)}{\partial \eta_i}$$
$$- \delta_{ik}(\nabla_\eta G(\eta, x) \cdot \nabla_\eta G(\eta, \xi)).$$

The tensor $T_{ik}(\eta; x, \xi)$ plays a central role in the variational theory of the Green's function. It may be called the Maxwell tensor of D; if we put $x = \xi$, we see in fact that $T_{ik}(\eta; x, x)$ is the classical Maxwell tensor of the electrostatic field created in the domain D with grounded conducting walls C by a unit charge at the point $x \in D$.

The tensor T_{ik} is obviously symmetric in its indices as well as in x and ξ. It satisfies the divergence condition

$$(3.5) \qquad \sum_{k=1}^{3} \frac{\partial}{\partial \eta_k} T_{ik}(\eta; x, \xi) = 0, \qquad\qquad i = 1, 2, 3.$$

This identity indicates that formula (3.4) can be simplified considerably by use of the divergence theorem; we can reduce the triple integral into an integral over the surface C of D provided that C is regular enough to admit the application of the divergence theorem. We have to observe in the transformation of the integral that T_{ik} becomes infinite for $\eta = x$ and $\eta = \xi$ and have to consider the residues from these singular points. Carrying out all the indicated steps, we arrive at

$$(3.6) \qquad \frac{\partial}{\partial \epsilon} G^*(x, \xi)\Big|_{\epsilon=0} = \int\int_C \frac{\partial G(\eta, x)}{\partial n} \frac{\partial G(\eta, \xi)}{\partial n} (S \cdot n) d\sigma$$

which is exactly Hadamard's variational formula (1.2). We obtain in this way a new and precise proof for Hadamard's classical formula under the weakest assumptions. On the other hand, (3.4) appears now clearly as the generalization of this formula to the case of arbitrarily bounded domains D.

A completely analogous formula holds also in the case of two-dimensional harmonic Green's functions. In this case it happens that the trace $\sum_i T_{ii}$ of the Maxwell tensor vanishes identically. Hence, T_{ik} has only two essential components, and we have

$$(3.7) \qquad\qquad T_{11} = -T_{22}, \qquad T_{12} = T_{21}.$$

If we choose now the vector field $S_i(x)$ in such a way that

$$(3.8) \qquad\qquad \frac{\partial S_1}{\partial x_1} = \frac{\partial S_2}{\partial x_2}, \qquad \frac{\partial S_1}{\partial x_2} = -\frac{\partial S_2}{\partial x_1}$$

we see that the integrand in (3.4) vanishes identically. This means, of course, the invariance of the Green's function under conformal mapping. It is now clear why the method of interior variation is particularly successful in the two-dimensional case; by choosing S to be an analytic vector field in the whole plane except for isolated singular points, we can express the variation of the Green's function in terms of its values and the values of its derivatives at these distinguished critical points. Extremum conditions on the Green's function which have, in general, the form of integro-differential equations reduce in this case to ordinary differential equations.

We illustrate the general variational method in a relatively simple case. We consider a plane, finite, and simply-connected domain D with boundary C and the eigenvalue problem

$$(3.9) \qquad\qquad \nabla^2 u + \lambda u = 0, \qquad\qquad u = 0 \text{ on } C.$$

Let z_0 be the complex coordinate of a point in the z-plane but not on C. We use the deformation

$$(3.10) \qquad\qquad z^* = z + \frac{\epsilon}{z - z_0},$$

with complex parameter ϵ, which is univalent in $|z - z_0| > |\epsilon|^{1/2}$ and which maps therefore, for small enough ϵ, D into a domain D^*. From the general variational procedure, we find that each eigenvalue $\lambda(D)$

which is nondegenerate corresponds to an eigenvalue $\lambda(D^*)$ by the formula [4]

$$(3.11) \quad \lambda(D^*) = \lambda(D) - \mathrm{Re}\ \left\{ 8\pi\epsilon\delta(z_0)\left(\frac{\partial u(z_0)}{\partial z_0}\right)^2 \right.$$

$$\left. - 2\lambda(D)\epsilon \iint_D \frac{u(z)^2}{(z-z_0)^2}\ o\omega \right\} + o(\epsilon).$$

Here $\delta(z_0) = 0$ or 1 if z_0 lies in the exterior or interior of D, respectively, and

$$(3.11') \qquad \frac{\partial}{\partial z_0} = \frac{1}{2}\left(\frac{\partial}{\partial x_0} - i\frac{\partial}{\partial y_0}\right).$$

$o(\epsilon)$ can be estimated uniformly for all domains D whose boundaries stay outside of a fixed circle $|z - z_0| = \alpha$. The above formula applies, in particular, always to the first eigenvalue $\lambda_1(D)$ which is known to be nondegenerate.

Under the same variation (3.10) the area $A(D)$ is transformed into

$$(3.12) \qquad A(D^*) = A(D) - \mathrm{Re}\ \left\{ 2\epsilon \iint \frac{d\omega}{(z-z_0)^2} \right\} + o(\epsilon).$$

We may now treat Rayleigh's problem to find the minimum for the product $A \cdot \lambda_1$. The existence of a minimum domain can be derived by use of the conformal mapping function which carries D into the unit circle and applying the normality of the family of univalent functions. We can characterize the extremum domain by the fact that the first variation of $\lambda_1 A$ has to vanish for each choice of ϵ and $z_0 \notin C$. This leads to the functional equation

$$(3.13) \qquad \delta(z_0)\left(\frac{\partial u(z_0)}{\partial z_0}\right)^2 = \frac{\lambda_1(D)}{4\pi A(D)}\iint_D \frac{A(D)u(z)^2 - 1}{(z-z_0)^2}\ d\omega.$$

By a careful analysis of this condition one can show that D has an analytic boundary curve C along which $\partial u/\partial n = \mathrm{const}$. We obtain by the variational method less than from the more elementary method of symmetrization which has been developed in great elegance by Pólya-Szegö [13]. However, the same variational technique can be applied in much more general problems where the extremum domain will not happen to be symmetric and the method will work as well.

In §2, we pointed out that the variational theory of domain functionals leads often to important existence theorems. We want to discuss here a problem which plays a role in applications. Let B be a body in three-space, considered immersed in an incompressible non-viscous fluid of density 1. Suppose that the fluid stream past the body is a stationary irrotational flow with velocity one at infinity and direction parallel to the x-axis. We denote the velocity potential of the flow by $\phi(x, y, z)$; it is a harmonic function in the domain D outside of B except at infinity and is at each fixed point in D a domain functional of B (or D). If we split off from $\phi(x, y, z)$ the singular term due to the source of the flow at infinity, we can write it in the form

$$(3.14) \qquad \phi(x, y, z) = - x + \varphi(x, y, z)$$

where φ is regular harmonic in D. The quantity

$$(3.15) \qquad M = \iiint_D (\nabla \varphi)^2 dx dy dz$$

is called the virtual mass of B and plays a central role in the hydromechanics of B; M is also a functional of B.

Suppose now that we deform the surface C of B into a surface C^* of B^* by a normal shift δn applied to all its points. It can be shown [22] that

$$(3.16) \qquad \delta M = \iint_C (\nabla \phi)^2 \delta n d\sigma - \delta V,$$

where at the same time the volume $V(B)$ varies according to

$$(3.17) \qquad \delta V = \iint_C \delta n d\sigma.$$

We apply these results to an important problem in fluid dynamics, namely, the free boundary problem. Helmholtz was the first to point out that the flow past a body B is by no means uniquely determined. The moving fluid may bypass the body B and leave also a part of the fluid itself adjacent to B at rest. The part of the fluid at rest together with the body B forms a larger body B' whose velocity potential $\phi'(x, y, z)$ determines the correct velocity field of the flow. However, since the hydrostatic pressure in the resting fluid must be constant, it is easily seen that along the surface dividing between resting and moving fluid, the so-called free boundary, the velocity has to be constant, that is, $(\nabla \phi')^2 = $ const. Hence, under a δn-deformation which affects only the free boundary of the body B', we should have, by (3.16) and (3.17), $\delta M' = \mu \delta V'$. Thus, a free boundary leads to a stationary value for $M - \mu V$. One can now invert the reasoning and pose the extremum problem for $M - \mu V$ [18]. If one can show that a solution exists and leads to a regular surface, one arrives at an existence proof for an interesting flow pattern. Such existence proof could be carried out indeed by interior variations in the case of plane flow [5] and of axially-symmetric flow [2]. The general case of a flow in three dimensions has not yet been solved satisfactorily.

The alternative approach to the free boundary problem for plane flows due to Leray [10] is based on fixed point theorems in function spaces and utilizes therefore also higher methods of functional analysis.

4. The second variation. The importance of the second variation for the theory of domain functionals is obvious. Consider a class of domains D and a functional $\phi[D]$. Suppose now that any two domains D_0 and D_1 of the class can be connected by a continuously varying set of domains D_t of the class with $0 \leq t \leq 1$. Then $\phi[D_t]$ becomes a function of the variable t. Suppose further that we can assert $d^2\phi_t/dt^2 > 0$ if $D_0 \neq D_1$. In this case, it is clear that there can exist only one domain in the class for which $\delta\phi = 0$ for all admissible variations and which yields, consequently, a minimum of ϕ. In many cases, it is more important to know that the functional equation implied by $\delta\phi \equiv 0$ has a unique solution than to know the exact value of the minimum. This is, for example, the case if the minimum problem has been set up artificially in order to obtain an existence proof as in the free boundary problem of the preceding section. In such cases, the knowledge of the second variation is most useful. Often it will even be sufficient to show that $d^2\phi_t/dt^2 > 0$ only in the case that $\delta\phi = 0$ in order to make analogous deductions.

As indicated already in §3, it is often easy to find the higher order variations for functionals if we use the method of interior variations. Returning to the special case of the three-dimensional harmonic Green's function, we know already that $g(x, \xi; \epsilon)$ is analytic in ϵ and that we can derive formally all coefficients in the power series development in ϵ, obtaining finally all desired variations for the Green's function. We may express, for example, the second variation of $G(x, \xi)$ in terms of a triple integral over the domain D involving the vector field $S_i(x)$ and the Green's function itself. We know, however, a priori that the variation of G must depend only on the shift of the boundary C of D and not on the interior values of the field; for the deformation of C alone determines the varied domain D^* with the Green's function G^*. Thus, it must be possible to express the second variation of the Green's function in the form of a surface integral over C, provided only that the boundary surface C of D is smooth enough. Similar arguments hold, of course, also in the case of variations of higher order.

We can give an elegant formula for the second variation of $G(x, \xi)$ in the case that the boundary surface C is three times continuously differentiable. We consider a fixed vector field S_i and denote by $G_\epsilon(x, \xi)$ the Green's function of the domain D_ϵ which arises from $D = D_0$ through the variation (3.1). We define $G(x, \xi) = G_0(x, \xi)$. Let

$$(4.1) \qquad\qquad N = S \cdot n;$$

N is defined on C and n is the exterior normal on C with respect to D. By Hadamard's formula (3.6), we have

$$(4.2) \quad H(x, \xi) = \frac{\partial}{\partial \epsilon} G_\epsilon(x, \xi)\bigg|_{\epsilon=0} = \int\int_C \frac{\partial G(\eta, x)}{\partial n} \frac{\partial G(\eta, \xi)}{\partial n} N(\eta) d\sigma.$$

Here, $H(x, \xi)$ is regular harmonic in D and a linear functional of the vector field S on the boundary C. We can prove [3]:

$$
\begin{aligned}
(4.3) \quad \frac{\partial^2}{\partial \epsilon^2} G_\epsilon(x, \xi)\bigg|_{\epsilon=0} = &-2 \int\int\int_D (\nabla_\eta H(\eta, x) \cdot \nabla_\eta H(\eta, \xi)) d\eta_1 d\eta_2 d\eta_3 \\
&- \int\int_C \frac{\partial G(\eta, x)}{\partial n} \frac{\partial G(\eta, \xi)}{\partial n} k(\eta) N^2(\eta) d\sigma,
\end{aligned}
$$

where $k(\eta) = 1/\rho_1 + 1/\rho_2$ is the mean curvature of C at the point η. This formula holds under the additional assumption that the vector field S has on C exact normal direction; otherwise some additional terms will come in depending on the tangential components of S. Such effect of the tangential deformation was to be expected in the theory of the second variation; for, if we displace each boundary point in direction of the tangent plane, we do not affect the domain in the first order, but we must expect an influence of the second order variation.

The above formula for the second variation can be simplified considerably in important special cases. Consider, for example, a closed surface T which encloses the boundary C of D. Let $V(x)$ be harmonic in the shell between C and T and let $V = 0$ on C, $V = 1$ on T. In other words, $V(x)$ is the potential of the conductor made up of the surfaces C and T. The level surfaces

$$4.4) \qquad\qquad V(x) = t, \qquad\qquad 0 \leqq t \leqq 1,$$

are a continuously varying sequence of surfaces C_t between C and T. Let $G_t(x, \xi)$ be the Green's function of the domain D_t interior to C_t. It is then easily shown that

$$\frac{\partial^2}{\partial t^2} G_t(x,\, \xi)$$

(4.5)

$$= -2 \iiint_{D_t} \left(\nabla_\eta \frac{\partial}{\partial t} G_t(\eta,\, x) \cdot \nabla_\eta \frac{\partial}{\partial t} G_t(\eta,\, \xi) \right) d\eta_1 d\eta_2 d\eta_3.$$

We observe that

(4.6) $G_t(x,\, \xi) = \dfrac{1}{4\pi r(x,\, \xi)} - \Gamma_t(x,\, \xi), \qquad r = \left(\sum_{i=1}^{3} (x_i - \xi_i)^2 \right)^{1/2},$

where $\Gamma_t(x,\, \xi)$ is regular harmonic in the entire domain D. Since the singularity term is independent of the domain,

(4.7) $\dfrac{\partial G_t}{\partial t} = -\dfrac{\partial \Gamma_t}{\partial t}, \qquad \dfrac{\partial^2 G_t}{\partial t^2} = -\dfrac{\partial^2 \Gamma_t}{\partial t^2}.$

Hence, we find

$$\frac{\partial^2}{\partial t^2} \left[\sum_{i,k=1}^{N} \alpha_i \alpha_k \Gamma_t(x^{(i)},\, x^{(k)}) \right]$$

(4.8)

$$= 2 \iiint_{D_t} \left(\nabla_\eta \sum_{i=1}^{N} \alpha_i \frac{\partial \Gamma_t(\eta,\, x^{(i)})}{\partial t} \right)^2 d\eta_1 d\eta_2 d\eta_3 \geqq 0$$

for any choice of the constants α_i and of the points $x^{(i)}$ in D. We arrive thus at a very general convexity theorem in potential theory and the applications indicated at the beginning of this section become possible.

Sometimes we can utilize the formula for the second variation more effectively than by just using the fact that it is positive. Take, for example, the functional $K[B]$, the electrostatic capacity of the body B; it is defined as follows: Let $\psi(x)$ be harmonic in the outside D of B and have the boundary value 1 on C. Then, $\psi(x)$ is called the conductor potential of B. It has at infinity the behavior

(4.9) $\psi = \dfrac{K[B]}{r} + o(1/r), \qquad r^2 = \sum_{i=1}^{3} x_i^2$

and this serves to define the electrostatic capacity $K[B]$.

Let now C_t be a sequence of surfaces connecting continuously the surface C of B with another enclosing surface T and being the level lines (4.4) of a harmonic function $V(x)$. The C_t bound bodies B_t with exterior D_t, capacity K_t, and conductor potential ψ_t. We can show from (4.5) that

$$(4.10) \qquad \frac{\partial^2}{\partial t^2} K_t = \frac{1}{2\pi} \iiint_{D_t} \left(\nabla \frac{\partial \psi_t}{\partial t} \right)^2 dx_1 dx_2 dx_3.$$

This shows that K_t is convex from below as a function of t. But we can estimate the right-hand integral better if we observe that $h(x) = \partial \psi_t / \partial t$ is harmonic in D_t. Then, we have by Green's identity

$$(4.11) \qquad \iiint_{D_t} (\nabla h \cdot \nabla \psi_t) dx_1 dx_2 dx_3 = -\iint_{C_t} \frac{\partial h}{\partial n} d\sigma = 4\pi a$$

if $\lim_{r \to \infty} hr = a$. Hence, by Schwarz's inequality

$$(4.12) \quad (4\pi a)^2 \leqq \iiint_{D_t} (\nabla h)^2 dx_1 dx_2 dx_3 \cdot \iiint_{D_t} (\nabla \psi_t)^2 dx_1 dx_2 dx_3.$$

Finally, by applying (4.11) for the case $h = \psi_t$ and using (4.9), we find

$$(4.13) \qquad \iiint_{D_t} (\nabla \psi_t)^2 dx_1 dx_2 dx_3 = 4\pi K_t.$$

From (4.9) we deduce that $\lim_{r \to \infty} r(\partial \psi_t / \partial t) = (\partial K_t / \partial t)$ and hence (4.12), (4.13) yield

$$(4.14) \qquad \iiint_{D_t} \left(\nabla \frac{\partial \psi_t}{\partial t} \right)^2 dx_1 dx_2 dx_3 \geqq \frac{4\pi}{K_t} \left(\frac{\partial K_t}{\partial t} \right)^2.$$

Hence, (4.10) leads to the differential inequality

$$(4.15) \qquad K_t \, \partial^2 K_t / \partial t^2 \geqq 2 \, (\partial K_t / \partial t)$$

which means that K_t^{-1} is convex from above as a function of t.

This result is the best possible; for in the case that T is a level surface $\psi(x) = \text{const.}$ of the conductor potential of B, it is easily seen that K_t^{-1} is linear in t.

As another application, we consider two convex curves C_0 and C_1 in the plane which are described by their supporting functions $p_0(\phi)$ and $p_1(\phi)$. We can connect both curves by a continuous sequence of convex curves C_t which are determined by the supporting functions,

$$(4.16) \qquad p_t(\phi) = (1 - t)p_0(\phi) + tp_1(\phi).$$

In this case, the theory of the second variation yields the formula

$$(4.17) \qquad \begin{aligned} &\frac{\partial^2}{\partial t^2} G_t(x, \xi) \\ &\leqq -2 \iint_{D_t} \left(\nabla_\eta \frac{\partial}{\partial t} G_t(\eta, x) \cdot \nabla_\eta \frac{\partial}{\partial t} G_t(\eta, \xi) \right) d\eta_1 d\eta_2 d\eta_3 \end{aligned}$$

where G_t is the Green's function of the interior D_t of C_t. This result leads again to numerous applications and convexity theorems in conformal mapping.

A main problem in applying the theory of the second variation is to find a convenient parametrization which permits us to connect any two domains of the class considered and which leads, on the other hand, to a simple second derivative of the functional in question with respect to that parameter.

5. Transplantation. Till now, we dealt mostly with functionals which arise from certain boundary value problems in partial differential equations. We consider now the important subclass of functionals which are defined as the minimum value of a given positive-definite integral extended over the domain considered and depending on functions of a prescribed function class with respect to this domain. The fact that the functional is an actual minimum allows certain estimates by use of competing functions which lead to interesting inequalities and even to variational formulas.

We explain the method for the case of the electrostatic capacity $K[B]$ of a body B with boundary surface C and exterior D. We can characterize $K[B]$ by the following extremum properties [13; 14]:

1. *Dirichlet's principle*: Consider all functions $U(x)$ which have continuous derivatives in D, are of order $O(1/r)$ at infinity, and have on C the boundary value one. Then

$$(5.1) \qquad K[B] = \min \frac{1}{4\pi} \iiint_D (\nabla U)^2 dx_1 dx_2 dx_3$$

and the minimum is attained for the conductor potential $\psi(x)$ of B.

Dirichlet's principle leads to easy upper bounds for $K[B]$ since we may choose any permissible function $U(x)$ and find an inequality for $K[B]$.

2. *Thomson's principle*: Consider all vector fields $q(x)$ in D with continuously differentiable components, which are solenoidal, that is, satisfy $\nabla q = 0$, are normalized by

$$(5.2) \qquad \iint_C q \cdot n d\sigma = 1,$$

and vanish at infinity such that $|q| = O(r^{-2})$ there. Then

$$(5.3) \qquad \frac{1}{4\pi K[B]} = \min \iiint_D q^2 dx_1 dx_2 dx_3$$

and the minimum is attained by the vector field $q = (1/4\pi K)\nabla \psi$.

Thomson's principle leads to lower bounds for $K[B]$ and is very useful in combination with the preceding Dirichlet principle.

3. *Gauss' principle*: Consider all functions $\mu(P)$ defined on the boundary surface C with the normalization

$$(5.4) \qquad \iint_C \mu d\sigma = 1.$$

Then

$$(5.5) \qquad \frac{1}{K} = \min \iint_C \iint_C \mu(x)\mu(\xi)r(x,\xi)^{-1}d\sigma_x d\sigma_\xi$$

and the minimum value is attained for $\mu(x) = -(1/4\pi K)\partial\psi/\partial n$.

Gauss' principle stresses the fact that the capacity is ultimately a functional of the surface C only. In physical interpretation, it expresses the capacity in terms of electric charges rather than in terms of electric fields.

We start now with a body $B = B(1)$ and consider the transformation of the x-space

$$(5.6) \qquad x_1' = tx_1, \qquad x_2' = x_2, \qquad x_3' = x_3;$$

this is a stretching in the x_1-direction in the ratio $t:1$. The body $B(1)$ is deformed continuously through a sequence of bodies $B(t)$ with exteriors $D(t)$, boundaries $C(t)$, conductor potentials $\psi_t(x)$, and capacities $K(t)$. Since $\psi_{t_0}(x)$ is defined in $D(t_0)$, clearly

$$(5.7) \qquad U(x) = \psi_{t_0}\left(\frac{t_0}{t} x_1, x_2, x_3\right)$$

is well-defined in $D(t)$. It has the boundary value 1 on $C(t)$ and the correct continuity and asymptotic behavior at infinity in order to be admissible in the Dirichlet principle (5.1) with respect to $D(t)$. Thus, we obtain

$$(5.8) \qquad K(t) \leqq \frac{1}{4\pi} \iiint_{D(t)} \left[\left(\frac{t_0}{t}\right)^2 \psi_{t_0;1}\left(\frac{t_0}{t} x_1, x_2, x_3\right)^2 \right. \\ \left. + \psi_{t_0;2}^2 + \psi_{t_0;3}^2 \right] dx_1 dx_2 dx_3$$

where $\psi_{t_0;i}$ denotes the partial derivative of ψ_{t_0} with respect to its ith variable. We may refer back the integration to $D(t_0)$ by replacing $(t_0/t)x_1$ by x_1. This will change also the volume element in the integral and (5.8) obtains the form

$$(5.9) \quad t^{-1}K(t) \leqq \frac{1}{4\pi t_0} \int\int\int_{D(t_0)} \left[\frac{t_0^2}{t^2} \psi_{t_0;1}^2 + \psi_{t_0;2}^2 + \psi_{t_0;3}^2 \right] dx_1 dx_2 dx_3$$

where the arguments of all $\psi_{t_0;i}$ are now x_1, x_2, x_3. Thus, we may put this inequality into the form

$$(5.9') \qquad\qquad t^{-1}K(t) \leqq t^{-2}A + B$$

where A and B depend on t_0 but not on t. Moreover

$$(5.9'') \qquad\qquad t_0^{-1}K(t_0) = t_0^{-2}A + B.$$

Hence, if we plot $t^{-1}K(t)$ versus $\tau = t^{-2}$, we see that the straight line $a\tau + B$ touches this curve at the point $\tau_0 = t_0^{-2}$ but lies elsewhere always above it. Hence, the curve $t^{-1}K(t)$ possesses at τ_0 a supporting line from above. Since t_0 is arbitrary, we proved that $t^{-1}K(t)$ is convex from above as a function of t^{-2}. If we did not know already from the general theory that $K(t)$ is analytic in t, we could deduce at least from the convexity result that $K(t)$ has a derivative in t almost everywhere. We can easily derive

$$(5.10) \quad K'(t) = \frac{1}{4\pi t} \int\int\int_{D(t)} [\psi_{t2}^2 + \psi_{t3}^2 - \psi_{t1}^2] dx_1 dx_2 dx_3;$$

and the differential inequality stating the above convexity

$$(5.11) \qquad\qquad t^2 K''(t) + t K'(t) \leqq K(t).$$

The basic idea in the above procedure is the transplantation of the correct extremum function from its domain $D(t_0)$ into another domain $D(t)$ where it serves as a comparison function in the extremum principle and leads thus easily to an inequality between the functionals of $D(t_0)$ and $D(t)$. This method of transplanting the extremum function has been used extensively in the following form. One uses one distinguished domain D_0, which is suspected to be the extremum domain, and its extremum function. Transplating this extremum function into all admissible domains D of the class considered one shows that the functional $\phi[D]$ by its minimum property is less than the value achieved by the transplanted function and moreover that this latter value is less than $\phi[D_0]$. This shows that $\phi[D_0] > \phi[D]$ and establishes the extremum property of D_0 [13].

The difference between this well known and useful device and the above method is that in the latter we transplant the extremum function of each domain into each other domain; in this way, we obtain convexity statements and differential inequalities [14].

We can repeat the procedure in the case of the Thomson principle. Here, we have to deal with an extremum vector field instead of an extremum function. Let $\mathbf{q} \equiv (q_1, q_2, q_3)$ be the extremum vector field for the exterior $D(t_0)$ of the body $B(t_0)$. Then

(5.12)
$$\mathbf{q}_t \equiv \left(q_1\left(\frac{t_0}{t} x_1, x_2, x_3\right), \frac{t_0}{t} q_2\left(\frac{t_0}{t} x_1, x_2, x_3\right), \right.$$
$$\left. \frac{t_0}{t} q_3\left(\frac{t_0}{t} x_1, x_2, x_3\right)\right)$$

will be defined in $D(t)$ and represent a solenoidal vector field there. It is easily seen that the normalization (5.2) is transplanted also. Hence, we may use $\mathbf{q}_t$ in order to obtain a lower bound for $K(t)$. An easy calculation shows that $tK(t)^{-1}$ is convex from above as a function of t^2. This leads to another differential inequality which combines with (5.11) to estimate $K''(t)$ from above and from below. We obtain the inequality

(5.9''') $$(at^2 + b)^{-1} \leqq t^{-1}K(t) \leqq At^{-2} + B$$

where equality holds on both sides for $t = t_0$. Thus, the curve $t^{-1}K(t)$ lies between two differentiable curves which touch for $t = t_0$. Clearly, $t^{-1}K(t)$ must be also differentiable there and since t_0 is quite arbitrary, we have proved that $K(t)$ is differentiable in t everywhere. Thus, combining the Dirichlet with the Thomson principle we obtained a much stronger result than could have been obtained from each of them.

Finally, we may utilize the Gauss principle by transplanting the charge density $\mu d\sigma$ in the deformation. This leads to the result that $tK(t)^{-1}$ is convex from above as a function of t. This result is an improvement on the convexity statement derived from the Thomson principle. It is also easy to read off from the Gauss principle that $K(t)$ increases with t; in fact, under transplantation corresponding charges are pulled away from each other and the energy of the charge distribution is clearly decreased and the capacity K is, consequently, increased. On the other hand, $K'(t)$ is given by (5.10) and it is by no means clear that this integral must be non-negative. But the combination of the two results leads to the inequality

(5.13) $$\iiint_D \psi_{x_1}^2 \, dx_1 dx_2 dx_3 \leqq \iiint_D [\psi_{x_2}^2 + \psi_{x_3}^2] \, dx_1 dx_2 dx_3$$

valid for every conductor potential and arbitrary choice of the coordinate system. Equality holds in (5.13) if the body B is a plate in a plane $x_1 = $ const. In fact, such a plate is not affected by a stretching in the x_1-direction and hence $K'(t) \equiv 0$.

It seems that the transformation by stretching is a very special kind of deformation and that the results obtained in this way are rather restricted. Before discussing other and more general deformations which admit a similar treatment, we want to show a slight generalization of the stretching deformation which leads to rather important applications. We consider the transformation

$$(5.14) \qquad x_1' = \begin{cases} tx_1 & \text{if} \quad x_1 \geq 0, \quad t > 0, \\ x_1 & \text{if} \quad x_1 \leq 0, \end{cases} \qquad x_2' = x_2, \; x_3' = x_3.$$

This is a continuous transformation of the x-space, but it is no longer continuously differentiable. We call it a partial stretching in the x_1-direction.

If we want to apply the transplantation method to the capacity $K[B]$ in the case of a partial stretching, we have to observe, at first, that the transplanted functions and vector fields have discontinuities or discontinuous derivatives along the plane $x_1 = 0$. However, it can be seen that Dirichlet's and Thomson's principles still hold for extended classes of functions or vector fields, which include the transplanted fields obtained. Hence, it can be shown, as before, that $t^{-1}K(t)$ is convex from above in t^{-2} while $tK(t)^{-1}$ is convex from above in t^2, even in the case of a partial stretching. The derivative of $K(t)$ with respect to t can be shown to be

$$(5.15) \qquad K'(t) = \frac{1}{4\pi t} \iiint_{D(t)^+} [\psi_{t2}^2 + \psi_{t3}^2 - \psi_{t1}^2] dx_1 dx_2 dx_3$$

where $D(t)^+$ is the intersect of $D(t)$ with the half-space $x_1 \geq 0$.

Since we may put the axes of reference arbitrarily with respect to the body considered, we obtain a great variety of deformations of the original body under partial stretching. In particular, we may put the plane $x_1 = 0$ in such a way that it cuts off only a very small part of the body B so that the partial stretching becomes a variation of a surface element of B. Using our above results, and in particular (5.15), we obtain a new approach to variational formulas of the Hadamard type.

Other interesting applications of the transplantation method arise in eigenvalue problems. In the membrane equation

$$(5.16) \qquad\qquad \nabla^2 u + \lambda u = 0, \qquad\qquad u = 0 \text{ on } C$$

we can characterize the eigenvalues by well known extremum principles. We restrict ourselves here to the lowest eigenvalue $\lambda_1(D)$ of a domain D in the (x, y)-plane. Let $u(x, y)$ be any continuously differentiable function in D which vanishes on the boundary C; then we have

$$(5.17) \qquad \lambda_1 = \min \frac{\iint_D (\nabla u)^2 dx dy}{\iint_D u^2 dx dy}.$$

The minimum is attained by the eigenfunction of (5.16) to the eigenvalue λ_1.

It is easy to find upper bounds for the eigenvalue λ_1 because of its minimum definition; it is, in general, not quite as easy to derive lower bounds. Here, convexity statements for the eigenvalue in dependence on some parameter t for a sequence of domains $D(t)$ may be of great value.

We can easily show that under a stretching

$$(5.18) \qquad x' = tx, \qquad y' = y$$

we obtain a sequence of domains $D(t)$ with the nth eigenvalue $\lambda_n(t)$ such that $\sum_{\nu=1}^{N} \lambda_\nu(t)$ is a monotonic function of t^{-2}, convex from above; this holds for every choice of the integer N. Observe that the sum considered is not necessarily continuously differentiable in t; it may have discontinuous derivatives for the values of the parameter t for which $\lambda_N(t)$ is a degenerate eigenvalue. The same result holds for the eigenvalues of the more general differential system

$$(5.19) \qquad \nabla^2 u + \lambda u = 0, \qquad \frac{\partial u}{\partial n} + ku = 0 \text{ on } C.$$

These convexity results have been applied to estimate the eigenvalues for isosceles triangles from the known eigenvalues for the equilateral triangle and their asymptotic behavior for $t \to 0$ and $t \to \infty$ [12].

Consider next the sequence of domains $D(t)$ which are obtained from a given domain $D = D(0)$ by the conformal mappings

$$(5.20) \qquad z^* = z + tf(z)$$

where $f(z)$ is analytic in $D(0)$. It is not necessary that all domains $D(t)$ lie schlicht over the complex plane; we may define their eigenvalues $\lambda_n(t)$ of the membrane problem even if they lie over a Riemann surface.

If $u(x, y)$ is a correct first eigenfunction of the membrane problem for the domain $D(0)$, we may consider $u(x, y)$ referred by $z(z^*)$ to $D(t)$ as a comparison function for the extremum problem. We may assume that u was normalized in $D(0)$ by the condition

$$(5.21) \qquad \iint_{D(0)} u^2 dx dy = 1.$$

Observe that the Dirichlet integral in the numerator of (5.17) is invariant under conformal transformation. Thus, (5.17) applied to $D(t)$ leads to

$$(5.22) \qquad \begin{aligned} \lambda_1(t)^{-1} &\geqq \lambda_1(0)^{-1} \iint_{D(0)} u^2(x, y) \left| 1 + t f'(z) \right|^2 dx dy \\ &\geqq \lambda_1(0)^{-1} + \lambda_1(0)^{-1} t \cdot 2 \operatorname{Re} \left\{ \iint_{D(0)} u^2 f'(z) dx dy \right\}. \end{aligned}$$

This shows that $\lambda_1(t)^{-1}$ is convex from below as a function of t, since we can deduce in the same way the existence of a supporting line for every value of t.

Using the extremum definition for the higher eigenvalues and the same transplantation argument, we can prove that $\sum_{\nu=1}^{N} \lambda_\nu^{-1}(t)$ is convex from below as a function of t, for every choice of the integer N. This result leads to interesting inequalities between the eigenvalues of the membrane problem and conformal moduli of the domain D [14].

An extension of the preceding convexity result to the more general system (5.19) seems not easy. The boundary conditions are only conformally invariant if $k = 0$ and, in this case, we have $\lambda_1 = 0$. Hence, no corresponding result is known in the more general case (5.19).

We do not want to multiply examples, but mention that the transplantation method proved also useful in the theory of torsional rigidity, virtual mass, conformal radius, etc. [14].

It may be of interest to quote a result from the theory of differential equations of higher than the second order. Consider the differential equation

$$(5.23) \qquad \nabla^4 u = \lambda u, \qquad u = \frac{\partial u}{\partial n} = 0 \text{ on } C,$$

for a plane domain D with boundary C. Then, it can be shown that under a stretching (5.18) of D, the functional $\lambda(D)$ is convex from above as a function of t^{-4}.

6. General method.

In the case that $\phi[D]$ is a functional defined

as the minimum value of a positive-definite Dirichlet integral $Q[u]$ which is quadratic in its argument function u, we can develop a very simple variational theory. We assume that the minimum in $Q[u]$ is understood with respect to a well-defined function class $\mathfrak{F}$ characterized by certain continuity and boundary conditions with respect to D. Suppose that a transformation

$$(6.1) \qquad\qquad x_i^* = x_i + \epsilon S_i(x)$$

is considered which carries $D = D(0)$ into a sequence of domains $D(\epsilon)$; we assume that the functions of the class $\mathfrak{F}$ are transformed by transplantation into $D(\epsilon)$ into functions of the corresponding function class $\mathfrak{F}_\epsilon$ of the new domain and vice versa. Thus, each admissible function $u^*(x^*)$ in $D(\epsilon)$ becomes admissible in $D(0)$ after transplantation.

The Dirichlet integrals $Q_\epsilon[u^*]$ for the various domains $D(\epsilon)$ become after transplantation into $D(0)$ well-defined Dirichlet integrals $Q[u; \epsilon]$ and are to be considered over the function class $\mathfrak{F}$; clearly $Q[u; 0] = Q[u]$.

Let $u(x; \epsilon)$ be the extremum function which yields the minimum

$$(6.2) \qquad \phi(\epsilon) = Q[u(x; \epsilon); \epsilon] \leqq Q[u_0; \epsilon], \qquad\qquad u_0 \in \mathfrak{F},$$

within the class $\mathfrak{F}$. Obviously

$$(6.3) \qquad\qquad \phi(\epsilon) = \phi[D(\epsilon)]$$

and the study of $\phi(\epsilon)$ in its dependence on the parameter ϵ leads to the complete variational theory of the functional $\phi[D]$.

Observe now that

$$(6.4) \qquad \begin{aligned} \phi(\epsilon) - \phi(0) &= Q[u(x; \epsilon); \epsilon] - Q[u(x; \epsilon); 0] \\ &\quad + Q[u(x; \epsilon); 0] - Q[u(x; 0); 0] \\ &= Q[u(x; \epsilon); \epsilon] - Q[u(x; 0); \epsilon] \\ &\quad + Q[u(x; 0); \epsilon] - Q[u(x; 0); 0]. \end{aligned}$$

Since $u(x; \epsilon)$ and $u(x; 0)$ both belong to the class $\mathfrak{F}$, we have by the minimum properties (6.2)

$$(6.5) \quad Q[u(x; \epsilon); \epsilon] \leqq Q[u(x; 0); \epsilon]; \quad Q[u(x; 0); 0] \leqq Q[u(x; \epsilon); 0].$$

Thus, (6.4) leads to the inequalities

$$(6.6) \quad \begin{aligned} & Q[u(x; \epsilon); \epsilon] - Q[u(x; \epsilon); 0] \\ & \qquad \leqq \phi(\epsilon) - \phi(0) \leqq Q[u(x; 0); \epsilon] - Q[u(x; 0); 0]. \end{aligned}$$

We have, until now, not made any assumption regarding the character of the quadratic Dirichlet integral $Q[u]$. We suppose that $Q[u; \epsilon]$ has for fixed $u(x)$ a continuous derivative $Q_1[u; \epsilon]$ with respect to ϵ which is again a quadratic functional of u. We have then by the mean value theorem

$$(6.7) \qquad Q[u; \epsilon_2] - Q[u; \epsilon_1] = (\epsilon_2 - \epsilon_1)Q_1[u; \eta]$$

where η is an intermediate value between ϵ_1 and ϵ_2. We restrict ourselves to the interval $|\epsilon| \leq 1$ and assume further that $Q_1[u; \epsilon]$ is there uniformly continuous as a functional of u in the $Q[u]$-metric and that there exist positive constants a, b, α, β such that

$$(6.8) \qquad aQ[u] \leq Q[u; \epsilon] \leq bQ[u]; \qquad \alpha Q[u] \leq |Q_1[u; \epsilon]| \leq \beta Q[u].$$

In many important cases, it is possible to verify that all these assumptions are fulfilled.

We derive now from (6.6) and (6.7) the inequality

$$(6.9) \qquad Q_1[u(x; \epsilon); \eta_1]\epsilon \leq \phi(\epsilon) - \phi(0) \leq Q_1[u(x; 0); \eta_2]\epsilon$$

where η_1 and η_2 lie between 0 and ϵ. From our assumptions (6.8) follows clearly that

$$(6.10) \qquad \phi(\epsilon) - \phi(0) = O(\epsilon),$$

that is, in particular, the continuity of $\phi(\epsilon)$ at $\epsilon = 0$.

Next, we observe that

$$(6.11) \qquad \begin{aligned} Q[u(x; \epsilon)] &- Q[u(x; 0)] \\ &= \phi(\epsilon) - \phi(0) - (Q[u(x; \epsilon); \epsilon] - Q[u(x; \epsilon); 0]) \end{aligned}$$

which leads by (6.7), (6.8), and (6.10) to

$$(6.12) \qquad Q[u(x; \epsilon)] - Q[u(x; 0)] = O(\epsilon).$$

But from the quadratic character of $Q[u]$ and the minimum property of $u(x; 0)$ with respect to this Dirichlet integral, we have

$$(6.13) \quad Q[u(x; \epsilon)] - Q[u(x; 0)] = Q[u(x; \epsilon) - u(x; 0)] = O(\epsilon).$$

Thus, $u(x; \epsilon)$ converges to $u(x; 0)$ in the Q-metric and, consequently, in view of the assumed continuity properties of $Q_1[u; \epsilon]$, we find from (6.9)

$$(6.14) \qquad \lim_{\epsilon \to 0} \frac{\phi(\epsilon) - \phi(0)}{\epsilon} = Q_1[u(x; 0); 0].$$

This proves the differentiability of the functional and yields an explicit variational formula.

It is interesting to observe the role which the actual minimum character of $\phi[D]$ plays in our reasoning. Terms like

$$\frac{1}{\epsilon} \left(Q[u(x; \epsilon); 0] - Q[u(x; 0); 0] \right),$$

which occur in (6.4) and are easily estimated from the minimum property, require a detailed investigation of the dependence of $u(x; \epsilon)$ on ϵ if the minimum property does not hold.

In important cases the value $u(x; \epsilon)$ itself can be characterized by a minimum problem of the above type. In this case, the above method leads to the derivative of $u(x; \epsilon)$ with respect to ϵ which, in turn, leads to the second derivative of $\phi[\epsilon]$. In the case of the harmonic Green's function, for example, such reasoning leads to the derivatives of all orders of $G_\epsilon(x, \xi)$ with respect to the parameter.

References

1. M. Biernacki, *Sur la représentation conforme des domaines linéairement accessibles*, Prace Math. Fiz. vol. 44 (1936) pp. 293–314.

2. P. R. Garabedian, H. Lewy and M. Schiffer, *Axially symmetric cavitational flow*, Ann. of Math. vol. 56 (1952) pp. 560–602.

3. P. R. Garabedian and M. Schiffer, *Convexity of domain functionals*, Journal d'Analyse Mathématique vol. 2 (1953) pp. 281–368.

4. ———, *Variational problems in the theory of elliptic partial differential equations*, Journal of Rational Mechanics and Analysis vol. 2 (1953) pp. 137–171.

5. P. R. Garabedian and D. C. Spencer, *Extremal methods in cavitational flow*, Journal of Rational Mechanics and Analysis vol. 1 (1952) pp. 359–409.

6. G. Golusin, *Method of variations in the theory of conformal mapping*, Mat. Sbornik N.S. vol. 19 (1946) pp. 203–236; vol. 21 (1947) pp. 83–117, 119–132.

7. J. Hadamard, *Mémoire sur le problème d'analyse relatif à l'équilibre des plaques élastiques encastrées*, Mémoires des Savants Etrangers vol. 33 (1908).

8. D. Hilbert, *Grundzüge einer allgemeinen Theorie der linearen Integralgleichungen*, Leipzig, 1912.

9. G. Julia, *Sur une équation aux dérivées fonctionelles liées à la représentation conforme*, Ann. École Norm. (3) vol. 39 (1922) pp. 1–28.

10. J. Leray, *Les problèmes de représentation conforme d'Helmholtz; théorie des sillages et des proues*, Comment. Math. Helv. vol. 8 (1935) pp. 149–180, 250–263.

11. K. Löwner, *Untersuchungen über schlichte konforme Abbildunges des Einheitskreises* I, Math. Ann. vol. 89 (1923) pp. 103–121.

12. G. C. Nooney, *On the vibrations of triangular membranes*, Technical Report No. 35, Office of Naval Research Contract N6ori-106 Task Order V, Stanford University.

13. G. Pólya and G. Szegö, *Isoperimetric inequalities of mathematical physics*, Princeton, 1951.

14. G. Pólya and M. Schiffer, *Convexity of functionals by transplantation*, Journal d'Analyse Mathématique, to appear.

15. H. Poincaré, *Figures d'équilibre d'une masse fluide*, Paris, 1902.

16. R. de Possel, *Zum Parallelschlitztheorem unendlich-vielfach zusammenhängender Gebiete*, Göttinger Nachr. (1931) pp. 199–202.

17. J. W. Rayleigh, *The theory of sound*, 2d ed., London, 1894.

18. D. Riabouchinsky, *Sur un problème de variation*, C. R. Acad. Sci. Paris vol. 185 (1927) pp. 840–841.

19. A. C. Schaeffer and D. C. Spencer, *Coefficient regions for schlicht functions*, Amer. Math. Soc. Colloquium Publications, vol. 35, New York, 1950.

20. M. Schiffer, *Variational methods in the theory of conformal mapping*, Proceedings of the International Congress of Mathematicians 1950, vol. II, pp. 233–240.

21. M. Schiffer and D. C. Spencer, *Functionals of finite Riemann surfaces*, Princeton, 1954.

22. M. Schiffer and G. Szegö, *Virtual mass and polarization*, Trans. Amer. Math. Soc. vol. 67 (1949) pp. 130–205.

23. O. Teichmüller, *Ungleichungen zwischen den Koeffizienten schlichter Funktionen*, Preuss. Akad. Wiss. Sitzungsber. Phys.-math. Klasse (1938) pp. 363–375.

STANFORD UNIVERSITY

Commentary on

[54] *Variation of domain functionals*, Bull. Amer. Math. Soc. **60** (1954), 303 328.

This survey article is a written version of an invited address given at a meeting of the American Mathematical Society. It can be viewed as a sequel to [17], which has a similar title. Comparing [54] with [17], one is amazed to see how far the variational method had progressed and how much Schiffer had accomplished, in the short span of eight years. In [17], he was just starting to look beyond univalent functions, broadening his outlook by adapting the variation of Green's function to problems involving transfinite diameter, harmonic measures, the Bergman kernel function, and Riemann surfaces. Eight years later, he had completed books on two of those topics, one with Bergman [51] on kernel functions and their application to boundary value problems for elliptic partial differential equations and the other with Spencer [55] applying variational methods to extremal problems for Riemann surfaces. He had written a series of important papers with Garabedian [26, 29, 44, 48, 49, 56] on application of variational techniques to existence theorems and three-dimensional problems of mathematical physics. He had also completed major papers with Szego [27] on virtual mass and polarization and with Pólya [53] on variation by transplantation, an ingenious method for verifying that a suspected extremal function is indeed extremal.

All of these topics are discussed in the survey paper [54], which lays heavy emphasis on the solution of extremal problems of classical physics. After a review of the variation of Green's function of a two-dimensional domain, Schiffer explains in detail how the method can be adapted to domains in three dimensions and discusses applications to electrostatics, eigenvalues of the Laplacian, and fluid dynamics. Adapting Hadamard's variational method [H], he then develops an expression for the second variation of Green's function of a smoothly bounded domain in $\mathbb{R}^3$, which he applies to prove the convexity of a certain domain functional defined in terms of electrostatic capacity. The paper concludes with an extended discussion of the method of transplantation, elucidated by applications to capacity, which occurs in the solution of three different problems of minimization: the classical principles of Dirichlet, Thomson, and Gauss. Similar problems are treated in the book of Pólya and Szego [PS].

References

[H] Jacques Hadamard, *Mémoire sur le problème d analyse relatif à l équilibre des plaques élastiques encastrées*, Mémoires présentés par divers savants à l'Académie de Sciences **33** (1908), N° 4, 1 128.

[PS] G. Pólya and G. Szego, *Isoperimetric Inequalities in Mathematical Physics*, Princeton University Press, 1951.

PETER DUREN

[59] (with P. R. Garabedian) A coefficient inequality for schlicht functions

[59] (with P. R. Garabedian) A coefficient inequality for schlicht functions. *Ann. of Math.* **61** (1955), 116–136.

ANNALS OF MATHEMATICS
Vol. 61, No. 1, January, 1955
Printed in U.S.A.

A COEFFICIENT INEQUALITY FOR SCHLICHT FUNCTIONS

BY P. R. GARABEDIAN AND M. SCHIFFER

(Received June 21, 1954)

1. Introduction

A great part of the theory of conformal mapping has been built around the study of the coefficients a_n of functions

$$(1) \qquad f(z) = z + a_2 z^2 + a_3 z^3 + a_4 z^4 + \cdots$$

schlicht in the interior $|z| < 1$ of the unit circle and the coefficients b_n of functions

$$(2) \qquad g(z) = z + b_0 + \frac{b_1}{z} + \frac{b_2}{z^2} + \frac{b_3}{z^3} + \cdots$$

schlicht in the exterior $|z| > 1$ of the unit circle. The estimates of these coefficients which have been obtained fall essentially into two classes, namely, those which follow in a natural way from the area theorem

$$(3) \qquad 1 \geq |b_1|^2 + 2|b_2|^2 + 3|b_3|^2 + \cdots$$

or its variants, and those, such as Loewner's theorem $|a_3| \leq 3$, which cannot be derived from such elementary considerations. While most of the useful distortion theorems of conformal mapping are consequences of the area theorem (3), there is nevertheless a great interest attached to the more remote class of inequalities because of the unanswered status of the Bieberbach conjecture $|a_n| \leq n$.

The principal result of the present paper is the sharp inequality

$$(4) \qquad |b_3| \leq \tfrac{1}{2} + e^{-6},$$

which belongs in this latter category. For earlier coefficients, the estimates $|b_1| \leq 1$ and $|a_2| \leq 2$ follow quickly from (3), while even the more difficult inequality $|b_2| \leq \tfrac{2}{3}$ can be deduced from a generalized area principle. Next in order of difficulty comes the theorem $|a_3| \leq 3$ due to Loewner [3], for which we shall give here a new and particularly simple proof. Thus the bound (4) on $|b_3|$ represents possibly the farthest point yet reached in estimating the higher coefficients of schlicht functions.

Our method of proving (4) is based on the differential equation for the schlicht function maximizing $|b_3|$ which results from an application of interior variations. We suppose that by now the precise derivation of such differential equations is familiar to the student of schlicht functions. Our contribution lies rather in determining the correct values of the parameters which appear in the differential equation, and this permits us to integrate the equation in closed form and find the largest value of $|b_3|$. Underlying our manipulations are a set of identities involving elliptic integrals which determine the parameters in the

116

differential equation in such a way that its solution is a schlicht function. The main difficulties of the investigation center about a successful analysis of these identities. It is because the corresponding identities for the case of higher coefficients involve hyperelliptic integrals that the Bieberbach conjecture $|a_n| \leqq n$ remains an unsettled problem.

A special significance attaches to the sharp estimate (4) because this result forces rejection of the earlier conjecture [9] that

$$(5) \qquad\qquad |b_n| \leqq \frac{2}{n+1}, \qquad\qquad n = 1, 2, 3, \cdots,$$

with equality holding for essentially only the function

$$(6) \qquad\qquad g(z) = (z^{n+1} + 2 + z^{-n-1})^{1/(n+1)}.$$

While this mapping function is a solution of the differential equation and the associated parameter relations, we succeed nevertheless in finding, for $n = 3$, another solution with a larger third coefficient. Proof of (4) consists merely in showing that this new solution and (6) are actually the only functions fulfilling the requirements upon an extremal mapping. The existence of superfluous solutions of the differential equations again illustrates the difficulties inherent in the coefficient problem for schlicht functions and indicates that a naive approach through conjectures based on familiar elementary maps is of no avail. We emphasize, however, that our advance here does not cast doubt on the Bieberbach conjecture, since we obtain an extremal function for (4) which has real coefficients and the Bieberbach conjecture has already been established for functions with real coefficients.

In the next section, we illustrate our fundamental technique by giving a new proof of Loewner's theorem, based on the differential equation for the extremal function. The sections following are devoted to the more tedious proof of the inequality (4). Closing portions of the paper take up corollaries of the principal theorem, such as the inequalities

$$(7) \qquad\qquad \mathrm{Re}\{b_3 - 3ib_1\} \leqq 3, \qquad \mathrm{Re}\{b_2 + 2b_1\} \leqq 2,$$

or indicate results which fit appropriately within the broader scope of our investigation.

2. Proof that $|a_3| \leqq 3$

Each schlicht function $f(z)$ of the form (1) generates a schlicht function $g(z)$ of the form (2) according to the rule

$$(8) \qquad\qquad g(z) = \frac{1}{f(1/z)} = z - a_2 + \frac{a_2^2 - a_3}{z} + \cdots.$$

If the behavior of $g(z)$ on the unit circle is sufficiently regular, we find by the residue theorem

$$(9) \qquad\qquad b_0 = \frac{1}{2\pi i} \oint_{|z|=1} g(z)\,\frac{dz}{z},$$

or, using (8) and setting $z = e^{i\theta}$,

$$(10) \qquad -a_2 = b_0 = \frac{1}{2\pi} \int_0^{2\pi} \frac{1}{f(1/z)} \, d\theta.$$

With $t = g(z)$, we can consider the image Γ in the t-plane of the unit circle $|z| = 1$ and we can interpret the measure

$$(11) \qquad d\mu = \frac{1}{2\pi} \, d\theta$$

in the usual electrostatic sense. Thus we think of $d\mu$ as the natural charge distribution on Γ. It is a non-negative distribution of total charge 1, and we therefore call the coefficient

$$(12) \qquad b_0 = \int_\Gamma t \, d\mu$$

the conformal centroid, or centroid, of the set Γ. Formula (10) shows that the coefficient a_2 is related to the centroid of Γ by

$$(13) \qquad -a_2 = \int_\Gamma t \, d\mu,$$

and thus it is clear that $-a_2$ is a point inside the convex hull of Γ.

With these preliminaries behind us, we proceed to the problem of maximizing $|a_3|$. We write

$$(14) \qquad w = f(z)$$

and we choose for $f(z)$, without loss of generality, the extremal function maximizing $|a_3|$ such that $a_3 > 0$. It follows from the method of interior variation that the extremal function (14) satisfies the ordinary differential equation [4, 5, 7, 8]

$$(15) \qquad \frac{dw^2}{w^2} \left(\frac{1}{w^2} + \frac{2a_2}{w} \right) = \frac{dz^2}{z^2} \left(\frac{1}{z^2} + \frac{2a_2}{z} + 2a_3 + 2\bar{a}_2 z + z^2 \right),$$

where the parenthesis on the right is non-negative for $|z| = 1$. Thus if we put

$$(16) \qquad t = \frac{1}{w} = \frac{1}{f(z)},$$

we find that the image Γ in the t-plane of the unit circle $|z| = 1$ consists of analytic arcs satisfying the differential equation

$$(17) \qquad \mathrm{Re} \left\{ \frac{(t + 2a_2)^{\frac{1}{2}}}{t^{\frac{1}{2}}} \, dt \right\} = 0.$$

The conformal transformation

$$(18) \qquad H = \int_0^t \frac{(t + 2a_2)^{\frac{1}{2}}}{t^{\frac{1}{2}}} \, dt$$

performs a univalent map of either of the half-planes bounded by the line L through the points 0 and $-2a_2$ in the t-plane onto a polygonal region R of the H-plane bounded by a linear ray from the origin, a finite line segment joining this ray at the origin under an angle of 90° with respect to the region, and a second infinite linear ray separating from the other end of the segment at an angle of 270° with respect to the region. The expression (18) is, in fact, merely a Schwarz-Christoffel transformation of a rotated half-plane. Now the arcs Γ in the t-plane correspond to a segment of the imaginary axis in the H-plane, according to the differential equation (17). Furthermore, if this segment, starting out from the origin, enters one of the above regions R, it must remain there, since R consists of the sum of two quadrants. Thus the curve Γ must either coincide with the line L between 0 and $-2a_2$, possibly forking at $-2a_2$, or else, if we overlook the origin, Γ must lie entirely interior to one of the half-planes bounded by L. We shall exclude this latter possibility.

Indeed, if Γ lies in the interior of one of the two half-planes bounded by L, then so does its centroid with respect to the natural charge distribution (11). But from the explicit calculation (13), the centroid lies on the line L, halfway between 0 and $-2a_2$. Thus Γ can lie in no such half-plane and must actually coincide with L between 0 and $-2a_2$, with a possible fork at the latter point.

To exclude the fork, we notice that such a configuration would entail two end-points of Γ corresponding to two double zeros of the right-hand side of the differential equation (15). Thus we would have

$$(19) \qquad \left(\frac{1}{z^2} + \frac{2a_2}{z} + 2a_3 + 2\bar{a}_2 z + z_2 \right) = \left(\frac{1}{z} + a_2 + z \right)^2,$$

whence

$$(20) \qquad 2a_3 = a_2^2 + 2.$$

Since $|a_2| \leqq 2$, this leads to the conclusion

$$(21) \qquad |a_3| \leqq \frac{|a_2|^2}{2} + 1 \leqq 3.$$

For equality to hold in (21) we must require $|a_2| = 2$, and this is true essentially only for the Koebe slit mapping

$$(22) \qquad w = \frac{z}{(1 - z)^2}.$$

Notice that (20) follows even when Γ does not fork at $-2a_2$, since in that case the right-hand side of (15) must have a quadruple root there.

This completes our proof of Loewner's theorem. It is based on an appropriate use of the identity (13), obtained from the schlicht character of the mapping (14), and it exploits in an essential way a geometrical analysis of the behavior of solutions of the differential equation (17).

3. The inequality for b_3

We proceed to the proof of (4) in several stages. Since the differential equation for the extremal function $g(z)$ maximizing $|b_3|$ and normalized so that $b_3 > 0$ is less familiar than the analogous differential equations for functions schlicht in the interior of the unit circle, we sketch a derivation of this equation. We stress that the derivation presented here is heuristic, and we refer to the literature for an exact treatment [7, 10].

The extremal function

$$(23) \qquad t = g(z)$$

maps the unit circle $|z| = 1$ onto a system of curves Γ. There is no loss of generality if we assume throughout that $b_0 = 0$, since this can be achieved simply by a translation of Γ. Let Γ_ρ be a small arc of Γ of outer mapping radius ρ and let t_0 be a point of Γ_ρ. Then there is a conformal mapping of the form

$$(24) \qquad \zeta = t - t_0 + C_0\rho + \frac{C_1\rho^2}{(t - t_0)} + \frac{C_2\rho^3}{(t - t_0)^2} + \cdots$$

taking the exterior of Γ_ρ into the exterior $|\zeta| > \rho$ of the circle of radius ρ in the ζ-plane. We introduce the special functions

$$(25) \qquad g^* = \zeta + \frac{B_1\rho^2}{\zeta},$$

with

$$(26) \qquad |B_1| \leqq 1,$$

which are schlicht in $|\zeta| > \rho$. The coefficient B_1 can be chosen arbitrarily except for the condition (26). It is well known that for each n the coefficients C_n are bounded uniformly in ρ.

By composition of the mappings (23), (24), and (25), we construct for $|z| > 1$ the schlicht function

$$g^*(z) = z + \frac{b_1}{z} + \frac{b_2}{z^2} + \frac{b_3}{z^3} + \cdots - t_0 + C_0\rho + \frac{(C_1 + B_1)\rho^2}{g(z) - t_0} + o(\rho^2)$$

$$(27) \qquad = z - t_0 + C_0\rho + \frac{b_1 + (C_1 + B_1)\rho^2}{z} + \frac{b_2 + t_0(C_1 + B_1)\rho^2}{z^2}$$

$$+ \frac{b_3 + (t_0^2 - b_1)(C_1 + B_1)\rho^2}{z^3} + \cdots + o(\rho^2).$$

From the extremal property of $g(z)$ it follows that

$$(28) \qquad |b_3 + (t_0^2 - b_1)(C_1 + B_1)\rho^2 + o(\rho^2)| \leqq |b_3|.$$

We let $\rho \to 0$ and we note that $C_1 \to -e^{2i\varphi}$, where φ is the angle of inclination of the tangent to Γ at t_0. Since $b_3 > 0$, we derive from (28) in the limit as $\rho \to 0$ the inequality

$$(29) \qquad \mathrm{Re}\{(t_0^2 - b_1)(B_1 - e^{2i\varphi})\} \leqq 0,$$

where B_1 is any complex number satisfying (26). Because of the freedom in the choice of B_1, the variational condition (29) yields the relation

$$(30) \qquad (t_0^2 - b_1)\, dt_0^2 \geqq 0$$

for the differential element dt_0 of Γ at the point t_0. This result is actually a differential equation for the system of arcs Γ.

We can derive from (30) a differential equation for the extremal function $g(z)$. Consider the expression

$$(31) \qquad z^2 g'(z)^2 [g(z)^2 - b_1] = z^4 - b_1 z^2 - 2b_2 z - 4b_3 + \cdots,$$

which is an analytic function of z in the exterior of the unit circle, except for the indicated pole at infinity. According to (30), this function must be real for $|z| = 1$, and hence we can continue it analytically into the interior of the unit circle by the Schwarz reflection principle. The function has a pole at the origin determined by the expansion (31), and hence we are able to calculate it explicitly and obtain

$$(32) \qquad z^2 \frac{dg^2}{dz^2} (g^2 - b_1) = z^4 - b_1 z^2 - 2b_2 z - 4b_3 - \frac{2\bar{b}_2}{z} - \frac{\bar{b}_1}{z^2} + \frac{1}{z^4}.$$

This is the desired differential equation for the schlicht function g, and we note only that, according to (30), the right-hand side must be non-positive for $|z| = 1$.

We turn to the rigorous integration of (32). There are three characteristically different cases to be considered. (i) The curves Γ contain both the square roots of b_1. In general, Γ will fork at these points and will have four end-points, each of which corresponds to a double root of the right-hand side of (32) on the unit circle $|z| = 1$. However, Γ might not fork and might even terminate at a critical point, but (32) will still have four double roots whenever Γ contains the two square roots of b_1. (ii) The curves Γ contain precisely one of the square roots of b_1; Γ will have, in general, three end-points, so that the right-hand side of (32) has only three double roots on $|z| = 1$ and the two remaining roots lie at inverse points inside and outside the unit circle. (iii) The set Γ does not contain either of the square roots of b_1 and hence consists of a simple arc without forks and with only two end-points, so that the right-hand side of (32) has two double roots and four simple roots. In this last case, the integration of (32) involves elliptic integrals, whereas in the first two cases only elementary integrals are required.

In later sections of the paper we shall prove that cases (ii) and (iii) can be excluded. We study in this section only case (i) and we determine the actual extremal function which maximizes $|b_3|$.

In case (i), the right-hand side of (32) has four double roots and is a perfect square, whence

$$(33) \qquad z^4 - b_1 z^2 - 2b_2 z - 4b_3 - \frac{2\bar{b}_2}{z} - \frac{\bar{b}_1}{z^2} + \frac{1}{z^4} = \left(z^2 - \frac{b_1}{2} - \frac{1}{z^2} \right)^2$$

and

(34) $$b_2 = 0, \qquad b_3 = \frac{1}{2} - \frac{b_1^2}{16}.$$

The coefficient b_1 must be pure imaginary, since both sides of (33) are non-positive. We can now take the square root of both sides of (32) and integrate to obtain

(35) $$\frac{g(g^2 - b_1)^{\frac{1}{2}}}{2} - \frac{b_1}{2} \log \frac{g + (g^2 - b_1)^{\frac{1}{2}}}{b_1^{\frac{1}{2}}} = \frac{z^2}{2} + \frac{1}{2z^2} - \frac{b_1}{2} \log z + K,$$

where K is a constant of integration.

In order to evaluate K, we expand (35) about the point at infinity in the z-plane, using (2) and remembering that we took $b_0 = 0$. By noting that the constant terms on both sides of the equation must be the same, we find that

(36) $$K = \frac{3}{4} b_1 - \frac{b_1}{4} \log \frac{4}{b_1}.$$

Since Γ passes through the point $b_1^{\frac{1}{2}}$, there exists a value z_0 of z on the unit circle $|z| = 1$ such that

(37) $$g(z_0) = b_1^{\frac{1}{2}}.$$

Substitution of this value of z into (35) yields the additional relation

(38) $$0 = \frac{z_0^2}{2} + \frac{1}{2z_0^2} - \frac{b_1}{2} \log z_0 + K.$$

The terms in (38) involving z_0 are real, and hence K is real. Therefore by (36) we have, since b_1 is pure imaginary,

(39) $$3b_1 + b_1 \log \frac{|b_1|}{4} = 0.$$

Using again the imaginary character of b_1, we deduce from equation (39) that either $b_1 = 0$ or else $|b_1| = 4e^{-3}$ and

(40) $$b_1 = 4ie^{-3}.$$

The root $b_1 = 0$ of (39) leads to the solution (6) of (32), with $n = 3$, and $b_3 = \frac{1}{2}$. On the other hand, the value (40) for b_1 leads by (34) to the value

(41) $$b_3 = \tfrac{1}{2} + e^{-6}$$

of the third coefficient of $g(z)$. This value is the larger of the two, and thus the function maximizing $|b_3|$ must be the one defined, according to (35), by the implicit relation

(42) $$g(g^2 - 4ie^{-3})^{\frac{1}{2}} - 4ie^{-3} \log \frac{g + (g^2 - 4ie^{-3})^{\frac{1}{2}}}{2z} = z^2 + \frac{1}{z^2} + 6ie^{-3}.$$

The extremal function (42) maps the exterior of the unit circle in the z-plane onto the exterior of a system of arcs Γ which consists of a line segment joining the two square roots of $4ie^{-3}$ and four analytic arcs forking from these square roots at angles of 120°. We remark that an extremal function with real coefficients can be obtained from the present one by rotation, and, indeed, $e^{(\pi i)/4}g(ze^{-(\pi i)/4})$ is such a function. However, its third coefficient is negative.

In order to establish that (41) is actually the largest value of b_3, we must exclude the above cases (ii) and (iii). This will be done in the next sections by a method based on the knowledge that for schlicht solutions of (32), the singular points of the differential equation in the z-plane must correspond to the singular points in the t-plane.

4. Exclusion of elliptic integrals

We establish in this section that there are no schlicht solutions of (32) in the case (iii) where the hypothesis is that the image Γ in the t-plane of the unit circle $|z| = 1$ consists of a single analytic arc and does not pass through the branch points of (30) at the square roots of b_1. Actual integration of the right-hand side of (32) in this case would involve the use of elliptic integrals, and the success of our treatment stems from the fact that we are able to avoid such a step and work only with the left-hand side of the equation, or, more precisely, with (30). Our first remark is that we do not lose any generality if we suppose that the z-plane and the t-plane have been rotated so that $b_1 \geq 0$, while b_3 is no longer necessarily real. This new normalization is more convenient for our study of (30), but we notice that (30) must now be replaced by the more general differential equation

$$(43) \qquad e^{2i\alpha}(t^2 - b_1)\, dt^2 > 0$$

for the arc Γ, where α is a fixed real parameter depending on the angle of rotation. The normalization $b_0 = 0$, made previously, is not altered by the rotation of coordinates.

We introduce the integral

$$(44) \qquad H = \int_0^t (t^2 - b_1)^{\frac{1}{2}}\, dt$$

and we point out that the differential equation (43) merely states that Γ is the image by the transformation (44) of a line segment L in the H-plane inclined at an angle $-\alpha$ with the real axis. This interpretation of (43) permits us to show that Γ cuts the real axis between $-b_1^{\frac{1}{2}}$ and $+b_1^{\frac{1}{2}}$. In fact (44), viewed as a Schwarz-Christoffel transformation, maps the upper half of the t-plane onto the exterior of a semi-infinite strip of the form

$$(45) \qquad \operatorname{Re} H > 0, \qquad -h < \operatorname{Im} H < h,$$

where

$$(46) \qquad h = \operatorname{Im} \int_0^{b_1^{\frac{1}{2}}} (t^2 - b_1)^{\frac{1}{2}}\, dt.$$

In order to establish that Γ cuts the real axis between $-b_1^{\frac{1}{2}}$ and $+b_1^{\frac{1}{2}}$, it suffices to prove that Γ cuts both the real axis and the imaginary axis. For if this is true, then the line segment L in the H-plane corresponding to Γ must cut both the boundary of the semi-infinite strip (45) and the negative real axis in the H-plane, since these map by (44) into the real and imaginary axes in the t-plane. Any line segment L with the above properties has to intersect the segment $\operatorname{Re} H = 0$, $-h < \operatorname{Im} H < h$, and since this segment corresponds to the interval between $-b_1^{\frac{1}{2}}$ and $+b_1^{\frac{1}{2}}$, the arc Γ must cut that interval.

It remains, then, to show that Γ cuts both the real axis and the imaginary axis. But, in the notation of formulas (9) and (12), we interpret the normalization $b_0 = 0$ to mean that the centroid b_0 of Γ lies at the origin, or in other words

$$(47) \qquad b_0 = \int_\Gamma t \, d\mu = 0.$$

This can only occur if Γ cuts both coordinate axes, and thus our lemma is established. A special application of this argument shows in addition that $b_1 \neq 0$.

We must consider all positions of the arc Γ which are consistent with the properties that it cuts the real axis between $-b_1^{\frac{1}{2}}$ and $+b_1^{\frac{1}{2}}$, that it has outer mapping radius 1, and that its centroid lies at the origin. We shall prove that these properties imply that Γ passes through the origin and is symmetric in the origin. It is evident that the converse is true, namely, that an arc which satisfies the differential equation (43) and is symmetric in the origin must have its centroid at the origin. Thus our proof can be carried out by establishing the uniqueness for each value of the parameter α of this solution of the equation (47).

As a preliminary, we derive variational formulas for the capacity γ and centroid b_0 of the curve Γ. We shall need such formulas for a shift of Γ corresponding to infinitesimal translation and magnification of the image segment L in the H-plane. We denote the shifted segment by L^* and we let Γ^* denote the correspondingly varied arc in the t-plane. We shall use addition of an asterisk to indicate all quantities associated with the varied configuration. The infinitesimal transformation carrying L into L^* can be written in the form

$$(48) \qquad H^* = (1 + \varepsilon_1)H + \varepsilon_2,$$

where ε_1 is a small real number and ε_2 is a small complex number. We set $\varepsilon = \max (|\,\varepsilon_1\,|, |\,\varepsilon_2\,|)$.

We denote by $p(t)$ the analytic function whose real part is the Green's function of the exterior of Γ with pole at infinity. Thus near infinity

$$(49) \qquad \operatorname{Re} p(t) = \log |\,t\,| - \gamma + O\left(\frac{1}{|\,t\,|}\right),$$

and e^γ is the outer mapping radius of Γ. We let $\varphi(t)$ be the function, analytic in the exterior of Γ, which has a pole of the form

$$(50) \qquad \varphi(t) = t + c_0 + \frac{c_1}{t} + \frac{c_2}{t^2} + \cdots$$

at infinity and which has real boundary values on Γ. Similarly, $\psi(t)$ is defined to be the function, analytic in the exterior of Γ, which has a pole of the form

$$(51) \qquad \psi(t) = t + d_0 + \frac{d_1}{t} + \frac{d_2}{t^2} + \cdots$$

at infinity and which has imaginary boundary values on Γ. The centroid b_0 of Γ can be expressed in terms of the expansions (50) and (51) by the formula

$$(52) \qquad b_0 = -\operatorname{Re} d_0 - i \operatorname{Im} c_0 = \beta_1 + i\beta_2,$$

since, in terms of the mapping (23),

$$(53) \qquad \varphi = z + \frac{1}{z} + \text{real const.},$$

$$(54) \qquad \psi = z - \frac{1}{z} + \text{imaginary const.}$$

Variational formulas for the capacity γ and centroid b_0 can be found using the domain functions p, φ, and ψ. We denote by $t^*(t)$ the infinitesimal transformation of Γ onto Γ^* induced by (44) and (48). By the residue theorem, we find

$$(55) \qquad \gamma^* - \gamma = \operatorname{Re} \frac{1}{2\pi i} \oint [p(t) - p^*(t)] \, dp(t),$$

where the path of integration is a closed curve surrounding Γ. Since $p(t)$ and $p^*(t^*(t))$ are pure imaginary on Γ, we can rewrite (55) in the form

$$(56) \qquad \gamma^* - \gamma = \operatorname{Re} \frac{1}{2\pi i} \oint [p^*(t^*) - p^*(t)] \, dp(t).$$

We wish to evaluate the integral over the corresponding path in the H-plane, and thus we replace t as the independent variable by H to obtain

$$(57) \qquad \gamma^* - \gamma = \operatorname{Re} \frac{1}{2\pi i} \oint [p^*(H^*) - p^*(H)] \, dp(H),$$

where we have used a loose notation that does not indicate explicitly the change in the functional form of p under the transformation from the t-plane to the H-plane.

From (48) and (57) we derive the variational formula

$$(58) \qquad \gamma^* - \gamma = \operatorname{Re} \frac{1}{2\pi i} \oint (\varepsilon_1 H + \varepsilon_2) p'(H)^2 \, dH + o(\varepsilon),$$

where the path of integration is a curve in the H-plane enclosing the line segment L. In a similar way, we derive from the expressions

$$(59) \qquad \beta_1^* - \beta_1 = \operatorname{Re} \frac{1}{2\pi i} \oint [\psi(t) - \psi^*(t)] \, dp(t),$$

$$(60) \qquad \beta_2^* - \beta_2 = \operatorname{Im} \frac{1}{2\pi i} \oint [\varphi(t) - \varphi^*(t)] \, dp(t)$$

the variational formulas

$$(61) \qquad \beta_1^* - \beta_1 = \operatorname{Re} \frac{1}{2\pi i} \oint (\varepsilon_1 H + \varepsilon_2)\psi'(H)p'(H) \, dH + o(\varepsilon),$$

$$(62) \qquad \beta_2^* - \beta_2 = \operatorname{Im} \frac{1}{2\pi i} \oint (\varepsilon_1 H + \varepsilon_2)\varphi'(H)p'(H) \, dH + o(\varepsilon)$$

for the real and imaginary parts β_1 and β_2 of the centroid b_0 of Γ.

Formula (58) can be simplified by evaluating the integral in the t-plane and by deforming the contour of integration into a path which consists of two small circles Ω_1 and Ω_2 about the end-points t_1 and t_2 of Γ and of the two edges of Γ joining these circles. In order to calculate the integrals over Ω_1 and Ω_2, we introduce the local uniformizers $\omega_1 = 2(t - t_1)^{\frac{1}{2}}$ and $\omega_2 = 2(t - t_2)^{\frac{1}{2}}$. We let the radii of the circles Ω_1 and Ω_2 tend to zero. The limit of the integral over Ω_k is $\operatorname{Re} \{\delta t_k (dp/d\omega_k)^2\}$, $k = 1, 2$, where δt_k is the displacement of t_k under the shift (48). The integral over the two edges of Γ has the limiting value

$$(2\pi)^{-1} \int_\Gamma (\partial p/\partial \nu)^2 \delta \nu \, ds,$$

where ν is the inner normal and s is the arc length along Γ, where the integration is carried out along both edges of Γ, and where $\delta \nu$ is the normal displacement of Γ under the shift (48). We denote by $\partial p/\partial \nu_k$ the normal derivative of the function p in the plane of the uniformizer ω_k and we denote by $\delta \nu_k$ the tangential projection with respect to Γ of the shift δt_k, $k = 1, 2$. Thus we obtain from (58) the Hadamard formula

$$(63) \qquad \delta \gamma = \frac{1}{2\pi} \int_\Gamma \left(\frac{\partial p}{\partial \nu}\right)^2 \delta \nu \, ds + \left(\frac{dp}{d\nu_1}\right)^2 \delta \nu_1 + \left(\frac{\partial p}{\partial \nu_2}\right)^2 \delta \nu_2 + o(\varepsilon),$$

and similarly, in an analogous notation, (61) and (62) yield

$$(64) \qquad \delta \beta_1 = \frac{1}{2\pi} \int_\Gamma \frac{\partial p}{\partial \nu} \frac{\partial \psi}{\partial \nu} \delta \nu \, ds + \frac{\partial p}{\partial \nu_1} \frac{\partial \psi}{\partial \nu_1} \delta \nu_1 + \frac{\partial p}{\partial \nu_2} \frac{\partial \psi}{\partial \nu_2} \delta \nu_2 + o(\varepsilon),$$

$$(65) \qquad \delta \beta_2 = \frac{1}{2\pi i} \int_\Gamma \frac{\partial p}{\partial \nu} \frac{\partial \varphi}{\partial \nu} \delta \nu \, ds + \frac{1}{i} \frac{\partial p}{\partial \nu_1} \frac{\partial \varphi}{\partial \nu_1} \delta \nu_1 + \frac{1}{i} \frac{\partial p}{\partial \nu_2} \frac{\partial \varphi}{\partial \nu_2} \delta \nu_2 + o(\varepsilon),$$

where $\delta \gamma = \gamma^* - \gamma$, $\delta \beta_1 = \beta_1^* - \beta_1$, and $\delta \beta_2 = \beta_2^* - \beta_2$. Notice that the derivatives $\partial p/\partial \nu$, $\partial \psi/\partial \nu$ and $-i\partial \varphi/\partial \nu$ are actually real numbers.

We attempt to choose the arc Γ as a solution of the differential equation (43) which cuts between $-b_1^{\frac{1}{2}}$ and $+b_1^{\frac{1}{2}}$ in such a way that the three equations

$$(66) \qquad \gamma = 0, \qquad \beta_1 = 0, \qquad \beta_2 = 0$$

are fulfilled. Given the upper end-point t_1 of Γ, the equation $\gamma = 0$ clearly determines the lower end-point t_2, since γ is a monotonic domain functional. We now establish that, along a prescribed curve solving (43), the equation

$\beta_2 = 0$ determines the upper end-point t_1 uniquely. This is a consequence of the variational formula (65) because, as we shall prove in a moment,

(67) $$\frac{1}{i}\frac{\partial\varphi}{\partial\nu_1} > 0, \qquad \frac{1}{i}\frac{\partial\varphi}{\partial\nu_2} < 0.$$

The inequalities (67) show that as t_1 rises along a solution of (43), and as t_2 follows t_1 so that $\gamma = 0$, the quantity β_2 increases monotonically. Thus, indeed, β_2 vanishes only once and t_1 is uniquely determined.

It remains to establish the inequalities (67). We prove first that our solution of (43) is a curve which intersects each horizontal line just once. Indeed, if such a curve intersects a horizontal line l more than once, it intersects it essentially at least three times, and between three intersections on the horizontal line l there will be two points where Im $\{e^{i\alpha}H\}$ is stationary, by Rolle's theorem. There is on the same horizontal line l an additional stationary value of Im $\{e^{i\alpha}H\}$ which occurs between the two intersections with the line l of one of the solutions of (43) which forks through $-b_1^{\frac{1}{2}}$ or $+b_1^{\frac{1}{2}}$. This accounts for at least three stationary values of Im $\{e^{i\alpha}H\}$ on the single horizontal line l, and at each such stationary point we find from differentiation of (44)

(68) $$\text{Im } e^{2i\alpha}(t^2 - b_1) = 0.$$

For t on the line l, (68) reduces to a real quadratic equation, and thus it can have at most two roots. Thus three stationary values could not appear, and we have proved that a solution of (43) which cuts the real axis between $-b_1^{\frac{1}{2}}$ and $+b_1^{\frac{1}{2}}$ must cut each horizontal line just once.

The inequalities (67) can now be deduced from the maximum principle. On the slit Γ the function Im $\{\varphi(t) - t + t_1\}$ is non-negative because Im $\{\varphi\} = 0$ and Im $\{t - t_1\} \leqq 0$ there. Hence, by the maximum principle, this function is positive in the exterior of Γ, and the first inequality (67) follows by differentiation when we note that at $t = t_1$ we have Im $\{\varphi(t) - t + t_1\} = 0$. Similarly, the harmonic function Im $\{\varphi(t) - t + t_2\}$ is negative in the exterior of Γ because it is non-positive on Γ, and since this function vanishes at $t = t_2$, we obtain the second inequality (67).

Thus we have shown that the equations $\gamma = 0$, $\beta_2 = 0$ determine a unique arc Γ on each level curve

(69) $$\text{Im } \{e^{i\alpha}H(t)\} = \lambda$$

cutting between $-b_1^{\frac{1}{2}}$ and $+b_1^{\frac{1}{2}}$. The problem is therefore to find all the values of λ such that $\beta_1 = 0$. It is clear from (44) that when $\lambda = 0$ the arc Γ satisfying $\gamma = \beta_2 = 0$ passes through the origin and is symmetric in the origin, so that $\beta_1 = 0$. We shall prove that this is the only choice for λ which gives $\beta_1 = 0$. Suppose, indeed, that there were another value λ_0 of λ for which we could find an arc of the level curve (69) satisfying all the equations (66). Then we could vary λ between 0 and λ_0 and consider for each intermediate value of λ the arc (69) with $\gamma = \beta_2 = 0$. By Rolle's theorem, there would exist an intermediate value λ_1 of λ for which β_1 would be stationary. Thus for the corresponding arc Γ and an appropriate variation of the type (48) we would obtain

$$(70) \qquad\qquad \delta\gamma = \delta\beta_1 = \delta\beta_2 = 0.$$

We shall establish that this is impossible, and therefore that the solution of (66) is unique.

For arbitrary real values of X, Y, and Z we consider the expression

$$(71) \qquad\qquad q(t) = Xp(t) + Y\psi(t) + Z\frac{\varphi(t)}{i}.$$

We can choose X, Y, and Z so that at the end-points t_1 and t_2 of the arc Γ satisfying (70) the conditions

$$(72) \qquad\qquad \frac{\partial q}{\partial \nu_1} = 0, \qquad \frac{\partial q}{\partial \nu_2} = 0$$

are fulfilled. This follows because $\partial q/\partial \nu$ is proportional to a linear combination of the functions 1, $\cos\theta$, and $\sin\theta$ on the unit circle $z = e^{i\theta}$ in the z-plane. By the argument principle, (72) accounts for all the zeros of the real quantity $\partial q/\partial \nu$.

With this choice of X, Y, and Z, the variational formulas (63), (64), and (65) yield for any shift (48) of Γ the simple relation

$$(73) \qquad X\delta\gamma + Y\delta\beta_1 + Z\delta\beta_2 = \frac{1}{2\pi}\int_\Gamma \frac{\partial p}{\partial \nu}\frac{\partial q}{\partial \nu} \delta\nu\, ds + o(\varepsilon).$$

However, the right-hand side of (73) does not vanish for the shift $\delta\nu$ which gave (70), because $\delta\nu$ is of one sign on one side of Γ and of the opposite sign on the other side of Γ for such a shift, and because by (72) the same is true of $\partial q/\partial \nu$. Thus we arrive at a contradiction.

Therefore, we conclude that the curve Γ must be symmetric in the origin. Next, if the parameter α of formula (69) lies in the interval $0 < \alpha < \pi/2$, we arrive at a contradiction because of the identity

$$(74) \qquad\qquad 2b_1 = \frac{1}{2\pi i}\int_{|z|=1} g(z)^2\,\frac{dz}{z} = \int_\Gamma t^2\, d\mu.$$

Indeed, the left-hand side of (74) is real and positive, whereas the right-hand side must have a positive imaginary part because Γ lies in the first and third quadrants. A similar conclusion holds when $-\frac{1}{2}\pi < \alpha < 0$, and when $\alpha = 0$ a contradiction is obtained because the two sides of (74) are real with opposite signs. Finally, the case $\alpha = \pi/2$ must be excluded also, because now in (74) we have $t^2 < b_1$ and the right-hand side is actually smaller than $2b_1$.

We thus obtain the final result that a single arc Γ without forks cannot occur in the solution of (32). We have therefore established that the case (iii) does not appear, and it remains only to exclude the case (ii) in order to prove the original inequality (4).

5. The equations for b_1 and b_2

Our treatment of case (ii) is based on direct integration of the differential equation (32). In the differential equation for any coefficient inequality, one can always find the correct number of conditions to determine all the coefficients which appear by expressing the fact that the singularities in the z-plane and the t-plane have to match up. In general, this procedure is not feasible because it involves hyperelliptic integrals, and even in the case (iii) it would have led to elliptic integrals. However, we are able to succeed with the method in case (ii), since the integrations can be executed in terms of elementary integrals.

The hypothesis in case (ii) is that the boundary Γ in the t-plane consists of three analytic arcs forking from one of the branch points $-b_1^{\frac{1}{2}}$ or $+b_1^{\frac{1}{2}}$ at angles of 120°. Since such a system of arcs has three end-points, the right-hand side of (32) must have three double zeros and it can be represented in the form

$$
(75) \quad
\begin{aligned}
z^4 - b_1 z^2 - 2b_2 z - 4b_3 - \frac{2\bar{b}_2}{z} - \frac{\bar{b}_1}{z^2} + \frac{1}{z^4} \\
= \left(E_1 z + E_2 + \frac{E_3}{z} + \frac{E_4}{z^2} \right)^2 (z^2 - 2rkz + k^2).
\end{aligned}
$$

Since the roots of the left-hand side of (75) lie at inverse points in the unit circle, we must have

$$
(76) \qquad |k| = 1, \qquad r > 1.
$$

Furthermore, the coefficients of z^4, z^3, z^{-3}, and z^{-4} on the left in (75) are known, so we obtain four conditions giving the coefficients E_1, E_2, E_3, and E_4 in terms of the two parameters r and k. Thus

$$
(77) \quad
\begin{aligned}
z^4 - b_1 z^2 - 2b_2 z - 4b_3 - \frac{2\bar{b}_2}{z} - \frac{\bar{b}_1}{z^2} + \frac{1}{z^4} \\
= \left(z + rk + \frac{r\bar{k}^2}{z} + \frac{\bar{k}}{z^2} \right)^2 (z^2 - 2rkz + k^2).
\end{aligned}
$$

By (77), we can express the three coefficients b_1, b_2, and b_3 in terms of the two real parameters r and arg k. In particular,

$$
(78) \qquad b_1 = 3r^2 k^2 - k^2 - 2r\bar{k}^2,
$$

and b_2 and b_3 can be expressed in terms of b_1.

In order to determine the unknown b_1, we have to integrate (32) explicitly. By (77), we can write (32) in the form

$$
(79) \qquad (g^2 - b_1)^{\frac{1}{2}} \, dg = \left(1 + \frac{rk}{z} + \frac{r\bar{k}^2}{z^2} + \frac{\bar{k}}{z^3} \right)(z^2 - 2rkz + k^2)^{\frac{1}{2}} \, dz.
$$

An equation for b_1 will result essentially from the fact that in the conformal transformation $t = g(z)$ one of the roots of $z^2 - 2rkz + k^2 = 0$ must map into a root of $t^2 - b_1 = 0$. We introduce the notation

(80) $$W = (z^2 - 2rkz + k^2)^{\frac{1}{2}},$$

and we integrate (79) to obtain

(81)
$$\frac{g(g^2 - b_1)^{\frac{1}{2}}}{2} - \frac{b_1}{2} \log [g + (g^2 - b_1)^{\frac{1}{2}}] + K$$
$$= -\frac{b_1}{2} \log [W + z - rk] + \frac{\bar{b}_1}{2} \log \left[r - \frac{W + k}{z} \right]$$
$$+ \frac{W}{2} \left(z + rk - \frac{r\bar{k}^2}{z} - \frac{k}{z^2} \right),$$

where K is a constant of integration. The reader can easily check (81) by direct differentiation.

The best way to evaluate K is to substitute for z in (81) a root of the equation $W = 0$; the corresponding value of g is $b_1^{\frac{1}{2}}$. This yields

(82) $$K = \frac{b_1}{4} \log \frac{b_1}{k^2(r^2 - 1)} + \frac{\bar{b}_1}{4} \log (r^2 - 1).$$

On the other hand, K can also be evaluated by expanding both sides of (81) about the point at infinity. We have thus

(83)
$$\frac{g^2}{2} - \frac{b_1}{4} - \frac{b_1}{2} \log 2g + K + O\left(\frac{1}{|g|}\right)$$
$$= \frac{z^2}{2} - \frac{3r^2k^2 - k^2 + 2r\bar{k}^2}{4} + \frac{\bar{b}_1}{2} \log (r - 1) - \frac{b_1}{2} \log 2z + O\left(\frac{1}{|z|}\right),$$

or, substituting for g the expansion (2) and letting $z \to \infty$,

(84)
$$K = -\frac{3b_1 + 3r^2k^2 - k^2 + 2r\bar{k}^2}{4} + \frac{\bar{b}_1}{2} \log (r - 1)$$
$$= -b_1 - r\bar{k}^2 + \frac{\bar{b}_1}{2} \log (r - 1).$$

The equations (78), (82), and (84) yield together a single equation for the determination of b_1. We prefer, however, to eliminate K and b_1 in order to obtain the one complex equation

(85)
$$\frac{k^2 + \bar{k}^2}{4} (3r + 1)(r - 1) \log (r - 1) + \frac{k^2 - \bar{k}^2}{4} (3r - 1)(r + 1)\log (r + 1)$$
$$= \frac{(3r^2 - 1)k^2 - 2r\bar{k}^2}{4} \log [3r^2 - 1 - 2r\bar{k}^4] + (3r^2 - 1)k^2 - r\bar{k}^2$$

for the two real unknowns r and arg k. Our objective is to prove that (85) has no solutions consistent with the hypothesis (76).

It is necessary to specify which branches of the logarithms are meant in equation (85). This question can be discussed by a more careful examination of our derivation of (85). We first perform rotations through 90° and reflections in the z-plane and the t-plane until b_1 lies in the first quadrant, while b_3 remains positive. Taking (82) into account, we then rewrite (81) in the form

$$(86) \quad g(g^2 - b_1)^{\frac{1}{2}} - b_1 \log \frac{[g + (g^2 - b_1)^{\frac{1}{2}}](r^2 - 1)^{\frac{1}{2}} k}{[W + z - rk] b_1^{\frac{1}{2}}}$$

$$= \bar{b}_1 \log \frac{rz - W - k}{z(r^2 - 1)^{\frac{1}{2}}} + W \left(z + rk - \frac{r\bar{k}^2}{z} - \frac{\bar{k}}{z^2} \right),$$

where it is now correct to take values of all the logarithms so that they vanish at the root z of the equations $W = 0$, $g = b_1^{\frac{1}{2}}$. In order to obtain (85), we let z increase from this root and become infinite along the ray $\arg z = \arg k$. A difficulty is encountered when we try to locate the corresponding trajectory of g.

The path covered by g obviously has the equation

$$(87) \qquad \operatorname{Im} p(g) = \arg k.$$

Also, we have the relation

$$(88) \qquad p'(g) = \int_\Gamma \frac{d\mu}{g - t},$$

so that the trajectory (87) is always directed away from the convex hull of Γ. Therefore, this trajectory cannot cross the ray $\arg g = \arg b_1^{\frac{1}{2}} + \pi$ beyond the point $-b_1^{\frac{1}{2}}$, since it starts out at the point $g = b_1^{\frac{1}{2}}$ and such a crossing would require an intermediate position in which its tangent does not cut Γ. Our analysis shows, then, that we are permitted to use only values of the logarithms in (86) which have an imaginary part lying between $-\pi$ and $+\pi$. It follows that in the final identity (85) the values of the logarithms should be chosen from this same principal branch. In particular, the logarithms on the left in (85) are real numbers.

Having determined the correct branch of the logarithms in (85), we proceed to pare down the region of variation of the parameters r and k which must be considered. We note that in addition to (78), the formula (77) yields the expressions

$$(89) \qquad b_2 = (r^3 - r)k^3 + (r^2 - 1)\bar{k},$$

$$(90) \qquad b_3 = r^3 - r^2 \frac{k^4 + \bar{k}^4}{4}$$

for the coefficients b_2 and b_3. Substituting these results into the area theorem (3), we find

$$(91) \quad | 3r^2 - 1 - 2r\bar{k}^4 |^2 + 2 | r^3 - r + (r^2 - 1)\bar{k}^4 |^2 + 3 \left| r^3 - r^2 \frac{k^4 + \bar{k}^4}{4} \right|^2 \leq 1.$$

From (76) and (91) we find easily that $\bar{k}^4$ lies in the right half-plane and that $4r^2[\operatorname{Im} \bar{k}^4]^2 + 3r^4/4 \leq 1$, whence

(92)
$$| \operatorname{Im} \bar{k}^4 | < \tfrac{1}{4}.$$

Thus we can replace (91) by the weaker estimate

$$(3r^2 - 2r - 1)^2 + 2r^2(r^2 - 1)^2 + 3r^4/4 \leq 1,$$

from which follows

$$(3r + 1)^2(r - 1)^2 + 2(r + 1)^2(r - 1)^2 \leq 1 - 3r^4/4 \leq \tfrac{1}{4},$$

or, finally,

(93)
$$(r - 1)^2[(3r + 1)^2 + 8] \leq \tfrac{1}{4}.$$

We write $r - 1 = \varepsilon$ and derive from (93) the inequality

(94)
$$\varepsilon^2 + \varepsilon^3 \leq \tfrac{1}{96},$$

whence

(95)
$$\varepsilon < \tfrac{1}{10}.$$

Rearranging the terms in (85) in a more suggestive notation, we must prove that the equation

(96)
$$(1 + \bar{k}^4)(4 + 3\varepsilon)\varepsilon \log \varepsilon + (1 - \bar{k}^4)(2 + 3\varepsilon)(2 + \varepsilon) \log (2 + \varepsilon)$$
$$= b_1\bar{k}^2 \log b_1\bar{k}^2 + 8 + 24\varepsilon + 12\varepsilon^2 - (4 + 4\varepsilon)\bar{k}^4$$

has no solutions in the region defined by (76), (92), and (95), when we use the principal branch of the logarithm. From (92) and (95) we obtain readily

(97)
$$\operatorname{Re} \{(1 + \bar{k}^4)(4 + 3\varepsilon)\varepsilon \log \varepsilon\} < 0,$$

(98)
$$\operatorname{Re} \{(1 - \bar{k}^4)(2 + 3\varepsilon)(2 + \varepsilon) \log (2 + \varepsilon)\} < \tfrac{1}{2},$$

and

(99)
$$| \arg b_1\bar{k}^2 | = | \arg [2 + 6\varepsilon + 3\varepsilon^2 - (2 + 2\varepsilon)\bar{k}^4] | \leq \pi/2.$$

By (99) and the estimate $| b_1 | \leq 1$, we have

(100)
$$| b_1 \bar{k}^2 \log b_1 \bar{k}^2 | < \frac{1}{e} + \frac{\pi}{2},$$

since

(101)
$$\max_{0 < x < 1} | x \log x | = \frac{1}{e}.$$

Finally, it is clear that

(102)
$$\operatorname{Re} \{8 + 24\varepsilon + 12\varepsilon^2 - (4 + 4\varepsilon)\bar{k}^4\} > 4.$$

Thus, combining (97), (98), (100), and (102), we derive from (96) the absurd inequality

$$(103) \qquad 4 < \frac{1}{e} + \frac{\pi}{2} + \frac{1}{2} < 2.6,$$

which shows that (85) has, indeed, no relevant solutions.

This finishes our proof that the case (ii) in the integration of the differential equation (32) does not actually arise, and thus, finally, we complete in every detail our proof of the original inequality (4). The tedious calculations required to bring us from the fundamental equation (85) determining b_1 to the contradiction (103) should not be allowed to obscure the basic value of the method.

6. Corollaries of the main theorem

Because we know that (42) is the extremal function for the inequality (4), we can make variations in the large and derive a set of further inequalities by composition of suitable mappings. We carry out one example of this type here and establish the inequalities (7).

We write the extremal function (42) in the form

$$(104) \qquad t = z + \frac{4ie^{-3}}{z} + \frac{\frac{1}{2} + e^{-6}}{z^3} + \cdots.$$

The boundary Γ in the t-plane includes for this mapping the segment from $-2 \exp (-\frac{3}{2} + (\pi i/4))$ to $+2 \exp (-\frac{3}{2} + (\pi i/4))$. The exterior of this segment is mapped onto the exterior of the circle of radius $\exp (-\frac{3}{2})$ in the w-plane by the transformation

$$(105) \qquad t = w + \frac{ie^{-3}}{w}.$$

In the region $|w| > \exp (-\frac{3}{2})$, we can consider the schlicht function

$$(106) \qquad g^* = w + \frac{b_1 e^{-3}}{w} + \frac{b_2 e^{-9/2}}{w^2} + \frac{b_3 e^{-6}}{w^3} + \cdots,$$

where the numbers b_n are arbitrary coefficients of a schlicht function of the form (2). Through composition of the transformations (104), (105), and (106) we obtain outside the unit circle in the z-plane the schlicht function

$$(107) \qquad g^* = z + \frac{b_1 e^{-3} + 3ie^{-3}}{z} + \frac{b_3 e^{-6} - 3ib_1 e^{-6} - 2e^{-6} + \frac{1}{2}}{z^3} + \cdots.$$

Since g^* is a competing function for the inequality (4), the expansion (107) yields

$$(108) \qquad \mathrm{Re} \ \{b_3 e^{-6} - 3ib_1 e^{-6} - 2e^{-6} + \frac{1}{2}\} \leq \frac{1}{2} + e^{-6},$$

and the first inequality (7) is an immediate consequence of (108).

In a similar fashion we can derive the second inequality (7) from the knowledge that (6), with $n = 2$, is the extremal function for the inequality $|b_2| \leq \frac{2}{3}$. The chief interest in the inequalities (7) stems from the fact that, in the sense of the substitution (8), they become equalities for the Koebe function.

The problem of maximizing b_3 when all the coefficients b_n of the schlicht function $g(z)$ are real has a somewhat unexpected solution, due to the fact that the extremal function (104) does not have real coefficients. In the case where the coefficients are real, routine application of variational methods leads to the same differential equation (32) which we obtained in the general case. One has only to make variations which are symmetric in the real axis to see this. But since we found in Sections 3, 4, and 5 all the schlicht solutions of (32), and since (104) does not have real coefficients, we deduce that (6), with $n = 3$, is actually the extremal function maximizing b_3 when all the coefficients b_n are real. On the other hand, rotation of (104) through 45° yields the function with real coefficients which minimizes b_3. Thus for real coefficients we have the peculiar result

$$(109) \qquad -\tfrac{1}{2} - e^{-6} \leq b_3 \leq \tfrac{1}{2}.$$

The exceptionally small difference in the size of the estimates on the left and on the right in (109) is quite remarkable.

After our discussion so far, it would appear that the next problem in order of difficulty is to settle the truth of the conjecture $|a_4| \leq 4$ for schlicht functions (1) inside the unit circle. Our success with the inequality (4) would indicate that the most promising approach to this question lies through the study of equations analogous to (85) for the earlier coefficients a_2 and a_3 of the extremal function. We are able here only to describe the nature of these equations.

From the differential equation for the extremal function $w = f(z)$ maximizing $|a_4|$ with $a_4 > 0$, we can derive the identity

$$(110) \qquad \begin{aligned} \int \left(\frac{1}{w^2} + \frac{3a_2}{w} + 2a_3 + a_2^2 \right)^{\tfrac{1}{2}} \frac{dw}{w^{\tfrac{1}{2}}} \\ = \int \left(\frac{1}{z^3} + \frac{2a_2}{z^2} + \frac{3a_3}{z} + 3a_4 + 3\bar{a}_3 z + 2\bar{a}_2 z^2 + z^3 \right)^{\tfrac{1}{2}} \frac{dz}{z}. \end{aligned}$$

With a suitable constant of integration and with the integrals interpreted as indefinite integrals, (110) defines the extremal map $w = f(z)$ implicitly. On the other hand, if we integrate over two corresponding closed paths in the z-plane and in the w-plane, (110) becomes merely a numerical equation. Furthermore, by Cauchy's theorem the closed paths need not correspond according to the map, provided their topology relative to the roots and the poles of the expressions in parentheses is the same. Let w_1 and w_2 be the zeros of the integrand on the left in (110) and let z_1, z_2, $\bar{z}_1^{-1}$, $\bar{z}_2^{-1}$, and $z_3 = \bar{z}_3^{-1}$ be the zeros of the integrand on the right, with $|z_1| < 1$, $|z_2| < 1$. We obtain two equations, independent of the conformal map $w = f(z)$, for a_2 and a_3 if we choose as the contours of integration in (110) loops around 0 and w_1 and around 0 and z_1, or loops around 0 and w_2 and around 0 and z_2. In order to discuss effectively these two complex equations for a_2 and a_3 it would first be necessary to generalize and refine preliminary estimates on a_2 and a_3 of the type (91).

Finally, we wish to point out that the geometrical analysis of Section 2 would lead rather easily to the result $|a_4| \le 4$ if it were possible to establish first even such a simple condition of symmetry for the extremal function as $\arg a_3 = \arg a_2^2$.

There is a problem related to coefficient inequalities for the schlicht functions (2) which is concerned with certain diameters D_n associated with a connected bounded closed set Γ. We define

$$(111) \qquad D_n = \max \left[\prod_{i<j} | t_i - t_j | \right]^{2/(n(n-1))}$$

for all choices of the n points t_j lying in the continuum Γ. The number D_2 is the usual diameter of the set Γ, and it can be shown that as $n \to \infty$ the nth diameter D_n approaches the outer mapping radius, or transfinite diameter, e^γ of Γ. We are interested in the problem of determining a set Γ which has, for a prescribed value of the outer mapping radius e^γ, the largest possible value of D_n [2, 6]. That such an extremal set Γ exists is an easy consequence of the theory of normal families of analytic functions.

It is well known that D_2 is a maximum when Γ is a line segment with endpoints t_1 and t_2. We shall give a new proof here that, for $e^\gamma = 1$, D_3 is a maximum when Γ consists of three equally spaced rays from the origin out to the three cube roots of 4, in which case t_1, t_2, and t_3 lie at these cube roots.

It takes only a routine application of variational methods to show that the analytic function $p(t)$ whose real part is the Green's function of the exterior of the extremal set Γ maximizing D_3 satisfies the differential equation [6]

$$(112) \qquad p'(t)^2 = \frac{t - \dfrac{t_1 + t_2 + t_3}{3}}{(t - t_1)(t - t_2)(t - t_3)}.$$

Furthermore, there is no loss of generality if we assume that Γ has been rotated and translated so that

$$(113) \qquad t_1 + t_2 + t_3 = 0, \qquad t_1 > 0.$$

Thus Γ consists of three analytic arcs which fork from the origin under angles of 120° and terminate at t_1, t_2, and t_3. The problem is to find t_1, t_2, and t_3.

We can find equations for t_1, t_2, and t_3 by noticing that the Green's function vanishes on Γ and hence vanishes at 0, t_1, t_2, and t_3. Thus, in particular,

$$(114) \qquad \text{Re} \int_0^{t_1} \frac{t^{\frac{1}{2}}\, dt}{[(t - t_1)(t - t_2)(t - t_3)]^{1/2}} = \text{Re}\,\{p(t_1) - p(0)\} = 0,$$

where we are allowed to integrate along the segment $0 < t < t_1$. It follows from (114) that the integrand must be pure imaginary for at least one value t_0 of t between 0 and t_1. Hence

$$(115) \qquad \text{Im}\,\{t_0^2 - (t_2 + t_3)t_0 + t_2 t_3\} = 0,$$

or, since $t_2 + t_3 = -t_1 < 0$,

(116)   Im $t_2 t_3 = 0$.

The relation (116), together with (113), implies that $t_2 = \bar{t}_3$, unless t_2 and t_3 are both real, a case which is easily excluded from the maximum problem by direct calculation. Thus t_1 lies on the perpendicular bisector of the segment joining t_2 and t_3, and since either of the points t_2 or t_3 could as easily have been chosen to be the one lying on the positive real axis, we deduce that t_1, t_2, and t_3 are the vertices of an equilateral triangle whose center lies at the origin.

From the normalization $e^\gamma = 1$ we now find that t_1, t_2, and t_3 are, indeed, the cube roots of 4, and (114) shows that Γ consists of three line segments joining these roots to the origin. This proof illustrates how simply conditions of the form (114) can sometimes be treated, even when they involve elliptic integrals.

It can be shown that the four rays from the origin to the fourth roots of 4 do not compose the extremal set maximizing D_4. The proof is too involved to present it profitably here, since the outcome is negative. The construction of a counter-example consists in applying the variation

(117) $$t^* = t + \frac{i\rho^2}{t}$$

to the fourth roots of 4 with a small positive value of ρ and developing in powers of ρ the outer mapping radius and the fourth diameter of the continuum through the varied points which has the smallest possible outer radius. Letting $\rho \to 0$, we find from this development that the varied continuum has a larger value of $D_4 e^{-\gamma}$ than that of the original symmetric fork.

STANFORD UNIVERSITY

REFERENCES

[1] M. FEKETE and G. SZEGÖ, *Eine Bemerkung über ungerade schlichte Funktionen*, J London Math. Soc., vol. 8 (1933), pp. 85–89.
[2] G. GOLUSIN, *Method of variations in the theory of conform representation. II*, Mat. Sbornik (Rec. Mathém.) (N.S.), vol. 21 (1947), pp. 83–117.
[3] K. LÖWNER, *Untersuchungen über schlichte konforme Abbildungen des Einheitskreises*, Math. Ann., vol. 89 (1923), pp. 103–121.
[4] A. C. SCHAEFFER and D. C. SPENCER, *The coefficients of schlicht functions*, Duke Math. J., vol, 10 (1943), pp. 611–635.
[5] A. C. SCHAEFFER and D. C. SPENCER, Coefficient regions for schlicht functions, Amer. Math. Soc. Colloquium Publications, vol. 35, New York, 1950.
[6] M. SCHIFFER, *Sur un problème d'extrémum de la représentation conforme*, Bull. Soc. Math. France, vol. 66 (1938), pp. 48–55.
[7] M. SCHIFFER, *A method of variation within the family of simple functions*, Proc. London Math. Soc., vol. 44 (1938), pp. 432–449.
[8] M. SCHIFFER, *Variation of the Green function and theory of the p-valued functions*, Amer. J. Math., vol., 65 (1943), pp. 341–360.
[9] D. C. SPENCER, *Some problems in conformal mapping*, Bull. Amer. Math. Soc., vol. 53 (1947), pp. 417–439.
[10] G. SPRINGER, *The coefficient problem for schlicht mappings of the exterior of the unit circle*, Trans. Amer. Math. Soc., vol. 70 (1951), pp. 421–450.

Commentary on

[59] P. R. Garabedian and M. M. Schiffer, *A coef"cient inequality for schlicht functions*, Ann. of Math. (2) **61** (1955), 116 136.

Let Σ_0 denote the class of analytic functions $g(z) = z + \sum_{n=1}^{\infty} b_n z^{-n}$ univalent in $|z| > 1$. The area theorem says that $\sum_{n=1}^{\infty} n|b_n|^2 \leq 1$ so that $|b_n| \leq 1/\sqrt{n}$, but the latter inequality is sharp only for $n = 1$. In 1938, Schiffer [4] proved that $|b_2| \leq \frac{2}{3}$, with equality only for $g(z) = z(1 + z^{-3})^{2/3}$ and its rotations. This result gave rise to the conjecture that $|b_n| \leq \frac{2}{n+1}$, with equality only for rotations of

$$g(z) = z\left(1 + z^{-(n+1)}\right)^{2/(n+1)}$$

$$= z + \frac{2}{n+1}z^{-n} + \cdots. \tag{1}$$

Springer [S] discussed the conjecture and showed that the function (1) is a critical point of $|b_n|$ in the sense that it satisfies the differential equation for an extremal function. In their paper [59], Garabedian and Schiffer use the technique of interior variation [13] to attack the conjecture for $n = 3$. They prove that the sharp bound for $|b_3|$ is not $\frac{1}{2}$ as the conjecture asserts, but $\frac{1}{2} + e^{-6}$. They arrive at this unexpected result by a long series of calculations involving elliptic integrals.

Following the publication of [59], Jenkins [J] gave another proof of the inequality $|b_3| \leq \frac{1}{2} + e^{-6}$ via his general coefficient theorem and Bombieri [Bo] found an easier approach along similar lines, with simplified analysis of the global structure of trajectories of the quadratic differential. Pommerenke ([P3, p. 123]) gave a relatively simple proof based upon the Garabedian Schiffer inequalities [91].

It later came to light, however, that the conjecture $|b_3| \leq \frac{1}{2}$ had been disproved as early as 1937. Bazilevich [Ba] had used Loewner's method [L] to prove that $|b_3| \leq \frac{1}{2} + e^{-6}$ is the sharp inequality for all *odd* functions $g \in \Sigma_0$. Actually, the result of Bazilevich follows easily from the Fekete Szego inequality [FS], also derived by Loewner's method, and the same argument gives the sharp bound

$$|b_{2k-1}| \leq \frac{1}{k}\left(1 + 2e^{-2(k+1)/(k-1)}\right),$$

$$k = 2, 3, \ldots,$$

for all functions $g \in \Sigma_0$ with k-fold symmetry (see [D, Sect. 4.7]). Waadeland [W] made a similar observation. In particular, the conjecture $|b_n| \leq \frac{2}{n+1}$ fails for every odd index $n \geq 3$, but the sharp bound in the full class Σ_0 is known only for $n = 3$.

Long after the publication of [59], Kubota [K] was able to disprove the conjecture $|b_4| \leq \frac{2}{5}$. Then Tsao [T] proved that the conjecture $|b_n| \leq \frac{2}{n+1}$ is false for all $n > 4$. She used a form of the second variation to show that the conjectured extremal function (1) does not provide even a local maximum for $|b_n|$, but is a saddle point. Chang, Schiffer, and Schober [123] also disproved the conjecture for $n > 4$, applying the second variation in a different way. The commentary on [123] gives further details.

In fact, the conjecture $|b_n| \leq \frac{2}{n+1}$ is false even in order of magnitude. Clunie [C] constructed a function of class Σ_0 with coefficients $b_n \neq O(1/n)$ as $n \to \infty$, and Pommerenke [P1, P2] improved this to $b_n \neq O(1/n^{0.83})$. (See also [P3, Sect. 5.2], or [D, Sect. 8.1].)

Finally, another result in the paper [59] is worthy of note. As a preliminary illustration of their method for proving $|b_3| \leq \frac{1}{2} + e^{-6}$, the authors give a purely variational proof of Loewner's result $|a_3| \leq 3$. Later, Charzyński and Schiffer [73] were able to adapt the same method to a "geometric proof" of $|a_4| \leq 4$.

References

[Ba] I. E. Bazilevich, *Supplement to the papers Zum Koef"zientenproblem der schlichten Funktionen and Sur les théorèmes de Koebe Bieberbach*, Mat. Sb. 2 (44) (1937), 689 698 (in Russian).

[Bo] Enrico Bombieri, *A geometric approach to some coef"cient inequalities for univalent functions*, Ann. Scuola Norm. Sup. Pisa (3) **22** (1968), 377 397.

[C] J. Clunie, *On schlicht functions*, Ann. of Math. (2) **69** (1959), 511 519.

[D] Peter L. Duren, *Univalent Functions*, Springer-Verlag, 1983.

[FS] M. Fekete and G. Szego, *Eine Bemerkung über ungerade schlichte Funktionen*, J. London Math. Soc. **8** (1933), 85 89.

[J] J. A. Jenkins, *On certain coef"cients of univalent functions*, Analytic Functions, Princeton University Press, 1960, pp. 159 194.

[K] Yoshihisa Kubota, *On the fourth coef"cient of meromorphic univalent functions*, Kōdai Math. Sem. Rep. **26** (1974/75), 267 288.

[L] Karl Löwner (Charles Loewner),*Untersuchungen über schlichte konforme Abbildungen des Einheitskreises, I* , Math. Ann. **89** (1923), 103 121.

[P1] Ch. Pommerenke, *On the coef"cients of univalent functions*, J. London Math. Soc. **42** (1967), 471 474.

[P2] Ch. Pommerenke, *Relations between the coef"cients of a univalent function*, Invent. Math. **3** (1967), 1 15.

[P3] Ch. Pommerenke, *Univalent Functions*, Vandenhoeck & Ruprecht, 1975.

[S] G. Springer, *The coef"cient problem for schlicht mappings of the exterior of the unit circle*, Trans. Amer. Math. Soc. **70** (1951), 421 450.

[T] Anna Tsao, *Disproof of a coef"cient conjecture for meromorphic univalent functions*, Trans. Amer. Math. Soc. **274** (1982), 783 796.

[W] Haakon Waadeland, *Über ein Koef"zientenproblem für schlichte Abbildungen des $|\zeta| > 1$*, Norske Vid. Selsk. Forh., Trondheim **30** (1957), 168 170.

PETER DUREN

Printed in the United States
By Bookmasters